New Models and Methods in
Dynamic Portfolio
Optimization

Series in Quantitative Finance   ISSN: 1756-1604

Series Editor: Shanjian Tang *(Fudan University, China)*

Published

Vol. 7 *New Models and Methods in Dynamic Portfolio Optimization*
by Lijun Bo and Xiang Yu

Vol. 6 *Simulating Copulas:
Stochastic Models, Sampling Algorithms and Applications
(Second Edition)*
by Jan-Frederik Mai and Matthias Scherer

Vol. 5 *Extreme Financial Risks and Asset Allocation*
by Olivier Le Courtois and Christian Walter

Vol. 4 *Simulating Copulas:
Stochastic Models, Sampling Algorithms, and Applications*
by Jan-Frederik Mai and Matthias Scherer

Vol. 3 *Option Pricing in Incomplete Markets:
Modeling Based on Geometric Lévy Processes and
Minimal Entropy Martingale Measures*
by Yoshio Miyahara

Vol. 2 *Advanced Asset Pricing Theory*
by Chenghu Ma

Vol. 1 *An Introduction to Computational Finance*
by Ömür Uğur

Series in Quantitative Finance – Vol. 7

New Models and Methods in Dynamic Portfolio Optimization

Lijun Bo
Xidian University, China

Xiang Yu
The Hong Kong Polytechnic University, Hong Kong, China

World Scientific

NEW JERSEY · LONDON · SINGAPORE · BEIJING · SHANGHAI · TAIPEI · CHENNAI

Published by

World Scientific Publishing Co. Pte. Ltd.
5 Toh Tuck Link, Singapore 596224
USA office: 27 Warren Street, Suite 401-402, Hackensack, NJ 07601
UK office: 57 Shelton Street, Covent Garden, London WC2H 9HE

Library of Congress Control Number: 2025013060

British Library Cataloguing-in-Publication Data
A catalogue record for this book is available from the British Library.

Series in Quantitative Finance — Vol. 7
NEW MODELS AND METHODS IN DYNAMIC PORTFOLIO OPTIMIZATION

Copyright © 2026 by World Scientific Publishing Co. Pte. Ltd.

All rights reserved. This book, or parts thereof, may not be reproduced in any form or by any means, electronic or mechanical, including photocopying, recording or any information storage and retrieval system now known or to be invented, without written permission from the publisher.

For photocopying of material in this volume, please pay a copying fee through the Copyright Clearance Center, Inc., 222 Rosewood Drive, Danvers, MA 01923, USA. In this case permission to photocopy is not required from the publisher.

ISBN 978-981-12-8056-6 (hardcover)
ISBN 978-981-12-8057-3 (ebook for institutions)
ISBN 978-981-12-8058-0 (ebook for individuals)

For any available supplementary material, please visit
https://www.worldscientific.com/worldscibooks/10.1142/13522#t=suppl

Desk Editors: Aanand Jayaraman/Kwong Lai Fun

Typeset by Stallion Press
Email: enquiries@stallionpress.com

Preface

The study of dynamic portfolio optimization has been a core research topic in quantitative finance since the pioneer work of Merton (1969, 1971). Abundant generalizations have been considered afterwards to address new demands and challenges coming from more realistic market models, performance measures, trading constraints, etc. This book aims to present some recent developments in models and methodology in the context of dynamic portfolio optimization, covering topics from risk-sensitive portfolio optimization with contagious risk, consumption habit formation to optimal tracking portfolio.

Chapter 1 starts with some introductory examples in optimal portfolio management that will be treated in the more general framework in the following chapters. This chapter also gives the perspective and overview of the book.

The first part (Chapters 2 and 3) of the book provides the fundamental introduction of stochastic differential equations (SDEs), hybrid diffusion system, Feymann–Kac's formula, backward stochastic differential equations (BSDEs) and important elements in stochastic control theory. In particular, the dynamic programming principle and stochastic maximum principle in Chapter 3 will be the key tools to derive the Hamilton–Jacobi–Bellman (HJB) equation (or the associated backward stochastic differential equations) and dual stochastic control problem in the remaining chapters of the book.

The second part of the book, Chapters 4 and 5, mainly focuses on the risk-sensitive optimal portfolio problem in the credit market with default contagion. The return rate and default intensity for the defaultable assets are modulated by a regime-switching process (continuous-time Markov chain) with countable states. Under the framework of reduce-form for credit risk, the multiple default indicator process of risky assets is formulated as a time-inhomogeneous Markov chain conditional on the regime-switching process. Then, the price dynamics for defaultable assets results in a class of hybrid systems introduced in Chapter 2 in the sense of jump-to-default. Chapter 4 studies the risk-sensitive portfolio optimization in the complete information case. The theory of monotone dynamical system and truncation of state space of the regime-switching process are developed and which are applied to solve the associated stochastic control problem. Chapter 5 deals with the partial information case where the regime-switching process is unobservable. The main results of this chapter include a martingale representation theorem under partial and phasing-out information and the well-posedness of a new type of quadratic BSDEs with jumps raised by the risk-sensitive control problem.

The third part (Chapter 6) of the book presents a recent study of the optimal portfolio-consumption problem under addictive habit formation preference in incomplete semimartingale markets. The methodology relies on the new extension of martingale and duality approach to the value function with two variables in semimartingale market models. Due to the path-dependence of the habit formation preference, we need to embed our original problem into a time-separable utility maximization problem with a shadow random endowment on the product space where the auxiliary dual process is not necessarily integrable. Some technical arguments developed in this problem might also be applicable to other types of path-dependent performance measures in semimartingale market models.

As the fourth part of the book, Chapter 7 proposes a type of optimal tracking problem using the capital injection. This problem is particularly motivated by the fund management when the fund manager can strategically inject capital to keep the total fund capital above a non-decreasing benchmark process at each intermediate time. The control problem involves the regular portfolio control and the singular capital injection control together with American type floor constraints. The optimality is attained when the cost of the accumulated capital injection is minimized. We reformulate the optimal tracking problem with dynamic floor constraints to a constraint-free stochastic control problem under a running maximum cost criterion. Then, we study the constraint-free stochastic control problem via the duality of the prime HJB equation with Neumann boundary condition in a complete market in which the dimension of Brownian risky factors coincides with that of risky assets.

As an extension to Chapter 7, Chapter 8 studies stochastic control problems motivated by optimal consumption with wealth benchmark tracking. This results in a type of extended Merton's problems in which the dynamical programming principle (DPP) does not hold. In the current chapter, the benchmark process is modeled by a combination of a geometric Brownian motion and a running maximum process, indicating its increasing trend in the long run. We provide a relaxed tracking formulation such that the wealth compensated by the injected capital always dominates the benchmark process. The stochastic control problem is to maximize the expected utility of consumption deducted by the cost of the capital injection under the dynamic floor constraint. By introducing two auxiliary state processes with reflections, an equivalent auxiliary control problem in which the DPP holds is formulated and studied, which results in the HJB equation with two non-homogeneous Neumann boundary conditions. We establish the well-posedness of the dual PDE in classical sense by using some novel probabilistic representations involving the local time of some dual processes together with a tailor-made decomposition-homogenization technique. The proof of the verification theorem on the optimal feedback control can be carried out by some stochastic flow analysis and technical estimations of the optimal control.

The last part (Chapter 9) studies the reinforcement learning for the infinite horizon optimal tracking portfolio problem using capital injection in incomplete market models. Here, the benchmark process is modeled by a geometric Brownian motion with zero drift driven by some unhedgeable risk. The relaxed tracking formulation is adopted where the fund account compensated by the injected capital needs to outperform the benchmark process at any time, and the goal is to minimize the cost of the discounted total capital injection. For the case where model parameters are known, we formulate the equivalent auxiliary control problem with reflected state dynamics, for which the classical solution of the HJB equation with Neumann boundary condition is obtained in a closed form. For the case with unknown model parameters, we introduce the exploratory formulation for the auxiliary control problem with entropy regularization and develop the continuous-time q-learning algorithm in models of reflected diffusion processes. In some illustrative numerical example, we show the satisfactory performance of the q-learning algorithm.

It is our hope that the reader can get inspired by some research problems presented in this book and may formulate and study other new and interesting problems in dynamic portfolio optimization and beyond.

L.J. Bo
X. Yu
Xi'an and Hong Kong
Spring 2025

About the Authors

Lijun Bo obtained his PhD from Nankai University. He is currently a Professor of Probability at School of Mathematics and Statistics, Xidian University, Xi'an, China. He serves as the vice director of *Shaanxi Society for Industrial and Applied Mathematics* and *China Society of Engineering Probability and Statistics*. He is an associate editor for the *Journal of Dynamics and Games* (AIMS) and *Chinese Journal of Applied Probability and Statistics*. His research interests are in applied probability, stochastic control, mean field games, and portfolio optimization. His research has been published in major journals of his field, including *Ann. Appl. Probab., Math. Finan., Math. Opers. Res., Product. Oper. Manag., SIAM J. Contr. Optim., SIAM J. Finan. Math., Stoch. Process. Appl.*. He has been funded by the Natural Science Foundation of China (3 grants), the Key Research Program of Frontier Sciences of the CAS and National Center for Applied Mathematics in Shaanxi.

Xiang Yu obtained his PhD from the University of Texas at Austin. He is currently an Associate Professor in the Department of Applied Mathematics at the Hong Kong Polytechnic University. His research interests lie primarily in mathematical finance, applied probability and stochastic analysis, stochastic control and optimization. He has publications in *Math. Finance, Finance & Stoch., Ann. Appl. Probab., Math. Opers. Res, SIAM J. Contr. Optim.*, and *Stoch. Process. Appl.*

Acknowledgments

Special thanks go to our collaborators Yijie Huang and Huafu Liao, and partial results of our collaborated research problems constitute the main content of this book. We are also indebted to Martin Larsson and Johannes Ruf for proposing the new index tracking problem using the capital injection during their visit to the Hong Kong Polytechnic University in 2018. We also would like to thank Wenyuan Wang and Yijie Huang for helping us to modify some proofs in Chapters 4 and 7 of the book.

The first author of the book is partially supported by National Natural Science Foundation of China (Program Nos. 12471451 and 11971368), Natural Science Basic Research Program of Shaanxi (Program No. 2023-JC-JQ-05) and the Fundamental Research Funds for the Central Universities (Program No. 20199235177). The second author of the book is partially supported by the Hong Kong Polytechnic University research grant (Program Nos. P0031417 and P0039251).

Notation

We need to introduce some of our notations which are used frequently throughout the book.

Notation	Definition
⊤	transpose operator
$\mathbb{N}$	$\{1, 2, \ldots\}$
$\mathbb{R}$	$(-\infty, \infty)$
$\mathbb{R}_+$ ($\overline{\mathbb{R}}_+$)	$(0, \infty)$ ($[0, \infty)$)
$|\cdot|$	absolute value or Euclidean norm
$\operatorname{Var}[\xi]$	variance of random variable ξ
$\operatorname{Cov}(\xi, \eta)$	covariance of two-dimensional random variable (ξ, η)
x^+	positive part of real number x, i.e., $x^+ = \max\{x, 0\}$
x^-	negative part of real number x, i.e., $x^- = \max\{-x, 0\}$
$x \in \mathbb{R}^n$	n-dimensional real column vector $x = (x_1, \ldots, x_n)^\top$
$x \leq y$	$x_i \leq y_i$ for all $i = 1, \ldots, n$, where $x, y \in \mathbb{R}^n$
$x < y$	$x \leq y$ and there is some $i \in \{1, \ldots, n\}$ such that $x_i < y_i$, where $x, y \in \mathbb{R}^n$
$x \ll y$	$x_i < y_i$ for all $i = 1, \ldots, n$, where $x, y \in \mathbb{R}^n$
$\mathcal{E}(X)$	stochastic exponential of a semimartingale X
a.e. (a.s.)	almost everywhere (almost surely)
$f'(t)$	first-order derivative of $f : \mathbb{R} \to \mathbb{R}$ w.r.t. variable $t \in \mathbb{R}$
∂_{x_i}	first-order partial derivative operator w.r.t. variable $x_i \in \mathbb{R}$
$\partial^2_{x_i x_j}$	second-order mixed partial derivative w.r.t variables (x_i, x_j)
∇_x	gradient operator $(\partial_{x_1}, \ldots, \partial_{x_n})^\top$

Notation	Definition
∇_x^2	Hessian matrix operator, i.e., $(\partial_{x_i x_j}^2)_{i,j=1}^n$
$a \cdot b$	dot product of $a, b \in \mathbb{R}^n$, i.e., $a \cdot b := \sum_{i=1}^n a_i b_i$
$a \wedge b$	$\min\{a, b\}$ for $a, b \in \mathbb{R}$
$a \vee b$	$\max\{a, b\}$ for $a, b \in \mathbb{R}$
$a := b$	b is denoted by a
$\mathrm{diag}(x)$	diagonal n-dimensional square matrix whose ith entry is x_i, where $x \in \mathbb{R}^n$
$\boldsymbol{I}_n$	n-dimensional identify matrix
$\boldsymbol{1}_n$	n-dimensional column vector whose elements are all one
$\boldsymbol{1}_A$	indicator function associated with the set A
$\mathrm{tr}[A]$	trace of the matrix A
A° ($\mathrm{int} A$)	interior of the set A
$\mathcal{A}_1 \otimes \mathcal{A}_2$	product σ-algebra of measurable spaces $(M_1, \mathcal{A}_1)$ and $(M_2, \mathcal{A}_2)$
(M, d_M)	metric space where d_M is a distance defined on M
$\mathcal{B}(M)$	Borel-σ-algebra generated by open sets of the metric space (M, d_M)
$\mathcal{P}(M)$	set of Borel probability measures on the metric space (M, d_M)
$\mathbb{P} \ll \mathbb{Q}$	$\mathbb{P}$ is absolutely continuous with respect to $\mathbb{Q}$, where $\mathbb{P}, \mathbb{Q} \in \mathcal{P}(M)$
$\mathbb{P} \sim \mathbb{Q}$	$\mathbb{P}$ is equivalent to $\mathbb{Q}$, i.e., $\mathbb{P} \ll \mathbb{Q}$ and $\mathbb{Q} \ll \mathbb{P}$, where $\mathbb{P}, \mathbb{Q} \in \mathcal{P}(M)$
δ_x	Dirac-delta measure concentrated on $x \in \mathbb{R}^n$, i.e., $\delta_x(A) = \boldsymbol{1}_A(x)$ for all $A \in \mathcal{B}(\mathbb{R}^n)$
$C(M)$	vector space of all continuous functions defined on M
$C_b(M)$	subset of $C(M)$, in which the elements are bounded on M
$C^k(\mathbb{R}^n)$	space of all-real valued continuous function which are continuously differentiable up to order k on $\mathbb{R}^n$ and whose derivatives up to order k are also continuous on $\mathbb{R}^n$
$\mathrm{Lip}(M)$	set of Lipschitz continuous functions on a metric space (M, d_M)
$\ell_{lip}(f)$	Lipschitz coefficient of a $\mathbb{R}^d$-valued Lipschitz function $f \in \mathrm{Lip}(M)$ that is, $\ell_{lip}(f) := \sup_{x \neq y \in M} \frac{\|f(x) - f(y)\|}{d_M(x,y)}$
BCT	Bounded convergence theorem
DCT	Dominated convergence theorem
MCT	Monotone convergence theorem

Contents

Preface   v

About the Authors   ix

Acknowledgments   xi

Notation   xiii

1. Introduction   1
   - 1.1 Simple Examples   1
     - 1.1.1 Merton's Investment-Consumption Problem   1
     - 1.1.2 Optimal Investment Problem with Default Risk .   4
     - 1.1.3 Optimal Tracking Portfolio with Infinite Horizon .   8
   - 1.2 Perspective   13
   - 1.3 Overview   13

2. Preliminary   15
   - 2.1 Stochastic Differential Equations   15
   - 2.2 Hybrid Diffusion System   24
   - 2.3 Feymann–Kac's Formula   33
   - 2.4 Backward Stochastic Differential Equations   38

3. Elements of Stochastic Control   43
   - 3.1 Formulation of Stochastic Control Problem   43
   - 3.2 Dynamic Programming Principle   47
   - 3.3 Hamilton–Jacobi–Bellman Equations   53
   - 3.4 Stochastic Maximum Principle   56
   - 3.5 Risk Sensitive Stochastic Control   61

4. Risk Sensitive Credit Portfolio Optimization: Complete Information   67

    4.1 The Market Model 70
    4.2 Formulation of Portfolio Optimization Problem 72
    4.3 Recursive HJB Equations 75
    4.4 Well-posedness of Recursive HJB Equations 76
        4.4.1 Finite State Case of Markov Chain 77
        4.4.2 Countably Infinite State Case of Markov Chain . . 91
    4.5 The Verification Result 101
    4.6 The Convergence Rate 105

5. Risk Sensitive Credit Portfolio Optimization: Partial Information   111

    5.1 The Market Model 113
    5.2 Partial Information and Filter Processes 114
    5.3 Martingale Representation Theorem 117
    5.4 General Correlation Volatility Matrix 124
    5.5 Risk Sensitive Control under Partial Information 126
    5.6 Quadratic BSDE with Jumps 130
        5.6.1 Formulation of Truncated BSDEs 132
        5.6.2 A Priori Estimates of Truncated Solutions 138
        5.6.3 Convergence of Solutions to Truncated BSDEs . . 145
    5.7 The Optimal Admissible Strategy 153
    5.8 Uniqueness of Solutions to BSDE 158

6. Portfolio Optimization with Consumption Habit in Incomplete Market   161

    6.1 The Market Model 164
        6.1.1 Absence of Arbitrage 165
        6.1.2 The Utility Function 166
    6.2 The Characterization of Financeable Consumption 168
        6.2.1 Path-dependence Reduction 169
        6.2.2 Embedding into Maximization Problem with Shadow Endowment 171
    6.3 The Dual Optimization Problem and Its Solvability 172
    6.4 Proofs of Main Results 175
        6.4.1 The Proof of First Main Result 175
        6.4.2 The Proof of Second Main Result 187

7. Optimal Tracking Problem in Portfolio Optimization 197
   7.1 The Market Model 199
   7.2 Formulation of Optimal Tracking Problem 201
   7.3 Auxiliary Stochastic Control Problem 204
   7.4 The Linearization of HJB Equation 207
   7.5 Probabilistic Solution of Linear Dual HJB Equation . . . 209
   7.6 Verification Results 220
   7.7 The GBM Benchmark Process 228
       7.7.1 Finite Horizon Case 229
       7.7.2 Infinite Horizon Case 232

8. Extended Merton's Problem with Relaxed Benchmark Tracking 235
   8.1 The Market Model 236
       8.1.1 The Benchmark Process 236
       8.1.2 Formulation of Extended Merton's Problem 238
       8.1.3 Unconstrained Control Problem 239
   8.2 Equivalent Stochastic Control Problem 240
   8.3 Solvability of the Dual PDE 243
   8.4 Verification Theorem 257

9. Reinforcement Learning for Optimal Tracking Portfolio
   in Incomplete Markets   271
   9.1 The Market Model 274
       9.1.1 The Value of Portfolio 274
       9.1.2 Formulation of Relaxed Tracking Problem 275
   9.2 The Equivalent Control Problem 276
   9.3 The Continuous Time Reinforcement Learning (RL) . . . 279
       9.3.1 q-Function and Martingale Characterization . . . 283
       9.3.2 Continuous-time q-Learning Algorithm 285
   9.4 Numerical Examples 288
   9.5 Proofs of Main Results 296

Bibliography   307

Index   321

Chapter 1

Introduction

This chapter first introduces three simple examples in the context of optimal consumption and investment problems, and then gives perspectives and an overview of the book.

1.1 Simple Examples

Before presenting the main context of this book, we introduce three simple examples for our motivation of the applications of stochastic control theory in dynamics portfolio optimization.

1.1.1 *Merton's Investment-Consumption Problem*

Let us first start with the seminal work by Merton (1969, 1971) on the optimal investment-consumption problem. More precisely, the price dynamics of a risky asset (e.g., stock) follows the following Black–Scholes model (c.f. Eq. (2.11) in Chapter 2):

$$S_t = S_0 \mathcal{E}_t(Z) = S_0 \exp\left(\left(\mu - \frac{\sigma^2}{2}\right)t + \sigma W_t\right), \quad t \geq 0, \quad (1.1)$$

where $S_0 > 0$ is the initial price, $\mu \in \mathbb{R}$ is the return rate, $\sigma > 0$ is the volatility, and $W = (W_t)_{t \geq 0}$ is a standard Brownian motion. Here, $Z_t := \mu t + \sigma W_t$ for $t \geq 0$, i.e., $Z = (Z_t)_{t \geq 0}$ is a drifted Brownian motion.

Let $T > 0$ be a finite terminal horizon. Now, there is an investor who may choose amount C_t of her wealth to consume, a fraction π_t of wealth to invest in the risky asset and the fraction $1 - \pi_t$ is invested in the risk-free asset with interest rate $r \geq 0$ at time $t \in [0, T]$. From the self-financing condition (c.f. Eq. (2.15) or (2.16) in Chapter 2), the wealth dynamics of this investor follows that

$$dX_t = rX_t dt + (\mu - r)\pi_t X_t dt + \sigma \pi_t X_t dW_t - C_t dt, \quad (1.2)$$

where $X_0 = x > 0$ is the investor's initial wealth level. The aim of the investor is to choose an optimal investment-consumption pair $(\pi, C) = (\pi_t, C_t)_{t \in [0,T]} \in \mathcal{U}$ (the admissible control set which will specified later) so to maximize the following objective functional given by

$$J(0, x; (\pi, C)) := \mathbb{E}\left[\int_0^T e^{-\rho s} U(C_s) ds + e^{-\rho T} U(X_T)\right], \quad (1.3)$$

where $\rho \geq 0$ is the subjective discount rate, and $U : [0, \infty) \to [0, \infty)$ is the utility function being of the constant relative risk aversion (CRRA) forms:

$$U(x) = \frac{1}{\gamma} x^\gamma, \quad x \geq 0. \quad (1.4)$$

Here, $\gamma \in (0, 1)$ and $\eta := 1 - \gamma$ expresses the investor's risk aversion (the higher γ, the more reluctance to own the asset).

The admissible control set $\mathcal{U}$ is defined as the set of adapted processes (π, C) with state space $U := \mathbb{R} \times [0, \infty)$, which satisfies that $\mathbb{E}[\int_0^T |\pi_t|^2 dt] < \infty$ and $\int_0^T C_t dt < \infty$, a.s. We will apply the dynamic program introduced in Chapter 3 to solve the problem. To this purpose, we define the dynamic version of the stochastic control problem by, for $(t, x) \in [0, T] \times \mathbb{R}_+$,

$$V(t, x) := \sup_{(\pi, C) \in \mathcal{U}} \mathbb{E}_{t,x}\left[\int_t^T e^{-\rho(s-t)} U(C_s) ds + e^{-\rho(T-t)} U(X_T)\right], \quad (1.5)$$

where $\mathbb{E}_{t,x}[\cdot] := \mathbb{E}[\cdot | X_t = x]$. Using the dynamic program, the value function formally satisfies the following HJB equation, for $(t, x) \in [0, T) \times \mathbb{R}_+$,

$$\partial_t V(t, x) + \mathcal{H}(t, x, \partial_x V(t, x), \partial_x^2 V(t, x)) = \rho V(t, x) \quad (1.6)$$

with the terminal condition $V(T, x) = U(x)$ for all $x > 0$. Here, $\mathcal{H}(t, x, p, q) : [0, T] \times \mathbb{R}_+ \times \mathbb{R} \times \mathbb{R} \to \mathbb{R}$ is called *Hamiltonian*, which is defined by

$$\mathcal{H}(t, x, p, q) \quad (1.7)$$
$$:= \sup_{(\pi, c) \in \mathbb{R} \times [0, \infty)} \left\{(rx + (\mu - r)\pi x - c) \cdot p + \frac{\sigma^2}{2} \pi^2 x^2 q + U(c)\right\}.$$

We next illustrate the procedure for solving the HJB equation (1.6):
Step 1. The first-order condition w.r.t. (π, c) yields that

$$\begin{cases} (\mu - r) x p + \sigma^2 x^2 q \pi = 0, \\ U'(c) = p. \end{cases} \Rightarrow \begin{cases} \hat{\pi}(x, p, q) = \frac{(r - \mu) p}{\sigma^2 x q}, & \text{if } q \neq 0, \\ \hat{c}(x, p, q) = p^{\frac{1}{\gamma - 1}}, & \text{if } p > 0. \end{cases}$$

Step 2. Assume that V is a classical solution to HJB equation (1.6) satisfying $\partial_x^2 V < 0$ which can be verified later. Then, we introduce the following optimal (feedback) investment-consumption candidate pair that

$$\begin{cases} \pi^*(t,x) = \hat{\pi}(x, \partial_x V(t,x), \partial_x^2 V(t,x)) = \dfrac{(r-\mu)\partial_x V(t,x)}{\sigma^2 x \partial_x^2 V(t,x)}, \\ c^*(t,x) = \hat{c}(x, \partial_x V(t,x), \partial_x^2 V(t,x)) = \partial_x V(t,x)^{\frac{1}{\gamma-1}}. \end{cases} \quad (1.8)$$

Step 3. Plugging (1.8) into the Hamiltonian, the HJB equation (1.6) becomes the following fully nonlinear PDE:

$$\partial_t V + rx \partial_x V - \frac{(r-\mu)^2}{2\sigma^2} \frac{(\partial_x V)^2}{\partial_x^2 V} + \frac{1-\gamma}{\gamma}(\partial_x V)^{\frac{\gamma}{\gamma-1}} = \rho V, \quad (1.9)$$

with $V(T,x) = U(x) = \frac{1}{\gamma} x^\gamma$. We make the following ansatz for the solution by $V(t,x) = f(t)x^\gamma$. Substitute it into Eq. (1.9) to have that $f(T) = \frac{1}{\gamma}$, and

$$f'(t) + \alpha f(t) + \beta f(t)^{\frac{\gamma}{\gamma-1}} = 0 \quad (1.10)$$

with $\alpha := r\gamma - \rho - \frac{(r-\mu)^2}{2\sigma^2}\frac{\gamma}{\gamma-1}$ and $\beta := \frac{1-\gamma}{\gamma}\gamma^{\frac{\gamma}{\gamma-1}}$. Eq. (1.10) is a Bernoulli equation, and hence we assume that $f(t) = u(t)^{1-\gamma}$. It follows from (1.10) that

$$u'(t) + \frac{\alpha}{1-\gamma}u(t) + \frac{\beta}{1-\gamma} = 0, \quad u(T) = \gamma^{\frac{1}{\gamma-1}}.$$

Solving the above linear ODE in an explicit way, we have

$$u(t) = v(T-t), \quad v(t) = e^{\frac{\alpha}{1-\gamma}t}\left[\gamma^{\frac{1}{\gamma-1}} + \frac{\beta}{1-\gamma}\int_0^t e^{-\frac{\alpha}{1-\gamma}s}ds\right]. \quad (1.11)$$

Thus, we conclude that the solution of HJB equation (1.6) has the following closed-form given by

$$V(t,x) = v(T-t)^{1-\gamma}x^\gamma, \quad \forall (t,x) \in [0,T] \times \mathbb{R}_+. \quad (1.12)$$

Having the above solution in hand, we can turn to **Step 2** and have the following optimal investment-consumption pair:

$$\begin{cases} \pi_t^* = \pi^*(t, X_t^*) = \dfrac{r-\mu}{(1-\gamma)\sigma^2}, \\ C_t^* = c^*(t, X_t^*) = \dfrac{\gamma^{\frac{1}{\gamma-1}}}{v(T-t)} X_t^*. \end{cases} \quad (1.13)$$

Here $X^* = (X_t^*)_{t \in [0,T]}$ satisfies the wealth dynamics (1.2) with (π, C) replaced by (π^*, C^*). That is, it satisfies that

$$\frac{dX_t^*}{X_t^*} = \left(r - \frac{(r-\mu)^2}{(1-\gamma)\sigma^2} - \frac{\gamma^{\frac{1}{\gamma-1}}}{v(T-t)}\right)dt + \frac{r-\mu}{(1-\gamma)\sigma}dW_t. \quad (1.14)$$

Note that $v(t) > 0$ for all $t \in [0,T]$. Hence, $C^* > 0$ a.s.. It will be seen in Section 7.4 of Chapter 7 that the fully nonlinear PDE (1.9) can be linearized by applying *Fenchel–Legendre* transform.

1.1.2 Optimal Investment Problem with Default Risk

We present an example on the optimal portfolio with a defaultable bond considered in Bielecki and Jang (2006) and Bo et al. (2019b).

In this example, the investor is allowed to invest her wealth in a stock, a defaultable bond and a riskless bond with interest rate $r \geq 0$. To price a defaultable bond, we first present the model of the default time for the bond by following a reduced-form formulation. Let $(\Omega, \mathcal{F}, \mathbb{P})$ be an original probability space and τ be a nontrivial random time (which is used to model the default time) on the space. Define the corresponding default indicator process as

$$H_t := \mathbf{1}_{\tau \leq t}, \quad \forall t \geq 0.$$

Consider a standard Brownian motion $W = (W_t)_{t \geq 0}$ on $(\Omega, \mathcal{F}, \mathbb{P})$ and $\mathbb{F} = (\mathcal{F}_t)_{t \geq 0}$ is defined as the natural filtration generated by W. Let $\mathcal{D}_t := \sigma(H_s; s \leq t)$ and $\mathcal{G}_t := \mathcal{F}_t \vee \mathcal{D}_t$ for $t \geq 0$ by the augmentation. Let $\mathbb{Q} \sim \mathbb{P}$ be a risk-neutral probability measure which will be specified later. Introduce the following conditional survival probability, for $t > 0$,

$$P_t := \mathbb{Q}(\tau > t | \mathcal{F}_t), \quad P_0 = 1. \tag{1.15}$$

Assume that $P_t > 0$ a.s. and $\mathbb{E}^{\mathbb{Q}}[P_t] > 0$ for all $t > 0$. This implies that there is always a chance that the firm defaults. Note that $P = (P_t)_{t \geq 0}$ is a $(\mathbb{Q}, \mathbb{F})$-supermartingale. It follows from Doob–Meyer theorem that, there is a unique compensator $K = (K_t)_{t \geq 0}$ to P such that $P + K$ becomes a $(\mathbb{Q}, \mathbb{F})$-martingale. Following Giesecke (2006), we can define the trend by

$$\Lambda_t = \int_0^t \frac{dK_s}{P_{s-}}, \quad \forall t \geq 0, \tag{1.16}$$

which is a non-decreasing $\mathbb{F}$-predictable process.

In the sequel, we consider the constant intensity-based case where $\Lambda_t = \lambda t$ for $t \geq 0$. Then, from Propositions 5.8 and 5.2 in Giesecke (2006), we get that $P_t = e^{-\lambda t}$ and

$$M_t^{\mathbb{Q}} := H_t - \int_0^{t \wedge \tau} \lambda ds, \quad t \geq 0 \tag{1.17}$$

is a $(\mathbb{Q}, \mathbb{G} = (\mathcal{G}_t)_{t \geq 0})$-martingale. Moreover, we have from Corollary 5.1.1 in Bielecki and Rutkowski (2002) that

$$\mathbb{E}^{\mathbb{Q}}\left[\mathbf{1}_{\tau > s} \xi | \mathcal{G}_t\right] = \mathbf{1}_{\tau > t} \mathbb{E}^{\mathbb{Q}}\left[e^{-\lambda(s-t)} \xi | \mathcal{F}_t\right], \quad \forall s \geq t, \ \xi \in \mathcal{F}_s. \tag{1.18}$$

Having the above default model in hand, we price a defaultable bond with coupon as in Bo et al. (2019b). Consider a coupon paying bonds with unit notional, underwritten by risky firms. The bond underwritten by the firm has maturity T_1 and generates the following dividend process, for $t \in [0, T_1]$,

$$D_t := \int_0^t c\mathbf{1}_{\tau>s}ds + R\mathbf{1}_{\tau \leq t} + \mathbf{1}_{\tau > T_1}\mathbf{1}_{t \geq T_1}$$

$$= \underbrace{\int_0^t c(1-H_s)ds}_{\text{coupon payments}} + \underbrace{RH_t}_{\text{recovery amount}} + \underbrace{(1-H_{T_1})\mathbf{1}_{t \geq T_1}}_{\text{terminal payoff}}, \quad (1.19)$$

where $c \geq 0$ is the coupon payment of the bond so that $\int_0^t c(1-H_s)ds$ is continuous stream of coupon payments received by the bond holder until the earliest of time t and the default of firm. The quality $R \in [0,1)$ is the recovery rate paid at the default time τ. The quantity $(1-H_{T_1})\mathbf{1}_{t \geq T_1}$ is unit notional payment received by the bond holder at the maturity T_1 if the firm has not defaulted.

Assuming that the bond market is free of arbitrage opportunities, the bond price is equal to the expected discounted payoff under the risk-neutral probability measure $\mathbb{Q}$. Then, the ex-dividend price of the bond at time $t \leq T_1$ given by

$$B_t := \mathbb{E}^{\mathbb{Q}}\left[\int_t^{T_1} e^{-r(s-t)}dD_s \Big| \mathcal{G}_t\right], \quad (1.20)$$

where $\mathbb{E}^{\mathbb{Q}}$ denotes the expectation operator under $\mathbb{Q}$. In terms of (1.19) and (1.20), it follows from (1.18) that

$$B_t = c\mathbb{E}^{\mathbb{Q}}\left[\int_t^{T_1} e^{-r(s-t)}\mathbf{1}_{\tau>s}ds \Big| \mathcal{G}_t\right] + R\mathbb{E}^{\mathbb{Q}}\left[\int_t^{T_1} e^{-r(s-t)}\mathbf{1}_{\tau \in ds} \Big| \mathcal{G}_t\right]$$
$$+ e^{-r(T_1-t)}\mathbb{E}^{\mathbb{Q}}\left[\mathbf{1}_{\tau>T_1}|\mathcal{G}_t\right]\mathbf{1}_{t \geq T_1}$$
$$= c\int_t^{T_1} e^{-r(s-t)}\mathbb{Q}(\tau > s|\mathcal{G}_t)ds + R\mathbb{E}^{\mathbb{Q}}\left[e^{-r(\tau-t)}\mathbf{1}_{t<\tau \leq T_1}|\mathcal{G}_t\right]$$
$$+ e^{-r(T_1-t)}\mathbb{Q}(\tau > T_1|\mathcal{G}_t)\mathbf{1}_{t \geq T_1}$$
$$= \mathbf{1}_{\tau>t}c\int_t^{T_1} e^{-(r+\lambda)(s-t)}ds + \mathbf{1}_{\tau>t}R\lambda\int_t^{T_1} e^{-(r+\lambda)(s-t)}ds$$
$$+ \mathbf{1}_{\tau>t}e^{-(r+\lambda)(T_1-t)}.$$

As a consequence, for all $t \in [0, T_1]$,

$$B_t = (1-H_t)\left\{(c+R\lambda)\int_t^{T_1} e^{-(r+\lambda)(s-t)}ds + e^{-(r+\lambda)(T_1-t)}\right\}$$
$$= (1-H_t)\left\{\frac{c+R\lambda}{r+\lambda} + \left(1 - \frac{c+R\lambda}{r+\lambda}\right)e^{-(r+\lambda)(T_1-t)}\right\}. \quad (1.21)$$

From above displays, we have $B_{T_1} = \mathbf{1}_{\tau > T_1}$. We also need the dynamics of the price process $t \to B_t$. By applying Itô's formula, we obtain

$$\begin{aligned} dB_t &= -\left\{\frac{c+R\lambda}{r+\lambda} + \left(1 - \frac{c+R\lambda}{r+\lambda}\right)e^{-(r+\lambda)(T_1-t)}\right\} dH_t \\ &\quad + (1-H_t)(r-c+(1-R)\lambda)\,e^{-(r+\lambda)(T_1-t)}dt \\ &= -B_{t-}dH_t + (r+\lambda)B_t dt - (c+R\lambda)(1-H_t)dt \\ &= -B_t dM_t^{\mathbb{Q}} - \lambda B_t dt + (r+\lambda)B_t dt - (c+R\lambda)(1-H_t)dt \\ &= -B_t dM_t^{\mathbb{Q}} + rB_t dt - (c+R\lambda)(1-H_t)dt. \end{aligned} \quad (1.22)$$

Let $W_t^{\mathbb{Q}} := W_t + \alpha t$ for $t \geq 0$ be a Brownian motion under the risk-neutral probability measure $\mathbb{Q}$ and $M_t := M_t^{\mathbb{Q}} - \int_0^{t \wedge \tau} \lambda \beta ds$ for $t \geq 0$ be a default martingale under $\mathbb{P}$. The coefficients $\alpha \in \mathbb{R}$ and $\beta \in (-1, \infty)$ represent, respectively, the market price of (diffusion) risk and the default risk premium. The single stock price follows the following Black–Scholes model under $\mathbb{Q}$, i.e.,

$$dS_t = rS_t dt + \sigma S_t dW_t^{\mathbb{Q}}. \quad (1.23)$$

The investor chooses the fraction π_t^1 of wealth to invest in the defaultable bond, the fraction π_t^2 of wealth to invest in the single stock, and the fraction $1 - \pi_t^1 - \pi_t^2$ is invested in the risk-free asset with interest rate $r \geq 0$ at time $t \in [0, T]$. We assume that π^1 and π^2 are predictable processes satisfying $\pi_t^1 = (1 - H_{t-})\pi_t^1$ (i.e., the investor will not invest in the default bond once it defaults). Then, using the self-financing condition, it follows that

$$\begin{aligned} \frac{dX_t}{X_{t-}} &= \pi_t^1 \frac{dB_t}{B_{t-}} + \pi_t^2 \frac{dS_t}{S_t} + (1 - \pi_t^1 - \pi_t^2)\frac{de^{rt}}{e^{rt}} \\ &= \pi_t^1 \left\{(r - \psi(t))dt - dM_t^{\mathbb{Q}}\right\} + \pi_t^2 \left\{rdt + \sigma dW_t^{\mathbb{Q}}\right\} + (1 - \pi_t^1 - \pi_t^2)rdt \\ &= rdt - \pi_t^1 \psi(t)dt + \sigma \pi_t^2 dW_t^{\mathbb{Q}} - \pi_t^1 dM_t^{\mathbb{Q}} \quad (1.24) \\ &= \left\{r + \sigma \alpha \pi_t^2 + (\lambda \beta - \psi(t))\pi_t^1\right\} dt + \sigma \pi_t^2 dW_t - \pi_t^1 dH_t, \quad (1.25) \end{aligned}$$

where $\psi(t) := \frac{(c+R\lambda)}{\frac{c+R\lambda}{r+\lambda} + (1 - \frac{c+R\lambda}{r+\lambda})e^{-(r+\lambda)(T_1-t)}}$. To keep the positive wealth process of the investor, it is required from (1.25) that $\pi_t^1 \in (-\infty, 1)$, a.s. for all $t \in [0, T]$.

We consider a risk-averse investor who wants to find the optimal admissible strategy, i.e., the one maximizing her expected utility from terminal wealth under the probability measure $\mathbb{P}$ that describes the actual distribution of risk factors. In other words, she wants to optimize the criterion $\mathbb{E}\left[U(X_T)\right]$ overall the all admissible trading strategies $(\pi^1, \pi^2) \in \mathcal{U}$. Here, $\mathbb{E}$ denotes the expectation operator under $\mathbb{P}$ and the utility function $x \to U(x)$ is given by (1.3).

To apply the dynamic program, define the value function by, for $(t, x, h) \in [0, T] \times \mathbb{R}_+ \times \{0, 1\}$,

$$V(t, x, h) := \sup_{\pi = (\pi^1, \pi^2) \in \mathcal{U}} \mathbb{E}_{t,x,h}[U(X_T)]$$

$$:= \sup_{\pi = (\pi^1, \pi^2) \in \mathcal{U}} \mathbb{E}[U(X_T) | X_t = x, H_t = h]. \qquad (1.26)$$

Then, the value function V formally satisfies the HJB equation:

$$\partial_t V + \mathcal{H}(t, x, h, \partial_x V, \partial_x^2 V) = 0 \qquad (1.27)$$

with $V(T, x, h) = U(x)$ for all $(x, h) \in \mathbb{R}_+ \times \{0, 1\}$. The Hamiltonian is defined as, for $(t, x, p, q) \in [0, T] \times \mathbb{R}_+ \times \mathbb{R} \times \mathbb{R}$,

$$\mathcal{H}(t, x, h, p, q) \qquad (1.28)$$

$$= \sup_{(\pi_1, \pi_2) \in (-\infty, 1) \times \mathbb{R}} \Big\{ (r + (\lambda \beta - \psi(t))(1 - h)\pi_1 + \sigma \alpha \pi_2) x \cdot p$$

$$+ \frac{\sigma^2}{2} \pi_2^2 x^2 q + \lambda (1 - h)(V(t, x(1 - (1 - h)\pi_1), 1 - h) - V(t, x, h)) \Big\}.$$

To highlight the default effect, we call $V_0(t, x) := V(t, x, 0)$ (resp. $V_1(t, x) := V(t, x, 1)$) as the *pre-default* value function (resp. the *post-default* value function). We then deal with these two cases:

- The post-default case ($h = 1$): In this case, the post-default value function $V_1(t, x)$ satisfies the HJB equation, for $(t, x) \in [0, T] \times \mathbb{R}_+$,

$$\partial_t V_1 + \mathcal{H}_1(t, x, \partial_x V_1, \partial_x^2 V_1) = 0 \qquad (1.29)$$

with $V_1(t, x) = U(x)$. The Hamiltonian is given by

$$\mathcal{H}_1(t, x, p, q) = \sup_{\pi_2 \in \mathbb{R}} \Big\{ (r + (\lambda \beta + \sigma \alpha \pi_2) x \cdot p + \frac{\sigma^2}{2} \pi_2^2 x^2 q \Big\}. \qquad (1.30)$$

We can apply the procedure for solving the HJB eqaution in the first example to obtain the closed-form solution to Eq. (1.29).
- The pre-default case ($h = 0$): In this case, the pre-default value function $V_0(t, x)$ satisfies the HJB equation, for $(t, x) \in [0, T] \times \mathbb{R}_+$,

$$\partial_t V_0 + \mathcal{H}_0(t, x, \partial_x V_0, \partial_x^2 V_0) = 0 \qquad (1.31)$$

with $V_0(t, x) = U(x)$. The Hamiltonian is given by

$$\mathcal{H}_0(t,x,h,p,q) = \sup_{(\pi_1,\pi_2)\in(-\infty,1)\times\mathbb{R}} \Big\{(r+(\lambda\beta-\psi(t))\pi_1+\sigma\alpha\pi_2)x\cdot p$$
$$+\frac{\sigma^2}{2}\pi_2^2 x^2 q + \lambda(\underbrace{V_1(t,x(1-\pi_1))}_{\text{post value function}}-V_0(t,x))\Big\}. \tag{1.32}$$

It can be seen from (1.32) that the pre-default HJB equation also depends on the post-default value function. Hence, Eq. (1.31) becomes more complex and it does not in general admit a closed-form solution apart from that the function $t \to \psi(t)$ is a constant (if $\psi(t) \equiv \psi \in \mathbb{R}$, then we can seek the solution form with $V_0(t,x) = f(t)x^\gamma$ as in the first example of Section 1.1.1). For the non-constant $\psi(t)$, we may study the viscosity solution and then improve its smoothness (thus technique has been adopted in Davis and Lleo (2013) for dealing with the well-posedness of partial-integro-HJB PDEs in classical sense). In addition, for the case with multiple default assets, Bo et al. (2019b) propose a self-exciting default model and study the associated recursive HJB equation (see also Chapter 4).

1.1.3 *Optimal Tracking Portfolio with Infinite Horizon*

The third example is slightly different from previous two examples on optimal investment-consumption. This example considers a dynamic optimal tracking portfolio problem over the infinite horizon. We first formulate the optimal tracking problem which is closely related to that discussed in Section 7.7.2 of Chapter 7.

Consider a financial market consists of d risky assets. The price process of the ith asset is described as the following Black–Scholes model, for $t \geq 0$,

$$\frac{dS_t^i}{S_t^i} = \mu_i dt + \sum_{j=1}^m \sigma_{ij} dW_t^j, \quad i=1,\ldots,d, \tag{1.33}$$

where $W = (W_t^1,\ldots,W_t^m)^\top_{t\geq 0}$ is an m-dimensional Brownian motion, $\mu_i \in \mathbb{R}$ and $\sigma_{ij} \in \mathbb{R}$. It is assumed that the interest rate $r=0$ that amounts to the change of numéraire. From this point onwards, all processes including the wealth process and the benchmark process are defined after the change of numéraire. In Section 7.7.2 of Chapter 7, it is assumed that $m = d$, but here m may be not same to d.

Let $B = (B_t)_{t\geq 0}$ be an another Brownian motion which is independent of W and $\mathbb{F} = (\mathcal{F}_t)_{t\geq 0}$ be the natural filtration generated by (W,B). Denote by $\theta^i = (\theta_t^i)_{t\geq 0}$ the amount of wealth (as an $\mathbb{F}$-adapted process) that the fund manager allocates in asset $S^i = (S_t^i)_{t\geq 0}$ at time t. The self-financing wealth process under the control $\theta = (\theta_t^1,\ldots,\theta_t^d)^\top_{t\geq 0}$ is given by

$$V_t^\theta = \mathrm{v} + \int_0^t \theta_s^\top \mu ds + \int_0^t \theta_s^\top \sigma dW_s, \tag{1.34}$$

where the initial wealth $V_0^\theta = \mathrm{v} \geq 0$, $\mu = (\mu_1,\ldots,\mu_d)^\top$ is the return vector, and $\sigma = (\sigma_{ij})_{d\times m}$ is the volatility matrix.

We consider the portfolio allocation by a fund manager that is to optimally track an exogenous capital benchmark process $Z = (Z_t)_{t\geq 0}$ satisfying a geometric Brownian motion (c.f. Example 2.2 in Chapter 2):

$$\frac{dZ_t}{Z_t} = \mu_Z dt + \sigma_Z dB_t \tag{1.35}$$

with the initial value $Z_0 = z > 0$, $\mu_Z \in \mathbb{R}$ and $\sigma_Z > 0$. We assume that the fund manager can inject capital to the fund account from time to time whenever it is necessary such that the total capital dynamically dominates the benchmark floor process Z. That is, the fund manager optimally tracks the process Z by choosing the regular control θ as the dynamic portfolio in risky assets and the singular control $C = (C_t)_{t\geq 0}$ as the cumulative capital injection such that $C_t + V_t^\theta \geq Z_t$ at each intermediate time $t \geq 0$.

The goal of the optimal tracking problem is to minimize the expected cost of the discounted cumulative capital injection under American-type floor constraints that

$$\begin{cases} u(\mathrm{v},z) := \inf_{C,\theta} \mathbb{E}\left[C_0 + \int_0^\infty e^{-\rho t} dC_t\right], \\ \text{s.t. } Z_t \leq C_t + V_t^\theta \text{ at each } t \geq 0, \end{cases} \tag{1.36}$$

where the constant $\rho > 0$ is the discount rate and $C_0 = (z - \mathrm{v})^+$ is the initial injected capital to match with the initial benchmark.

To study problem (1.36) with dynamic floor constraints, our first step is to reformulate it based on the observation that, for a fixed control θ, the optimal C is always the smallest adapted right-continuous and non-decreasing process that dominates $Z - V^\theta$. Let $\mathcal{U}$ be the set of regular $\mathbb{F}$-adapted control processes $\theta = (\theta_t)_{t\geq 0}$ such that (1.34) is well-defined. By applying Lemma 7.1 in Chapter 7, we obtain an equivalent representation for problem (1.36) given by

$$u(\mathrm{v},z) = (z-\mathrm{v})^+ + \inf_{\theta\in\mathcal{U}} \mathbb{E}\left[\int_0^\infty e^{-\rho t} d\left(0 \vee \sup_{s\in[0,t]} (Z_s - V_s^\theta)\right)\right]. \tag{1.37}$$

As in Section 7.2 of Chapter 7, we define the difference process by $D_t := Z_t - V_t^\theta + \mathrm{v} - a$ for $t > 0$, and $D_0 = 0$. For any $x \geq 0$, consider its running maximum process by $L_t := x \vee \sup_{s \in [0,t]} D_s \geq 0$ for $t \geq 0$. Thus, we have that $(z - \mathrm{v})^+ - u(\mathrm{v}, z)$ with $u(\mathrm{v}, z)$ given in (1.37) is equivalent to the auxiliary stochastic control problem:

$$\sup_{\theta \in \mathcal{U}} \mathbb{E}\left[-\int_0^\infty e^{-\rho s} dL_s\right], \quad L_0 = x = (\mathrm{v} - z)^+. \tag{1.38}$$

We also introduce the new controlled state process $X = (X_t)_{t \geq 0}$ for problem (1.38), which is defined as the reflected process $X_t := L_t - D_t$ for $t \geq 0$ that satisfies

$$X_t = -\int_0^t \mu_Z Z_s ds - \int_0^t \sigma_Z Z_s dB_s + \int_0^t \theta_s^\top \mu ds$$
$$+ \int_0^t \theta_s^\top \sigma dW_s + L_t, \tag{1.39}$$

where $X_0 = L_0 = x \geq 0$. The running maximum process L_t increases if and only if $X_t = 0$, i.e., $L_t = D_t$. Hereafter, we use L^X to highlight the dependence of L on X, and the benchmark process $Z = (Z_t)_{t \geq 0}$ defined in (1.35) is chosen as another state process. We define the optimal value function for the auxiliary problem (1.38) by

$$\mathrm{w}(x, z) := \sup_{\theta \in \mathcal{U}} J(\theta; x, z) := \sup_{\theta \in \mathcal{U}} \mathbb{E}\left[-\int_0^\infty e^{-\rho s} dL_s^X\right]. \tag{1.40}$$

Hence, we have the equivalence that $\mathrm{w}(z, (\mathrm{v} - z)^+) = (z - \mathrm{v})^+ - u(\mathrm{v}, z)$.

In the sequel, we apply the change of measure to reduce the dimension of problem (1.40). Recall the dynamics satisfied by X, Z given by

$$\begin{cases} dX_t = -\mu_Z Z_t dt - \sigma_Z Z_t dB_t + \theta_t^\top \mu dt + \theta_t^\top \sigma dW_t + dL_t^X, \\ dZ_t = \mu_Z Z_t dt + \sigma_Z Z_t dB_t. \end{cases}$$

Define the new process $Y_t = \frac{X_t}{Z_t}$ for all $t \geq 0$. Since $dZ_t^{-1} = Z_t^{-1}((\sigma_Z^2 - \mu_Z)dt - \sigma_Z dB_t)$, we obtain $d[Z^{-1}, X]_t = \sigma_Z^2 dt$. Then, the Itô's rule yields that

$$dY_t = (\sigma_Z^2 - \mu_Z)(Y_t + 1)dt - \sigma_Z(Y_t + 1)dB_t + \tilde{\theta}_t^\top \mu dt$$
$$+ \tilde{\theta}_t^\top \sigma dW_t + dL_t^Y, \tag{1.41}$$

where, for all $t \geq 0$,

$$\tilde{\theta}_t := \frac{\theta_t}{Z_t} \quad \text{and} \quad L_t^Y := \int_0^t \frac{dL_s^X}{Z_s}. \tag{1.42}$$

We deduce from (1.42) that, for all $t \geq 0$,

$$\int_0^t \mathbf{1}_{Y_s=0} dL_s^Y = \int_0^t \mathbf{1}_{X_s=0} \frac{dL_s^X}{Z_s} = \int_0^t \frac{dL_s^X}{Z_s} dL_s^X = L_t^Y, \quad a.s.$$

This yields that the non-decreasing (non-negative) process increases only on $\{t \geq 0; Y_t = 0\}$. That is, $L^Y = (L_t^Y)_{t \geq 0}$ is the local time process for $Y = (Y_t)_{t \geq 0}$. As a consequence, it follows from (1.41) that

$$dY_t = (\sigma_Z^2 - \mu_Z)(Y_t + 1)dt - \sigma_Z(Y_t + 1)dB_t + \tilde{\theta}_t^\top \mu dt + \tilde{\theta}_t^\top \sigma dW_t + dL_t^Y$$
$$= -\mu_Z(Y_t + 1)dt - \sigma_Z(Y_t + 1)d\tilde{B}_t + \tilde{\theta}_t^\top \mu dt + \tilde{\theta}_t^\top \sigma dW_t + dL_t^Y, \quad (1.43)$$

where $\tilde{B}_t := B_t - \sigma_Z t$ for $t \geq 0$. We introduce the change of measure given by the density process $\frac{d\mathbb{Q}}{d\mathbb{P}}|_{\mathcal{F}_t} = \xi_t$ with $\xi_t := e^{-\mu_Z t} Z_t$. Hence, $\tilde{B} = (\tilde{B}_t)_{t \geq 0}$ is a $\mathbb{Q}$-Brownian motion, which is also independent of $W = (W)_{t \geq 0}$.

We next consider the following stochastic control problem, for $(Y_0, Z_0) = (y, z) \in [0, \infty) \times \mathbb{R}_+$,

$$u(y) := \sup_{\tilde{\theta} \in \tilde{\mathcal{U}}} \mathbb{E}\left[-\int_0^\infty e^{-\rho s} \frac{Z_s}{Z_0} L_s^Y\right]$$
$$= \sup_{\tilde{\theta} \in \tilde{\mathcal{U}}} \mathbb{E}^{\mathbb{Q}}\left[-\int_0^\infty e^{-(\rho - \mu_Z)s} dL_s^Y\right]. \quad (1.44)$$

In virtue of (1.40), we have from (1.42) that, for $(X_0, Z_0) = (x, z) \in [0, \infty) \times \mathbb{R}_+$ (and hence $Y_0 = \frac{x}{z}$),

$$w(x, z) = \sup_{\theta \in \mathcal{U}} \mathbb{E}\left[-\int_0^\infty e^{-\rho s} dL_s^X\right] = \sup_{\tilde{\theta} \in \tilde{\mathcal{U}}} \mathbb{E}\left[-\int_0^\infty e^{-\rho s} Z_s dL_s^Y\right]$$
$$= z \sup_{\tilde{\theta} \in \tilde{\mathcal{U}}} \mathbb{E}\left[-\int_0^\infty e^{-\rho s} \frac{Z_s}{Z_0} dL_s^Y\right] = zu\left(\frac{x}{z}\right). \quad (1.45)$$

Above, $\tilde{\mathcal{U}}$ is the admissible control set for $\tilde{\theta}$. It follows from (1.44) and (1.45) that, to solve problem (1.40), it suffices to solve the problem in RHS of (1.44), i.e.,

$$\begin{cases} u(y) = \sup_{\tilde{\theta} \in \tilde{\mathcal{U}}} \mathbb{E}^{\mathbb{Q}}\left[-\int_0^\infty e^{-\tilde{\rho} s} dL_s^Y\right], \quad \tilde{\rho} := \rho - \mu_Z, \\ \text{s.t. } dY_t = -\mu_Z(Y_t + 1)dt - \sigma_Z(Y_t + 1)d\tilde{B}_t + \tilde{\theta}_t^\top \mu dt \\ \qquad\qquad + \tilde{\theta}_t^\top \sigma dW_t + dL_t^Y, \end{cases} \quad (1.46)$$

where we choose $\rho > \mu_Z$, and hence $\tilde{\rho} > 0$. Using the dynamic program, the value function u formally satisfies the following stationary HJB equation, for $y > 0$,

$$\begin{cases} \dfrac{\sigma_Z^2}{2}(y+1)^2 u''(y) - \mu_Z(y+1)u'(y) \\ \quad + \sup\limits_{\theta \in \mathbb{R}^n} \left\{ u'(y)\theta^\top \mu \dfrac{u''(y)}{2} \theta^\top \sigma \sigma^\top \theta \right\} = \tilde{\rho} u(y), \\ u'(0) = 1. \end{cases} \quad (1.47)$$

If $\sigma\sigma^\top$ is invertible, the (feedback) optimal control is obtained by

$$\tilde{\theta}^*(y) = -\left(\sigma\sigma^\top\right)^{-1} \frac{u'(y)\mu}{u''(y)}, \quad \forall y \geq 0. \quad (1.48)$$

Then, plugging (1.48) into the HJB equation (1.47) to have that

$$-\alpha \frac{(u'(y))^2}{u''(y)} + \frac{\sigma_Z^2}{2}(y+1)^2 u''(y) - \mu_Z(y+1)u'(y) = \tilde{\rho} u(y) \quad (1.49)$$

with $\alpha := \tfrac{1}{2}\mu^\top \left(\sigma\sigma^\top\right)^{-1} \mu$. We can obtain the closed-form solution of Eq. (1.49) with $u'(0) = 1$ given by

$$u(y) = \frac{\gamma - 1}{\gamma}(1+y)^{\frac{\gamma}{\gamma-1}}, \quad \forall y \geq 0. \quad (1.50)$$

It follows from (1.49) that

$$\tilde{\theta}^*(y) = -(\gamma-1)(1+y)\left(\sigma\sigma^\top\right)^{-1}\mu, \quad \forall y \geq 0. \quad (1.51)$$

Here, $\gamma \in (0,1)$ is the unique solution to the following equation (since $\tilde{\rho} = \rho - \mu_Z > 0$):

$$\ell(\gamma) := \alpha\gamma(\gamma-1)^2 + \rho(\gamma-1)^2 - \frac{1}{2}\sigma_Z^2 \gamma + \mu_Z(\gamma-1) = 0. \quad (1.52)$$

In fact, since $\tilde{\rho} > 0$, we can easily check that

$$\lim_{\gamma \to -\infty} \gamma(\gamma) = -\infty, \quad \gamma(0) = \tilde{\rho} > 0, \quad \gamma(1) = -\frac{1}{2}\sigma_Z^2 < 0, \quad \lim_{\gamma \to +\infty} \gamma(\gamma) = +\infty.$$

This yields that Eq. (1.52) has a unique root $\gamma \in (0,1)$.

Since the HJB equation (1.47) is solvable in a closed-form, one can apply this closed-form solution to prove the verification result (transversality condition) associated with the control problem (1.46).

1.2 Perspective

The purpose of this book is to introduce graduate students and young researchers some recent studies in financial models and methodology motivated by different problems related to dynamic portfolio optimization. The book is comprised of main chapters (Chapters 4–9) to represent different stochastic control problems and their financial applications and implications, which may help readers to understand some new focuses in financial decision making and inspire them to formulate other new research problems in dynamic portfolio optimization and beyond.

On the other hand, some topics on dynamical portfolio optimization documented in this book deserve further extensions. For instance, the market model with default contagion studied in Chapters 4 and 5 can also be applied to the optimal reinsurance problems in the field of actuarial science. One can incorporate the consumption strategy into the optimal tracking portfolio problem discussed in Chapter 7, which leads to a type of extended Merton's problems with relaxed benchmark tracking partially discussed in Chapter 8. Moreover, for some research problems introduced in this book, one can also generalize the single investor to a large number of competitive agents where the competition among agents occurs in a mean field sense. That is, some previous optimal portfolio problems may be formulated and studied as mean field game problems.

1.3 Overview

Chapters 2 and 3 provide an introduction to the fundamentals of stochastic analysis and stochastic control problems. The main body of the book focuses on four research topics in dynamical portfolio optimization in continuous time framework. The first topic discusses the risk-sensitive credit asset management problem with contagious default risk in Chapters 4 and 5. The market risk is modeled as a continuous time Markov chain with countable infinitely states and the contagious risk is established as a time-inhomogeneous Markov chain in which the generator is coupled with the market risk factor. Chapter 4 deals with the complete information case, while Chapter 5 studies the partial information case. The second topic studies the optimal portfolio-consumption problem in incomplete semimartingale markets when the consumption is measured by the habit formation preference in Chapter 6. The martingale and duality approach is developed in the general semimartingale framework when the value function has two state variables and the dual process may not be integrable. The third topic is documented in Chapter 7 and examines a type of relaxed benchmark tracking scheme when the wealth process compensated by the fictitious capital injection dominates the non-decreasing benchmark process at all times. By introducing the auxiliary controlled state process with reflection, the dynamic floor constraint can be hidden and we can equivalently study the associated HJB equation with Neumann boundary condition. The fourth topic is concerned with an extension to the optimal tracking problem discussed in Chapter 7 by introducing the optimal consumption as in Merton (1969, 1971). This results in a type of so-called extended Merton's problem which is studied in Chapter 8. This chapter establishes the well-posedness of the resulting stochastic control problem by proposing a tailor-made decomposition-homogenization technique. The last topic in Chapter 9 concerns the reinforcement learning for the infinite horizon optimal tracking portfolio problem using capital injection in incomplete market with unknown model parameters. The exploratory formulation for the equivalent control problem with entropy regularization is introduced. This chapter develops the continuous-time q-learning algorithm based on reflected diffusion processes and shows the satisfactory performance of the q-learning algorithm in an illustrative numerical example.

Chapter 2

Preliminary

This chapter presents a basic knowledge of elementary stochastic differential equations (SDEs) and some previous exposures to Feymann–Kac's formula and backward stochastic differential equations (BSDEs).

2.1 Stochastic Differential Equations

Let n, m be two positive integers. Consider two Borel functions $b : \mathbb{R}^n \to \mathbb{R}^n$ and $\sigma : \mathbb{R}^n \to \mathbb{R}^{n \times m}$. More precisely, we write $b = (b_1, \ldots, b_n)^\top$ and $\sigma = (\sigma_{ij}; i = 1, \ldots, n, j = 1, \ldots, m)$. For a complete probability space $(\Omega, \mathcal{F}, \mathbb{P})$ with a reference filtration $\mathbb{F} = (\mathcal{F}_t)_{t \geq 0}$ satisfying the usual conditions, let $W = (W_t)_{t \geq 0}$ be an m-dimensional $(\mathbb{P}, \mathbb{F})$-Brownian motion. Then, an Itô SDE can be described as the following integral equation:

$$X_t = X_0 + \int_0^t b(X_s)ds + \int_0^t \sigma(X_s)dW_s, \quad t \geq 0, \tag{2.1}$$

where $A_t^X := \int_0^t b(X_s)ds$ for $t \geq 0$ is the absolutely continuous part and $M_t^X := \int_0^t \sigma(X_s)dW_s$ for $t \geq 0$ is the Itô's stochastic integral part as a local martingale.

For the given Brownian motion W on the filtered probability space $(\Omega, \mathcal{F}, \mathbb{F}, \mathbb{P})$, an $\mathbb{F}$-adapted (right continuous with left limits) process $X = (X_t)_{t \geq 0}$ is called a *strong* solution of SDE (2.1) if it satisfies (2.1), $\mathbb{P}$-a.s. with $A^X = (A_t^X)_{t \geq 0}$ and $M^X = (M_t^X)_{t \geq 0}$ being well-defined. In this case, $X_t = X_0 + A_t^X + M_t^X$ for $t \geq 0$ gives the Doob–Meyer decomposition of the solution process X. Hence, X is a continuous $(\mathbb{P}, \mathbb{F})$-semimartingale. A *weak* solution of SDE (2.1) is defined as sextuple $(\Omega, \mathcal{F}, \mathbb{F}, \mathbb{P}, W = (W_t)_{t \geq 0}, X = (X_t)_{t \geq 0})$, where W is an m-dimensional $(\mathbb{P}, \mathbb{F})$-Brownian motion and X is an $\mathbb{F}$-adapted (right continuous with left limits) process satisfying (2.1), $\mathbb{P}$-a.s.

It is obvious to have that a strong solution is also a weak solution, not vice-versa. We next give a fundamental result on existence and uniqueness of strong solutions to SDE (2.1):

Theorem 2.1. *Suppose that the coefficients (b,σ) in SDE (2.1) satisfy the following Lipschitz condition with linear growth:*

(**A**$_{lip}$) *for all $x,y \in \mathbb{R}^n$, there exist constants $L, K > 0$ which are independent of x, y such that*

$$|b(x) - b(y)| + |\sigma(x) - \sigma(y)| \leq L|x - y|,$$
$$|b(x)| + |\sigma(x)| \leq K(1 + |x|).$$

Let $\mathbb{E}[|X_0|^p] < +\infty$ for some $p \geq 1$. Then, for any $T > 0$, there exists a unique $\mathbb{F}^W := (\sigma(W_s;\ s \leq t))_{t \geq 0}$-adapted, continuous solution X of SDE (2.1) satisfying

$$\|X\|_{p,T}^p := \mathbb{E}\left[\sup_{t \in [0,T]} |X_t|^p\right] < +\infty. \tag{2.2}$$

Proof. The proof of the theorem is standard, see Karatzas and Shreve (1991, Chapter 5), by using the Picard iteration and Borel–Cantelli lemma. Here, we give a fixed point argument. More precisely, for $p \geq 1$, let $\mathcal{X}^p$ be the set of $\mathbb{F}^W$-adapted, continuous processes $X = (X_t)_{t \in [0,T]}$ satisfying $\|X\|_{p,T} < +\infty$ for any $T > 0$. Then, it is easy to see that $(\mathcal{X}^p, \|\cdot\|_T)$ is a Banach space. We then define a mapping $\mathcal{T}$ on $\mathcal{X}^p$ as: for any $X \in \mathcal{X}^p$,

$$(\mathcal{T}X)_t := X_0 + \int_0^t b(X_s)ds + \int_0^t \sigma(X_s)dW_s, \quad \forall t \in [0,T].$$

Obviously, $(\mathcal{T}X)_0 = X_0$. It follows from the linear growth of (b,σ) in Assumption (**A**$_{lip}$) that $A^X = (A_t^X)_{t \in [0,T]}$ is an $\mathbb{F}^W$-adapted absolutely continuous process and $M^X = (M_t^X)_{t \in [0,T]}$ is a continuous $(\mathbb{P}, \mathbb{F}^W)$-local martingale. Hence $t \to (\mathcal{T}X)_t$ is continuous and $\|\mathcal{T}X\|_{p,T} < \infty$ using BDG inequality. This includes that $\mathcal{T}X \in \mathcal{X}^p$.

For any $X, Y \in \mathcal{X}^p$, using Assumption (**A**$_{lip}$) and the similar argument above, we have

$$\|\mathcal{T}X - \mathcal{T}Y\|_T \leq C_{T,p}\|X - Y\|_T,$$

where $C_{T,p}$ is a positive constant (which may depend on L, K, p and T) such that $T \to C_{T,p}$ is a non-decreasing continuous function satisfying $\lim_{T \downarrow 0} C_{T,p} = 0$. Choose $T = t_0$ small enough such that $C_{T,p} < 1$, we have a unique fixed point $X^* \in \mathcal{X}^p$ such that $X^* = \mathcal{T}X^*$ on $[0, t_0]$. Since $C_{T,p}$

depends on L, K, T, p only (it is independent of X_0), we can divide $[0, T]$ into infinitely many small time intervals. In each small interval, we have a unique fixed point and fit them together on $[0, T]$. □

In many cases, the coefficients (b, σ) may be *not globally* Lipschitz continuous. Let us impose the following conditions:

($\mathbf{A}_{loc}$) b and σ are *locally* Lipschitz continuous on $\mathbb{R}^n$.
($\mathbf{A}_{lya}$) there exists a function $V : \mathbb{R}^n \to \mathbb{R}_+$ which satisfies that (i) let $q_R := \inf_{|x|>R} V(x)$, then $\lim_{R \to \infty} q_R = +\infty$; (ii) there exists a constant $C > 0$ such that, for all $(t, x) \in [0, T] \times \mathbb{R}^n$,

$$\mathbb{E}_x[V(X_{t \wedge \tau_R \wedge T})] \leq V(x) + C \int_0^t \{1 + \mathbb{E}_x[V(X_{s \wedge \tau_R \wedge T})]\} ds,$$

where $\tau_R := \inf\{t \geq 0; |X_t| \geq R\}$ with $\inf \emptyset = +\infty$ by convention and $\mathbb{E}_x[\cdot] := \mathbb{E}[\cdot | X_0 = x]$.

Then, we have

Theorem 2.2. *Let Assumptions* ($\mathbf{A}_{loc}$) *and* ($\mathbf{A}_{lya}$) *be satisfied. Then, for any $T > 0$, there exists a unique $\mathbb{F}^W$-adapted and continuous solution $X = (X_t)_{t \in [0,T]}$ of SDE* (2.1).

Proof. By Theorem 2.1, using Assumption ($\mathbf{A}_{loc}$), we have, for any $T > 0$ and $R > 0$, SDE (2.1) has a unique continuous strong solution on $[0, \tau_R \wedge T]$. We next prove $\tau_R \wedge T \to T$ as $R \to \infty$, $\mathbb{P}_x$-a.s.. In fact, by the condition (ii) of Assumption ($\mathbf{A}_{lya}$), we have

$$\mathbb{E}_x[V(X_{t \wedge \tau_R \wedge T})] \leq e^{Ct}(1 + V(x)), \quad \forall t \in [0, T].$$

Therefore, for all $t \in [0, T]$, and $R > 0$,

$$\mathbb{P}_x(\tau_R \wedge T < t) \leq \frac{1}{q_R} \mathbb{E}_x\left[\mathbf{1}_{\tau_R \wedge T < t} V(X_{\tau_R \wedge T})\right] = \frac{1}{q_R} \mathbb{E}_x\left[\mathbf{1}_{\tau_R \wedge T < t} V(X_{t \wedge \tau_R \wedge T})\right]$$
$$\leq \frac{e^{Ct}}{q_R}(1 + V(x)).$$

This yields that, for all $t \in [0, T]$, $\mathbb{P}_x(\tau_R \wedge T < t) \to 0$ as $R \to \infty$, using the condition (i) of Assumption ($\mathbf{A}_{loc}$). Thus, we complete the proof of the theorem. □

The linear SDE may have a closed-form solution, which is given by the following lemma:

Lemma 2.1. *For any $T > 0$, let $Y = (Y_t)_{t\in[0,T]}$ and $Z = (Z_t)_{t\in[0,T]}$ be one-dimensional continuous $(\mathbb{P},\mathbb{F})$-semimartingales. Consider the following linear SDE given by*

$$X_t = Y_t + \int_0^t X_s dZ_s, \quad \forall t \in [0,T]. \tag{2.3}$$

Then, the linear SDE (2.3) admits a closed-form solution given by, for all $t \in [0,T]$,

$$X_t = \mathcal{E}_t(Z)\left(Y_0 + \int_0^t \mathcal{E}_s(Z)^{-1}(dY_s - d[Y,Z]_s)\right), \tag{2.4}$$

where recall that the process $\mathcal{E}(Z) = (\mathcal{E}_t(Z))_{t\in[0,T]}$ is Doléans–Dade exponential of the semimartingale Z, i.e., it satisfies that $\frac{d\mathcal{E}_t(Z)}{\mathcal{E}_t(Z)} = dZ_t$ with $\mathcal{E}_0(Z) = 1$.

Proof. Let us first consider the following simple linear SDE:

$$X_t^1 = 1 + \int_0^t X_s^1 dZ_s, \quad \forall t \in [0,T].$$

Then, the solution is Doléans–Dade exponential of Z, i.e., $X_t^1 = \mathcal{E}(Z)_t$ for $t \in [0,T]$. Consider the solution of the linear SDE admitting the form:

$$X_t = X_t^1 L_t, \quad \forall t \in [0,T], \tag{2.5}$$

where $L = (L_t)_{t\in[0,T]}$ is a continuous semimartingale with $L_0 = Y_0$. Using integration by parts, it holds that

$$dX_t = d(X_t^1 L_t) = L_t dX_t^1 + X_t^1 dL_t + d[X^1, L]_t$$
$$= X_t^1 L_t dZ_t + X_t^1 dL_t + d[X^1, L]_t$$
$$= X_t dZ_t + X_t^1 dL_t + d[X^1, L]_t.$$

Compare the above equation with (2.3), i.e., $dX_t = X_t dZ_t + dY_t$, we obtain that

$$dY_t = X_t^1 dL_t + d[X^1, L]_t, \quad L_0 = Y_0.$$

This yields that

$$dL_t = (X_t^1)^{-1} dY_t - (X_t^1)^{-1} d[X^1, L]_t, \quad L_0 = Y_0. \tag{2.6}$$

Note that $\int_0^\cdot (X_s^1)^{-1} d[X^1, L]_t$ is a finite variation (FV) process. Therefore

$$d[X^1, L]_t = d\left[\int_0^\cdot X_s^1 dZ_s, \int_0^\cdot (X_s^1)^{-1} dY_s\right]_t = d[Y, Z]_t.$$

We get from (2.6) get

$$dL_t = (X_t^1)^{-1}(dY_t - d[Y,Z]_t) = \mathcal{E}_t(Z)^{-1}(dY_t - d[Y,Z]_t), \quad L_0 = Y_0.$$

Then, we arrive at

$$L_t = Y_0 + \int_0^t \mathcal{E}_s(Z)^{-1}(dY_s - d[Y,Z]_s), \quad \forall t \in [0,T].$$

Thus, the proof of the lemma is complete. $\square$

The linear SDE (2.3) in Lemma 2.1 involves many important examples in the practical modeling for physics and finance. We next introduce several related examples as special one-dimensional linear SDEs.

Example 2.1 (OU process). Let $W = (W_t)_{t \geq 0}$ be a scalar $(\mathbb{P}, \mathbb{F})$-Brownian motion, i.e., $m = 1$. We consider the special coefficient (b, σ) of SDE (2.1) with $n = 1$ as follows: for $x \in \mathbb{R}$, let $b(x) = \alpha(\beta - x)$ and $\sigma(x) \equiv \sigma$ with $\alpha, \beta, \sigma > 0$. This results in the following linear SDE (it is called Ornstein–Uhlenbeck (OU) process):

$$dX_t = \alpha(\beta - X_t)dt + \sigma dW_t. \tag{2.7}$$

By taking the semimartingales Y, Z in Lemma 2.1 as $Y_t = X_0 + \alpha\beta t + \sigma W_t$ and $Z_t = -\alpha t$ for $t \in [0, T]$, we have $[Y, Z] \equiv 0$. Then, it follows from (2.4) that

$$X_t = X_0 e^{-\alpha t} + \int_0^t e^{-\alpha(t-s)}(\alpha\beta ds + \sigma dW_s)$$
$$= X_0 e^{-\alpha t} + \beta(1 - e^{-\alpha t}) + \sigma \int_0^t e^{-\alpha(t-s)} dW_s. \tag{2.8}$$

The OU process is a class of important stochastic models in physics (Langevin equation) and finance (Vasicek interest rate model). It is also both continuous Gaussian process and semimartingale. Moreover, by using (2.8), the mean and covariance functions are respectively given by, for $x \in \mathbb{R}$,

$$\mathbb{E}_x[X_t] = xe^{-\alpha t} + \beta(1 - e^{-\alpha t}), \quad \forall t \geq 0,$$
$$\mathrm{Cov}_x(X_t, X_s) = \frac{\sigma^2}{2\alpha}\left(e^{-\alpha(t-s)} - e^{-\alpha(t+s)}\right), \quad \forall s, t \geq 0.$$

In view of the expression of the above mean function, we have

$$\lim_{t \to \infty} \mathbb{E}_x[X_t] = \beta, \quad \forall x \in \mathbb{R}, \tag{2.9}$$

which is called the mean-reverting property satisfied by the OU process (2.7).

In particular, the Langevin equation is given by $dV_t = -\alpha V_t dt + dW_t$ (i.e., the case where the mean reverting level $\beta = 0$ in (2.7)). Then, the OU process (2.8) reduces to

$$V_t = e^{-\alpha t}\left(V_0 + \int_0^t e^{\alpha s} dW_s\right), \quad t \geq 0.$$

The above Langevin equation is founded by French physicist, Paul Langevin (1872–1946), who is one of students of French physicist Pierre Curie (1859–1906). In the above Langevin equation, $\int_0^t V_s ds$ is used by physicists as a model of physical Brownian motion. More recently, Langevin equation and its variation have been used to improve the performance of SGD algorithm in the machine learning and the non-convex optimization problem (c.f. Cheng et al. (2020)).

Example 2.2 (GBM). Let $W = (W_t)_{t \geq 0}$ be a scalar $(\mathbb{P}, \mathbb{F})$-Brownian motion, i.e., $m = 1$. We consider the special coefficient (b, σ) of SDE (2.1) with $n = 1$ as follows: for $x > 0$, let $b(x) = \mu x$ and $\sigma(x) = \sigma x$ with $\mu \in \mathbb{R}$ and $\sigma > 0$. This leads to the following SDE which is called geometric Brownian motion (GBM):

$$\frac{dX_t}{X_t} = \mu dt + \sigma dW_t, \quad X_0 > 0. \tag{2.10}$$

By taking the semimartingales Y, Z in Lemma 2.1 as $Y_t \equiv X_0$ and $Z_t = \mu t + \sigma W_t$ for $t \geq 0$. Then, $[Y, Z] \equiv 0$. Hence, it follows from (2.4) that

$$X_t = X_0 \mathcal{E}_t(Z) = X_0 \exp\left(\left(\mu - \frac{\sigma^2}{2}\right)t + \sigma W_t\right), \quad t \geq 0. \tag{2.11}$$

In finance, GBM is used to model the stock price which is known as Black–Scholes (BS) model. Myron Samuel Scholes (1941–) is Canadian–American financial economist, Frank E. Buck Professor of Finance, Emeritus, at the Stanford Graduate School of Business and the winner of 1997 Nobel Prize winner in economics. The main motivation of the stock price model by using GBM is based on the following discretization of GBM:

$$\underbrace{\frac{X_{t_{i+1}} - X_{t_i}}{X_{t_i}}}_{\text{Stock Return}} = \underbrace{\mu \Delta t_i + \sigma \xi_i}_{\text{Return+Risk}}, \quad \xi_i := W_{t_{i+1}} - W_{t_i} \sim N(0, \Delta t_i),$$

where $\Delta t_i := t_{i+1} - t_i$.

Let us consider an investor who will invest her wealth in a stock with price process $X = (X_t)_{t \geq 0}$ and a money market account with a spot interest rate $r \geq 0$. The initial wealth of the investor is given by $v > 0$. For $t \geq 0$, denote by θ_t the number of shares (as an $\mathbb{F}$-adapted process) that the investor allocates in asset X at time t. Let V_t be the wealth level of the investor at time t, and κ_t the number of money market accounts invested in the money market account with price $B_t = e^{rt}$. Then, we have

$$V_t = \theta_t X_t + \kappa_t B_t, \quad t \geq 0. \tag{2.12}$$

By the *self-financing* assumption (i.e., in an enough small time interval, the change of the investor's wealth is only determined by the price change of assets), it follows that

$$dV_t = \theta_t dX_t + \kappa_t dB_t, \quad V_0 = v > 0. \tag{2.13}$$

Note that $dB_t = rB_t dt$. Then, using (5.33) and (2.10), we arrive at
$$dV_t = \theta_t X_t \frac{dX_t}{X_t} + r\kappa_t B_t dt = \theta_t X_t(\mu dt + \sigma dW_t) + r(V_t - \theta_t X_t)dt.$$
Therefore, we have the wealth dynamics of the investor given by
$$dV_t = rV_t dt + (\mu - r)\theta_t X_t dt + \sigma \theta_t X_t dW_t, \quad V_0 = v > 0. \tag{2.14}$$

The investor will choose $\theta = (\theta_t)_{t \geq 0}$ (as a portfolio strategy) such that $V_t > 0$, $\mathbb{P}$-a.s.. We may introduce the amount ϕ_t (resp. the fraction π_t) of wealth that the investor allocates in stock at time t. Then, it holds that
$$\phi_t = \theta_t X_t, \quad \pi_t = \frac{\phi_t}{V_t}, \quad \forall t \geq 0.$$
Hence, the wealth dynamics (2.14) can be rewritten as follows:
$$dV_t = rV_t dt + (\mu - r)\phi_t dt + \sigma \phi_t dW_t, \quad V_0 = v > 0, \tag{2.15}$$
$$dV_t = rV_t dt + (\mu - r)\pi_t V_t dt + \sigma \pi_t V_t dW_t, \quad V_0 = v > 0. \tag{2.16}$$
If the consumption of the investor is additionally considered, then the wealth dynamics (2.15)–(2.16) with a consumption rate strategy $C = (C_t)_{t \geq 0}$ is given by
$$dV_t = rV_t dt + (\mu - r)\phi_t dt + \sigma \phi_t dW_t - C_t dt, \quad V_0 = v > 0, \tag{2.17}$$
$$dV_t = rV_t dt + (\mu - r)\pi_t V_t dt + \sigma \pi_t V_t dW_t - C_t dt, \quad V_0 = v > 0. \tag{2.18}$$

Example 2.3 (GARCH). Let $W = (W_t)_{t \geq 0}$ be a scalar $(\mathbb{P}, \mathbb{F})$-Brownian motion, i.e., $m = 1$. We consider the special coefficient (b, σ) of SDE (2.1) with $n = 1$ as follows: for $x > 0$, $b(x) = \theta - ax$ and $\sigma(x) = \sigma x$ with $\theta, a, \sigma > 0$. Then, we have the following linear SDE:
$$dX_t = (\theta - aX_t)dt + \sigma X_t dW_t, \quad X_0 > 0. \tag{2.19}$$
By taking the semimartingales Y, Z in Lemma 2.1 as $Y_t = X_0 + \theta t$ and $Z_t = -at + \sigma W_t$ for $t \geq 0$. Then, we have $[Y, Z] \equiv 0$. Hence, it follows from (2.4) that
$$X_t = \mathcal{E}_t(Z)\left(X_0 + \theta \int_0^t \mathcal{E}_s(Z)^{-1} ds\right), \quad t \geq 0. \tag{2.20}$$
Obviously, if $\theta = 0$ in (2.20), then X becomes GBM given by (2.11). Therefore, we also call it inhomogeneous Brownian motion. In Lewis (2000), it is also referred to as a GARCH model. Related to the GARCH model, the solution of the following SDE: for $a, \sigma > 0$ and $\theta > \sigma^2$,
$$dY_t = (\theta - aY_t)Y_t dt + \sigma Y_t dW_t, \quad Y_0 > 0 \tag{2.21}$$
is called an inverse GARCH model. This is because that $Y^{-1} = (Y_t^{-1})_{t \geq 0}$ will be a GARCH model.

In some one-dimensional case, the coefficients (b, σ) of SDE (2.1) is even not locally Lipschitz continuous. However, one may expect to have a unique strong solution for the equation. Let us first consider the following example which is introduced by Itô and Watanabe (1978):

$$X_t = \int_0^t 3X_t^{\frac{1}{3}} dt + \int_0^t 3X_s^{\frac{2}{3}} dW_s, \quad t \geq 0, \qquad (2.22)$$

where $W = (W_t)_{t \geq 0}$ is a scalar $(\mathbb{P}, \mathbb{F})$-Brownian motion. It can be seen that the coefficients $b(x) = 3x^{\frac{1}{3}}$ and $\sigma(x) = 3x^{\frac{2}{3}}$ for $x \in \mathbb{R}$, although continuous in x, are not smooth at $x = 0$. They are *not* locally Lipschitz continuous. It is not difficult to verify by Itô's formula that $X_t = 0$ and $X_t = W_t^3$ for $t \geq 0$ are two different strong solutions of SDE (2.22), i.e., (strong) uniqueness does not hold. The strong uniqueness for SDE can be extended to the so-called *pathwise uniqueness* of SDE which is defined as:

Definition 2.1 (Pathwise Uniqueness of Solutions to SDE). Let $(\Omega, \mathcal{F}, \mathbb{F}, \mathbb{P}, W = (W_t)_{t \geq 0}, X = (X_t)_{t \geq 0})$ and $(\Omega, \mathcal{F}, \tilde{\mathbb{F}}, \mathbb{P}, W = (W_t)_{t \geq 0}, \tilde{X} = (\tilde{X}_t)_{t \geq 0})$ be two weak solutions of SDE (2.1). If $\mathbb{P}(X_0 = \tilde{X}_0) = 1$ and $\mathbb{P}(X_t = \tilde{X}_t; \forall t \geq 0) = 1$ (i.e., X and $\tilde{X}$ are indistinguishable), then we call the pathwise uniqueness holds for SDE (2.1).

An important implication of the pathwise uniqueness is that *the pathwise uniqueness and the weak existence jointly imply the strong existence* (c.f. Yamada and Watanabe (1971)).

We next introduce the following so-called Yamada–Watanabe SDE given in the following theorem:

Theorem 2.3 (Yamada and Watanabe, 1971). *Let the coefficients (b, σ) in SDE (2.1) with $n = m = 1$ satisfy that*

$$|b(x) - b(y)| \leq C|x - y|, \quad |\sigma(x) - \sigma(y)| \leq \rho_\sigma(|x - y|), \quad \forall x, y \in \mathbb{R}, \quad (2.23)$$

where $C > 0$ and $\rho_\sigma : [0, \infty) \to [0, \infty)$ is strictly increasing function with $\rho_\sigma(0) = 0$ and

$$\int_{(0,\epsilon)} \frac{ds}{\rho_\sigma^2(s)} = +\infty, \quad \forall \epsilon > 0. \qquad (2.24)$$

Let $X = (X_t)_{t \geq 0}$ and $Y = (Y_t)_{t \geq 0}$ be two strong solutions with $X_0 - Y_0$ being integrable. Then, there exists a constant $C > 0$ such that

$$\mathbb{E}\left[|X_t - Y_t|\right] \leq e^{Ct} \mathbb{E}[|X_0 - Y_0|], \quad \forall t \geq 0. \qquad (2.25)$$

Proof. In light of (2.24), there exists a sequence of strictly decreasing $(a_k)_{k\geq 1} \subset (0,1]$ such that $a_0 = 1$, $a_\infty = 0$, and
$$\int_{a_k}^{a_{k-1}} \rho_\sigma^{-2}(x)dx = k, \quad \forall k \geq 1.$$
For each $k \geq 1$, one can construct a $C(\mathbb{R})$-probability density function ρ_k with support (a_k, a_{k-1}) satisfying $0 \leq \rho_k(x) \leq \frac{2}{k\rho_\sigma^2(x)}$ for all $x > 0$. Define a sequence of auxiliary $C^2(\mathbb{R})$-functions by, for $k \geq 1$,
$$\psi_k(x) := \int_0^{|x|} \int_0^y \rho_k(s) ds dy, \quad \forall x \in \mathbb{R}. \tag{2.26}$$
Then, $|\psi_k'(x)| \leq 1$, $\lim_{k\to\infty} \psi_k(x) = |x|$ for $x \in \mathbb{R}$, and $(\psi_k)_{k\geq 1}$ is non-decreasing. Note that $\mathbb{E}[\int_0^t \psi_k''(X_s - Y_s)\rho_\sigma^2(|X_s - Y_s|)] \leq \frac{2t}{k}$. Using those properties, Itô formula and the localization argument, it follows from (2.23) that, for all $k \geq 1$,
$$\mathbb{E}[\psi_k(X_t - Y_t)] \leq \mathbb{E}[|X_0 - Y_0|] + C \int_0^t \mathbb{E}[|X_s - Y_s|]ds + \frac{t}{k}, \quad \forall t \geq 0.$$
By taking limit as $k \to \infty$ in both sides of the above display, we conclude the estimate (2.25) by using Gronwall's lemma. $\square$

Since the solution of Yamada–Wanatabe SDE is continuous a.s., it implies from (2.25) that the pathwise uniqueness of the equation holds. On the other hand, it is not difficult to show that Yamada-Wanatabe SDE admits a weak solution using approximating argument, tightness and Skorokhod's representation theorem (see also Trutnau (2010)). Then, it follows from Theorem 2.3 that the equation has a unique strong solution.

Let us look back to the condition (2.23) satisfied by the volatility function $\sigma(\cdot)$. It is obvious to see that $b(x) = \alpha(\beta - x)$ with $\alpha, \beta > 0$ and $\sigma(x) = \sigma|x|^p$ with $\sigma > 0$ and $p \geq \frac{1}{2}$ satisfy (2.23). In particular, in this case, if $X_0 \geq 0$, a.s., we also have the solution $X_t \geq 0$ for $t \geq 0$, a.s. using the comparison theorem (Proposition 5.2.18 in Karatzas and Shreve (1991)). If p is also assumed to be less one, then the equation is also called the *constant elasticity of variance* (CEV) model with p being referred to as the elasticity of variance. The CEV model was first considered by Cox (1996). For $p = \frac{1}{2}$, the CEV model becomes the so-called Cox–Ingersoll–Ross (CIR) process (c.f. Cox et al. (1985)) or Feller diffusion process (c.f. Feller (1951)):
$$X_t = X_0 + \int_0^t \alpha(\beta - X_s)ds + \sigma \int_0^t \sqrt{X_s}dW_s, \quad X_0 > 0. \tag{2.27}$$

Let us talk more about the CIR process because it has many important applications in the modeling of finance and queueing system. Note that $\sigma^2(x) = \sigma^2 x$ is also a linear function in x, then the CIR process is an *affine process* introduced in Duffie et al. (2003). In addition, the CIR process is also related to the OU process given in Example 2.1. In fact, let $W = (W^1, \ldots, W^m)$ be an m-dimensional Brownian motion. For $\beta, \sigma > 0$, consider the following OU process with mean-reverting level zero:

$$dX_t^j = -\frac{\beta}{2} X_t^j dt + \frac{\sigma}{2} dW_t^j, \quad X_0^j \in \mathbb{R}, \quad j = 1, \ldots, m.$$

Define $X_t := \sum_{j=1}^m (X_t^j)^2$ for $t \geq 0$. Then, it follows from Itô's formula that

$$dX_t = \left(\frac{\sigma^2 m}{4} - \beta X_t\right) dt + \sigma \sum_{j=1}^m X_t^j dW_t^j, \quad t \geq 0.$$

Define $B_t := \sum_{j=1}^m \int_0^t \frac{X_s^j}{\sqrt{X_s}} dW_s^j$ for $t \geq 0$, which is continuous local martingale. Moreover, we have $\langle B, B \rangle_t = t$ for $t \geq 0$. This implies from the Lévy theorem that $B = (B_t)_{t \geq 0}$ is a Brownian motion. Thus, we show that $X_t := \sum_{j=1}^m (X_t^j)^2$ for $t \geq 0$ satisfies (2.27) with $\alpha = \frac{\sigma^2 m}{4}$. Finally, let us introduce the so-called Feller's condition given by

$$\alpha \beta \geq \frac{\sigma^2}{2}, \tag{2.28}$$

under which, we have $\mathbb{P}(X_t > 0, \ \forall \ t \geq 0) = 1$. Namely, under Feller's condition (2.28), the CIR process will never hit boundary zero if the initial date $X_0 > 0$.

2.2 Hybrid Diffusion System

This section introduces a class of stochastic differential equations with regime-switching. This class of equations is usually called a hybrid diffusion system, which can be described as follows:

$$dX_t = b(X_t, I_t) dt + \sigma(X_t, I_t) dW_t, \quad X_0 \in \mathbb{R}^n, \tag{2.29}$$

where the process $I = (I_t)_{t \geq 0}$ is a pure jump process with a discrete state space $S := \{1, 2, \ldots, d\}$ with $d \geq 1$ being a finite integer or $+\infty$, and meanwhile it can couple with the solution process $X = (X_t)_{t \geq 0}$ such that, for $t \geq 0$ and $(i, j) \in S$,

$$\mathbb{P}(I_{t+\delta} = j | I_t = i, X_t = x) = \begin{cases} q_{ij}(x)\delta + o(\delta), & i \neq j, \\ 1 + q_{ii}(x)\delta + o(\delta), & i = j \end{cases} \tag{2.30}$$

uniformly in $x \in \mathbb{R}^n$ as $\delta \downarrow 0$. We further assume that

(A$_q$) (i) for all $x \in \mathbb{R}^n$, $q_{ij}(x) \geq 0$ for $i \neq j$ and $\sum_{j \in S} q_{ij}(x) = 0$ for all $i \in S$. (ii) there exists a constant $C > 0$ such that, for all $i \in S$ and $x \in \mathbb{R}^n$,

$$q_i(x) := -q_{ii}(x) = \sum_{j \in S \setminus \{i\}} q_{ij}(x) \leq C(1+i),$$

$$\sum_{j \in S \setminus \{i\}} (f(j) - f(i))q_{ij}(x) \leq C(1 + f(i) + |x|^2),$$

where $f : S \to [0, \infty)$ is a nondecreasing function satisfying $\lim_{k \to \infty} f(k) = \infty$. (iii) there exists a constant $\gamma \in (0, 1]$ such that, for all $i \in S$ and $x, y \in \mathbb{R}^n$,

$$\sum_{j \in S \setminus \{i\}} |q_{ij}(x) - q_{ij}(y)| \leq C|x - y|^\gamma.$$

If I is a continuous-time Markov chain (i.e., $q_{ij} = q_{ij}(x)$), then Assumption (A$_q$) holds. We next impose the assumptions on the coefficients (b, σ) of SDE (2.29):

(A$_{lip,i}$) for any $i \in S$ and $x, y \in \mathbb{R}^n$,

$$|b(x, i) - b(y, i)| + |\sigma(x, i) - \sigma(y, i)| \leq C|x - y|,$$
$$|b(x, i)| + |\sigma(x, i)| \leq C(1 + |x|),$$

where $C > 0$ is a constant which is independent of i, x, y.

Then, we have the following existence and uniqueness of strong solutions of the hybrid diffusion system (2.29)–(2.30):

Theorem 2.4 (Xi and Zhu (2017)). *Let Assumptions (A$_q$) and (A$_{lip,i}$) hold. Then, the hybrid diffusion system (2.29)–(2.30) has a unique strong solution with $(X_0, I_0) = (x, i)$ for $(x, i) \in \mathbb{R}^n \times S$.*

Proof. The proof of the theorem relies on a classical "interlacing procedure" proposed in Ikeda and Watanabe (1989). We here give a sketch of the proof (see Xi and Zhu (2017) for more details). First of all, we define a sequence of stopping times as follows, for any $(x, i) \in \mathbb{R}^n \times S$, consider the following SDE:

$$X_t^i = x + \int_0^t b(X_s^i, i)ds + \int_0^t \sigma(X_s^i, i)dW_s, \quad t \geq 0. \tag{2.31}$$

With Assumption (A$_{lip,i}$), it follows from Theorem 2.1 that SDE (2.31) admits a unique strong solution. Let $(\xi_k)_{k \geq 1}$ be a sequence of standard exponential random variables which are independent of W. Then, we define

$$\tau_1 = \theta_1 := \inf\left\{t \geq 0; \int_0^t q_i(X_s^i)ds > \xi_1\right\}. \tag{2.32}$$

Above, we set $\inf \emptyset = +\infty$ by convention. Therefore, for all $t \geq 0$,

$$\mathbb{P}(\tau_1 > t | \mathcal{F}_t) = \exp\left(-\int_0^t q_i(X_s^i) ds\right).$$

Using Assumption $(\mathbf{A}_{lip,i})$-(ii), we have $\mathbb{P}(\tau_1 > t) \geq e^{-C(1+i)t}$. This yields that $\mathbb{P}(\tau_1 > 0) = 1$ using the continuity of $A \to \mathbb{P}(A)$. Thus, one can define the process (X, I) on $[0, \tau_1]$ as follows:

$$X_t = X_t^i, \ \forall \, t \in [0, \tau_1], \text{ and } I_t = i, \ \forall \, t \in [0, \tau_1),$$

and define $I_{\tau_1} \in S$ according to the probability law given by

$$\begin{cases} \mathbb{P}(I_{\tau_1} = j | \mathcal{F}_{\tau_1 -}) = \dfrac{q_{ij}(X_{\tau_1 -})}{q_i(X_{\tau_1 -})} \mathbf{1}_{q_i(X_{\tau_1 -}) > 0}, & i \neq j, \\ \mathbb{P}(I_{\tau_1} = j | \mathcal{F}_{\tau_1 -}) = \mathbf{1}_{q_i(X_{\tau_1 -}) = 0}, & i = j. \end{cases} \quad (2.33)$$

We inductively define, for $l \geq 0$ ($\tau_0 = 0$),

$$\tau_{l+1} := \tau_l + \theta_{l+1}, \quad \theta_{l+1} := \inf\left\{t \geq 0; \int_0^t q_{I_{\tau_l}}(X_s^{I_{\tau_l}}) ds > \xi_{l+1}\right\}, \quad (2.34)$$

where the state process $X_t^{I_{\tau_l}}$ for $t \geq 0$ is given by

$$X_t^{I_{\tau_l}} = X_{\tau_l} + \int_0^t b(X_s^{I_{\tau_l}}, I_{\tau_l}) ds + \int_0^t \sigma(X_s^{I_{\tau_l}}, I_{\tau_l}) dW_s, \quad t \geq 0. \quad (2.35)$$

It follows from Theorem 2.1 and Assumption $(\mathbf{A}_{lip,i})$ that SDE (2.35) admits a unique strong solution. We also have that

$$\mathbb{P}(\theta_{l+1} > t | \mathcal{F}_{\tau_l + t}) = \exp\left(-\int_0^t q_{I_{\tau_l}}(X_s^{I_{\tau_l}}) ds\right), \quad \forall \, t \geq 0. \quad (2.36)$$

Assumption $(\mathbf{A}_{lip,i})$-(ii) also yields that $\mathbb{P}(\theta_{l+1} > 0) = 1$. Thus, one can define the process (X, I) on $[\tau_l, \tau_{l+1}]$ as follows:

$$X_t = X_{t-\tau_l}^{I_{\tau_l}}, \ \forall \, t \in [\tau_l, \tau_{l+1}], \text{ and } I_t = I_{\tau_l}, \ \forall \, t \in [\tau_l, \tau_{l+1}),$$

and define $I_{\tau_{l+1}} \in S$ according to the probability law given by

$$\begin{cases} \mathbb{P}(I_{\tau_{l+1}} = j | \mathcal{F}_{\tau_{l+1} -}) = \dfrac{q_{I_{\tau_l} j}(X_{\tau_{l+1} -})}{q_{I_{\tau_l}}(X_{\tau_{l+1} -})} \mathbf{1}_{q_{I_{\tau_l}}(X_{\tau_{l+1} -}) > 0}, & I_{\tau_l} \neq j, \\ \mathbb{P}(I_{\tau_{l+1}} = j | \mathcal{F}_{\tau_{l+1} -}) = \mathbf{1}_{q_{I_{\tau_l}}(X_{\tau_{l+1} -}) = 0}, & I_{\tau_l} = j. \end{cases} \quad (2.37)$$

Repeating the above procedure, we can construct a strong Markov process (X_t, I_t) satisfying (2.29)–(2.30) for $t \in [0, \tau_\infty)$ with $\tau_\infty := \lim_{l \to \infty} \tau_l$ (note that $l \to \tau_l$ is strictly increasing).

For the proof of the global solution, it suffices to prove $\tau_\infty \to \infty$, a.s.. Let $(X_0, I_0) = (x, i) \in \mathbb{R}^n \times S$. For any $k \geq i+1$, denote by ρ_k the first exit time of I from the finite set $S_k := \{1, \ldots, k-1\}$. Namely, $\rho_k := \inf\{t \geq 0; I_t \geq k\}$ and $\rho_k = +\infty$ if $\{t \geq 0; I_t \geq k\} = \emptyset$. We then define
$$A_i := \{\omega \in \Omega; \tau_\infty(\omega) \leq \rho_{k_0}(\omega), \exists\, k_0 \geq i+1\}.$$
Then $\mathbb{P}(\tau_\infty = +\infty) = \mathbb{P}(\tau_\infty = +\infty|A_i)\mathbb{P}(A_i) + \mathbb{P}(\tau_\infty = +\infty|A_i^c)\mathbb{P}(A_i^c)$.

In the sequel, we first verify $\mathbb{P}(\tau_\infty = \infty|A_i) = 1$. In fact, for all $\omega \in A_i$, we have $\tau_\infty(\omega) \leq \rho_{k_0}(\omega)$ for some $k_0 \geq i+1$. This yields that $I_{\tau_d}(\omega) \in S_{k_0}$ for all $d \geq 0$. It follows from Assumption $(\mathbf{A}_q)$-(ii) that, on A_i, for all $l \geq 0$,
$$q_{I_{\tau_l}}(X_t^{I_{\tau_l}}) \leq C(I_{\tau_l} + 1) \leq Ck_0, \quad \forall t \geq 0.$$
By using (2.36), we arrive at, for all $l \geq 0$,
$$\mathbb{P}(\theta_{l+1} > t | \mathcal{F}_{\tau_l + t}) \geq \mathbf{1}_{A_i} \exp\left(-\int_0^t q_{I_{\tau_l}}(X_s^{I_{\tau_l}})ds\right) \geq \mathbf{1}_{A_i} e^{-Ck_0 t}.$$
This gives that $\mathbb{P}(\theta_{l+1} > t) \geq e^{-Ck_0 t}\mathbb{P}(A_i)$. For any $t > 0$, by Fatou's lemma
$$\mathbb{P}(\tau_\infty = \infty) = \mathbb{P}\left(\lim_{l \to \infty} \tau_l = \infty\right) \geq \mathbb{P}\left(\bigcap_{k=1}^{\infty} \bigcup_{\ell=k}^{\infty} \{\theta_\ell > t\}\right)$$
$$= \mathbb{P}\left(\limsup_{k \to \infty}\{\theta_k > t\}\right)$$
$$\geq \limsup_{k \to \infty} \mathbb{P}(\theta_k > t) \geq e^{-Ck_0 t}\mathbb{P}(A_i). \tag{2.38}$$
Therefore $\mathbb{P}(\tau_\infty = \infty|A_i) \geq e^{-Ck_0 t}$ for all $t > 0$. Thus, by letting $t \downarrow 0$, $\mathbb{P}(\tau_\infty = \infty|A_i) = 1$.

We next consider $\mathbb{P}(A_i^c) > 0$ (for $\mathbb{P}(A_i^c) = 0$, it is clear to have that $\mathbb{P}(\tau_\infty = \infty) = \mathbb{P}(\tau_\infty = \infty|A_i)\mathbb{P}(A_i) = 1 \times (1 - 0) = 1$). Since $k \to \rho_k$ is increasing, denote by $\rho_\infty := \lim_{k \to \infty} \rho_k$. Then, $A_i^c = \{\tau_\infty \geq \rho_\infty\}$. This gives that $\mathbb{P}(\tau_\infty = \infty|A_i^c) \geq \mathbb{P}(\rho_\infty = \infty|A_i^c)$. Hence, if we can prove
$$\mathbb{P}(\rho_\infty = \infty|A_i^c) = 1, \quad (\text{i.e. } \mathbb{P}(\rho_\infty < \infty|A_i^c) = 0), \tag{2.39}$$
then we would have $\mathbb{P}(\tau_\infty = \infty|A_i^c) = 1$. We next prove (2.39) by contradiction and hence assume that there exists a constant $v > 0$ such that $\alpha := \mathbb{P}(\rho_\infty \leq v, A_i^c) > 0$. Let $f : S \to [0, \infty)$ be the function given in Assumption $(\mathbf{A}_q)$. Then, by (2.40) in Lemma 2.2 below, for any $k \geq i+1$, we have
$$f(i) = \mathbb{E}_i\left[e^{-Cv \wedge \tau_\infty \wedge \rho_k} f(I_{v \wedge \tau_\infty \wedge \rho_k})\right]$$
$$+ \mathbb{E}_i\left[\int_0^{v \wedge \tau_\infty \wedge \rho_k} e^{-Cs}\left(Cf(I_s) - \sum_{j \in S} q_{I_s, j}(X_s)(f(j) - f(I_s))\right)ds\right]$$

$$\geq \mathbb{E}_i \left[e^{-Cv \wedge \tau_\infty \wedge \rho_k} f(I_{v \wedge \tau_\infty \wedge \rho_k}) \right]$$
$$+ \mathbb{E}_i \left[\int_0^{v \wedge \tau_\infty \wedge \rho_k} e^{-Cs} \left(Cf(I_s) - C(1 + I_s^2 + f(I_s)) \right) ds \right]$$
$$\geq \mathbb{E}_i \left[e^{-Cv \wedge \tau_\infty \wedge \rho_k} f(I_{v \wedge \tau_\infty \wedge \rho_k}) \right].$$

Hence, one can get

$$e^{Cv} f(i) \geq \mathbb{E}_i \left[f(I_{v \wedge \tau_\infty \wedge \rho_k}) \right] \geq \mathbb{E}_i \left[f(I_{\rho_k}) \mathbf{1}_{\rho_k \leq v \wedge \tau_\infty} \right] \geq f(k) \mathbb{P}(\rho_k \leq v \wedge \tau_\infty)$$
$$\geq f(k) \mathbb{P}(\rho_k \leq v \wedge \tau_\infty, A_i^c) \geq f(k) \mathbb{P}(\rho_\infty \leq v \wedge \tau_\infty, A_i^c).$$

Note that $A_i^c = \{\rho_\infty \leq \tau_\infty\}$. Then

$$\mathbb{P}(\rho_\infty \leq v \wedge \tau_\infty, A_i^c) = \mathbb{P}(\rho_\infty \leq v \wedge \tau_\infty, \rho_\infty \leq \tau_\infty) \geq \mathbb{P}(\rho_\infty \leq v, \rho_\infty \leq \tau_\infty)$$
$$= \mathbb{P}(\rho_\infty \leq v, A_i^c) = \alpha > 0.$$

Consequently, for all $k \geq i + 1$,

$$\infty > e^{Cv} f(i) \geq \alpha f(k) \xrightarrow{k \to \infty} \infty,$$

since $\lim_{k \to \infty} f(k) = \infty$. This results in a contradiction.

We omit the proof of the pathwise uniqueness of (2.29) under Assumption ($\mathbf{A}_{lip,i}$) since the proof is standard (see also Xi and Zhu (2017) for details). Thus, we complete the proof of the theorem. $\square$

The Dykin formula on Markov process (X, I) is given in the following lemma.

Lemma 2.2. *Let Assumptions ($\mathbf{A}_q$) and ($\mathbf{A}_{lip,i}$) hold. Then, for any $f \in (\cdot, i) \in C^2(\mathbb{R}^n)$ with $i \in S$,*

$$\mathbb{E}_{x,i} \left[f(X_{t \wedge \tau_\infty}, I_{t \wedge \tau_\infty}) \right] = f(x, i) + \mathbb{E}_{x,i} \left[\int_0^{t \wedge \tau_\infty} \mathcal{L} f(X_s, I_s) ds \right], \quad (2.40)$$

where, for $(x, i) \in \mathbb{R}^n \times S$, the conditional expectation $\mathbb{E}_{x,i}[\cdot] := \mathbb{E}[\cdot | X_0 = x, I_0 = i]$. The operator $\mathcal{L}$ is defined as, for $f(\cdot, i) \in C^2(\mathbb{R}^n)$ with $i \in S$,

$$\mathcal{L} f(x, i) := \frac{1}{2} \mathrm{tr}[\sigma \sigma^\top (x, i) \nabla_x^2 f(x, i)] + b(x, i)^\top \nabla_x f(x, i)$$
$$+ \sum_{j \neq i} q_{ij}(x)(f(x, j) - f(x, i)). \quad (2.41)$$

Proof. We first reconstruct the regime-switching process I in terms of stochastic integrals w.r.t. a class of Poisson random measures. To this purpose, we define the following sequence of intervals:

$$\begin{cases} R_{01}(x) := [0, q_{01}(x)), \ldots, R_{0d}(x) := \left[\sum_{j=1}^{d-1} q_{0j}(x), \sum_{j=1}^{d} q_{0j}(x)\right), \ldots, \\ R_{10}(x) := [q_0(x), q_0(x) + q_{10}(x)), \\ R_{12}(x) := [q_0(x) + q_{10}(x), q_0(x) + q_{10}(x) + q_{12}(x)), \ldots, \\ R_{20}(x) := [q_0(x) + q_1(x), q_0(x) + q_1(x) + q_{20}(x)), \ldots, \\ \quad \vdots \qquad \qquad \vdots \end{cases} \quad (2.42)$$

where we set $\sum_{j=1}^{0} \cdot = 0$. In addition, if $q_{ij}(x) = 0$ with $i \neq j$, we set $R_{ij}(x) = \emptyset$. It follows from (2.42) that the length of the interval $R_{ij}(x)$ is given by $q_{ij}(x)$ and $(R_{ij}(x))_{i,j \in S}$ are disjoint for any $x \in \mathbb{R}^n$. Then, we define the following function $\phi : \mathbb{R}^n \times S \times [0, \infty) \to \mathbb{R}$ as, for all $(x, i, z) \in \mathbb{R}^n \times S \times [0, \infty)$,

$$\phi(x, i, z) := \sum_{j \in S} (j - i) \mathbf{1}_{R_{ij}(x)}(z). \quad (2.43)$$

For $t \in [0, \tau_\infty)$, define $\lambda_t := \int_0^t q_{I_s}(X_s) ds$, and

$$C_t := \max\left\{ k \geq 1; \sum_{j=1}^{k} \xi_j \leq \lambda_t \right\},$$

where recall that $(\xi_j)_{j \geq 1}$ is a sequence of i.i.d. standard exponential random variables. Then $K := (K_{t \wedge \tau_\infty})_{t \geq 0}$ is a point process that counts the number of switches for the component I. We may view K as a non-homogeneous Poisson process with a random intensity $q_{I_t}(X_t)$ for $t \in [0, \tau_\infty)$.

For any $0 \leq s < t < \tau_\infty$ and $B \in \mathcal{B}(S)$, define

$$p((s, t] \times B) := \sum_{\ell \in (s,t]} \mathbf{1}_{I_\ell \neq I_{\ell-},\ I_\ell \in A}, \quad p(t, B) := p((0, t] \times B). \quad (2.44)$$

Then $p(t \wedge \tau_\infty, S) = K_{t \wedge \tau_\infty}$ for $t \geq 0$, and

$$I_{t \wedge \tau_\infty} = I_0 + \sum_{d=1}^{\infty} (I_{\tau_d} - I_{\tau_d-}) \mathbf{1}_{\tau_d \leq t \wedge \tau_\infty}$$

$$= I_0 + \sum_{j \in S} \int_0^{t \wedge \tau_\infty} (j - I_{s-}) p(ds, j). \quad (2.45)$$

We next define the Poisson random measure N on $[0,\infty) \times [0,\infty)$ as, for all $(t, A) \in [0,\infty) \times \mathcal{B}([0,\infty))$,

$$N(t \wedge \tau_\infty, A) := \sum_{j \in S \cap A} p(t \wedge \tau_\infty, j). \quad (2.46)$$

Additionally, $\mathbb{E}[N(t, A)] = m(A)t$ with m being Lebesgue measure on $\mathbb{R}$. Thus, we may rewrite (2.45) as follows:

$$I_{t \wedge \tau_\infty} = I_0 + \int_0^{t \wedge \tau_\infty} \int_0^\infty \phi(X_{s-}, I_{s-}, z) N(ds, dz). \quad (2.47)$$

Note that, for any $(x, i) \in \mathbb{R}^n \times S$ and $j \neq i$, we have

$$\begin{cases} m(\{z \geq 0;\ \phi(x, i, z) = 0\}) = q_i(x), \\ m(\{z \geq 0;\ \phi(x, i, z) = j - i\}) = q_{ij}(x), \end{cases} \quad (2.48)$$

and

$$\sum_{j \neq i} q_{ij}(x)(f(x, j) - f(x, i))$$
$$= \int_0^\infty (f(x, i + \phi(x, i, z)) - f(x, i)) m(dz). \quad (2.49)$$

Thus, the equality (2.40) follows from Itô's formula. $\square$

The hybrid diffusion system (2.29) can result in a class of important credit risk models with default contagion (c.f. Bo et al. (2019)). More precisely, let $S := \{0, 1\}^p$ be the state space of the so-called default indicator process $H = (H^1, \ldots, H^p)$ of p firms. In other words, if $H_t^i = 1$ (resp. $H_t^i = 0$), it is equivalent to saying that the i-th firm has (resp. does not) defaulted before time $t \geq 0$. Hence, $\tau_i := \inf\{t \geq 0;\ H_t^i = 1\}$ represents the corresponding default time of the i-th firm (by convention $\tau_i = \infty$ if $\{t \geq 0;\ H_t^i = 1\} = \emptyset$). We assume that a joint process (X, H) is a continuous Markov process with state space $\mathbb{R}^n \times S$. The joint process (X, H) satisfies the following hybrid diffusion system:

$$dX_t = b(X_t, H_t)dt + \sigma(X_t, H_t)dW_t, \quad (2.50)$$

and for $(x, h) \in \mathbb{R}^n \times S$,

$$\mathbb{P}(H_{t+\delta} = h^i | H_t = h, X_t = x, (X_s, H_s)_{s \leq t})$$
$$= \begin{cases} \delta \lambda_i(x, h) \mathbf{1}_{h_i = 0} + o(\delta), & i = 1, \ldots, p; \\ 1 + \delta \lambda_0(x, h) + o(\delta), & i = 0, \end{cases} \quad (2.51)$$

uniformly in $x \in \mathbb{R}^n$ as $\delta \downarrow 0$. Here $\lambda_0(x,h) = -\sum_{i=1}^{p} \lambda_i(x,h)\mathbf{1}_{h_i=0}$ and $(x,h) \to \lambda_i(x,h)$ is a measurable function taking values on $(0,\infty)$ for all $(x,h) \in \mathbb{R}^n \times S$. In (2.51), we have introduced the notation, for $h \in S$,

$$h^i = (h_1, \ldots, h_{i-1}, 1 - h_i, h_{i+1}, \ldots, h_p) \in S, \quad (2.52)$$

where we set $h^0 := h$. If the coefficients $(b, \sigma, \lambda_1, \ldots, \lambda_p)$ satisfy Assumptions $(\mathbf{A}_q)$ and $(\mathbf{A}_{lip,i})$, then (X, H) exists and it is unique by using Theorem 2.4.

Example 2.4 (Default Contagion with Two Names). We consider an example for the default indicator process H with two firms, i.e., $p = 2$ and $H_t = (H_t^1, H_t^2)$ for $t \geq 0$. Then, the default intensity matrix is given by

states	(0,0)	(0,1)	(1,0)	(1,1)
(0,0)	$-\sum_{i=1}^{2} \lambda_i(x,(0,0))$	$\lambda_2(x,(0,0))$	$\lambda_1(x,(0,0))$	0
(0,1)	0	$-\lambda_1(x,(0,1))$	0	$\lambda_1(x,(0,1))$
(1,0)	0	0	$-\lambda_2(x,(1,0))$	$\lambda_2(x,(1,0))$
(1,1)	0	0	0	0

Observe the above intensity matrix, we can see that the default state $(1,1)$ is an absorbing state. By Corollary 2.5 in Xi and Zhu (2017), the extended generator of (X, H) is given by, for all $f(\cdot, h) \in C^2(\mathbb{R}^n)$ with $h \in S$,

$$\mathcal{L}f(x,h) := \frac{1}{2}\mathrm{tr}[\sigma\sigma^\top(x,h)\nabla_x^2 f(x,h)] + b(x,h)^\top \nabla_x f(x,h)$$
$$+ \underbrace{\sum_{i=1}^{p}(1-h_i)\lambda_i(x,h)(f(x,h^i) - f(x,h))}_{\text{default contagion effect}}. \quad (2.53)$$

Then, with (2.53), using the Dynkin's formula by taking $f(x,h) = h_i$ for $i = 1, \ldots, p$, the following process

$$M_t^i := H_t^i - \int_0^t \mathcal{L}f(X_s, H_s)ds$$
$$= H_t^i - \int_0^t (1 - H_s^i)\lambda_i(X_s, H_s)(1 - H_s^i - H_s^i)ds$$
$$= H_t^i - \int_0^t (1 - H_s^i)\lambda_i(X_s, H_s)ds, \quad t \geq 0 \quad (2.54)$$

is a martingale w.r.t. the natural filtration generated by (X, H). In (2.54), we also used the fact that $(1 - H_t^i)H_t^i = 0$ and $(1 - H_t^i)^2 = 1 - H_t^i$, a.s.

The vector of price processes of the p stocks issued by the p firms is denoted by $\tilde{P} = (\tilde{P}^i_t; i = 1, \ldots, p)^\top_{t \geq 0}$. For $t \geq 0$, the price process of the i-th defaultable stock is then given by

$$\tilde{P}^i_t = (1 - H^i_t)P^i_t, \quad i = 1, \ldots, p. \tag{2.55}$$

In other words, the price of the i-th stock is given by the *pre-default price* $P^i = (P^i_t)_{t \geq 0}$ up to τ_i-, and jumps to 0 at time τ_i, where it remains forever afterwards (c.f. Figure 2.1). We refer to the price dynamics (2.55) as the

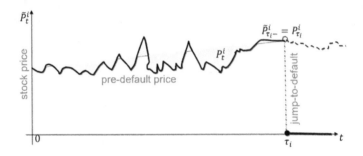

Fig. 2.1 The path of price process of the i-th stock.

jump-to-default (c.f. Linetsky (2006)). The dynamics of the pre-default price process $P = (P^i_t; i = 1, \ldots, p)^\top_{t \geq 0}$ of the p defaultable stocks is given by

$$dP_t = \mathrm{diag}(P_t)\left\{(\mu(X_t) + \lambda(X_t, H_t))dt + a(X_t)dW_t\right\}. \tag{2.56}$$

Above, we recall that $\mathrm{diag}(P_t)$ is the diagonal p-dimensional square matrix whose i-th entry is P^i_t. Recall that W is an m-dimensional Brownian motion. The vector μ is a $\mathbb{R}^p$-valued function, the matrix a is a $\mathbb{R}^{p \times m}$-valued function and $\lambda(x, h) = (\lambda_i(x, h); i = 1, \ldots, p)^\top$ for $(x, h) \in \mathbb{R}^n \times S$. The process $X = (X_t)_{t \geq 0}$ can be viewed as the stochastic factor which coupled by the default indicator process $H = (H_t)_{t \geq 0}$. Equation (2.56) indicates that the investor holding the credit sensitive stock is compensated for the incurred default risk at the premium rate (X_t, H_t) (i.e., the hazard rate needs to be added to the drift rate of the pre-default price process, e.g., Davis and Lischka (2002)). In terms of (2.56), we can apply Itô's rule to conclude that the price dynamics of p defaultable stocks is given by

$$d\tilde{P}_t = \mathrm{diag}(\tilde{P}_t)\left\{\mu(X_t)dt + a(X_t)dW_t - dM_t\right\}, \tag{2.57}$$

where $M_t = (M^1_t, \ldots, M^p_t)^\top$ for $t \geq 0$ is the pure jump (default) martingale defined by (2.54). The dynamics (2.56)-(2.57) will be a fundamental model of the price dynamics of multiple risky assets used in Chapter 4 and Chapter 5.

2.3 Feymann–Kac's Formula

The classical Feynman–Kac's formula, named after Richard Feynman (1918–1988, who won 1965 Nobel Prize in Physics) and Mark Kac (1914–1984), establishes a link between (linear) partial differential equations (PDEs) and stochastic analysis by providing stochastic representations for classical solutions of linear Kolmogorov PDEs.

In this section, we provide a general Feynman–Kac's formula proved in Heath and Schweizer (2000) under weak enough conditions to be satisfied in some typical examples from finance. Let $D \subseteq \mathbb{R}^n$ be a domain (i.e., an open connected subset of $\mathbb{R}^n$) which is not necessarily bounded (for instant, $D = \mathbb{R}^n$). We first make the following assumption on the domain D:

(**A**$_D$) there exists a sequence $(D_k)_{k \geq 1}$ of bounded domains with C^2-boundary and $\overline{D_k} \subset D$ s.t. $\bigcup_{k=1}^{\infty} D_k = D$.

Let $b : D \to \mathbb{R}^{n \times 1}$ and $\sigma : D \to \mathbb{R}^{n \times m}$ be two matrix-valued measurable functions. Consider the following SDE, for $(t, x) \in [0, \infty) \times D$,

$$X_s^{t,x} = x + \int_t^s b(X_r^{t,x}) dr + \int_t^s \sigma(X_r^{t,x}) dW_r, \quad \forall s \geq t. \tag{2.58}$$

The generator of $X^{0,x} = (X_t^{0,x})_{t \geq 0}$ acted on $C^2(D)$ is given by: for $f \in C^2(D)$,

$$\mathcal{A}f(x) := b(x)^\top \nabla_x f(x) + \frac{1}{2} \text{tr}[a(x) \nabla_x^2 f(x)], \quad \forall x \in D, \tag{2.59}$$

with $a(x) := \sigma \sigma^\top(x)$ for $x \in D$.

We impose the following assumption on SDE (2.58).

(**A**$_X$) for any $T > 0$, and $(t, x) \in [0, T] \times D$, the solution process $X^{t,x} = (X_s^{(t,x)})_{s \in [t,T]}$ of SDE (2.58) neither explodes nor leaves D before T, i.e.,

$$\mathbb{P}\left(\sup_{s \in [t,T]} |X_s^{t,x}| < \infty \right) = \mathbb{P}\left(X_s^{t,x} \in D, \, \forall \, s \in [t, T] \right) = 1.$$

Assumption (**A**$_X$) yields that $\sigma_T^t := \inf\{s \in [t, T); X_s^{t,x} \notin D\} = \inf \emptyset = T$. For $(t, x) \in [0, T] \times D$, we define

$$u(t, x) := \mathbb{E}\left[\phi(X_T^{t,x}) \exp\left(\int_t^T g(X_s^{t,x}) ds \right) \right]$$

$$- \mathbb{E}\left[\int_t^T f(s, X_s^{t,x}) \exp\left(\int_t^s g(X_r^{t,x}) dr \right) ds \right], \tag{2.60}$$

where $\phi : D \to \mathbb{R}$, $f : [0,T] \times D \to \mathbb{R}$ and $g : D \to \mathbb{R}$ are measurable functions.

The following lemma provides sufficient conditions under which the function u defined by (2.60) is well-defined.

Lemma 2.3 (Continuity of $(t,x) \to u(t,x)$). *Let Assumption* **(A_{lip})** *(imposed in Section 2.1), Assumption* **(A_X)** *and the following additional assumptions hold:*

(HSC) *the mapping $g : D \to \mathbb{R}$ is continuous and is bounded from above, $f : [0,T] \times D \to \mathbb{R}$ and $\phi : D \to \mathbb{R}$ are continuous. Moreover, there exist $C, p > 0$ such that, for all $(t,x) \in [0,T] \times D$,*
$$|f(t,x)| \vee |\phi(x)| \leq C(1 + |x|^p).$$

(HSX) *there exists a constant $C > 0$ such that*
$$\mathbb{E}\left[\sup_{s \in [t,T]} |X_s^{t,x}|^q\right] \leq C(1 + |x|^q), \quad \forall q \geq 1.$$

Then, the function $u : [0,T] \times D \to \mathbb{R}$ defined by (2.60) is continuous.

Remark 2.1. If $p = 0$ in Assumption **(HSC)** of Lemma 2.3 (i.e., f, ϕ are all bounded), then Lemma 2.3 holds without Assumption **(HSX)**. This is also the case considered in Heath and Schweizer (2000).

Proof of Lemma 2.3. Denoted by $V_{t,x}$ the term in the expectation of (2.60). Define $H_{t,x}^\epsilon := \{(s,y) \in [0,T] \times D; |s-t| + |y-x| < \epsilon\}$ for $\epsilon > 0$. Then, for $k > 1$ such that $pk \geq 1$, using Assumptions **(HSC)** and **(HSX)**, it follows that
$$\sup_{(s,y) \in H_{t,x}^\epsilon} \mathbb{E}[|V_{s,y}|^k] \leq C_{k,T}\left\{1 + \mathbb{E}\left[\sup_{s \in [t,T]} |X_s^{t,x}|^{pk}\right]\right\} \leq C_{k,T}(1 + |x|^{pk}).$$

This implies that the family $(V_{r,y})_{(r,y) \in H_{t,x}^\epsilon}$ is uniformly integrable. On the other hand, Assumptions **(A_{lip})** and **(A_X)** yield that $(s,t,x) \to X_s^{t,x}$ is $\mathbb{P}$-a.s. continuous. As a consequence, $(t,x) \to \phi(X_T^{t,x})$ is $\mathbb{P}$-a.s. continuous, $(s,t,x) \to g(X_s^{t,x})$ and $(s,t,x) \to f(s, X_s^{t,x})$ are $\mathbb{P}$-a.s. uniformly continuous and bounded on any compact set of $[0,T] \times [0,T] \times D$. Then $(t,x) \to \int_t^T g(X_s^{t,x})ds$ and $(t,x) \to \int_t^T f(s, X_s^{t,x})ds$ are $\mathbb{P}$-a.s. continuous. In summary, $(t,x) \to V_{t,x}$ is $\mathbb{P}$-a.s. continuous. Thus, the uniform integrability of $(V_{r,y})_{(r,y) \in H_{t,x}^\epsilon}$ implies that $(t,x) \to u(t,x) = \mathbb{E}[V_{t,x}]$ is continuous. $\square$

Building upon the continuity of $(t,x) \to u(t,x)$ verified in Lemma 2.3, for each $k \geq 1$, consider the following initial-boundary problem on $[0,T] \times D_k$:

$$\begin{cases} (\partial_t + \mathcal{A} + g)v_k(t,x) = f(t,x), & \text{in } (t,x) \in [0,T) \times D_k, \\ v_k(T,x) = u(T,x) = \phi(x), & \text{on } D_k, \\ v_k(t,x) = u(t,x), & \text{on } [0,T) \times \partial D_k, \end{cases} \quad (2.61)$$

where $\mathcal{A}$ is the operator defined by (2.59), and the function $u(t,x)$ is given by (2.60). It is, in general, more reasonable to call (2.61) a terminal-boundary problem. However, we can change t to $T-t$ and then transfer it into an initial-boundary problem.

We further assume that, for each $k \geq 1$,

(A$_{\text{HS}}$)

  (AHSba): the operator $\mathcal{A}$ is uniformly elliptic in D_k, i.e., there is a $l_k > 0$ s.t. $\xi^\top a(x)\xi \geq l_k|\xi|^2$ for all D_k and $\xi \in \mathbb{R}^n$;
  (AHSfg): the mapping g is Hölder continuous on $\overline{D}_k$ and f is uniformly Hölder continuous on $[0,T] \times \overline{D}_k$.

Then, we have

Theorem 2.5 (Feynman–Kac's Formula). *Let Assumptions* **(A$_{lip}$)**, **(A$_X$)**, **(A$_D$)**, **(A$_{\text{HS}}$)**, **(HSC)** *and* **(HSX)** *hold. Then, $u(t,x)$ defined by the probabilistic representation* (2.60) *belongs to* $C^{1,2}([0,T) \times D) \cap C([0,T] \times D)$, *and satisfies the following Cauchy problem:*

$$\begin{cases} (\partial_t + \mathcal{A} + g)u(t,x) = f(t,x), & \text{in } (t,x) \in [0,T) \times D, \\ u(T,x) = \phi(x), & \text{on } D. \end{cases} \quad (2.62)$$

Furthermore, there exists a unique classical solution of Cauchy problem (2.62).

Proof. We give a sketch of the proof (c.f. Heath and Schweizer (2000) for details). We first note that, for any $k \geq 1$, Assumption **(A$_{lip}$)** implies that b, a are Lipschitz continuous on the bounded set $\overline{D}_k$. By Lemma 2.3, Assumptions **(HSC)** and **(HSX)** yield that $u(t,x)$ is continuous on $[0,T] \times \partial D_k \cup \{T\} \times \overline{D}_k$ (since $\overline{D}_k \subset D$). Combine above two claims and Assumption **(A$_{\text{HS}}$)** to obtain that the conditions imposed in Theorem 6.5.2 of Friedman (1975), page 147, are satisfied on $[0,T] \times D_k$. Then, using Friedman (1975), the initial-boundary problem (2.61) admits a unique classical solution $v_k(t,x)$. Now, for any $(t,x) \in [0,T] \times D$, the assumption $(\mathbf{A}_D)$ implies that one can find a $k \geq 1$ s.t. $x \in D_k$. Define σ_k as the exit time of $X^{t,x}$ from D_k from t before T, i.e., $\sigma_k := \inf\{s \in [t,T); X_s^{t,x} \notin D_k\}$ and $\inf \emptyset = T$. Since the path $s \to X_s^{t,x}$ is continuous, we have $(\sigma_k, X_{\sigma_k}^{t,x}) \in (0,T) \times \partial D_k \cup \{T\} \times D_k$. By Lemma 2.3, we obtain $u(\sigma_k, X_{\sigma_k}^{t,x}) < \infty$. Therefore, we can also apply the stochastic representation of $v_k(t,x)$ given in Theorem 6.5.2 of Friedman (1975), one has

$$v_k(t,x) = \mathbb{E}\left[u(\sigma_k, X_{\sigma_k}^{t,x}) \exp\left(\int_t^{\sigma_k} g(X_s^{t,x})ds\right) \mathbf{1}_{\sigma_k < T}\right]$$
$$+ \mathbb{E}\left[\phi(X_T^{t,x}) \exp\left(\int_t^T g(X_s^{t,x})ds\right) \mathbf{1}_{\sigma_k = T}\right]$$
$$- \mathbb{E}\left[\int_t^{\sigma_k} f(s, X_s^{t,x}) \exp\left(\int_t^s g(X_r^{t,x})dr\right) ds\right]$$
$$\overset{u(T,x)=\phi(x)}{=\!=\!=\!=\!=} \mathbb{E}\left[u(\sigma_k, X_{\sigma_k}^{t,x}) \exp\left(\int_t^{\sigma_k} g(X_s^{t,x})ds\right)\right]$$
$$- \mathbb{E}\left[\int_t^{\sigma_k} f(s, X_s^{t,x}) \exp\left(\int_t^s g(X_r^{t,x})dr\right) ds\right].$$

We next prove that $u(t,x) = v_k(t,x)$ if $(t,x) \in [0,T] \times D_k$. Since $T \geq \sigma_k \geq t$, we get that

$$\phi(X_T^{t,x}) \exp\left(\int_t^T g(X_s^{t,x})ds\right) - \int_t^T f(s, X_s^{t,x}) \exp\left(\int_t^s g(X_r^{t,x})dr\right) ds$$
$$= \exp\left(\int_t^{\sigma_k} g(X_s^{t,x})ds\right) \phi(X_T^{t,x}) \exp\left(\int_{\sigma_k}^T g(X_s^{t,x})ds\right)$$
$$- \exp\left(\int_t^{\sigma_k} g(X_r^{t,x})dr\right) \int_{\sigma_k}^T f(s, X_s^{t,x}) \exp\left(\int_{\sigma_k}^s g(X_r^{t,x})dr\right) ds$$
$$- \int_t^{\sigma_k} f(s, X_s^{t,x}) \exp\left(\int_t^s g(X_r^{t,x})dr\right) ds.$$

This gives that

$$\mathbb{E}\left[\phi(X_T^{t,x}) \exp\left(\int_t^T g(X_s^{t,x})ds\right)\right.$$
$$\left. - \int_t^T f(s, X_s^{t,x}) \exp\left(\int_t^s g(X_r^{t,x})dr\right) ds \bigg| \mathcal{F}_{\sigma_k}\right]$$
$$= \exp\left(\int_t^{\sigma_k} g(X_s^{t,x})ds\right) \mathbb{E}\left[\phi(X_T^{t,x}) \exp\left(\int_{\sigma_k}^T g(X_s^{t,x})ds\right)\right.$$

$$-\int_{\sigma_k}^{T} f(s, X_s^{t,x}) \exp\left(\int_{\sigma_k}^{s} g(X_r^{t,x}) dr\right) ds \bigg| \mathcal{F}_{\sigma_k}\right]$$
$$-\int_{t}^{\sigma_k} f(s, X_s^{t,x}) \exp\left(\int_{t}^{s} g(X_r^{t,x}) dr\right) ds. \tag{2.63}$$

Using the strong Markov property of $X^{t,x}$, we have

$$\mathbb{E}\left[\phi(X_T^{t,x}) \exp\left(\int_{\sigma_k}^{T} g(X_s^{t,x}) ds\right)\right.$$
$$\left.-\int_{\sigma_k}^{T} f(s, X_s^{t,x}) \exp\left(\int_{\sigma_k}^{s} g(X_r^{t,x}) dr\right) ds \bigg| \mathcal{F}_{\sigma_k}\right] = u(\sigma_k, X_{\sigma_k}^{t,x}).$$

Taking expectation on both sides of (2.63), for each $k \geq 1$ and $(t, x) \in [0, T] \times D_k$,

$$u(t,x) = \mathbb{E}\left[u(\sigma_k, X_{\sigma_k}^{t,x}) \exp\left(\int_{t}^{\sigma_k} g(X_s^{t,x}) ds\right)\right]$$
$$- \mathbb{E}\left[\int_{t}^{\sigma_k} f(s, X_s^{t,x}) \exp\left(\int_{t}^{s} g(X_r^{t,x}) dr\right) ds\right] = v_k(t,x).$$

From Assumption (**A**$_D$) and (2.61), it follows that $u(t,x)$ satisfies Cauchy problem (2.62). The uniqueness follows from the probabilistic representation (2.60) for $u(t,x)$ since $X^{t,x}$ is unique. □

Example 2.5 (Option Pricing under Stochastic Volatility Model).
Consider the option pricing problem under a stochastic volatility model, i.e., the stock price dynamics under a risk-neutral pricing measure $\mathbb{Q}$ is given by

$$dS_t = rS_t dt + \sqrt{V_t} S_t dW_t^1. \tag{2.64}$$

The stochastic volatility process in (2.64) is described as a CIR process:

$$dV_t = \alpha(\beta - V_t)dt + \sigma\sqrt{V_t} d(\rho W_t^1 + \sqrt{1-\rho^2} W_t^2), \tag{2.65}$$

where $W = (W^1, W^2)^\top$ is a two-dimensional Brownian motion under $\mathbb{Q}$, $\alpha\beta \geq \frac{\sigma^2}{2}$ (i.e., Feller's condition given by (2.28)), and $\rho \in (-1, 1)$.

The risk-neutral price at time $t \in [0, T]$ of a European put option on $S = (S_t)_{t \in [0,T]}$ with maturity $T > 0$ and strike $K > 0$ is then given by

$$\mathbb{E}^{\mathbb{Q}}\left[e^{-r(T-t)}(K - S_T)^+ \big| \mathcal{F}_t\right] = v(t, S_t, V_t), \tag{2.66}$$

where $r \geq 0$ is the interest rate, $\mathbb{E}^{\mathbb{Q}}[\cdot]$ is the expectation operator under $\mathbb{Q}$, and $\mathbb{F} = (\mathcal{F}_t)_{t \in [0,T]}$ is the natural filtration generated by W representing the available market information. For $t \in [0,T]$ and $x = (x_1, x_2) \in D := (0, \infty)^2$, we introduce $X_t = (S_t, V_t)^\top$, $b(x) = (rx_1, \alpha(\beta - x_2))^\top$ and

$$\sigma(x) := \begin{bmatrix} x_1\sqrt{x_2} & 0 \\ \sigma\rho\sqrt{x_2} & \sigma\sqrt{1-\rho^2}\sqrt{x_2} \end{bmatrix}.$$

By taking $f(t,x) \equiv 0$, $g(x) \equiv -r$ and $\phi(x) = (K-x_1)^+$ (which is continuous and bounded), it is not difficult to verify that Assumptions in Theorem 2.5 can be satisfied. It follows from Theorem 2.5 that the following pricing equation admits a unique classical solution:

$$\begin{cases} (\partial_t + \mathcal{A})v(t,x) = rv(t,x), & \text{in } (t,x) \in [0,T) \times D, \\ v(T,x) = (K-x_1)^+, & \text{on } D. \end{cases} \quad (2.67)$$

Note that the payoff function $\phi(x) = (K-x_1)^+$ for $x \in D$ are automatically continuous and bounded. Then, by Remark 2.1, Lemma 2.3 holds without Assumption **(HSX)**. However, if one considers the European call option, then the corresponding payoff function is continuous, but not bounded. In this case, we need to apply Lemma 2.3 via the verification of Assumption **(HSX)** therein. This case on the pricing of call options was not discussed in Heath and Schweizer (2000).

2.4 Backward Stochastic Differential Equations

This section provides a brief introduction to the backward stochastic differential equations (BSDEs). In essence, BSDE is a nonlinear extension to the martingale representation theorem which is used to prove the well-posedness of BSDEs under Lipschitz drivers.

We start with the filtered probability space $(\Omega, \mathcal{F}, \mathbb{F}, \mathbb{P})$ introduced in Section 2.1. Let $W = (W_t)_{t \geq 0}$ be an m-dimensional $(\mathbb{P}, \mathbb{F})$-Brownian motion. Then, for convenience, we assume that $\mathbb{F} = \mathbb{F}^W = (\mathcal{F}_t^W)_{t \geq 0}$ with $\mathcal{F}_t^W = \sigma(W_s; s \leq t)$ for $t \geq 0$, which is augmented by all the $\mathbb{P}$-null sets. The one-dimensional BSDE can be written as follows, for a terminal horizon $T > 0$,

$$dY_t = -f(t, Y_t, Z_t)dt + Z_t^\top dW_t, \quad Y_T = \xi, \quad (2.68)$$

where ξ is an $\mathcal{F}_T$-measurable random variable under $\mathbb{P}$, and $f(t, \omega, y, z) : [0,T] \times \Omega \times \mathbb{R} \times \mathbb{R}^m \to \mathbb{R}$ is a random function. We call ξ the terminal condition and f the driver of the BSDE (2.68). The solution of this BSDE is in fact is a pair of $\mathbb{F}$-adapted ($\mathbb{R} \times \mathbb{R}^m$-valued) processes $(Y, Z) = (Y_t, Z_t)_{t \in [0,T]}$

satisfying (2.68), or equivalently the following integral form given by, for all $t \in [0,T]$,

$$Y_t = \xi + \int_t^T f(s, Y_s, Z_s) ds - \int_t^T Z_s^\top dW_s. \tag{2.69}$$

Note that the solution component Z is in fact uniquely determined by Y and W (formally, Z is the derivative of Y w.r.t. W, c.f. Zhang (2017)). Moreover, the role of the solution component Z can be used to guarantee the $\mathbb{F}$-adapted property of the solution component Y.

We next give the well-posedness of the BSDE (2.68) (or equivalently (2.69)). To this purpose, we define the following spaces for stochastic processes: for $p \geq 1$,

- The space S_T^p: the set of real-valued progressively measurable processes $Y = (Y_t)_{t \in [0,T]}$ such that

$$\mathbb{E}\left[\sup_{t \in [0,T]} |Y_t|^p\right] < +\infty.$$

- The space $H_{m,T}^p$: the set of $\mathbb{R}^m$-valued progressively measurable processes $Z = (Z_t)_{t \in [0,T]}$ such that

$$\mathbb{E}\left[\int_0^T |Z_t|^p dt\right] < +\infty.$$

We also impose the following assumptions on the terminal condition ξ and the driver f:

(BSDE-ξ) $\xi \in L^2(\mathbb{P}; \mathcal{F}_T)$, i.e., it is a square-integrable $\mathcal{F}_T$-adapted random variable.

(BSDE-f) write $f(t, y, z) := f(t, \cdot, y, z)$, it is progressively measurable for all $(y,z) \in \mathbb{R} \times \mathbb{R}^m$; $f(t,0,0) \in H_{1,T}^p$; $f(t,\cdot,\cdot)$ is Lipschitz in (y,z) uniformly in (t,ω), i.e., there exists a constant $K > 0$ such that, $dt \otimes \mathbb{P}(d\omega)$-a.s.

$$|f(t, y_1, z_1) - f(t, y_2, z_2)| \leq K(|y_1 - y_2| + |z_1 - z_2|),$$

for all $(y_i, z_i) \in \mathbb{R} \times \mathbb{R}^m$ with $i = 1, 2$.

Then, the well-posedness of the BSDE (2.68) (or equivalently (2.69)) is given as follows:

Theorem 2.6. *Let Assumptions* **(BSDE-ξ)** *and* **(BSDE-f)** *hold. Then, the BSDE* (2.68) *admits a unique solution* $(Y, Z) \in S_T^2 \times H_{m,T}^2$.

Proof. The proof is standard. We here provide a sketch of the proof based on a fixed point argument (see the proof of Theorem 4.3.1 in Zhang (2017), which is based on a Picard iteration). Let $\mathcal{X} := S_T^2 \times H_{m,T}^2$. For any $(U, V) \in \mathcal{X}$, we define

$$M_t := \mathbb{E}\left[\xi + \int_0^T f(s, U_s, V_s)ds \Big| \mathcal{F}_t\right], \quad \forall t \in [0, T].$$

Then, from Assumptions **(BSDE-ξ)** and **(BSDE-f)**, it follows that $M = (M_t)_{t \in [0,T]}$ is a square-integrable $(\mathbb{P}, \mathbb{F})$-martingale. The martingale representation theorem yields that, there exists a $\mathbb{R}^m$-valued process $Z \in H_{m,T}^2$ such that

$$M_t = M_0 + \int_0^t Z_s^\top dW_s, \quad \forall t \in [0, T]. \tag{2.70}$$

Given $Z \in H_{m,T}^2$ above, we then define that, for all $t \in [0, T]$,

$$Y_t := \mathbb{E}\left[\xi + \int_t^T f(s, U_s, V_s)ds \Big| \mathcal{F}_t\right] = M_t - \int_0^t f(s, U_s, V_s)ds. \tag{2.71}$$

It is easy from (2.71) to see that $Y_T = \xi$, and it holds from (2.70) that

$$Y_t - Y_T = M_t - M_T + \int_t^T f(s, U_s, V_s)ds$$
$$= -\int_t^T Z_s^\top dW_s + \int_t^T f(s, U_s, V_s)ds.$$

This is, we have

$$Y_t = \xi - \int_t^T Z_s^\top dW_s + \int_t^T f(s, U_s, V_s)ds. \tag{2.72}$$

We then define a mapping Φ on $\mathcal{X}$ via (2.72) as follows:

$$(Y, Z) = \Phi(U, V), \quad \forall (U, V) \in \mathcal{X}. \tag{2.73}$$

By applying BDG's inequality, we obtain $(Y, Z) \in \mathcal{X}$, i.e., $\Phi : \mathcal{X} \to \mathcal{X}$. Consequently, (Y, Z) is a solution of the BSDE (2.69) if and only if (Y, Z) is a fixed point of $\Phi : \mathcal{X} \to \mathcal{X}$. Then, $\Phi : \mathcal{X} \to \mathcal{X}$ is a contraction mapping on Banach space $\mathcal{X}$ with norm given by

$$\|(Y, Z)\|_\lambda := \left\{\mathbb{E}\left[\int_0^T e^{\lambda s}(|Y_s|^2 + |Z_s|^2)ds\right]\right\}^{\frac{1}{p}},$$

by taking a suitable parameter $\lambda \in \mathbb{R}$. We omit the remaining proof. $\square$

Similar to that the linear Itô SDE (2.3) admits a closed-form solution, the linear BSDE will also have a closed-form solution, which is documented in the following lemma.

Lemma 2.4. *Consider the following linear BSDE given by*

$$dY_t = -(A_t Y_t + Z_t^\top B_t + C_t)dt + Z_t^\top dW_t, \quad Y_T = \xi, \qquad (2.74)$$

where $A = (A_t)_{t \in [0,T]}$ and $B = (B_t)_{t \in [0,T]}$ are respectively bounded progressively measurable $\mathbb{R}$-valued and $\mathbb{R}^m$-valued processes, and $C = (C_t)_{t \in [0,T]} \in H_{1,T}^2$. Then, we have

$$Y_t = X_t^{-1} \mathbb{E}\left[X_T \xi + \int_t^T X_s C_s ds \Big| \mathcal{F}_t \right], \quad \forall t \in [0,T]. \qquad (2.75)$$

Here $X_t = \mathcal{E}_t(\int_0^\cdot A_s ds + \int_0^\cdot B_s^\top dW_s)$ for $t \in [0,T]$.

We next introduce forward-backward SDEs (FBSDEs). To this purpose, let us recall the (forward) SDE given by (2.1):

$$X_t = X_0 + \int_0^t b(X_r)dr + \int_0^t \sigma(X_r)dW_r, \quad \forall t \in [0,T],$$

where $b : \mathbb{R}^n \to \mathbb{R}^n$ and $\sigma : \mathbb{R}^n \to \mathbb{R}^{n \times m}$ are two matrix-valued measurable functions satisfying Assumption **(A_{lip})**. Consider the following BSDE associated to the above SDE:

$$Y_t = \phi(X_T) + \int_t^T f(s, X_s, Y_s, Z_s) ds - \int_t^T Z_s^\top dW_s, \quad \forall t \in [0,T], \quad (2.76)$$

where $\phi : \mathbb{R}^n \to \mathbb{R}^d$, $f : [0,T] \times \mathbb{R}^n \times \mathbb{R}^d \times \mathbb{R}^{d \times m} \to \mathbb{R}^d$, and $Y_s \in \mathbb{R}^d$, $Z_s \in \mathbb{R}^{d \times m}$ for $s \in [0,T]$. The system (2.76) is called a (decoupled) forward-backward SDE (FBSDE). The solution of FBSDE (2.76) will be connected to a solution to a class of semi-linear PDEs.

We impose the following assumptions:

(NFKf) the deterministic function $f(t,x,y,z) : [0,T] \times \mathbb{R}^n \times \mathbb{R}^d \times \mathbb{R}^{d \times m} \to \mathbb{R}^d$ is continuous and satisfies the linear growth condition in (x,y,z) uniformly in t, and Lipschitz condition in (y,z) uniformly in (t,x).

(NFKϕ) the mapping $\phi : \mathbb{R}^n \to \mathbb{R}^d$ is continuous and it satisfies the linear growth condition.

Assumptions **(NFKf)**, **(NFKϕ)** and **(A_{lip})** can guarantee the well-posedness of the FBSDE (2.76). For $(t,x) \in [0,T] \times \mathbb{R}^n$, let $X^{t,x} = (X_s^{t,x})_{s \in [t,T]}$ be the strong solution to SDE (2.1) with $X_t = x$ (i.e., (2.58) with $D = \mathbb{R}^n$). Let $(Y_s^{t,x}, Z_s^{t,x})_{s\in[t,T]}$ be the solution of FBSDE (2.76) with $X_s = X_s^{t,x}$ for $s \in [t,T]$. Then, $u(t,x) := Y_t^{t,x}$ for $(t,x) \in [0,T] \times \mathbb{R}^n$ is a deterministic function mapping from $(t,x) \in [0,T] \times \mathbb{R}^n$ to $\mathbb{R}^d$. It is obvious to have that $u(T,x) = Y_T^{T,x} = \phi(X_T^{T,x}) = \phi(x)$. Using Markov property of X and the uniqueness of the solution to BSDE, we have

$$Y_t = u(t, X_t), \quad \forall t \in [0,T]. \tag{2.77}$$

Now, we assume that $u \in C^2([0,T] \times \mathbb{R}^n)$ and satisfies that

$$|u(t,x)| + |\nabla_x u(t,x)^\top \sigma(x)| \leq C(1 + |x|^p), \quad \forall (t,x) \in [0,T] \times \mathbb{R}^n,$$

for some $C > 0$ and $p \geq 1$. By applying Itô's formula to $u(T, X_T)$, we arrive at

$$u(T, X_T) = u(t, X_t) + \int_t^T (\partial_t + \mathcal{A})u(r, X_r)dr$$
$$+ \int_t^T \nabla_x u(r, X_r)^\top \sigma(X_r) dW_r. \tag{2.78}$$

Here, $\mathcal{A}$ is the operator defined by (2.59), and $\nabla_x u$ with $u = (u_1, \ldots, u_d)$ is an $n \times d$ matrix which is defined by

$$\nabla_x u = \begin{bmatrix} \partial_{x_1} u_1 & \partial_{x_1} u_2 & \cdots & \partial_{x_1} u_d \\ \partial_{x_2} u_1 & \partial_{x_2} u_2 & \cdots & \partial_{x_2} u_d \\ \vdots & \vdots & \ddots & \vdots \\ \partial_{x_n} u_1 & \partial_{x_n} u_2 & \cdots & \partial_{x_n} u_d \end{bmatrix}_{n \times d}. \tag{2.79}$$

It follows from (2.78) and (2.77) that

$$Y_t = \phi(X_T) - \int_t^T (\partial_t + \mathcal{A})u(r, X_r)dr - \int_t^T \nabla_x u(r, X_r)^\top \sigma(X_r) dW_r.$$

Compare it with (2.76) and by applying the uniqueness of solutions to (2.76), we obtain

$$Z_t = \sigma(X_t)^\top \nabla_x u(t, X_t), \quad t \in [0,T], \tag{2.80}$$

and $u(t,x)$ satisfies that, for $(t,x) \in [0,T] \times \mathbb{R}^n$,

$$(\partial_t + \mathcal{A})u(t,x) + f(t, x, u(t,x), \sigma(x)^\top \nabla_x u(t,x)) = 0,$$
$$u(T,x) = \phi(x), \quad \forall x \in \mathbb{R}^n. \tag{2.81}$$

It can be proved that $u(t,x)$ for $(t,x) \in [0,T] \times \mathbb{R}^n$ is continuous and it is a viscosity solution of the semilinear Cauchy problem (2.81) (c.f. Pham (2009)).

Chapter 3

Elements of Stochastic Control

This chapter provides a simple review on the formulation of stochastic control problems and the methodology for solving the stochastic control problem.

3.1 Formulation of Stochastic Control Problem

Let $(\Omega, \mathcal{F}, \mathbb{F}, \mathbb{P})$ be a filtered probability space with the filtration $\mathbb{F} = (\mathcal{F}_t)_{t \geq 0}$ satisfying the usual conditions. A stochastic control problem consists of the following elements:

(i) the admissible control set $\mathcal{U}$. It is defined as the set of $\mathbb{F}$-adapted U-valued process $u = (u_t)_{t \geq 0}$ satisfying some constraints, where U is s subset of $\mathbb{R}^p$. We call $u \in \mathcal{U}$ an admissible control process.

(ii) the controlled state process $X^u = (X^u_t)_{t \geq 0}$ for $u \in \mathcal{U}$. The controlled state process X^u is assumed to satisfy the following SDE:

$$dX^u_t = b(X^u_t, u_t)dt + \sigma(X^u_t, u_t)dW_t, \qquad (3.1)$$

where $W = (W_t)_{t \geq 0}$ be an m-dimensional $(\mathbb{P}, \mathbb{F})$-Brownian motion. The coefficients $b : \mathbb{R}^n \times U \to \mathbb{R}^n$ and $\sigma : \mathbb{R}^n \times U \to \mathbb{R}^{n \times m}$ are continuous.

Moreover, we assume that

($\mathbf{A}_{clip}$) for all $x, y \in \mathbb{R}^n$ and $u \in U$, there exist constants $L, K > 0$ which are independent of x, y, u such that

$$|b(x, u) - b(y, u)| + |\sigma(x, u) - \sigma(y, u)| \leq L|x - y|,$$
$$|b(x, u)| + |\sigma(x, u)| \leq K(1 + |x|).$$

Assumption ($\mathbf{A}_{clip}$) guarantees the existence and uniqueness of strong solutions of SDE (3.1) for each $u \in \mathcal{U}$. For any $(t,x) \in [0,T] \times \mathbb{R}^n$, it also from ($\mathbf{A}_{clip}$) holds that

$$\sup_{u \in \mathcal{U}} \mathbb{E}_{t,x}\left[\sup_{s \in [t,T]} |X_s^u|^p\right] \leq C(1+|x|^p), \quad \forall p \geq 1, \qquad (3.2)$$

where $\mathbb{E}_{t,x}[\cdot] := \mathbb{E}[\cdot | X_t^u = x]$, and $C > 0$ is a finite constant which is independent of (t,x).

For $(t,x) \in [0,T] \times \mathbb{R}^n$, we may rewrite (3.1) as in an integral form given by, for all $s \in [t,T]$,

$$X_s^{t,x} = x + \int_t^s b(X_r^{t,x}, u_r)dr + \int_t^s \sigma(X_r^{t,x}, u_r)dW_r. \qquad (3.3)$$

where we have omitted the dependence of X on the control process u.

(iii) the objective functional $J(x;u)$. It is in general defined by, for all $x \in \mathbb{R}^n$ and $u \in \mathcal{U}$,

$$J(x;u) := \mathbb{E}\left[g(X_T^u) + \int_0^T f(t, X_t^u, u_t)dt \Big| X_0^u = x\right], \qquad (3.4)$$

where $g : \mathbb{R}^n \to \mathbb{R}$ and $f : [0,T] \times \mathbb{R}^n \times U$ are respectively referred to as the terminal function and the running function.

We assume that

($\mathbf{A}_{TR}$) the terminal function $g : \mathbb{R}^n \to \mathbb{R}$ is continuous in x; The running function $f : [0,T] \times \mathbb{R}^n \times U \to \mathbb{R}$ is continuous in (t,x) for all $u \in U$; g, f satisfy the quadratic growth condition in x, i.e., for any $(t,x,u) \in [0,T] \times \mathbb{R}^n \times U$,

$$|g(x)| \vee |f(t,x,u)| \leq C(1+|x|^2),$$

where $C > 0$ is independent of (t,x,u).

Using the ingredients (i)-(iii) provided above, the stochastic control problem is formulated as follows, for $x \in \mathbb{R}^n$,

$$V(x) = \sup_{u \in \mathcal{U}} J(x;u) = \sup_{u \in \mathcal{U}} \mathbb{E}\left[g(X_T^u) + \int_0^T f(t, X_t^u, u_t)dt \Big| X_0^u = x\right]. \qquad (3.5)$$

Here, we explain g (resp. f) as the terminal revenue function (resp. running revenue function) in (3.5). If g (resp. f) is interpreted as the terminal cost function (resp. running cost function), then we may replace "$\sup_{u \in \mathcal{U}}$" with

"$\inf_{u \in \mathcal{U}}$" in (3.5). Our aim is to find a so-called *optimal* control process $u^* \in \mathcal{U}$ (if exists) such that $V(x) = J(x; u^*)$ for $x \in \mathbb{R}^n$.

To solve the stochastic control problem (3.5), we introduce the following so-called "dynamic version" of value function defined by, for all $(t, x) \in [0, T] \times \mathbb{R}^n$,

$$V(t,x) := \sup_{u \in \mathcal{U}} J(t, x; u)$$

$$:= \sup_{u \in \mathcal{U}} \mathbb{E}_{t,x} \left[g(X_T^u) + \int_t^T f(s, X_s^u, u_s) ds \right]$$

$$= \sup_{u \in \mathcal{U}} \mathbb{E} \left[g(X_T^{t,x}) + \int_t^T f(s, X_s^{t,x}, u_s) ds \right]. \quad (3.6)$$

The following lemma proves that the value function is well-defined and it satisfies the following properties:

Lemma 3.1. *Let Assumptions* ($\mathbf{A}_{clip}$) *and* ($\mathbf{A}_{TR}$) *hold. Then, the value function $V(t, x)$ for $(t, x) \in [0, T] \times \mathbb{R}^n$ satisfies that*

(i) *it satisfies the quadratic growth condition, i.e., $|V(t,x)| \leq C(1 + |x|^2)$ for some constant $C > 0$ which is independent of (t, x).*

(ii) *if g and f are Lipschitz continuous w.r.t. x uniformly in (t, u), then V is continuous on $[0, T] \times \mathbb{R}^n$.*

Proof. It follows from Assumptions ($\mathbf{A}_{clip}$) and ($\mathbf{A}_{TR}$) that

$$\sup_{u \in \mathcal{U}} \left| \mathbb{E}_{t,x} \left[g(X_T^u) + \int_t^T f(s, X_s^u, u_s) ds \right] \right|$$

$$\leq C \sup_{u \in \mathcal{U}} \mathbb{E}_{t,x} \left[|X_T^u|^2 + \int_t^T |X_s^u|^2 ds \right] \leq C(1+T) \sup_{u \in \mathcal{U}} \mathbb{E}_{t,x} \left[\sup_{s \in [t,T]} |X_s^u|^2 \right]$$

$$\leq C_T(1 + |x|^2),$$

where we used the moment estimate (3.2), and $C_T > 0$ is a finite constant which is depending on T only.

Let $(t_k, x_k)_{k \geq 1} \subset [0, T] \times \mathbb{R}^n$ be a sequence converging to $(t, x) \in [0, T] \times \mathbb{R}^n$. Take $\epsilon > 0$ and let $u^\epsilon = u^\epsilon(t, x) \in \mathcal{U}$ be an ϵ-optimal control process with value function $V(t, x)$. Namely, it holds that

$$V(t, x) \leq \epsilon + J(t, x; u^\epsilon). \quad (3.7)$$

Since $V(t_k, x_k) \geq J(t_k, x_k; u^\epsilon)$, we obtain

$$V(t, x) - V(t_k, x_k) \leq \epsilon + J(t, x; u^\epsilon) - J(t_k, x_k; u^\epsilon). \quad (3.8)$$

Using Assumptions ($\mathbf{A}_{clip}$), ($\mathbf{A}_{TR}$) and DCT with (3.2), we get

$$\lim_{k\to\infty} V(t_k, x_k) \geq V(t, x) - \epsilon. \tag{3.9}$$

Similarly, taking ϵ-control process $u_k^\epsilon = u^\epsilon(t_k, x_k)$ with value function $V(t_k, x_k)$. Then, we have

$$V(t_k, x_k) - V(t, x) \leq \epsilon + J(t_k, x_k; u_k^\epsilon) - J(t, x; u_k^\epsilon). \tag{3.10}$$

First of all, it holds that

$$|J(t_k, x_k; u_k^\epsilon) - J(t, x; u_k^\epsilon)| \leq \mathbb{E}\left[|g(X_T^{t_k, x_k}) - g(X_T^{t, x})|\right]$$
$$+ \mathbb{E}\left[\left|\int_{t_k}^T f(s, X_s^{t_k, x_k}, u_{k,s}^\epsilon)ds - \int_t^T f(s, X_s^{t, x}, u_{k,s}^\epsilon)ds\right|\right].$$

We next prove that $\mathbb{E}[|g(X_T^{t_k, x_k}) - g(X_T^{t, x})|] \to 0$ as $k \to \infty$. We can also apply the similar argument to show that

$$\lim_{k\to\infty} \mathbb{E}\left[\left|\int_{t_k}^T f(s, X_s^{t_k, x_k}, u_{k,s}^\epsilon)ds - \int_t^T f(s, X_s^{t, x}, u_{k,s}^\epsilon)ds\right|\right] = 0.$$

On the other hand, since g is Lipschitz continuous w.r.t. x uniformly in (t, u), we arrive at

$$\mathbb{E}\left[|g(X_T^{t_k, x_k}) - g(X_T^{t, x})|\right] \leq \ell_{lip}(g)\mathbb{E}\left[|X_T^{t_k, x_k} - X_T^{t, x}|\right],$$

where

$$X_T^{t_k, x_k} - X_T^{t, x} = x_k - x + \int_{t_k}^T b(X_s^{t_k, x_k}, u_{k,s}^\epsilon)ds - \int_t^T b(X_s^{t, x}, u_{k,s}^\epsilon)ds$$
$$+ \int_{t_k}^T \sigma(X_s^{t_k, x_k}, u_{k,s}^\epsilon)dW_s - \int_t^T \sigma(X_s^{t, x}, u_{k,s}^\epsilon)dW_s.$$

For the case with $t_k \uparrow t$ as $k \to \infty$, it holds that

$$X_T^{t_k, x_k} - X_T^{t, x} = x_k - x + \int_{t_k}^t b(X_s^{t_k, x_k}, u_{k,s}^\epsilon)ds$$
$$+ \int_t^T \{b(X_s^{t_k, x_k}, u_{k,s}^\epsilon) - b(X_s^{t, x}, u_{k,s}^\epsilon)\}ds + \int_{t_k}^t \sigma(X_s^{t_k, x_k}, u_{k,s}^\epsilon)dW_s$$
$$+ \int_t^T \{\sigma(X_s^{t_k, x_k}, u_{k,s}^\epsilon) - \sigma(X_s^{t, x}, u_{k,s}^\epsilon)\}dW_s. \tag{3.11}$$

Using the linear growth of b and σ, it follows from (3.2) and BDG's inequality that

$$\mathbb{E}\left[\int_{t_k}^t |b(X_s^{t_k, x_k}, u_{k,s}^\epsilon)|ds\right]$$

$$\leq K(t-t_k)\mathbb{E}\left[\left(1+\sup_{s\in[t_k,T]}\left|X_s^{t_k,x_k}\right|\right)\right] \to 0, \quad k\to\infty,$$

and

$$\mathbb{E}\left[\sup_{\theta\in[t_k,t]}\left|\int_{t_k}^{\theta}\sigma(X_s^{t_k,x_k},u_{k,s}^\epsilon)dW_s\right|^2\right] \leq K\mathbb{E}\left[\int_{t_k}^{t}\sigma^2(X_s^{t_k,x_k},u_{k,s}^\epsilon)ds\right]$$

$$\leq C\mathbb{E}\left[\int_{t_k}^{t}\left(1+|X_s^{t_k,x_k}|^2\right)ds\right]$$

$$\leq C(t-t_k)\mathbb{E}\left[\left(1+\sup_{s\in[t_k,T]}\left|X_s^{t_k,x_k}\right|^2\right)\right] \to 0, \quad k\to\infty.$$

By applying the Lipschitz continuous property of b and σ, it holds from BDG's inequality that

$$\mathbb{E}\left[\int_t^T \left|b(X_s^{t_k,x_k},u_{k,s}^\epsilon) - b(X_s^{t,x},u_{k,s}^\epsilon)\right| ds\right]$$

$$\leq L\int_t^T \mathbb{E}\left[\left|X_s^{t_k,x_k} - X_s^{t,x}\right|\right] ds \to 0, \quad k\to\infty,$$

and

$$\mathbb{E}\left[\left|\int_t^T \sigma(X_s^{t_k,x_k},u_{k,s}^\epsilon) - \sigma(X_s^{t,x},u_{k,s}^\epsilon)dW_s\right|^2\right]$$

$$\leq C\int_t^T \mathbb{E}\left[\left|X_s^{t_k,x_k} - X_s^{t,x}\right|^2\right] ds \to 0, \quad k\to\infty.$$

Similarly, the above convergences hold for the case with $t_k \downarrow t$ as $k\to\infty$. Thus, in view of (3.11), we have that $\mathbb{E}[|g(X_T^{t_k,x_k}) - g(X_T^{t,x})|] \to 0$, as $k\to\infty$. □

3.2 Dynamic Programming Principle

This section presents a core method for solving (stochastic) control problem, i.e., dynamic programming principle (DPP for short). The term dynamic programming was first used in the 1940s by Richard Bellman (1920–1984)[1]

---

[1] Richard Bellman was an American applied mathematician, who introduced dynamic programming in 1953, and made important contributions in other fields of mathematics, such as biomathematics. He founded the leading biomathematical journal *Mathematical Biosciences*.

to describe problems where one needs to find the best decisions one after another. The idea of DPP in essence is to divide a control problem into smaller nested subproblems, and then combine the solutions to reach an overall solution. This concept is known as the principle of optimality, and a more formal exposition is provided in this chapter. The mathematical statement of principle of optimality is remembered in his name as the Bellman equation or Hamilton-Jacobi-Bellman equation (c.f. Section 3.3).

We first start with the deterministic control problem and introduce the DPP. To be precise, consider a deterministic control problem described as follows:

- Time horizon: $T \in (0, \infty)$, time variable: $t \in [0, T]$;
- State function of a controlled system: $X^{t,x,u} = (X_s^{t,x,u})_{s \in [t,T]}$, control function $u = (u_t)_{t \in [0,T]}$.

The state dynamics of the controlled system is described as, for $(t, x) \in [0, T] \times \mathbb{R}^n$,

$$X_s^{t,x,u} = x + \int_t^s b(X_r^{t,x,u}, u_r) dr, \quad \forall s \in [t, T]. \tag{3.12}$$

The control function u_s for $s \in [t, T]$ is called a "control" or "strategy", which is assumed to take values in U as a compact subset of $\mathbb{R}^m$. We assume that $b(x, u) : \mathbb{R}^n \times U \to \mathbb{R}^n$ satisfies a uniform Lipschitz condition in U. The set of admissible controls $\mathcal{U}_t^T$ is specified as:

$$\mathcal{U}_t^T := \{u_s : [t, T] \to U; \ s \to u_s \text{ is measurable}\}.$$

A deterministic control problem can be described as follows, for $(t, x) \in [0, T] \times \mathbb{R}^n$,

$$J(t, x; u^*) = \inf_{u \in \mathcal{U}_t^T} J(t, x; u)$$

$$:= \inf_{u \in \mathcal{U}_t^T} \left[g(X_T^{t,x,u}) + \int_t^T f(X_s^{t,x,u}, u_s) ds \right], \tag{3.13}$$

where $u^* \in \mathcal{U}_t^T$ is the optimal control (strategy) and $g : \mathbb{R}^n \to \mathbb{R}$ (resp. $f : \mathbb{R}^n \times U \to \mathbb{R}$) is the terminal cost function (resp. running cost function).

As in (3.6), the corresponding "dynamic version" of value function is defined by, for all $(t, x) \in [0, T] \times \mathbb{R}^n$,

$$V(t, x) := \inf_{u \in \mathcal{U}_t^T} \left[g(X_T^{t,x,u}) + \int_t^T f(X_s^{t,x,u}, u_s) ds \right]. \tag{3.14}$$

Obviously, the value function satisfies the terminal condition $V(T,x) = g(x)$ for all $x \in \mathbb{R}^n$. Then, the DPP is stated as follows:

Theorem 3.1 (DPP for Deterministic Control Problem). *For the value function $V(t,x)$ defined by (3.17), and $t < s \leq T$, it holds that*

$$V(t,x) = \inf_{u \in \mathcal{U}_t^T} \left[V(s, X_s^{t,x,u}) + \int_t^s f(X_r^{t,x,u}, u_r) dr \right]. \quad (3.15)$$

It is saying that if one knows the value function at time $s > t$, one may determine the value function at time t by optimizing from time t to time s and using $V(\cdot, s)$ as the terminal cost. The DPP described as (7.55) means that $V(t,x)$ satisfies a "semigroup" property, but, which is running "backwards".

Proof of Theorem 3.1. The proof is based on the following observation on the admissible control set: for $t < s \leq T$, we have $\mathcal{U}_t^T = \mathcal{U}_t^s \oplus \mathcal{U}_s^T$. Here, $\oplus$ means that if $u^1 : [t,s] \to U \in \mathcal{U}_t^s$ and $u^2 : [s,T] \to U \in \mathcal{U}_s^T$, then $u^t \oplus u^s$ is defined as, for $r \in [t,T]$,

$$u^t \oplus u^s := \begin{cases} u_r^1, & r \in [t,s], \\ u_r^2, & r \in [s,T]. \end{cases}$$

Consequently $u := u^1 \oplus u^2 \in \mathcal{U}_t^T$ if $u^1 \in \mathcal{U}_t^s$ and $u^2 \in \mathcal{U}_s^T$. On the other hand, if $u \in \mathcal{U}_t^T$, by restricting the domain of u to $[t,s]$, we obtain an admissible control in $\mathcal{U}_t^s$. Similarly, by restricting the domain of u to $[s,T]$, we obtain an admissible control in $\mathcal{U}_s^T$. Therefore, we proved that $\mathcal{U}_t^T = \mathcal{U}_t^s \oplus \mathcal{U}_s^T$. Using the definition (3.17) of the value function, it holds that

$$V(t,x) := \inf_{u \in \mathcal{U}_t^T} \left[g(X_T^{t,x,u}) + \int_t^T f(X_s^{t,x,u}, u_s) ds \right]$$

$$= \inf_{u \in \mathcal{U}_t^T} \left[g(X_T^{t,x,u}) + \int_t^s f(X_r^{t,x,u}, u_r) dr + \int_s^T f(X_r^{t,x,u}, u_r) dr \right]$$

$$= \inf_{u = u^1 \oplus u^2;\ u^1 \in \mathcal{U}_t^s,\ u^2 \in \mathcal{U}_s^T} \left[g(X_T^{t,x,u}) + \int_t^s f(X_r^{t,x,u}, u_r) dr + \int_s^T f(X_r^{t,x,u}, u_r) dr \right].$$

Moreover, we decompose the state process $X^{t,x,u}$ into the ones in the time interval $[t,s]$ and $[s,T]$, i.e., $X = X^1 \oplus X^2$:

$$\begin{cases} dX_r^{1,u^1} = b(X_r^{1,u^1}, u_r^1) dr, & r \in (t,s];\ X_t^{1,u^1} = x; \\ dX_r^{2,u^2} = b(X_r^{2,u^2}, u_r^2) dr, & r \in (s,T];\ X_s^{2,u^2} = X_s^{1,u^1} = X_s^{t,x,u}. \end{cases}$$

Then, it holds that
$$V(t,x) = \inf_{u^1 \in \mathcal{U}_t^s} \inf_{u=u^1 \oplus u^2;\ u^2 \in \mathcal{U}_s^T,\ X_s^{2,u^2}=X_s^{1,u^1}} \left[g(X_T^{2,u^2}) + \int_t^s f(X_r^{1,u^1}, u_r^1)dr + \int_s^T f(X_r^{2,u^2}, u_r^2)dr \right].$$

Note that X^{1,u^1} depends only on x and u^1, not on X^{2,u^2} or u^2. Since the first integral depends only on X^{1,u^1} and u^1, this may be rearranged as:

$$V(t,x) = \inf_{u^1 \in \mathcal{U}_t^s} \left\{ \int_t^s f(X_r^{1,u^1}, u_r^1)dr \right.$$
$$\left. + \inf_{u^2 \in \mathcal{U}_s^T,\ X_s^{2,u^2}=X_s^{1,u^1}} \left[g(X_T^{2,u^2}) + \int_s^T f(X_r^{2,u^2}, u_r^2)dr \right] \right\}$$
$$= \inf_{u^1 \in \mathcal{U}_t^s} \left[\int_t^s f(X_r^{1,u^1}, u_r^1)dr + V(X_s^{1,u^1}, s) \right]$$
$$= \inf_{u \in \mathcal{U}_t^T} \left[\int_t^s f(X_r^{t,x,u}, u_r)dr + V(X_s^{t,x,u}, s) \right].$$

Thus, we complete the proof of the theorem. □

We next turn to the DPP for stochastic control problem. To do it, we reformulate the stochastic control problem. Consider the following controlled diffusion process described as, given $(t,x) \in [0,T] \times \mathbb{R}^n$,

$$X_s^{t,x} = x + \int_t^s b(X_r^{t,x}, u_r)dr + \int_t^s \sigma(X_r^{t,x}, u_r)dW_r, \quad \forall s \in [t,T]. \quad (3.16)$$

Here, $b(x,u): \mathbb{R}^n \times U \to \mathbb{R}^n$ and $\sigma(x,u): \mathbb{R}^n \times U \to \mathbb{R}^{n \times m}$. The control $(u_t)_{t \in [0,T]}$ is a progressively measurable process, valued in $U \subset \mathbb{R}^m$. We impose the assumption on the coefficients (b, σ):

(**A**$_c$) the coefficients (b, σ) satisfy a uniform Lipschitz condition in U, for any $x, y \in \mathbb{R}^n$ and $u \in U$,
$$|b(x,u) - b(y,u)| + |\sigma(x,u) - \sigma(y,u)| \le L|x-y|,$$
for some constant $L > 0$.

Denote by $\mathcal{U}$ the set of control processes $(u_t)_{t \in [0,T]}$ such that
$$\mathbb{E}\left[\int_0^T (|b(0, u_t)|^2 + |\sigma(0, u_t)|^2)dt \right] < +\infty.$$

In the sequel, we describe the objective functional of the stochastic control problem. First, we consider that

(**A$_g$**) the terminal revenue function: $g : \mathbb{R}^n \to \mathbb{R}$ is measurable and satisfies quadratic growth condition: i.e., $|g(x)| \leq K(1 + |x|^2)$ for all $x \in \mathbb{R}^n$.

(**A$_f$**) the running revenue function: $f : \mathbb{R}^n \times U \to \mathbb{R}$ is measurable and satisfies quadratic growth condition in x: i.e., $|f(x,u)| \leq K(1 + |x|^2) + l(u)$ for all $(x,u) \in \mathbb{R}^n \times U$, where $l : U \to \mathbb{R}_+$ is a positive function.

Now, we specify the *admissible control set*. For any $(t,x) \in [0,T] \times \mathbb{R}^n$, denote by $\mathcal{U}_{t,x}$ the subset of controls $(u_t)_{t \in [0,T]} \in \mathcal{U}$ satisfying

$$\mathbb{E}\left[\int_t^T |f(s, X_s^{t,x}, u_s)| ds\right] < +\infty.$$

Then, under Assumption (**A$_f$**), we have $\mathcal{U}_{t,x} \neq \emptyset$. The dynamic version of value function is defined as, for $(t,x) \in [0,T] \times \mathbb{R}^n$,

$$V(t,x) := \sup_{u \in \mathcal{U}_{t,x}} J(t,x,u)$$

$$:= \sup_{u \in \mathcal{U}_{t,x}} \mathbb{E}\left[g(X_T^{t,x}) + \int_t^T f(s, X_s^{t,x}, u_s) ds\right]. \quad (3.17)$$

For any $(t,x) \in [0,T] \times \mathbb{R}^n$, if there exists $u^* \in \mathcal{U}_{t,x}$ such that $V(t,x) = J(t,x,u^*)$, then u^* is called an optimal control. A Markovian control is defined as: for any $u \in \mathcal{U}_{t,x}$, if there exists a measurable mapping $\mathrm{u} : [0,T] \times \mathbb{R}^n \to U$ such that $u_s = \mathrm{u}(s, X_s^{t,x})$ for all $s \in [t,T]$. Correspondingly, the optimal Markovian (feedback) control is given by $u_s^* = \mathrm{u}^*(s, X_s^{t,x,*})$ with $s \in [t,T]$. Here, $X^{t,x,*} = (X_s^{t,x,*})_{s \in [t,T]}$ obeys SDE (3.13) with u replaced by u^*.

We start with the following DPP associated with the stochastic control problem (3.17):

Theorem 3.2 (DPP for Stochastic Control Problem). *Let $(t,x) \in [0,T] \times \mathbb{R}^n$. Then, it holds that*

$$V(t,x) = \sup_{u \in \mathcal{U}_{t,x}} \sup_{\tau \in \mathcal{T}_t^T} \mathbb{E}\left[V(\tau, X_\tau^{t,x}) + \int_t^\tau f(s, X_s^{t,x}, u_s) ds\right]$$

$$= \sup_{u \in \mathcal{U}_{t,x}} \inf_{\tau \in \mathcal{T}_t^T} \mathbb{E}\left[V(\tau, X_\tau^{t,x}) + \int_t^\tau f(s, X_s^{t,x}, u_s) ds\right],$$

where $\tau \in \mathcal{T}_t^T$ is the set of $\mathbb{F}$-stopping times taking values on $[t,T]$.

Proof. The proof is standard (c.f. Pham (2009)). To be complete, we provide a sketch of the proof. For any $u \in \mathcal{U}_{t,x}$, i.e., u is an admissible control, using the pathwise uniqueness of SDE (3.13), we arrive at

$$X_s^{t,x} = X_s^{\tau, X_\tau^{t,x}}, \quad \forall s \geq \tau \in \mathcal{T}_t^T. \tag{3.18}$$

Then, for any $\tau \in \mathcal{T}_t^T$, it follows from (3.18) that

$$J(t,x,u) = \mathbb{E}\left[g(X_T^{t,x}) + \int_\tau^T f(s, X_s^{t,x}, u_s)ds + \int_t^\tau f(s, X_s^{t,x}, u_s)ds\right]$$

$$= \mathbb{E}\left\{\mathbb{E}\left[g(X_T^{t,x}) + \int_\tau^T f(s, X_s^{t,x}, u_s)ds + \int_t^\tau f(s, X_s^{t,x}, u_s)ds \Big| \mathcal{F}_\tau\right]\right\}$$

$$= \mathbb{E}\left\{\mathbb{E}\left[g(X_T^{t,x}) + \int_\tau^T f(s, X_s^{t,x}, u_s)ds \Big| \mathcal{F}_\tau\right] + \int_t^\tau f(s, X_s^{t,x}, u_s)ds\right\}$$

$$= \mathbb{E}\left[J(\tau, X_\tau^{t,x}, u) + \int_t^\tau f(s, X_s^{t,x}, u_s)ds\right].$$

Since $J(t,x,u) \leq V(t,x)$, it holds that

$$J(t,x,u) \leq \mathbb{E}\left[V(\tau, X_\tau^{t,x}) + \int_t^\tau f(s, X_s^{t,x}, u_s)ds\right], \quad \forall \tau \in \mathcal{T}_t^T.$$

Hence, for all $u \in \mathcal{U}_{t,x}$,

$$J(t,x,u) \leq \inf_{\tau \in \mathcal{T}_t^T} \mathbb{E}\left[V(\tau, X_\tau^{t,x}) + \int_t^\tau f(s, X_s^{t,x}, u_s)ds\right]$$

$$\leq \sup_{u \in \mathcal{U}_{t,x}} \inf_{\tau \in \mathcal{T}_t^T} \mathbb{E}\left[V(\tau, X_\tau^{t,x}) + \int_t^\tau f(s, X_s^{t,x}, u_s)ds\right].$$

This yields that

$$V(t,x) \leq \sup_{u \in \mathcal{U}_{t,x}} \inf_{\tau \in \mathcal{T}_t^T} \mathbb{E}\left[V(\tau, X_\tau^{t,x}) + \int_t^\tau f(s, X_s^{t,x}, u_s)ds\right].$$

On the other hand, fix an arbitrary admissible control $u \in \mathcal{U}_{t,x}$ and $\tau \in \mathcal{T}_t^T$. Using the definition of the value function, for any $\epsilon > 0$ and $\omega \in \Omega$-a.s., there exists $u^\epsilon(\omega) \in \mathcal{U}_{\tau(\omega), X_{\tau(\omega)}^{t,x}(\omega)}$, which is an ϵ-optimal control for $V(\tau(\omega), X_{\tau(\omega)}^{t,x}(\omega))$, i.e.,

$$V(\tau(\omega), X_{\tau(\omega)}^{t,x}(\omega)) - \epsilon \leq J(\tau(\omega), X_{\tau(\omega)}^{t,x}(\omega), u^\epsilon(\omega)). \tag{3.19}$$

Let us define

$$\tilde{u}_t(\omega) := \begin{cases} u_t(\omega), & t \in [0, \tau(\omega)], \\ u_t^\epsilon(\omega), & t \in [\tau(\omega), T]. \end{cases} \tag{3.20}$$

Note that there is the measurability issue on $\tilde{u} \in \mathcal{U}_{t,x}$. However, it can be shown by the measurable selection theorem. Then, we have from (3.19) that

$$V(t,x) \geq J(t,x,\tilde{u}) = \mathbb{E}\left[J(\tau, X_\tau^{t,x}, u^\epsilon) + \int_t^\tau f(s, X_s^{t,x}, u_s)ds\right]$$

$$\geq \mathbb{E}\left[V(\tau, X_\tau^{t,x}) + \int_t^\tau f(s, X_s^{t,x}, u_s)ds\right] - \epsilon.$$

Using the arbitrariness of $u \in \mathcal{U}_{t,x}$, $\tau \in \mathcal{T}_t^T$ and $\epsilon > 0$, we obtain

$$V(t,x) \geq \sup_{u \in \mathcal{U}_{t,x}} \sup_{\tau \in \mathcal{T}_t^T} \mathbb{E}\left[V(\tau, X_\tau^{t,x}) + \int_t^\tau f(s, X_s^{t,x}, u_s)ds\right].$$

Thus, we finish the proof of the theorem. □

3.3 Hamilton–Jacobi–Bellman Equations

The DPP given by Theorem 3.1 (deterministic version) and Theorem 3.2 (stochastic version) in the previous section can yield a class of important PDEs: Hamilton–Jacobi–Bellman (HJB) equations.

We start with the derivation of the HJB equation via the deterministic version of DPP described by Theorem 3.1. To do it, we assume that the value function $V(t,x)$ defined by (3.17) belongs to $C^{1,1}([0,T] \times \mathbb{R}^n) \cap C([0,T] \times \mathbb{R}^n)$. For a sufficient small $h > 0$, it follows from Theorem 3.1 that

$$V(t,x) = \inf_{u \in \mathcal{U}_t^{t+h}} \left[V(t+h, X_{t+h}^{t,x}) + \int_t^{t+h} f(X_r^{t,x}, u_r)dr\right]. \quad (3.21)$$

It holds from the chain rule that

$$V(t+h, X_{t+h}^{t,x}) = V(t+h, x) + \int_t^{t+h} b(X_r^{t,x}, u_r)^\top \nabla_x V(r, X_r^{t,x})dr.$$

Inserting the above display into (3.21), we obtain

$$V(t,x) = \inf_{u \in \mathcal{U}_t^{t+h}} \left[V(t+h, x) + \int_t^{t+h} (b(X_r^{t,x}, u_r)^\top \nabla_x V(r, X_r^{t,x}) + f(X_r^{t,x}, u_r))dr\right].$$

Divide h on both sides of the above equality, one has

$$0 = \inf_{u \in \mathcal{U}_t^{t+h}} \left[\frac{V(t+h, x) - V(t, x)}{h}\right.$$

$$+ \frac{1}{h}\int_{t}^{t+h}(b(X_r^{t,x}, u_r)^\top \nabla_x V(r, X_r^{t,x}) + f(X_r^{t,x}, u_r))dr\bigg].$$

Letting $h \to 0$ and we arrive at the following so-called *HJB equation* given by

$$\begin{cases} \partial_t V(t,x) + \mathcal{H}(\nabla_x V(t,x), x) = 0, & \forall (t,x) \in [0,T) \times \mathbb{R}^n, \\ V(T,x) = g(x), & \forall x \in \mathbb{R}^n, \end{cases} \quad (3.22)$$

where $\mathcal{H}(p,x)$ with $(p,x) \in \mathbb{R}^n \times \mathbb{R}^n$ is called *Hamiltonian*, which is defined as follows:

$$\mathcal{H}(p,x) := \inf_{u \in U}\left\{b(x,u)^\top p + f(x,u)\right\}. \quad (3.23)$$

For the infinite-horizon case, we have the following remark:

Remark 3.1 (Infinite-Horizon Control Problem). When $T \to \infty$, the finite horizon control problem becomes an infinite horizon control problem, which is formulated as, for all $x \in \mathbb{R}^n$,

$$V(x) := \inf_{u \in \mathcal{U}_0^\infty}\left[\int_0^\infty e^{-\rho s} f(X_s^x, u_s)ds\right], \quad (3.24)$$

where $\rho > 0$ is the discount factor and $X_s^x := X_s^{0,x}$ for all $s \geq 0$. The corresponding DPP is formulated as, for any $s > 0$,

$$V(x) = \inf_{u \in \mathcal{U}_0^\infty}\left[e^{-\rho s}V(X_s^x) + \int_0^s e^{-\rho r} f(X_r^x, u_r)dr\right]. \quad (3.25)$$

Formally, the HJB equation for this infinite-horizon control problem is

$$\mathcal{H}(\nabla_x V(x), x) = \rho V(x), \quad \forall x \in \mathbb{R}^n, \quad (3.26)$$

where $\mathcal{H}(p,x)$ with $(p,x) \in \mathbb{R}^n \times \mathbb{R}^n$ is defined by (3.23).

We next illustrate the HJB equation approach in terms of *calculus of variations* rather than optimal control. Let $L(q,x)$ be the Lagrangian, which is a sufficiently smooth function in $q, x \in \mathbb{R}^n$. Fix two points $x, y \in \mathbb{R}^n$, consider the class of admissible trajectories connecting the two points: for $t > 0$,

$$\mathcal{U}_t^{x,y} := \left\{\ell \in C^1([0,t]; \mathbb{R}^n); \ \ell(0) = x, \ \ell(t) = y\right\}.$$

The mathematical problem of the calculus of variations is to find the optimal curve $\ell^* \in \mathcal{U}_t^{x,y}$ such that

$$J(\ell^*) = \inf_{\ell \in \mathcal{U}_t^{x,y}} J(\ell) := \inf_{\ell \in \mathcal{U}_t^{x,y}}\int_0^t L(\ell'(s), \ell(s))ds.$$

Now, assume that ℓ^* exists and we want to see what is the property satisfied by $\ell^* \in \mathcal{U}_t^{x,y}$. The result is stated as follows:

Proposition 3.1 (Euler–Lagrange Equation). *The optimal curve $\ell^* \in \mathcal{U}_t^{x,y}$ satisfies that, for all $r \in [0,t]$,*

$$\frac{d}{dr}[\nabla_q L(\ell^{*,'}(r), \ell^*(r))] = \nabla_x L(\ell^{*,'}(r), \ell^*(r)). \tag{3.27}$$

Proof. Let $\psi \in C^1([0,t]; \mathbb{R}^n)$ with $\psi(0) = \psi(t) = 0$, i.e., $\psi \in \mathcal{U}_t^{0,0}$. Define $\ell^s(r) := \ell^*(r) + s\psi(r)$ for $r \in [0,t]$ and $s \in \mathbb{R}$. Hence $\ell^s \in \mathcal{U}_t^{x,y}$. We also define $\Phi(s) := J(\ell^s)$. Then, we next compute $\Phi'(s)$ for $s \in \mathbb{R}$. Recall that

$$\Phi(s) = \int_0^t L(\ell^{*,'}(r) + s\psi'(r), \ell^*(r) + s\psi(r))dr.$$

By setting $r_s := \ell^{*,'}(r) + s\psi'(r)$, we deduce that

$$\Phi'(s) = \int_0^t [\nabla_q L(r_s)^\top \psi'(r) + \nabla_x L(r_s)^\top \psi(r)]dr.$$

Using integration by parts, we arrive at

$$\int_0^t \nabla_q L(r_s)^\top \psi'(r) dr = \nabla_q L(r_s)^\top \psi(r) \Big|_{r=0}^{r=t} - \int_0^t \frac{d}{dr}[\nabla_q L(r_s)]^\top \psi(r) dr$$

$$= -\int_0^t \frac{d}{dr}[\nabla_q L(r)]^\top \psi(r) dr.$$

Thus, it holds that

$$\Phi'(s) = \int_0^t \left\{ \nabla_x L(r_s)^\top - \frac{d}{dr}[\nabla_q L(r_s)]^\top \right\} \psi(r) dr.$$

Since $\ell^* \in \mathcal{U}_t^{x,y}$ is a minimizer of $J(\ell)$ over $\ell \in \mathcal{U}_t^{x,y}$, we have $\Phi'(0) = 0$. This yields the Euler–Lagrange equation (3.27). □

In the sequel, we connect the Euler–Lagrange equation to the so-called Hamilton equation. To do it, let $p \in \mathbb{R}^n$ and $x \in \mathbb{R}^n$ be given. Suppose that the equation $\nabla_q L(q, x) = p$ in the unknown q has a unique smooth solution $q(p, x) \in \mathbb{R}^n$. Define the Hamiltonian as:

$$\mathcal{H}(p, x) := pq(p, x) - L(q(p, x), x), \quad \forall (p, x) \in \mathbb{R}^n \times \mathbb{R}^n. \tag{3.28}$$

Let $\ell^*(r)$ for $r \in [0,t]$ be the solution of the Euler–Lagrange equation (3.27). Then, we can prove that ℓ^* satisfies the Hamilton equation: for all $r \in [0,t]$,

$$\begin{cases} \ell^{*,'}(r) = \nabla_p \mathcal{H}(p(r), \ell^*(r)), \\ p'(r) = -\nabla_x \mathcal{H}(p(r), \ell^*(r)). \end{cases} \tag{3.29}$$

Similar to the derivation of the HJB equation (3.22) for the deterministic control problem, the DPP described in Theorem 3.2 can yield the HJB equation for the stochastic control problem (3.17) (we may apply Itô's formula instead of the chain rule). More precisely, if the value function $V \in C^{1,2}([0,T) \times \mathbb{R}^n) \cap C([0,T] \times \mathbb{R}^n)$, then V solves the following HJB equation given by, for $(t,x) \in [0,T) \times \mathbb{R}^n$,

$$\partial_t V(t,x) + \mathcal{H}(x, \nabla_x V(t,x), \nabla_x^2 V(t,x)) = 0 \qquad (3.30)$$

with terminal condition $V(T,x) = g(x)$ for all $x \in \mathbb{R}^n$. Here, the Hamiltonian $\mathcal{H}(x,p,M)$ for $(x,p,M) \in \mathbb{R}^n \times \mathbb{R}^n \times \mathbb{R}^{n \times n}$ is defined as:

$$\mathcal{H}(x,p,M) := \sup_{u \in U} \left\{ b(x,u)^\top p + \frac{1}{2}\mathrm{tr}[\sigma\sigma^\top(x,u)M] + f(x,u) \right\}. \qquad (3.31)$$

When $\sigma(x,u) = \sigma(x)$ (i.e., the volatility function of the controlled state process is independent of the control u) and the following assumptions hold:

(a) the policy space $U \subset \mathbb{R}^p$ is compact;
(b) the coefficient $b(x,u) = \tilde{b}(x) + \sigma(x)\theta(x,u)$;
(c) the mappings $\tilde{b}, \sigma \in C^2(\mathbb{R}^n)$; σ, σ^{-1} and $\nabla_x \sigma, \nabla_x \tilde{b}$ are bounded in $\mathbb{R}^n$; $\theta \in C^1(\mathbb{R}^n \times U)$, $\theta, \nabla_x \theta$ are bounded;
(d) the mapping $f \in C^1(\mathbb{R}^n \times U)$ and $f, \nabla_x f$ satisfy the polynomial growth condition;
(e) the mapping $g \in C^2(\mathbb{R}^n)$ and $g, \nabla_x g$ satisfy the polynomial growth condition,

we then have that the following well-posedness of the second-order HJB equation (3.30):

Theorem 3.3 (Fleming and Rishel (1975), Theorem VI.6.2, p. 169). *Under Assumptions* (a)–(e), *the HJB equation* (3.30) *admits a classical solution with polynomial growth.*

3.4 Stochastic Maximum Principle

The (deterministic) optimal control theory and the maximum principle originate from Pontryagin et al. (1962). It is known that the stochastic maximum principle (SMP) provides some necessary conditions for optimality for stochastic optimal control problems. For the stochastic control problem in which the control domain is convex or diffusive coefficient does not depend on the control strategy, the corresponding SMP was developed by

Bismut (1976, 1978) where the first-order expansion is only needed. For the general stochastic control problem where the control domain need not to be convex and the diffusive coefficient can depend on the control variable, Peng (1990) made a breakthrough by establishing the corresponding SMP via the second-order expansion and second-order BSDE.

In what follows, we first provide a heuristic derivation of the Pontryagin's maximum principle for the deterministic control problem using a perturbation method. Recall the deterministic control problem described as (3.13), i.e., at time $t = 0$,

$$J(0, x; u^*) = \inf_{u \in \mathcal{U}_0^T} J(0, x; u) := \inf_{u \in \mathcal{U}_0^T} \left[g(X_T) + \int_0^T f(X_s, u_s) ds \right]. \quad (3.32)$$

In (3.32), we used the notation $X = (X_t)_{t \in [0,T]}$ to represent $X^{0,x,u} = (X_t^{0,x,u})_{t \in [0,T]}$ for the notational convenience. Also, recall that the state process X satisfies the SDE (3.12), i.e., $X_t' := \frac{dX_t}{dt} = b(X_t, u_t) dt$.

Next, we want to derive necessary conditions for optimality. It follows from the Lagrangian multiplier that, we minimize the following Lagrangian functional given by, with $X = (X_t)_{t \in [0,T]}$, $u = (u_t)_{t \in [0,T]}$ and $p = (p_t)_{t \in [0,T]}$,

$$L(X, u, p) := g(X_T) + \int_0^T \left(f(X_s, u_s) + p_s^\top (b(X_s, u_s) - X_s') \right) ds. \quad (3.33)$$

Let $\mathcal{H}(x, u, p)$ for $(x, u, p) \in \mathbb{R}^n \times U \times \mathbb{R}^n$ be Hamiltonian with the form defined by

$$\mathcal{H}(x, u, p) := p^\top b(x, u) + f(x, u). \quad (3.34)$$

Then, the Lagrangian functional becomes that

$$L(X, u, p) = g(X_T) + \int_0^T \left(\mathcal{H}(X_s, u_s, p_s) - p_s^\top X_s' \right) ds. \quad (3.35)$$

Let $u^* = (u_t^*)_{t \in [0,T]}$ be a minimizer of L associated with X^* and p. By performing a first-order Taylor expansion with perturbation $(\delta X, \delta u, \delta p) = (\delta X_t, \delta u_t, \delta p_t)_{t \in [0,T]}$ such that $\delta X_0 = 0$ and

$$\delta L(X, u, p) := L(X^* + \delta X, u^* + \delta u, p + \delta p) - L(X^*, u^*, p)$$
$$= \int_0^T \left(\mathcal{H}(X^* + \delta X, u_s^* + \delta u_s, p_s + \delta p_s) - (p_s + \delta p_s)^\top (X_s^* + \delta X_s)' \right) ds$$
$$- \int_0^T \left(\mathcal{H}(X_s^*, u_s^*, p_s) - p_s^\top (X_s^*)' \right) ds + g(X_T^* + \delta X_T^*) - g(X_T^*)$$

$$\approx \int_0^T (\nabla_x \mathcal{H}_s^\top \delta X_s + \nabla_u \mathcal{H}_s^\top \delta u_s + \nabla_p \mathcal{H}_s^\top \delta p_s - p_s^\top (\delta X_s)' - \delta p_s^\top (X_s^*)'$$
$$- (\delta p_s)^\top (\delta X_s)')ds + \delta X_T^\top \nabla_x g,$$

where $\nabla_x \mathcal{H}_s := \nabla_x \mathcal{H}(X_s^*, u_s^*, p_s)$ and $\nabla_x g := \nabla_x g(X_T^*)$. We have assumed that the policy set U is *linear*, i.e., $u + \delta u \in U$ for all $u, \delta u \in U$. By Integrating by parts, we arrive at

$$\delta L(X, u, p) \approx \nabla_x g^\top \delta X_T + \int_0^T ((\nabla_x \mathcal{H}_s + p_s')^\top \delta X_s + \nabla_u \mathcal{H}_s^\top \delta u_s$$
$$+ (\nabla_p \mathcal{H}_s - (X^*)')^\top \delta p_s - (p_s^\top \delta X_s)')ds.$$

Note that $\delta L(X, u, p) = 0$ and $\delta X_0 = 0$. Then, we deduce that

$$0 = \int_0^T ((\nabla_x \mathcal{H}_s + p_s')^\top \delta X_s + \nabla_u \mathcal{H}_s^\top \delta u_s + (\nabla_p \mathcal{H}_s - (X^*)')^\top \delta p_s)ds$$
$$+ (\nabla_x g - p_T)^\top \delta X_T.$$

This in turn suggests that the optimum (X^*, u^*, p) satisfies the so-called *Hamiltonian system* associated with the deterministic optimal control problem (3.32):

$$\begin{cases} dX_t^* = \nabla_p \mathcal{H}(X_t^*, u_t^*, p_t)dt, & X_0^* = x & \text{(controlled dynamics)}, \\ dp_t = -\nabla_x \mathcal{H}(X_t^*, u_t^*, p_t)dt, & p_T = \nabla_x g(X_T^*) & \text{(adjoint equation)}, \\ \nabla_u \mathcal{H}(X_t^*, u_t^*, p_t) = 0 & & (u^* \text{ is extremal for } \mathcal{H}). \end{cases} \quad (3.36)$$

Then, the Pontryagin's maximum principle is stated as: if (X^*, u^*) is an optimal solution of the deterministic control problem (3.32), there exists a solution p to the adjoint equation such that $u_t^* \in \arg\min_{u \in U} \mathcal{H}(X_t^*, u, p_t)$ for all $t \in [0, T]$.

We stress that the above perturbation method assumes that the policy set U is linear. However, in many practical examples, the space U is rather convex or a general metric space. Hence, a new perturbation method we will develop should be compatible with the structure of the set of controls. If the policy set U is convex, then a natural perturbation can be a convex perturbation of a given optimal control $u^* = (u_t^*)_{t \in [0,T]} \in \mathcal{U}_0^T$:

$$\delta u_t := \epsilon(u_t - u_t^*), \quad u_t \in U, \quad (3.37)$$

where $\epsilon \downarrow 0$. Therefore, for the optimal strategy u^* for J,

$$J(0, x; u^* + \delta u) = J(0, x; \epsilon u + (1-\epsilon)u^*) \geq J(0, x; u^*).$$

Thus, we have $(J'(0, x; u^*), u - u^*) \geq 0$ for all $u \in \mathcal{U}_0^T$ provided that the objective functional $J(0, x; \cdot)$ is Gâteaux differentiable.

We next describe the stochastic maximum principle for the stochastic control problem (3.16) and (3.17) with $t = 0$, but we here minimizes the objective functional over $u \in \mathcal{U}_{0,x}$. That is, we consider

$$V(x) := \inf_{u \in \mathcal{U}_{0,x}} J(x, u) := \inf_{u \in \mathcal{U}_{0,x}} \mathbb{E}\left[g(X_T) + \int_0^T f(s, X_s, u_s) ds\right]. \quad (3.38)$$

We first impose the following assumptions:

- ($\mathbf{A_{SMPc}}$) the coefficients b, σ and f are continuously differentiable with respect to $(x, u) \in \mathbb{R}^n \times U$ with Lipschitz continuous and bounded partial derivatives. The terminal function g is continuously differentiable with respect to $x \in \mathbb{R}^n$ with Lipschitz continuous and bounded partial derivatives.
- ($\mathbf{A_{SMPu}}$) the policy set $U \subset \mathbb{R}^m$ is a convex set and the admissible control set $\mathcal{U}_{0,x}$ is convex (i.e., a control process $u \in \mathcal{U}_{0,x}$ to be admissible are stable under convex combinations).

Furthermore, we introduce the so-called *generalized Hamiltonian* $\mathcal{H}$: $[0, T] \times \mathbb{R}^n \times \mathbb{R}^n \times \mathbb{R}^{n \times m} \times U \to \mathbb{R}$, which is defined as:

$$\mathcal{H}(t, x, y, z, u) := b(x, u)^\top y + \text{tr}[\sigma(x, u)^\top z] + f(t, x, u). \quad (3.39)$$

Thus, the controlled diffusion process can be rewritten as follows:

$$dX_t = \nabla_y \mathcal{H}(t, X_t, Y_t, Z_t, u_t) dt + \sigma(X_t, u_t) dW_t, \quad t \in (0, T].$$

Hence, the *adjoint equation* is the following BSDE given by, for $u \in \mathcal{U}_{0,x}$,

$$\begin{cases} dY_t = -\nabla_x \mathcal{H}(t, X_t, Y_t, Z_t, u_t) dt + Z_t dW_t, \quad t \in [0, T), \\ Y_T = \nabla_x g(X_T). \end{cases} \quad (3.40)$$

In summary, the stochastic maximum principle (necessary condition) can be described as:

Theorem 3.4. *Let Assumptions ($\mathbf{A_{SMPc}}$) and ($\mathbf{A_{SMPu}}$) hold. Let $u^* \in \mathcal{U}_{0,x}$ be an optimal control strategy of the stochastic control problem (3.16) and (3.38). Correspondingly, let $X^* = (X_t^*)_{t \in [0,T]}$ be the associated (optimally) controlled state (i.e., the solution to SDE (3.16) with u replaced with u^*) and $(Y^*, Z^*) = (Y_t^*, Z_t^*)_{t \in [0,T]}$ be the solution to the adjoint equation (3.40) with X replaced with X^*. Then, it holds that, for all $u \in U$,*

$$\nabla_u \mathcal{H}(t, X_t^*, Y_t^*, Z_t^*, u_t^*)^\top (u - u_t^*) \geq 0, \quad a.e. \ t \in [0, T]. \quad (3.41)$$

Proof. Given $v \in \mathcal{U}_{0,x}$, we take the perturbation $u_t^\epsilon := u_t^* + \epsilon(v_t - u_t^*)$ for any $\epsilon \in [0,1]$. It follows from Assumption ($\mathbf{A}_{\text{SMPu}}$) that $u^\epsilon \in \mathcal{U}_{0,x}$. Since $u^* \in \mathcal{U}_{0,x}$ is optimal, we arrive at the following Gâteaux derivative of the objective functional:

$$\frac{d}{d\epsilon}J(x; u^\epsilon)|_{\epsilon=0} = \mathbb{E}\left[\int_0^T \nabla \mathcal{H}_u(t, X_t^*, Y_t^*, Z_t^*, u_t^*)^\top (v_t - u_t^*) dt\right] \geq 0.$$

This proves (3.41), and hence the proof of the theorem is complete. □

The next theorem gives a sufficient condition for optimality (sufficient condition):

Theorem 3.5. *Let Assumptions ($\mathbf{A}_{\text{SMPc}}$) and ($\mathbf{A}_{\text{SMPu}}$) hold. Additionally, we assume that g is convex and $\mathbb{R}^n \times U \ni (x, u) \to \mathcal{H}(t, x, Y_t^*, Z_t^*, u)$ is convex, $\mathbb{P}$-a.s.. Let $u^* \in \mathcal{U}_{0,x}$ and $X^* = (X_t^*)_{t\in[0,T]}$ be the controlled diffusion process with control u^*. Assume that there is a solution $(Y^*, Z^*) = (Y_t^*, Z_t^*)_{t\in[0,T]}$ to the adjoint equation (3.40) such that, for $t \in [0, T]$,*

$$\mathcal{H}(t, X_t^*, Y_t^*, Z_t^*, u_t^*) = \inf_{u \in \mathcal{U}_{0,x}} \mathcal{H}(t, X_t^*, Y_t^*, Z_t^*, u), \quad \mathbb{P}\text{-a.s..} \qquad (3.42)$$

Then u^* is an optimal control, i.e., $J(x; u^*) = \inf_{u \in \mathcal{U}_{0,x}} J(x; u)$.

Proof. For an arbitrary control strategy $\hat{u} \in \mathcal{U}_{0,x}$, denote by $\hat{X} = (\hat{X}_t)_{t\in[0,T]}$ the solution to the SDE (3.16) with u replaced by $\hat{u}$. Since g is convex, we deduce from Assumption ($\mathbf{A}_{\text{SMPu}}$) that

$$g(x) - g(y) \leq (x - y)^\top \nabla_x g(x), \quad \forall x, y \in \mathbb{R}^n.$$

Then, it holds from (3.40) and Itô's formula that

$$\mathbb{E}[g(X_T^*) - g(\hat{X}_T)] \leq \mathbb{E}[(X_T^* - \hat{X}_T)^\top \nabla_x g(X_T^*)] = \mathbb{E}[(X_T^* - \hat{X}_T)^\top Y_T^*]$$

$$= \mathbb{E}\left[\int_0^T (X_s^* - \hat{X}_s)^\top dY_s^*\right] + \mathbb{E}\left[\int_0^T (Y_s^*)^\top d(X_s^* - \hat{X}_s)\right]$$

$$+ \mathbb{E}\left[\int_0^T \text{tr}[(\sigma(X_s^*, u_s^*) - \sigma(\hat{X}_s, \hat{u}_s))^\top Z_s^*] ds\right]$$

$$= -\mathbb{E}\left[\int_0^T (X_s^* - \hat{X}_s)^\top \nabla_x \mathcal{H}(s, X_s^*, Y_s^*, Z_s^*, u_s^*) ds\right]$$

$$+ \mathbb{E}\left[\int_0^T (Y_s^*)^\top (b(X_s^*, u_s^*) - b(\hat{X}_s, \hat{u}_s)) ds\right]$$

$$+ \mathbb{E}\left[\int_0^T \mathrm{tr}[(\sigma(X_s^*, u_s^*) - \sigma(\hat{X}_s, \hat{u}_s))^\top Z_s^*]ds\right].$$

Furthermore, using the definition (3.39) for generalized Hamiltonian, we have

$$\mathbb{E}\left[\int_0^T (f(s, X_s^*, u_s^*) - f(s, \hat{X}_s, \hat{u}_s))ds\right]$$
$$= \mathbb{E}\left[\int_0^T (\mathcal{H}(s, X_s^*, Y_s^*, Z_s^*, u_s^*) - \mathcal{H}(s, \hat{X}_s, \hat{Y}_s, \hat{Z}_s, \hat{u}_s))ds\right]$$
$$- \mathbb{E}\left[\int_0^T (Y_s^*)^\top (b(X_s^*, u_s^*) - b(\hat{X}_s, \hat{u}_s))ds\right]$$
$$- \mathbb{E}\left[\int_0^T \mathrm{tr}[(\sigma(X_s^*, u_s^*) - \sigma(\hat{X}_s, \hat{u}_s))^\top Z_s^*]ds\right].$$

Make a summation of the above two displays, it follows from (3.42) and the convexity assumption that

$$J(x, u^*) - J(x, \hat{u})$$
$$= \mathbb{E}[g(X_T^*) - g(\hat{X}_T)] + \mathbb{E}\left[\int_0^T (f(s, X_s^*, u_s^*) - f(s, \hat{X}_s, \hat{u}_s))ds\right]$$
$$\leq \mathbb{E}\left[\int_0^T (\mathcal{H}(s, X_s^*, Y_s^*, Z_s^*, u_s^*) - \mathcal{H}(s, \hat{X}_s, \hat{Y}_s, \hat{Z}_s, \hat{u}_s)\right.$$
$$\left. - (X_s^* - \hat{X}_s)^\top \nabla_x \mathcal{H}(s, X_s^*, Y_s^*, Z_s^*, u_s^*))ds\right]$$
$$\leq 0.$$

Thus, we complete the proof of the theorem. □

3.5 Risk Sensitive Stochastic Control

The risk sensitive control in discrete time framework is first studied by Howard and Matheson (1972). After it, Jacobson (1973) considers a risk sensitive stochastic control problem with linear quadratic Gaussian in the continuous time. In fact, the risk sensitive stochastic control generalizes the conventional stochastic control problem. In essence, the latter can be viewed as a special case of the former with a so-called risk neutral parameter. The application of the risk sensitive stochastic control to the dynamic portfolio optimization is first discussed in Bielecki and Pliska (1999). This section mainly introduces the risk sensitive stochastic control problem in continuous time and the path integral method for solving the optimal problem.

We start with the controlled state process given by (3.16), i.e., for all $(t,x) \in [0,T] \times \mathbb{R}^n$,

$$X_s^{t,x} = x + \int_t^s b(X_r^{t,x}, u_r)dr + \int_t^s \sigma(X_r^{t,x}, u_r)dW_r, \quad \forall s \in [t,T].$$

The performance of the control problem is evaluated by the following payoff function given by, for all $(t,x) \in [0,T] \times \mathbb{R}^n$,

$$F(t,x; X|_{[t,T]}) := g(X_T^{t,x}) + \int_t^T f(s, X_s^{t,x})ds. \qquad (3.43)$$

For any $\gamma \in \mathbb{R}$, we define the following value function given by

$$V_\gamma(t,x) := \sup_{u \in \mathcal{U}_{t,x}} J_\gamma(t,x;u)$$

$$:= \sup_{u \in \mathcal{U}_{t,x}} \begin{cases} \mathbb{E}\left[F(t,x; X|_{[t,T]})\right], & \gamma = 0, \\ \dfrac{1}{\gamma} \ln \mathbb{E}\left[e^{\gamma F(t,x; X|_{[t,T]})}\right], & \gamma \neq 0. \end{cases} \qquad (3.44)$$

It can be seen from (3.44) that $V_0(t,x)$ is the value function for the conventional stochastic control problem discussed in the previous sections. For thus case with $\gamma = 0$, we call $V_0(t,x)$ the value function of the risk-neutral control. For the non-zero γ, by applying Taylor expansion of $V_\gamma(t,x)$ around $\gamma = 0$, we deduce that

$$J_\gamma(t,x;u) = \mathbb{E}\left[F(t,x; X|_{[t,T]})\right] + \frac{\gamma}{2}\mathrm{Var}[F(t,x; X|_{[t,T]})] + o(\gamma^2). \qquad (3.45)$$

By taking $\gamma = -\frac{\theta}{2}$ with $\theta > 0$ and $F(t,x; X|_{[t,T]}) = \mathbb{E}[\ln X_T^{t,x}]$ with $X^{t,x} = (X_s^{t,x})_{s \in [0,T]}$ being the self-financing wealth process of an investor in (3.45), the expansion (3.45) can be read as follows:

$$-\frac{2}{\theta}\mathbb{E}[e^{-\frac{\theta}{2}\ln X_T^{t,x}}] = \mathbb{E}\left[\ln X_T^{t,x}\right] - \frac{\theta}{4}\mathrm{Var}[\ln X_T^{t,x}] + o(\theta^2). \qquad (3.46)$$

This expansion has been mentioned in Bielecki and Pliska (1999). Therein, the objective functional $J_{-\frac{\theta}{2}}$ is interpreted as the expected growth rate minus a penalty term (since $\theta > 0$), with an error that is proportional to θ^2. Here, the penalty term is proportional to the variance of the logarithm of wealth which is captured in Konno et al. (1993). Note that, the penalty term $\frac{\theta}{4}\mathrm{Var}[\ln X_T^{t,x}]$ in (3.46) is also proportional to the positive parameter

θ. We then explain θ as a risk-sensitive parameter or risk-averse parameter. The special case $\theta = 0$ is referred to as the risk-neutral case. This corresponds to a risk-neutral investor. It also follows from (3.46) that $J_{-\frac{\varrho}{2}}$ admits a large deviation type form w.r.t. the logarithm of wealth. Therefore, we want to maximize the objective functional $J_{-\frac{\varrho}{2}}$ for $\theta > 0$. This means that it protects an investor interested in maximizing the expected growth rate of the logarithm of wealth against large deviations of the actually realized rate from the expectation.

We next provide a formal derivation of the HJB equation corresponding to the risk sensitive control problem (3.44). In fact, using the DPP, it follows that, for any $\ell \in [t, T]$,

$$V_\gamma(t,x) = \sup_{u \in \mathcal{U}_{t,x}} \ln \frac{1}{\gamma} \mathbb{E}\left[\exp\left(\gamma F(\ell, X_\ell^{t,x}; X|_{[\ell,T]}) + \gamma \int_t^\ell f(s, X_s^{t,x}) ds\right)\right]$$

$$= \sup_{u \in \mathcal{U}_{t,x}} \ln \frac{1}{\gamma} \mathbb{E}\left[\exp\left(\gamma V_\gamma(\ell, X_\ell^{t,x}) + \gamma \int_t^\ell f(s, X_s^{t,x}) ds\right)\right].$$

For $\gamma > 0$, the HJB equation satisfied by $\hat{V}_\gamma(t,x) = \exp(\gamma V_\gamma(t,x))$ with $(t,x) \in [0,T] \times \mathbb{R}^n$ can be formally described as follows:

$$\sup_{u \in U} \{\partial_t + \mathcal{A}^u + \gamma f\}\hat{V}_\gamma(t,x) = 0, \quad \hat{V}_\gamma(T,x) = e^{\gamma g(x)}, \tag{3.47}$$

where, for any $u \in U$, the operator $\mathcal{A}^u$ acted on $C^2(\mathbb{R}^n)$ is defined by

$$\mathcal{A}^u := b(x,u)^\top \nabla_x + \frac{1}{2} \text{tr}[\sigma \sigma^\top(x,u) \nabla_x^2]. \tag{3.48}$$

Note that $\partial_t \hat{V}_\gamma(t,x) = \gamma \hat{V}_\gamma(t,x) \partial_t V_\gamma(t,x)$, and it also holds that

$$\mathcal{A}^u \hat{V}_\gamma(t,x) = \hat{V}_\gamma(t,x)\{\gamma b(x,u)^\top \nabla_x V_\gamma(t,x) + \frac{\gamma}{2}\text{tr}[\sigma\sigma^\top(x,u)\nabla_x^2 V_\gamma(t,x)]\}$$

$$+ \frac{\gamma^2}{2}\left|\sigma(x,u)^\top \nabla_x V_\gamma(t,x)\right|^2$$

$$= \gamma \hat{V}_\gamma(t,x) \mathcal{A}^u V_\gamma(t,x) + \frac{\gamma^2}{2}\left|\sigma(x,u)^\top \nabla_x V_\gamma(t,x)\right|^2.$$

Plugging them into Eq. (3.47) to obtain that, the value function $V_\gamma(t,x)$ satisfies the HJB equation given by

$$\sup_{u \in U} \left\{\{\partial_t + \mathcal{A}^u\}V_\gamma(t,x) + f(t,x) + \frac{\gamma}{2}\left|\sigma(x,u)^\top \nabla_x V_\gamma(t,x)\right|^2\right\} = 0 \tag{3.49}$$

with terminal condition $V_\gamma(T,x) = g(x)$ for all $x \in \mathbb{R}^n$.

In the case with $\gamma < 0$, the HJB equation (3.49) remain valid, and it is derived in a similar fashion. However, observe (3.49), it can be seen that a quadratic form of the gradient of the value function additionally appears in the equation compared with the HJB equation (3.30) of the conventional stochastic control problem. To handle this term, one in general applies the so-called *Cole–Hopf* transformation of the value function $V_\gamma(t,x)$ (i.e, $\hat{V}_\gamma(t,x) = \exp(\gamma V_\gamma(t,x))$) to transfer the HJB equation (3.49) into a standard HJB equation (3.47). Consequently, it is more convenient to study the transformed equation (3.47) instead of (3.49). This technique has been used in Chapters 4 and 5 of the book.

Lastly, we consider the so-called *path integral method* to solve the risk sensitive stochastic control problem (van den Broek et al. (2010)) under some conditions. To apply path integral method, the controlled state process should admit the following form given by, for all $s \in [t,T]$,

$$X_s^{t,x} = x + \int_t^s \mu(X_r^{t,x})dr + \int_t^s c(X_r^{t,x})(u(r,X_r^{t,x})dr + vdW_r). \quad (3.50)$$

Here, $\mu : \mathbb{R}^n \to \mathbb{R}^n$, $c : \mathbb{R}^n \to \mathbb{R}^{n \times p}$, $v \in \mathbb{R}^{p \times m}$ and $u : [0,T] \times \mathbb{R}^n \to U = \mathbb{R}^p$ is the feedback control function. That is, we choose that

$$b(x,u) = \mu(x,u) + c(x)u, \quad \sigma(x,u) = c(x)v, \quad (3.51)$$

where $\sigma(x,u) = c(x)v$ which is independent of the control strategy in (3.16). Additionally, we take the running payoff function having the form:

$$f(t,x) = -\frac{1}{2}\|Ru(t,x)\|^2 + q(t,x), \quad (3.52)$$

where $\rho > 0$, $R \in \mathbb{R}^{p \times p}$ with $R^\top R$ being invertible and $q : [0,T] \times \mathbb{R}^n \to \mathbb{R}$.

We also assume that $\gamma < 0$ (in the context of the portfolio optimization, it corresponds to a risk-sensitive investor). In this setup with $U = \mathbb{R}^p$, the HJB equation (3.49) becomes that

$$\{\partial_t + \mathcal{A}^0\}V_\gamma + q(t,x) + \frac{\gamma}{2}(\nabla_x V_\gamma)^\top cvv^\top c^\top \nabla_x V_\gamma$$
$$+ \sup_{u \in \mathbb{R}^p}\left\{u^\top c(x)^\top \nabla_x V_\gamma - \frac{1}{2}u^\top R^\top Ru\right\} = 0.$$

The optimal feedback control function implied in the above equation is given by

$$u^*(t,x) = (R^\top R)^{-1}c(x)^\top \nabla_x V_\gamma(t,x). \quad (3.53)$$

Plugging the optimal (feedback) strategy (3.53) into the above HJB equation, we deduce that

$$\{\partial_t + \mathcal{A}^0\}V_\gamma + q(t,x)$$

$$+ \frac{1}{2}(\nabla_x V_\gamma)^\top c(x)(\gamma vv^\top + (R^\top R)^{-1})c(x)^\top \nabla_x V_\gamma = 0, \quad (3.54)$$

where $V_\gamma(T,x) = g(x)$ for all $x \in \mathbb{R}^n$. We next assume that $vv^\top$ admits the following form given by (for the one-dimensional case, it trivially holds):

$$vv^\top = -\frac{1}{\gamma}(R^\top R)^{-1}. \quad (3.55)$$

In light of (3.55), Eq. (3.54) is reduced to the following *linear* PDE, for all $(t,y) \in [0,T] \times \mathbb{R}^n$,

$$\partial_t V_\gamma(t,y) + \mathcal{A}^0 V_\gamma(t,y) + q(t,y) = 0. \quad V_\gamma(T,y) = g(y). \quad (3.56)$$

Let us introduce the following SDE, for all $(t,y) \in [0,T] \times \mathbb{R}^n$,

$$Y_s^{t,y} = y + \int_t^s \mu(Y_r^{t,y})dr + \int_t^s c(Y_r^{t,y})v dW_r, \quad \forall s \in [t,T]. \quad (3.57)$$

Namely, the operator $\mathcal{A} := \mathcal{A}^0$ is the generator of $Y^{t,y} = (Y_s^{t,y})_{s \in [0,T]}$. Then, if Assumptions of Theorem 2.5 (*Feynman–Kac's formula*) in Section 2.3 remain valid for the coefficients of PDE (3.56), it has a unique classical solution admitting the following probabilistic representation:

$$V_\gamma(t,y) = \mathbb{E}\left[g(Y_T^{t,y}) + \int_t^T q(s, Y_s^{t,y})ds\right]. \quad (3.58)$$

We here stress that the process $Y^{t,y} = (Y_s^{t,y})_{s \in [0,T]}$ is depending on the risk-sensitive parameter γ (c.f. (3.55)). Denote by $p(t,x)$ the probability density of the process Y, i.e., $p(t, z-y)dz = \mathbb{P}(Y_t^{0,y} \in dz)$. Then, $p(t,x)$ satisfies the forward Kolmogorov (Fokker-Planck-Kolmogorov) equation:

$$\begin{cases} \partial_t p(t,x) = \mathcal{A}^* p(t,x), & \forall (t,x) \in (0,T] \times \mathbb{R}^n, \\ p(0,x) = \delta_x, & \forall x \in \mathbb{R}^n, \end{cases} \quad (3.59)$$

where $\mathcal{A}^*$ is the adjoint operator of $\mathcal{A}$. Thus, it holds that

$$V_\gamma(t,y) = \int_{\mathbb{R}^n} g(z) p(T, z-y) dy$$

$$+ \int_t^T \left(\int_{\mathbb{R}^n} q(s,z) p(s, z-y) dz\right) ds. \quad (3.60)$$

We next consider the case that, for a positive constant $\alpha \neq -\frac{1}{\gamma}$,

$$vv^\top = \alpha_0 (R^\top R)^{-1}. \quad (3.61)$$

In this case, let us define an exponential transformation for V_γ by

$$E_\gamma(t,x) := \exp\left(\frac{V_\gamma(t,x)}{\alpha_\gamma}\right) \iff V_\gamma(t,x) = \alpha_\gamma \ln E_\gamma(t,x)$$

with the constant $\alpha_\gamma := \frac{\alpha_0}{1+\alpha_0\gamma}$ (i.e., $\ell_\gamma := \alpha_0(\gamma\alpha_\gamma - 1) + \alpha_\gamma = 0$). It follows from (3.49) that

$$\{\partial_t + \mathcal{A}\}E_\gamma + \alpha_\gamma^{-1}q(t,x)E_\gamma$$
$$+ \ell_\gamma E_\gamma^{-1}(\nabla_x E_\gamma)^\top c(x)(R^\top R)^{-1}c(x)^\top \nabla_x E_\gamma = 0$$

with terminal condition $E_\gamma(T,x) = e^{\alpha_\gamma^{-1}g(x)}$ for all $x \in \mathbb{R}^n$. The corresponding optimal feedback strategy is given by

$$u^*(t,x) = \frac{\alpha_\gamma}{E_\gamma(t,x)}(R^\top R)^{-1}c(x)^\top \nabla_x E_\gamma(t,x). \tag{3.62}$$

Since $\ell_\gamma = 0$, we deduce the following *linear* PDE:

$$\{\partial_t + \mathcal{A}\}E_\gamma + \alpha_\gamma^{-1}q(t,x)E_\gamma = 0, \quad E_\gamma(T,x) = e^{\alpha_\gamma^{-1}g(x)}. \tag{3.63}$$

If Assumptions of Theorem 2.5 in Section 2.3 remain valid for the coefficients of PDE (3.63), then

$$E_\gamma(t,y) = \mathbb{E}\left[\exp\left(\frac{1}{\alpha_\gamma}g(Y_T^{t,y}) + \frac{1}{\alpha_\gamma}\int_t^T q(s,Y_s^{t,y})ds\right)\right]. \tag{3.64}$$

Therefore, it holds that

$$V_\gamma(t,y) = \ln\left\{\mathbb{E}\left[\exp\left(g(Y_T^{t,y}) + \int_t^T q(s,Y_s^{t,y})ds\right)^{\alpha_\gamma^{-1}}\right]\right\}^{\alpha_\gamma}. \tag{3.65}$$

Similarly to (3.60), we can use the transition probability density $p(t,x)$ of Y to represent the value function V given by (3.65).

An alternative approach to solve the risk-sensitive control problem with the state process X specified by (3.50) is to apply the change of measure. However, to apply the "change of measure", it is required that the control strategy u may be in a more smaller admissible control set. It may raise the additional difficulty in the proof of the corresponding verification result.

Chapter 4

Risk Sensitive Credit Portfolio Optimization: Complete Information*

This chapter presents a class of risk sensitive portfolio optimization problems with default contagion under a complete information framework. In Chapter 5, we will discuss the similar risk sensitive portfolio optimization problem but under an incomplete information setting.

Credit portfolio with contagion. It is known that, the dynamic credit portfolio optimization is the process of selecting the optimal portfolio associated with credit assets dynamically, out of the set of portfolios being considered, according to the investor's objective. The importance of considering the defaultable assets has attracted a lot of attention, especially after the systemic failure caused by some global financial crisis. Some recent developments extend the early model of single defaultable security to default contagion effects on portfolio allocations. The research of these mutual contagion influence opens the door to provide possible answers to some empirical puzzles like the high mark-to-market variations in prices of credit sensitive assets. Kraft and Steffensen (2009) discuss the contagion effects on defaultable bonds. Callegaro et al. (2012) consider an optimal investment problem with multiple defaultable assets which depend on a partially observed exogenous factor process. Jiao et al. (2013) study the model in which multiple jumps and default events are allowed. Bo and Capponi (2016) examine the optimal portfolio problem of a power utility investor who allocates the wealth between credit default swaps and a money market for which the contagion risk is modeled via interacting default intensities.

Regime-switching model. The market trend is often described as some random factors such as Markov chains. In particular, the so-called regime-switching model is widely accepted and usually proposed to capture the

---

*This chapter is based on the work Bo et al. (2019).

influence on the behavior of the market caused by transitions in the macroeconomic system or the macroscopic readjustment and regulation. For instance, the empirical results by Ang and Bekaert (2002) illustrate the existence of two regimes characterized by different levels of volatility. It is well known that default events modulated by the regime-switching process have an impact on the distress state of the surviving securities in the portfolio. More specifically, by an empirical study of the corporate bond market over 150 years, Giesecke et al. (2011) suggest the existence of three regimes corresponding to high, middle, and low default risk. With finitely many economical regimes, Capponi and Figueroa-López (2014) study the classical utility maximization problem from terminal wealth based on a defaultable security, and Capponi et al. (2014) obtain a Poisson series representation for the arbitrage-free price process of vulnerable contingent claims. Back to the practical implementation in financial markets with stochastic factors, the regime-switching model or continuous time Markov chain is frequently used to approximate the dynamics of time-dependent market parameter or factors. The continuous state space of the parameter or factor is usually discretized which lead to infinite states of the approximating Markov chain (c.f. Ang and Timmermann (2012)). This mainly motivates us to consider the countable regime states in this work and it is shown that this technical difficulties can eventually be reconciled using an appropriate approximation by counterparts with finite states. Therefore, our analytical conclusions for regime-switching can potentially provide theoretical foundations for numerical treatment of risk sensitive portfolio optimization with defaults and stochastic factor processes.

Risk-sensitive asset management. Apart from the celebrated Merton's model on utility maximization, there has been an increasing interest in the risk-sensitive stochastic control criterion in the portfolio management during recent years, see, e.g., Davis and LIeo (2014) for an overview of the theory and practice of risk-sensitive asset management. In a typical risk sensitive portfolio optimization problem, the investor maximizes the long run growth rate of the portfolio, adjusted by a measure of volatility. In particular, the classical utility maximization from terminal wealth can be transformed to the risk-sensitive control criterion by introducing a change of measure and a so-called risk-sensitive parameter which characterizes on the degree of risk tolerance of investors, see, e.g., Bielecki and Pliska (1999) and Nagai and Peng (2002). We will only name a small portion of the vast literature, for instance, the risk sensitive criterion can be linked to the dynamic version of Markowitz's mean-variance optimization by Bielecki and Pliska (1999), to differential games by Fleming (2006) and more recently by Bayraktar and Yao (2013) for the connection to zero-sum stochastic differential games using BSDEs and the weak DPP. Hansen et al. (2006) further connect the risk-sensitive objective to a robust criteria in which perturbations are characterized by the relative entropy. Bayraktar and Cohen (2018) later examine a risk sensitive control version of the lifetime ruin probability problem.

Even if many existing work on the risk-sensitive control, optimal investment with credit risk or regime switching respectively, it remains an open problem of the risk-sensitive portfolio allocation with both scenarios of default risk and regime-switching. This chapter aims to fill this gap and considers an extended situation when the default contagion effect can depend on regime states, possibly infinitely many. For some recent related work, it is worth noting that in the default-free market with finite regime states, Andruszkiewicz et al. (2016) explore the existence and uniqueness of the solution to the risk-sensitive maximization problem, and provide an ODE for the optimal value function, which may be efficiently solved numerically. Meanwhile, Das et al. (2018) consider a risk-sensitive portfolio optimization problem with multiple stocks modeled as a multi-dimensional jump diffusion whose coefficients are modulated by an age-dependent semi-Markov process. They also establish the existence and uniqueness of classical solutions to the corresponding HJB equations. In the context of theoretical stochastic control, we also note that Kumar and Pal (2013) derive the DPP for a class of risk-sensitive control problem of pure jump process with near monotone cost. To model hybrid diffusions, Nguyen and Yin (2016) propose a switching diffusion system with countably infinite states. The existence and uniqueness of the solution to the hybrid diffusion with past-dependent switching are obtained.

Aims of this chapter. The aims of this chapter are twofold. From the modeling perspective, it is considered that the correlated stocks are subject to credit events, and in particular, the dynamics of defaultable stocks, namely the drift, the volatility and the default intensity coefficients, all depend on the macroeconomic regimes. As defaults can occur sequentially, the default contagion is modeled in the sense that default intensities of surviving names are affected simultaneously by default events of other stocks as well as on current regimes states. This set up in our model enables us to analyze the joint complexity rooted in the investor's risk sensitivity, the regime changes and the default contagion among stocks. From the mathematical perspective, the resulting HJB equation can be viewed as a recursive infinite-dimensional nonlinear dynamical system in terms of default states. The depth of the recursion equals the number of stocks in the portfolio. Our recipe to study this new type of recursive dynamical system can be summarized in the following scheme: First, it is proposed to truncate the countably infinite state space of the regime switching process and consider the recursive HJB equation only with a finite state space. Second, for the finite state case, the existence and uniqueness of solutions to the recursive HJB equation are analyzed based upon a backward recursion, namely from the state in which all stocks are defaulted toward the state in which all stocks are alive. It is worth noting that no bounded constraint is reinforced on the trading strategies of securities or control variables as in Andruszkiewicz et al. (2016) and Kumar and Pal (2013). As a price to pay, the nonlinearities of the HJB systems are not globally Lipschitz continuous. To overcome this new challenge, we develop a truncation technique by proving a comparison theorem based on the theory of monotone dynamical systems documented in Smith (2008). Then, we establish a unique classical solution to the recursive HJB equation by showing that the solution of truncated system has a uniform (strictly positive) lower bound independent of the truncation level. This also enables us to characterize the optimal admissible (feedback) trading strategy in the verification theorem. Next, when the states are relaxed to be countably infinite, the results in the finite state case can be applied to construct a sequence of approximating risk sensitive control problems to the original problem and obtain elegant uniform estimates to conclude that the sequence of associated smooth value functions will successfully converge to the classical solution of the original recursive HJB equation. We also contribute to the existing literature by exploring the possible construction and approximation of the optimal (feedback) strategy in some rigorous verification theorems.

The rest of the chapter is organized as follows. Section 4.1 describes the credit market model with default contagion and regime switching. Section 4.2 formulates the risk-sensitive control problem. Section 4.3 derives the recursive HJB equation in terms of default states. Section 4.4 analyzes the well-posedness of global solution of recursive infinite-dimensional HJB equations in classical sense. Section 4.5 develops a rigorous verification theorem. Section 4.6 establishes the convergence rate for approximating risk-sensitive control problems.

4.1 The Market Model

In this section, we consider a financial market consisting of n defaultable stocks and a risk-free money market account on a given filtered probability space $(\Omega, \mathcal{G}, \mathbb{G}, \mathbb{P})$. Let $T > 0$ be a terminal horizon and $I = (I_t)_{t \in [0,T]}$ be a regime-switching process which is described as a continuous-time (conservative) Markov chain $I = (I_t)_{t \in [0,T]}$ with countable state space $S_I = \mathbb{N}$. The generator of the Markov chain I is given by the Q-matrix $Q = (q_{il})_{i,l \in S_I}$. This implies that $q_{ii} \leq 0$ for $i \in S_I$, $q_{il} \geq 0$ for $i \neq l$, and $\sum_{l=1}^{\infty} q_{il} = 0$ for $i \in S_I$ (i.e., $\sum_{l \neq i} q_{il} = -q_{ii}$ for $i \in S_I$). The global filtration $\mathbb{G} = \mathbb{F} \vee \mathbb{H}$ augmented by all $\mathbb{P}$-null sets satisfies the usual conditions. The filtration $\mathbb{F} = (\mathcal{F}_t)_{t \in [0,T]}$ is jointly generated by I and a d-dimensional Brownian motions (which is independent of the Markov chain) denoted by $W = (W_t^j; j = 1, \ldots, d)^{\top}_{t \in [0,T]}$. The price process of the money market account B_t satisfies $dB_t = r(I_t)B_t dt$, where $r(I_t) \geq 0$ is interest rate modulated by the Markov chain I. The filtration $\mathbb{H}$ is generated by an n-dimensional default indicator process $H = (H_t^j; j = 1, \ldots, n)_{t \in [0,T]}$ taking values in $\mathcal{S} := \{0,1\}^n$.

We assume that (I, H) is a hybrid process introduced by Section 2.2 with the underlying diffusion process X replaced by the Markov chain I. More precisely, in terms of (2.51) in Chapter 2, we assume that, for $(i,h) \in S_I \times \mathcal{S}$,

$$\mathbb{P}(H_{t+\delta} = h^j | H_t = h, I_t = i, (I_s, H_s)_{s \leq t})$$
$$= \begin{cases} \delta \lambda_j(i,h) \mathbf{1}_{h_j = 0} + o(\delta), & j = 1, \ldots, n, \\ 1 + \delta \lambda_0(i,h) + o(\delta), & j = 0, \end{cases} \quad (4.1)$$

uniformly in $i \in S_I$ as $\delta \downarrow 0$. Here, $\lambda_0(i,h) = -\sum_{j=1}^{n} \lambda_j(i,h) \mathbf{1}_{h_j=0}$. Recall that, for $h \in \mathcal{S}$, h^j is defined by (2.52) for $j = 1, \ldots, n$. We set $\lambda(i,h) = (\lambda_j(i,h); j = 1, \ldots, n)^{\top}$ for $(i,h) \in S_I \times \mathcal{S}$. The default indicator process H links to the default times of the n defaultable stocks via $\tau_j := \inf\{t \geq 0; H_t^j = 1\}$ for $j = 1, \ldots, n$. The filtration $\mathbb{H} = (\mathcal{H}_t)_{t \in [0,T]}$ is defined by $\mathcal{H}_t = \bigvee_{j=1}^{n} \sigma(H_s^j; s \leq t)$. Hence, $\mathbb{H}$ contains all information about default events until the terminal time T.

Let the vector process $\tilde{P} = (\tilde{P}_t^j; j = 1, \ldots, n)^{\top}_{t \in [0,T]}$ be the price process of the n defaultable stocks, which is described as in (2.55) of Section 2.2 in Chapter 2. The dynamics of the *pre-default* price process $P = (P_t^i; i = 1, \ldots, n)^{\top}_{t \geq 0}$ of the n defaultable stocks is given by

$$dP_t = \text{diag}(P_t)\{(\mu(I_t) + \lambda(I_t, H_t))dt + \sigma(I_t)dW_t\}, \quad (4.2)$$

where, for each $i \in S_I$, the vector $\mu(i)$ is $\mathbb{R}^n$-valued column vector and $\sigma(i)$ is $\mathbb{R}^{n \times d}$-valued matrices such that $\sigma(i)\sigma(i)^{\top}$ is positive definite. Using (2.57) in Chapter 2, the price dynamics of defaultable stocks satisfies that

$$d\tilde{P}_t = \text{diag}(\tilde{P}_{t-})\{\mu(I_t)dt + \sigma(I_t)dW_t - dM_t\}. \quad (4.3)$$

Here, in view of (2.54) with $\mathbb{G} = (\mathcal{G}_t)_{t\in[0,T]}$, the vector of processes $M = (M_t^j;\ j = 1,\ldots,n)_{t\in[0,T]}^\top$ is a pure jump (n-dimensional) $\mathbb{G}$-martingale given by

$$M_t^j := H_t^j - \int_0^t (1 - H_s^j)\lambda_j(I_s, H_s)ds$$
$$= H_t^j - \int_0^{t\wedge\tau_j} \lambda_j(I_s, H_s)ds, \quad \forall t \in [0, T]. \quad (4.4)$$

In terms of the construction of the default indicator process H in Bo and Capponi (2018), it can be seen that W is also a $\mathbb{G}$-Brownian motion using the condition (M.2a) in Section 6.1.1 of Chapter 6 in Bielecki and Rutkowski (2002).

4.2 Formulation of Portfolio Optimization Problem

This sections formulates the optimal credit portfolio problem with default contagion.

For $j = 1,\ldots,n$, let $\tilde{\pi}_t^j$ be the fraction of wealth invested in the j-th stock at time $t \geq 0$. Note that the price of the j-th stock jumps to zero when the j-th stock defaults, the fraction of wealth held by the investor in this stock is zero after it defaults. In particular, as a $\mathbb{G}$-predictable process $\tilde{\pi}^i = (\tilde{\pi}_t^j)_{t\in[0,T]}$, the equality holds $\tilde{\pi}_t^j = (1 - H_{t-}^j)\tilde{\pi}_t^j$ for $j = 1,\ldots,n$. Then, in light of (2.16) in Chapter 2 under the self-financing assumption, the investor's wealth process becomes that, $X_0^{\tilde{\pi}} = x \in \mathbb{R}_+$, and

$$dX_t^{\tilde{\pi}} = X_t^{\tilde{\pi}}\{r(I_t) + \tilde{\pi}_t^\top(\mu(I_t) - r(I_t)\mathbf{1}_n)\}dt$$
$$+ X_{t-}^{\tilde{\pi}}\tilde{\pi}_t^\top\{\sigma(I_t)dW_t - dM_t\} \quad (4.5)$$

with the portfolio $\tilde{\pi}_t = (\tilde{\pi}_t^j;\ j = 1,\ldots,n)^\top$.

We next introduce the definition of the set of all admissible controls used in this chapter.

Definition 4.1 (Admissible Control Set). The admissible control set $\tilde{\mathcal{U}}$ is a class of $\mathbb{G}$-predictable feedback strategies $\tilde{\pi}_t$ for $t \in [0,T]$, given by $\tilde{\pi}_t^j = \pi_j(t, X_{t-}^{\tilde{\pi}}, I_{t-}, H_{t-})$ such that SDE (4.5) admits a unique positive (strong) solution for $X_0^{\tilde{\pi}} = x \in \mathbb{R}_+$ (i.e., the feedback strategies $\tilde{\pi}_t$ should take values in $U := (-\infty, 1)^n$). Moreover, the control $\tilde{\pi} = (\tilde{\pi}_t)_{t\in[0,T]}$ is required to make the positive process $\Gamma^{\tilde{\pi}} = (\Gamma_t^{\tilde{\pi}})_{t\in[0,T]}$ defined later by (4.9) to be a $\mathbb{P}$-martingale.

The martingality of $\Gamma^{\tilde\pi}$ for any $\tilde\pi \in \tilde{\mathcal{U}}$ imposed in Definition 4.1 can guarantee that we can apply the change of measure to transform our credit portfolio optimization problem into a risk-sensitive control problem with form as in Section 3.5. We will prove the martingality of $\Gamma^{\tilde\pi^*}$ for a candidate optimal strategy $\tilde\pi^*$ by verifying the generalized Novikov's condition in Section 4.4. As in Section 3.5, we consider the following finite-horizon objective functional as in Bielecki and Pliska (1999). For $\tilde\pi \in \tilde{\mathcal{U}}$, and given the initial values $(X_0, I_0, H_0) = (x, i, h) \in \mathbb{R}_+ \times S_I \times S$, we define

$$\mathcal{J}(\tilde\pi; T, x, i, h) := -\frac{2}{\theta} \ln \mathbb{E}\left[(X_T^{\tilde\pi})^{-\frac{\theta}{2}}\right]$$

$$= -\frac{2}{\theta} \ln \mathbb{E}\left[\exp\left(-\frac{\theta}{2} \ln X_T^{\tilde\pi}\right)\right]. \qquad (4.6)$$

The investor aims to maximize the objective functional $\mathcal{J}$ over all strategies $\tilde\pi \in \tilde{\mathcal{U}}$. Let us only focus on the case when $\theta \in (0, \infty)$ for a risk-sensitive investor. The case $\theta \in (-2, 0)$ is ignored as it is associated to a risk-seeking behavior which is less encountered in practise.

We next rewrite the objective functional as the exponential of an integral criterion (c.f. Nagai and Peng (2002)) in order to apply the change of measure to simplify the objective functional $\mathcal{J}$. For any $\tilde\pi \in \tilde{\mathcal{U}}$, the wealth process satisfied by SDE (4.5) is given by

$$\begin{aligned}X_T^{\tilde\pi} =&\, x \exp\bigg\{\int_0^T (r(I_s) + \tilde\pi_s^\top(\mu(I_s) - r(I_s)\mathbf{1}_n))ds + \int_0^T \tilde\pi_s^\top \sigma(I_s)dW_s \\ & - \frac{1}{2}\int_0^T |\sigma(I_s)^\top \tilde\pi_s|^2 ds + \sum_{j=1}^n \int_0^T \ln(1-\tilde\pi_s^j)dM_s^j \\ & + \sum_{j=1}^n \int_0^{T\wedge\tau_j} \lambda_j(I_s, H_s)[\tilde\pi_s^j + \ln(1-\tilde\pi_s^j)]ds\bigg\}.\end{aligned}$$

Therefore, it holds that

$$\exp\left(-\frac{\theta}{2}\ln X_T^{\tilde\pi}\right) = x^{-\frac{\theta}{2}} \Gamma_T^{\tilde\pi} \exp\left(\frac{\theta}{2}\int_0^T L(\tilde\pi_s; I_s, H_s)ds\right), \qquad (4.7)$$

where, for $(\pi, i, h) \in U \times S_I \times \mathcal{S}$, the risk sensitive function $L(\pi; i, h)$ is defined by

$$\begin{aligned}L(\pi; i, h) :=&\, -r(i) - \pi^\top(\mu(i) - r(i)\mathbf{1}_n) + \frac{1}{2}\left(1 + \frac{\theta}{2}\right)|\sigma(i)^\top \pi|^2 \\ & - \sum_{j=1}^n (1-h_j)\left[\frac{2}{\theta} + \pi_j - \frac{2}{\theta}(1-\pi_j)^{-\frac{\theta}{2}}\right]\lambda_j(i, h).\end{aligned} \qquad (4.8)$$

Here, the positive density process is given by

$$\Gamma_t^{\tilde{\pi}} := \mathcal{E}_t(\Pi^{\tilde{\pi}}), \quad \forall t \in [0,T], \tag{4.9}$$

$$\Pi_t^{\tilde{\pi}} := -\frac{\theta}{2}\int_0^t \tilde{\pi}_s^\top \sigma(I_s)dW_s + \sum_{j=1}^n \int_0^t \{(1-\tilde{\pi}_s^j)^{-\frac{\theta}{2}} - 1\}dM_s^j,$$

where we recall that $\mathcal{E}(\cdot)$ denotes the stochastic exponential (Doléans–Dade exponential).

In light of Definition 4.1, for any $\tilde{\pi} \in \tilde{\mathcal{U}}$, we have that $\Gamma^{\tilde{\pi}} = (\Gamma_t^{\tilde{\pi}})_{t\in[0,T]}$ is a $\mathbb{P}$-martingale. Then, we may define the following change of measure given by

$$\frac{d\mathbb{P}^{\tilde{\pi}}}{d\mathbb{P}}\bigg|_{\mathcal{G}_t} = \Gamma_t^{\tilde{\pi}}, \cdot \quad \forall t \in [0,T], \tag{4.10}$$

under which

$$W_t^{\tilde{\pi}} := W_t + \frac{\theta}{2}\int_0^t \sigma(I_s)^\top \tilde{\pi}_s ds, \quad t \in [0,T] \tag{4.11}$$

is a d-dimensional Brownian motion; while under $\mathbb{P}^{\tilde{\pi}}$, for $j=1,\ldots,n$,

$$M_t^{\tilde{\pi},j} := H_t^j - \int_0^{t\wedge\tau_j}(1-\tilde{\pi}_s^j)^{-\frac{\theta}{2}}\lambda_j(I_s,H_s)ds, \quad t \in [0,T] \tag{4.12}$$

is a martingale. Thus, by applying the change of measure, we deduce from (4.7) that

$$J(\tilde{\pi};T,x,i,h) = -\frac{2}{\theta}\ln\mathbb{E}\left[x^{-\frac{\theta}{2}}\Gamma_T^{\tilde{\pi}}\exp\left(\frac{\theta}{2}\int_0^T L(\tilde{\pi}_s;I_s,H_s)ds\right)\right]$$

$$= \ln x - \frac{2}{\theta}\ln\mathbb{E}^{\tilde{\pi}}\left[\exp\left(\frac{\theta}{2}\int_0^T L(\tilde{\pi}_s;I_s,H_s)ds\right)\right]$$

$$=: \ln x + \bar{J}(\tilde{\pi};T,i,h).$$

Here, $\mathbb{E}^{\tilde{\pi}}[\cdot]$ represents the expectation w.r.t. $\mathbb{P}^{\tilde{\pi}}$ defined by (4.10). In terms of the relation between J and $\bar{J}$, our original problem is equivalent to maximize $\bar{J}$ over $\tilde{\pi} \in \tilde{\mathcal{U}}$. We can therefore reformulate the value function of the risk-sensitive control problem as

$$V(T,i,h) := \sup_{\tilde{\pi}\in\tilde{\mathcal{U}}} \bar{J}(\tilde{\pi};T,i,h)$$

$$= -\frac{2}{\theta}\inf_{\tilde{\pi}\in\tilde{\mathcal{U}}}\ln\mathbb{E}^{\tilde{\pi}}\left[\exp\left(\frac{\theta}{2}\int_0^T L(\tilde{\pi}_s;I_s,H_s)ds\right)\right]. \tag{4.13}$$

4.3 Recursive HJB Equations

This section will derive the recursive HJB equation in terms of default states satisfied by the value function (4.13) using heuristic arguments in Birge et al. (2018).

We first define the following dynamical version of the optimization problem (4.13) as, for $(t, i, h) \in [0, T] \times S_I \times S$,

$$\bar{V}(t, i, h) := -\frac{2}{\theta} \inf_{\tilde{\pi} \in \tilde{\mathcal{U}}_t} \ln J(\tilde{\pi}; t, i, h)$$

$$:= -\frac{2}{\theta} \inf_{\tilde{\pi} \in \tilde{\mathcal{U}}_t} \ln \mathbb{E}^{\tilde{\pi}}_{t,i,h} \left[\exp\left(\frac{\theta}{2} \int_t^T L(\tilde{\pi}_s; I_s, H_s) ds\right) \right], \quad (4.14)$$

where $\tilde{\mathcal{U}}_t$ is the admissible control set in Definition 4.1 with starting time t, and $\mathbb{E}^{\tilde{\pi}}_{t,i,h}[\cdot] := \mathbb{E}^{\tilde{\pi}}[\cdot | I_t = i, H_t = h]$. This yields the relation $V(T, i, h) = \bar{V}(0, i, h)$. For $0 \leq t < s \leq T$, the DPP results in

$$\bar{V}(t, i, h) \qquad (4.15)$$
$$= -\frac{2}{\theta} \inf_{\tilde{\pi} \in \tilde{\mathcal{U}}_t} \ln \mathbb{E}^{\tilde{\pi}}_{t,i,h} \left[\exp\left(-\frac{\theta}{2} \bar{V}(s, I_s, H_s) + \frac{\theta}{2} \int_t^s L(\tilde{\pi}_u; I_u, H_u) du\right) \right].$$

Using heuristic arguments as in Birge et al. (2018), we have the following recursive HJB equation satisfied by the value function, for $(t, i, h) \in [0, T) \times S_I \times S$,

$$0 = \partial_t \bar{V}(t, i, h) - \frac{2}{\theta} \sum_{l \neq i} q_{il} \left[\exp\left(-\frac{\theta}{2}(\bar{V}(t, l, h) - \bar{V}(t, i, h))\right) - 1 \right]$$
$$+ \sup_{\pi \in U} \mathcal{H}\left(\pi; i, h, (\bar{V}(t, i, h^j); \ j = 0, 1, \ldots, n)\right) \qquad (4.16)$$

with terminal condition $\bar{V}(T, i, h) = 0$ for all $(i, h) \in S_I \times S$. The function $\mathcal{H}$ is defined by, for $(\pi, i, h) \in U \times S_I \times S$,

$$\mathcal{H}(\pi; i, h, \bar{f}(h))$$
$$:= -\frac{2}{\theta} \sum_{j=1}^n \left[\exp\left(-\frac{\theta}{2}(f(h^j) - f(h))\right) - 1 \right] (1 - h_j)(1 - \pi_j)^{-\frac{\theta}{2}} \lambda_j(i, h)$$
$$+ r(i) + \pi^\top(\mu(i) - r(i)\mathbf{1}_n) - \frac{1}{2}\left(1 + \frac{\theta}{2}\right) |\sigma(i)^\top \pi|^2$$
$$+ \sum_{j=1}^n \left[\frac{2}{\theta} + \pi_j - \frac{2}{\theta}(1 - \pi_j)^{-\frac{\theta}{2}} \right] (1 - h_j) \lambda_j(i, h). \qquad (4.17)$$

Here, $\bar{f}(h) := (f(h^j); \ j = 0, 1, \ldots, n)$ for some function $f(h)$. Eq. (4.16) is in fact a *recursive HJB system*.

We introduce the *Cole-Hopf* transform of the solution given by, for all $(t, i, h) \in [0, T] \times S_I \times S$,

$$\varphi(t, i, h) := \exp\left(-\frac{\theta}{2}\bar{V}(t, i, h)\right). \tag{4.18}$$

Then, we get that

$$0 = \partial_t \varphi(t, i, h) + \sum_{l \neq i} q_{il} \left[\varphi(t, l, h) - \varphi(t, i, h)\right]$$
$$+ \inf_{\pi \in U} \tilde{\mathcal{H}}\left(\pi; i, h, (\varphi(t, i, h^j); j = 0, 1, \ldots, n)\right) \tag{4.19}$$

with terminal condition $\varphi(T, i, h) = 1$ for all $(i, h) \in S_I \times S$. The function $\tilde{\mathcal{H}}$ is specified as

$$\tilde{\mathcal{H}}(\pi; i, h, \bar{f}(h))$$
$$:= \left\{ -\frac{\theta}{2} r(i) - \frac{\theta}{2} \pi^\top (\mu(i) - r(i)\mathbf{1}_n) + \frac{\theta}{4}\left(1 + \frac{\theta}{2}\right) |\sigma(i)^\top \pi|^2 \right.$$
$$+ \sum_{j=1}^n \left(-1 - \frac{\theta}{2}\pi_j\right)(1 - h_j)\lambda_j(i, h) \right\} f(h)$$
$$+ \sum_{j=1}^n f(h^j)(1 - h_j)(1 - \pi_j)^{-\frac{\theta}{2}} \lambda_j(i, h). \tag{4.20}$$

The well-posedness of recursive HJB system (4.19) will be addressed in next section 4.4.

4.4 Well-posedness of Recursive HJB Equations

This section analyzes the existence of global solutions to the recursive HJB system (4.19) in a two-step procedure:

- we first study the existence and uniqueness of (classical) solutions to Eq. (4.19) as a dynamical system when the Markov chain I takes values in the finite state space;
- we then proceed to explore the countably infinite state case using approximation arguments.

Before presenting the main results, we introduce some notations which will be used frequently in this section. For the general default state $h \in S$, we write this default state as follows:

$$h = 0^{j_1, \ldots, j_k} \tag{4.21}$$

for indices $j_1 \neq \cdots \neq j_k$ belonging to $\{1,\ldots,n\}$, and $k \in \{0,1,\ldots,n\}$. Here, we recall that n is the number of risky assets. Such a vector h is obtained by flipping the entries $j_1,\ldots,j_k$ of the zero vector to one, i.e., $h_{j_1} = \cdots = h_{j_k} = 1$, and $h_j = 0$ for $j \notin \{j_1,\ldots,j_k\}$ (if $k = 0$, we set $h = 0^{j_1,\ldots,j_k} = 0$). It is obvious to have $0^{j_1,\ldots,j_n} = \mathbf{1}_n^\top$.

4.4.1 Finite State Case of Markov Chain

This subsection focuses on the case where the Markov chain I has a finite state space:
$$D_m := \{1,\ldots,m\}, \quad \forall m \in S_I. \tag{4.22}$$

The corresponding Q-matrix of the Markov chain I is given by $Q_m = (q_{il})_{i,l \in D_m}$ satisfying $\sum_{l \in D_m} q_{il} = 0$ for $i \in D_m$ and $q_{il} \geq 0$ when $i \neq l$. Note that q_{il}, $i,l \in D_m$ here may be different from what is given in Section 4.1. With slight abuse of notation, we still use q_{il} only for notational convenience.

Let $\varphi(t,h) := (\varphi(t,i,h); \ i = 1,\ldots,m)^\top$ be a column vector of the solution for $(t,h) \in [0,T] \times S$. Then, we can rewrite Eq. (4.19) as the following dynamical system:
$$\begin{cases} \partial_t \varphi(t,h) + (Q_m + \mathrm{diag}(\nu(z)))\varphi(t,h) \\ \qquad\qquad + G(t,\varphi(t,h),h) = 0, \quad \forall t \in [0,T) \times S, \\ \varphi(T,h) = \mathbf{1}_m, \qquad\qquad\qquad \forall h \in S, \end{cases} \tag{4.23}$$

where $G(t,x,h) = (G_i(t,x,h); \ i = 1,\ldots,m)^\top$ is given by, for $i \in D_m$, and $(t,x,h) \in [0,T] \times \mathbb{R}^m \times S$,

$$G_i(t,x,h) := \inf_{\pi \in U} \left\{ \sum_{j=1}^n \varphi(t,i,h^j)(1-h_j)(1-\pi_j)^{-\frac{\theta}{2}} \lambda_j(i,h) \right.$$
$$+ \left[\frac{\theta}{4}\left(1+\frac{\theta}{2}\right)|\sigma(i)^\top \pi|^2 - \frac{\theta}{2}\pi^\top(\mu(i) - r(i)\mathbf{1}_n) \right.$$
$$\left. \left. - \frac{\theta}{2}\sum_{j=1}^n \pi_j(1-h_j)\lambda_j(i,h) \right] x_i \right\}.$$

The vector of coefficients $\nu(h) = (\nu_i(h); \ i = 1,\ldots,m)^\top$ with $h \in S$ is defined as follows:
$$\nu_i(h) := -\frac{\theta}{2} r(i) - \sum_{j=1}^n (1-h_j)\lambda_j(i,h), \quad \forall i \in D_m. \tag{4.24}$$

For $k = 0, 1, \ldots, n$, Eq. (4.23) is a recursive differential system in terms of default states $h = 0^{j_1, \ldots, j_k} \in \mathcal{S}$. Hence, the solvability can be in fact analyzed in the recursive form on default states. Our strategy for analyzing the system is based on a recursive procedure, starting from the default state $h = \mathbf{1}_n^\top$ (i.e., all stocks have defaulted) and proceeding backward to the default state $h = 0$ (i.e., all stocks are alive).

(i) $k = n$ (i.e., all stocks have defaulted). In this default state, the investor will not invest in stocks, and hence the optimal fraction strategy in stocks for this case is given by $\pi_1^* = \cdots = \pi_n^* = 0$ by Definition 4.1. Let $\varphi(t, \mathbf{1}_n^\top) = (\varphi(t, i, \mathbf{1}_n^\top); \ i = 1, \ldots, n)^\top$. As a consequence, the dynamical system (4.23) can be written as:

$$\begin{cases} \varphi'(t, \mathbf{1}_n^\top) = -A^{(n)} \varphi(t, \mathbf{1}_n^\top), & \text{in } [0, T), \\ \varphi(T, \mathbf{1}_n^\top) = \mathbf{1}_m. \end{cases} \quad (4.25)$$

The matrix of coefficients $A^{(n)} := Q_m + \mathrm{diag}(\nu(\mathbf{1}_n^\top))$. To establish the unique positive solution of the dynamical system (4.25), we need the following auxiliary result.

Lemma 4.1. *Let $g(t) = (g_i(t); \ i = 1, \ldots, m)^\top$ satisfy the following dynamical system $g'(t) = Bg(t)$ in $(0, T]$ and $g(0) = \xi$. If $B = (b_{ij})_{m \times m}$ satisfies $b_{ij} \geq 0$ for $i \neq j$, and $\xi \gg 0$, then $g(t) \gg 0$ for all $t \in [0, T]$.*

Proof. Set $f(x) := Bx$ for $x \in \mathbb{R}^m$. In lieu of Proposition 1.1 of Chapter 3 in Smith (2008), it suffices to verify that $f : \mathbb{R}^m \to \mathbb{R}^m$ is of type K, i.e., for any $x, y \in \mathbb{R}^m$ satisfying $x \leq y$ and $x_i = y_i$ for some $i = 1, \ldots, m$, then $f_i(x) \leq f_i(y)$. Note that $b_{ij} \geq 0$ for all $i \neq j$. Then, we have

$$f_i(x) = (Bx)_i = \sum_{j=1}^m b_{ij} x_j = b_{ii} x_i + \sum_{j=1, j \neq i}^m b_{ij} x_j$$

$$= b_{ii} y_i + \sum_{j=1, j \neq i}^m b_{ij} x_j \leq b_{ii} y_i + \sum_{j=1, j \neq i}^m b_{ij} y_j = f_i(y), \quad (4.26)$$

and hence f is of type K. Thus, we complete the proof of the lemma. $\square$

The following result is consequent on the above lemma.

Lemma 4.2. *The dynamical system (4.25) has a unique solution which is given by, for all $t \in [0, T]$,*

$$\varphi(t, \mathbf{1}_n^\top) = e^{A^{(n)}(T-t)} \mathbf{1}_m = \sum_{i=0}^\infty \frac{(A^{(n)})^i (T-t)^i}{i!} \mathbf{1}_m, \quad (4.27)$$

where $A^{(n)} = Q_m + \text{diag}(\nu(\mathbf{1}_n^\top)) = Q_m - \frac{\theta}{2}\text{diag}(r)$ with $r = (r(i); i = 1,\ldots,m)^\top$, which is an $m \times m$-dimensional matrix. Moreover, it holds that $\varphi(t, \mathbf{1}_n^\top) \gg 0$ for all $t \in [0,T]$.

Proof. The solution representation (4.27) for $\varphi(t, \mathbf{1}_n^\top)$ is obvious. Note that $\mathbf{1}_m \gg 0$ and $q_{il} \geq 0$ for all $i \neq l$ since $Q_m = (q_{il})_{m \times m}$ is the generator of the Markov chain. Then, to prove $\varphi(t, \mathbf{1}_n^\top) \gg 0$ for all $t \in [0,T]$, using Lemma 4.1, it suffices to verify $[A^{(n)}]_{ij} \geq 0$ for all $i \neq j$. However, $[A^{(n)}]_{ij} = q_{ij}$ for all $i \neq j$ and the condition given in Lemma 4.1 is therefore verified which implies that $\varphi(t, \mathbf{1}_n^\top) \gg 0$ for all $t \in [0,T]$. $\square$

We next consider the general default case with $h = 0^{j_1,\ldots,j_k}$ for $0 \leq k \leq n-1$, i.e., the stocks $j_1,\ldots,j_k$ have defaulted but the stocks $\{j_{k+1},\ldots,j_n\} := \{1,\ldots,n\} \setminus \{j_1,\ldots,j_k\}$ remain alive. Then, we have

(ii) since the stocks $j_1,\ldots,j_k$ have defaulted, the optimal fraction strategies for the stocks $j_1,\ldots,j_k$ are given by $\pi_j^{(k,*)} = 0$ for $j \in \{j_1,\ldots,j_k\}$ via Definition 4.1.

Through introducing the notation:
$$\varphi^{(k)}(t) = (\varphi(t, i, 0^{j_1,\ldots,j_k}); i = 1,\ldots,n)^\top,$$
$$\lambda_j^{(k)}(i) = \lambda_j(i, 0^{j_1,\ldots,j_k}), \quad i = 1,\ldots,m, \ j \notin \{j_1,\ldots,j_k\},$$

Eq. (4.23) to this default case is given by
$$\begin{cases} \dfrac{d}{dt}\varphi^{(k)}(t) = -A^{(k)}\varphi^{(k)}(t) - G^{(k)}(t, \varphi^{(k)}(t)), & \text{in } [0,T), \\ \varphi^{(k)}(T) = \mathbf{1}_m, \end{cases} \quad (4.28)$$

where, the $m \times m$-dimensional matrix $A^{(k)}$ is given by
$$A^{(k)} = \text{diag}\left[\left(-\frac{\theta}{2}r(i) - \sum_{j \notin \{j_1,\ldots,j_k\}} \lambda_j^{(k)}(i); i = 1,\ldots,m\right)\right] + Q_m. \quad (4.29)$$

The coefficient $G^{(k)}(t,x) = (G_i^{(k)}(t,x); i = 1,\ldots,m)^\top$ for $(t,x) \in [0,T] \times \mathbb{R}^m$ is given by, for $i \in D_m$,

$$G_i^{(k)}(t,x) \quad (4.30)$$
$$:= \inf_{\pi^{(k)} \in U^{(k)}} \left\{ \sum_{j \notin \{j_1,\ldots,j_k\}} \varphi^{(k+1),j}(t,i)\left(1 - \pi_j^{(k)}\right)^{-\frac{\theta}{2}} \lambda_j^{(k)}(i) + \mathcal{H}^{(k)}(\pi^{(k)}; i)x_i \right\}.$$

Here, for $(\pi^{(k)}, i) \in U^{(k)} \times D_m$, the function $\mathcal{H}^{(k)}$ is given by

$$\mathcal{H}^{(k)}(\pi^{(k)}; i) := \frac{\theta}{4}\left(1 + \frac{\theta}{2}\right)\left|\sigma^{(k)}(i)^\top \pi^{(k)}\right|^2 \qquad (4.31)$$

$$- \frac{\theta}{2}(\pi^{(k)})^\top (\mu^{(k)}(i) - r(i)\mathbf{1}_{n-k}) - \frac{\theta}{2} \sum_{j \notin \{j_1,\ldots,j_k\}} \pi_j^{(k)} \lambda_j^{(k)}(i).$$

The policy space related to this state is $U^{(k)} = (-\infty, 1)^{n-k}$, and $\varphi^{(k+1),j}(t,i) := \varphi(t, i, 0^{j_1,\ldots,j_k,j})$ for $j \notin \{j_1,\ldots,j_k\}$ corresponds to the i-th element of the positive solution vector of Eq. (4.23) at the default state $h = 0^{j_1,\ldots,j_k,j}$. Here, for each $i = 1, \ldots, m$, we have also used the following notations:

$$\begin{cases} \pi^{(k)} = (\pi_j^{(k)}; \ j \notin \{j_1, \ldots, j_k\})^\top, \quad \theta^{(k)}(i) = (\theta_j(i); \ j \notin \{j_1, \ldots, j_k\})^\top, \\ \sigma^{(k)}(i) = (\sigma_{j\kappa}(i); \ j \notin \{j_1, \ldots, j_k\}, \kappa \in \{1, \ldots, d\}), \qquad (4.32) \\ \mu^{(k)}(i) = (\mu_j(i); \ j \notin \{j_1, \ldots, j_k\})^\top. \end{cases}$$

It follows from (4.30) that, the solution $\varphi^{(k)}(t)$ for $t \in [0,T]$ of Eq. (4.23) at the default state $h = 0^{j_1,\ldots,j_k}$ in fact depends on the solution $\varphi^{(k+1),j}(t)$ for $t \in [0,T]$ of (4.23) at the default state $h = 0^{j_1,\ldots,j_k,j}$ for $j \notin \{j_1,\ldots,j_k\}$. In particular, when $k = n-1$, the solution $\varphi^{(k+1),j}(t) = \varphi(t, \mathbf{1}_n^\top) \gg 0$ corresponds to the solution of (4.23) at the default state $h = \mathbf{1}_n$ (i.e., $k = n$), which has been obtained by Lemma 4.2. This suggests us to solve Eq. (4.23) backward recursively in terms of default states $h = 0^{j_1,\ldots,j_k}$. Thus, in order to study the existence and uniqueness of positive (classical) solutions to the system (4.28), we first assume that (4.23) has a positive unique (classical) solution $\varphi^{(k+1),j}(t)$ for $t \in [0,T], j \notin \{j_1,\ldots,j_k\}$.

We can first obtain an estimate on $G^{(k)}(t,x)$, which is presented in the following lemma.

Lemma 4.3. *For each $k = 0, 1, \ldots, n-1$, assume that Eq. (4.23) has a positive unique (classical) solution $\varphi^{(k+1),j}(t)$ on $t \in [0,T]$ for $j \notin \{j_1,\ldots,j_k\}$. Then, for any $x, y \in \mathbb{R}^m$ satisfying $x, y \geq \varepsilon\mathbf{1}_m$ with $\varepsilon > 0$, there exists a positive constant $C = C(\varepsilon)$ which only depends on $\varepsilon > 0$ such that*

$$\left|G^{(k)}(t,x) - G^{(k)}(t,y)\right| \leq C |x - y|. \qquad (4.33)$$

Proof. It suffices to prove that, for each $i = 1, \ldots, m$, $|G_i^{(k)}(t,x) - G_i^{(k)}(t,y)| \leq C(\varepsilon)|x-y|$ for all $x, y \in \mathbb{R}^m$ satisfying $x, y \geq \varepsilon e_m$ with $\varepsilon > 0$, where $C(\varepsilon) > 0$ is independent of time t. By the recursive assumption, $\varphi^{(k+1),j}(t)$ on $t \in [0,T]$ is the unique positive (classical) solution to (4.23) for $j \notin \{j_1,\ldots,j_k\}$. Then, it is continuous on $[0,T]$ which implies the existence of a constant $C_0 > 0$ independent of t such that $\sup_{t \in [0,T]} |\varphi^{(k+1),j}(t)| \leq C_0$ for $j \notin \{j_1, \ldots, j_k\}$. Thus, by (4.30), and thanks to $\mathcal{H}^{(k)}(0;i) = 0$ for all $i \in D_n$ using (4.31), it follows that, for all $(t,x) \in [0,T] \times \mathbb{R}^m$,

$$G_i^{(k)}(t,x)$$

$$\leq \left[\sum_{j \notin \{j_1,\ldots,j_k\}} \varphi^{(k+1),j}(t,i)(1-\pi_j^{(k)})^{-\frac{\theta}{2}} \lambda_j^{(k)}(i) + \mathcal{H}^{(k)}(\pi^{(k)};i)x_i \right]\bigg|_{\pi^{(k)}=0}$$

$$= \sum_{j \notin \{j_1,\ldots,j_k\}} \varphi^{(k+1),j}(t,i) \lambda_j^{(k)}(i) \leq C_0 \sum_{j \notin \{j_1,\ldots,j_k\}} \lambda_j^{(k)}(i). \tag{4.34}$$

On the other hand, since $\sigma^{(k)}(i)^\top \sigma^{(k)}(i)$ is positive-definite, there exists a positive constant $\delta > 0$ such that $\|\sigma^{(k)}(i)^\top \pi^{(k)}\|^2 \geq \delta \|\pi^{(k)}\|^2$ for all $i \in D_m$. Hence, the following estimate holds:

$$\mathcal{H}^{(k)}(\pi^{(k)};i) \geq \frac{\theta}{4}\left(1+\frac{\theta}{2}\right)\delta \left|\pi^{(k)}\right|^2 \tag{4.35}$$

$$- \frac{\theta}{2}\left(\left|\mu^{(k)}(i) - r(i)\mathbf{1}_{n-k}\right| + \sum_{j \notin \{j_1,\ldots,j_k\}} \lambda_j^{(k)}(i) \right) \left|\pi^{(k)}\right|.$$

We next take the positive constant defined as:

$$C_1 := 2 \frac{\left|\mu^{(k)}(i) - r(i)\mathbf{1}_{n-k}\right| + \sum_{j \notin \{j_1,\ldots,j_k\}} \lambda_j^{(k)}(i)}{(1+\frac{\theta}{2})\delta}.$$

For all $\pi^{(k)} \in \{\pi^{(k)} \in U^{(k)}; |\pi^{(k)}| \geq C_1\}$, it holds that

$$\mathcal{H}^{(k)}(\pi^{(k)};i) \geq 0, \quad i \in D_m. \tag{4.36}$$

This implies that, for all $\pi^{(k)} \in \{\pi^{(k)} \in U^{(k)}; |\pi^{(k)}| \geq C_1\}$ and all $x \geq \varepsilon e_n$, we deduce from (4.35) and (4.36) that

$$\sum_{j \notin \{j_1,\ldots,j_k\}} \varphi^{(k+1),j}(t,i)(1-\pi_j^{(k)})^{-\frac{\theta}{2}} \lambda_j^{(k)}(i) + H^{(k)}(\pi^{(k)};i)x_i$$

$$\geq \mathcal{H}^{(k)}(\pi^{(k)};i)x_i \geq \mathcal{H}^{(k)}(\pi^{(k)};i)\varepsilon \geq \varepsilon \left[\frac{\theta}{4}\left(1+\frac{\theta}{2}\right)\delta \left|\pi^{(k)}\right|^2\right.$$

$$\left. - \frac{\theta}{2}\left(\left|\mu^{(k)}(i) - r(i)\mathbf{1}_{n-k}\right| + \sum_{j \notin \{j_1,\ldots,j_k\}} \lambda_j^{(k)}(i) \right) \left|\pi^{(k)}\right| \right].$$

We choose another positive constant depending on $\varepsilon > 0$ as

$$C_2(\varepsilon) := \frac{C_1}{2} + \sqrt{\frac{C_1^2}{4} + \frac{8}{\varepsilon\theta(2+\theta)\delta}C_0 \sum_{j \notin \{j_1,\ldots,j_k\}} \lambda_j^{(k)}(i)}.$$

Then, for all $\pi^{(k)} \in \{\pi \in U^{(k)}; |\pi| \geq C_2(\varepsilon)\}$ and all $x \geq \varepsilon e_m$, it holds that

$$\sum_{j \notin \{j_1,\ldots,j_k\}} \varphi^{(k+1),j}(t,i)(1-\pi_j^{(k)})^{-\frac{\theta}{2}} \lambda_j^{(k)}(i) + \mathcal{H}^{(k)}(\pi^{(k)};i)x_i$$
$$\geq C_0 \sum_{j \notin \{j_1,\ldots,j_k\}} \lambda_j^{(k)}(i). \tag{4.37}$$

It follows from (4.34) that $G_i^{(k)}(t,x) \leq C_0 \sum_{j \notin \{j_1,\ldots,j_k\}} \lambda_j^{(k)}(i)$ for all $(t,x) \in [0,T] \times \mathbb{R}^m$. Thus, in light of (4.37),

$$G_i^{(k)}(t,x) \tag{4.38}$$
$$= \inf_{\pi^{(k)} \in U^{(k)}} \left\{ \sum_{j \notin \{j_1,\ldots,j_k\}} \varphi^{(k+1),j}(t,i)(1-\pi_j^{(k)})^{-\frac{\theta}{2}} \lambda_j^{(k)}(i) + \mathcal{H}^{(k)}(\pi^{(k)};i)x_i \right\}$$
$$= \inf_{\substack{\pi^{(k)} \in \{\pi \in U^{(k)}: \\ |\pi| \leq C_2(\varepsilon)\}}} \left\{ \sum_{j \notin \{j_1,\ldots,j_k\}} \varphi^{(k+1),j}(t,i)(1-\pi_j^{(k)})^{-\frac{\theta}{2}} \lambda_j^{(k)}(i) + \mathcal{H}^{(k)}(\pi^{(k)};i)x_i \right\}.$$

In view of (4.38), it holds that

$$G_i^{(k)}(t,x) = \inf_{\substack{\pi^{(k)} \in \{\pi \in U^{(k)}: \\ |\pi| \leq C_2(\varepsilon)\}}} \left\{ \sum_{j \notin \{j_1,\ldots,j_k\}} \varphi^{(k+1),j}(t,i)(1-\pi_j^{(k)})^{-\frac{\theta}{2}} \lambda_j^{(k)}(i) \right.$$
$$\left. + \mathcal{H}^{(k)}(\pi^{(k)};i)y_i + \mathcal{H}^{(k)}(\pi^{(k)};i)(x_i - y_i) \right\}$$
$$\leq \inf_{\substack{\pi^{(k)} \in \{\pi \in U^{(k)}: \\ |\pi| \leq C_2(\varepsilon)\}}} \left\{ \sum_{j \notin \{j_1,\ldots,j_k\}} \varphi^{(k+1),j}(t,i)(1-\pi_j^{(k)})^{-\frac{\theta}{2}} \lambda_j^{(k)}(i) \right.$$
$$\left. + \mathcal{H}^{(k)}(\pi^{(k)};i)y_i \right\} + C(\varepsilon)|x_i - y_i|$$
$$= G_i^{(k)}(t,y) + C(\varepsilon)|x_i - y_i|. \tag{4.39}$$

Here, the finite positive constant $C(\varepsilon) = \max_{i=1,\ldots,n} C^{(i)}(\varepsilon)$, where, for all $i \in D_m$,

$$C^{(i)}(\varepsilon) := \sup_{\substack{\pi^{(k)} \in \{\pi \in U^{(k)}: \\ |\pi| \leq C_2(\varepsilon)\}}} \mathcal{H}^{(k)}(\pi^{(k)};i). \tag{4.40}$$

Note that $C^{(i)}(\varepsilon) \in [0,\infty)$ for any $i \in D_m$. Using the estimate (4.39), we have $|G_i^{(k)}(t,x) - G_i^{(k)}(t,y)| \leq C(\varepsilon)|x-y|$ for any $x,y \in \mathbb{R}^m$ satisfying $x,y \geq \varepsilon 1_m$ with $\varepsilon > 0$, which completes the proof of the lemma. $\square$

We next prove the existence and uniqueness of global (classical) solutions to the system (4.28). Toward this end, we prepare the following comparison results:

Lemma 4.4 (Comparison Result). *Let $g_\kappa(t) = (g_{\kappa i}(t); \ i = 1,\ldots, m)^\top$ with $\kappa = 1, 2$ satisfy the following dynamical systems on $[0, T]$, respectively*

$$\begin{cases} g_1'(t) = f(t, g_1(t)) + \tilde{f}(t, g_1(t)), \\ g_1(0) = \xi_1, \end{cases} \qquad \begin{cases} g_2'(t) = f(t, g_2(t)), \\ g_2(0) = \xi_2. \end{cases}$$

Here, $f(t, x), \tilde{f}(t, x) : [0, T] \times \mathbb{R}^m \to \mathbb{R}^m$ are assumed to be Lipschitz continuous w.r.t. $x \in \mathbb{R}^m$ uniformly in $t \in [0, T]$. The function $f(t, \cdot)$ satisfies the type K condition for each $t \in [0, T]$ (i.e., for any $x, y \in \mathbb{R}^m$ satisfying $x \leq y$ and $x_i = y_i$ for some $i = 1,\ldots, m$, it holds that $f_i(t, x) \leq f_i(t, y)$ for each $t \in [0, T]$). If $\tilde{f}(t, x) \geq 0$ for $(t, x) \in [0, T] \times \mathbb{R}^m$ and $\xi_1 \geq \xi_2$, then $g_1(t) \geq g_2(t)$ for all $t \in [0, T]$.

Proof. For $p > 0$, let $g_1^{(p)}(t) = (g_{1i}^{(p)}(t); \ i = 1,\ldots, m)^\top$ be the solution to the following dynamical system given by

$$\begin{cases} \dfrac{d}{dt} g_1^{(p)}(t) = f(t, g_1^{(p)}(t)) + \tilde{f}(t, g_1^{(p)}(t)) + \dfrac{1}{p} \mathbf{1}_m^\top, \quad \text{in } (0, T], \\ g_1^{(p)}(0) = \xi_1 + \dfrac{1}{p} \mathbf{1}_m^\top. \end{cases}$$

Then, for all $t \in (0, T]$, we have

$$|g_1^{(p)}(t) - g_1(t)| \leq |g_1^{(p)}(0) - g_1(0)| + \int_0^t |f(s, g_1^{(p)}(s)) - f(s, g_1(s))| ds$$

$$+ \int_0^t |\tilde{f}(s, g_1^{(p)}(s)) - \tilde{f}(s, g_1(s))| ds + \frac{1}{p} \int_0^t |\mathbf{1}_m| ds$$

$$\leq \frac{1+T}{p} |\mathbf{1}_m| + (C + \tilde{C}) \int_0^t |g_1^{(p)}(s) - g_1(s)| ds,$$

where $C > 0$ and $\tilde{C} > 0$ are Lipschitz constant coefficients for $f(t, x)$ and $\tilde{f}(t, x)$, respectively. The Gronwall's lemma yields that $g_1^{(p)}(t) \to g_1(t)$ for all $t \in [0, T]$ as $p \to \infty$. We claim that $g_1^{(p)}(t) \gg g_2(t)$ for all $t \in [0, T]$. Assume that the claim does not hold, the fact that $g_1^{(p)}(0) \gg g_2(0)$, and $g_1^{(p)}(t), g_2(t)$ are continuous on $[0, T]$ jointly imply that there exists a $t_0 \in (0, T]$ such that $g_1^{(p)}(s) \geq g_2(s)$ on $s \in [0, t_0]$ and $g_{1i}^{(p)}(t_0) = g_{2i}(t_0)$ for some $i \in \{1,\ldots, m\}$. Since, for $t_0 > 0$, $g_1^{(p)}(t)$ and $g_2(t)$ are differentiable on $(0, T]$, it follows that

$$\frac{d}{dt} g_{1i}^{(p)}(t)\Big|_{t=t_0} = \lim_{\epsilon \to 0} \frac{g_{1i}^{(p)}(t_0) - g_{1i}^{(p)}(t_0 - \epsilon)}{\epsilon}$$

$$\leq \lim_{\epsilon \to 0} \frac{g_{2i}(t_0) - g_{2i}(t_0 - \epsilon)}{\epsilon} = \frac{d}{dt} g_{2i}(t)\Big|_{t=t_0}.$$

On the other hand, since $f(t, \cdot)$ satisfies the type K condition for each $t \in [0, T]$ and $\tilde{f}(t, x) \geq 0$ for all $(t, x) \in [0, T] \times \mathbb{R}^m$, for the above i, we also have

$$\frac{d}{dt} g_{1i}^{(p)}(t)\Big|_{t=t_0} = f_i(t_0, g_{1i}^{(p)}(t_0)) + \tilde{f}_i(t_0, g_1^{(p)}(t_0)) + \frac{1}{p}$$

$$> f_i(t_0, g_{1i}^{(p)}(t_0)) \geq f_i(t_0, g_2(t_0)) = \frac{d}{dt} g_{2i}(t)\Big|_{t=t_0}.$$

We obtain a contradiction, and hence $g_1^{(p)}(t) \gg g_2(t)$ for all $t \in [0, T]$. Therefore $g_1(t) \geq g_2(t)$ for all $t \in [0, T]$ by passing p to infinity. □

Now, we are ready to present the following existence and uniqueness result for positive (classical) solutions of Eq. (4.28).

Theorem 4.1. *For each $k = 0, 1, \ldots, n-1$, assume that Eq. (4.23) has a positive unique solution $(\varphi^{(k+1),j}(t))_{t \in [0,T]}$ with $j \notin \{j_1, \ldots, j_k\}$. Then, there exists a unique positive solution $(\varphi^{(k)}(t))_{t \in [0,T]}$ of (4.23) at the default state $h = 0^{j_1, \ldots, j_k}$ (i.e., Eq. (4.28) has a unique positive solution).*

Proof. For any $a > 0$, we consider the following truncated dynamical system given by

$$\begin{cases} \dfrac{d}{dt} \varphi_a^{(k)}(t) + A^{(k)} \varphi_a^{(k)}(t) + G_a^{(k)}(t, \varphi_a^{(k)}(t)) = 0, & \text{in } [0, T), \\ \varphi_a^{(k)}(T) = \mathbf{1}_m, \end{cases} \quad (4.41)$$

where $\varphi_a^{(k)}(t) = (\varphi_a^{(k)}(t, i); \ i \in D_m)^\top$ is the vector-valued solution and the $m \times m$-dimensional matrix $A^{(k)}$ is given by (4.29). The vector-valued function $G_a^{(k)}(t, x)$ is defined by

$$G_a^{(k)}(t, x) := G^{(k)}(t, x \vee a\mathbf{1}_m), \quad \forall (t, x) \in [0, T] \times \mathbb{R}^m. \quad (4.42)$$

Thanks to Lemma 4.3, there exists $C = C(a) > 0$ which only depends on $a > 0$ such that

$$\sup_{t \in [0,T]} |G_a^{(k)}(t, x) - G_a^{(k)}(t, y)| \leq C|x - y|, \quad \forall x, y \in \mathbb{R}^m, \quad (4.43)$$

i.e., $x \to G_a^{(k)}(t, x)$ is globally Lipschitz continuous uniformly in $t \in [0, T]$. By reversing the time $\tilde{\varphi}_a^{(k)}(t) := \varphi_a^{(k)}(T - t)$ for $t \in [0, T]$, we have that $\tilde{\varphi}_a^{(k)}(t)$ satisfies the system given by

$$\begin{cases} \dfrac{d}{dt} \tilde{\varphi}_a^{(k)}(t) = A^{(k)} \tilde{\varphi}_a^{(k)}(t) + G_a^{(k)}(T - t, \tilde{\varphi}_a^{(k)}(t)), & \text{in } (0, T], \\ \tilde{\varphi}_a^{(k)}(0) = \mathbf{1}_m. \end{cases} \quad (4.44)$$

Using the globally Lipschitz continuity condition (4.43), it follows that the system (4.44) has a unique (classical) solution $\tilde{\varphi}_a^{(k)}(t)$ for $[0, T]$.

Let us consider the following linear system given by

$$\begin{cases} \dfrac{d}{dt}\psi^{(k)}(t) = f^{(k)}(\psi^{(k)}(t)), & \text{in } (0, T], \\ \psi^{(k)}(0) = \mathbf{1}_m. \end{cases} \quad (4.45)$$

Here, the vector-valued function $f^{(k)}(x) = (f_i^{(k)}(x); \ i \in D_m)^\top$ is defined by, for $i \in D_m$,

$$f_i^{(k)}(x) = \sum_{j=1}^n q_{ij} x_j - \left(\frac{\theta}{2} r(i) + \sum_{j \notin \{j_1, \ldots, j_k\}} h_j^{(k)}(i) \right) x_i - \beta_i x_i. \quad (4.46)$$

The constants β_i for $i \in D_m$ are given by

$$\beta_i = - \inf_{\pi^{(k)} \in U^{(k)}} \mathcal{H}^{(k)}(\pi^{(k)}; i), \quad (4.47)$$

where $\mathcal{H}^{(k)}(\pi^{(k)}; i)$ for $i \in D_m$ is defined by (4.31). It is not difficult to see that $\beta_i \in [0, \infty)$ for each $i \in D_m$ using (4.31).

We next verify that the vector-valued function $f^{(k)}(x)$ given by (4.54) is of type K. Namely, we need to verify that, for any $x, y \in \mathbb{R}^m$ satisfying $x \leq y$ and $x_{i_0} = y_{i_0}$ for some $i_0 = 1, \ldots, m$, $f_{i_0}^{(k)}(x) \leq f_{i_0}^{(k)}(y)$. In fact, we have from (4.54) that, for all $x, y \in \mathbb{R}^m$ satisfying $x \leq y$ and $x_{i_0} = y_{i_0}$ for some $i_0 = 1, \ldots, m$,

$$f_{i_0}^{(k)}(x) = \sum_{j=1}^m q_{i_0 j} x_j - \left(\frac{\theta}{2} r(i_0) + \sum_{j \notin \{j_1, \ldots, j_k\}} \lambda_j^{(k)}(i_0) \right) x_{i_0} - \beta_i x_{i_0}$$

$$= q_{i_0 i_0} x_{i_0} - \left(\frac{\theta}{2} r(i_0) + \sum_{j \notin \{j_1, \ldots, j_k\}} \lambda_j^{(k)}(i_0) \right) x_{i_0} - \beta_{i_0} x_{i_0}$$

$$+ \sum_{j \neq i_0} q_{i_0 j} x_j$$

$$= q_{i_0 i_0} y_{i_0} - \left(\frac{\theta}{2} r(i_0) + \sum_{j \notin \{j_1, \ldots, j_k\}} \lambda_j^{(k)}(i_0) \right) y_{i_0} - \beta_{i_0} y_{i_0}$$

$$+ \sum_{j \neq i_0} q_{i_0 j} x_j$$

$$\leq q_{i_0 i_0} y_{i_0} - \left(\frac{\theta}{2} r(i_0) + \sum_{j \notin \{j_1, \ldots, j_k\}} \lambda_j^{(k)}(i_0) \right) y_{i_0} - \beta_{i_0} y_{i_0}$$

$$+ \sum_{j \neq i_0} q_{i_0 j} y_j$$
$$= f_{i_0}^{(k)}(y), \qquad (4.48)$$

where, we used the fact that, for all $j \neq i_0$, $q_{i_0 j} \geq 0$ since $Q_m = (q_{il})_{m \times m}$ is the generator of the Markov chain I, and hence $\sum_{l \neq i_0} q_{i_0 j} x_l \leq \sum_{l \neq i_0} q_{i_0 j} y_l$ for all $x \leq y$. Hence, using Proposition 1.1 of Chapter 3 in Smith (2008) and Lemma 4.1, we deduce that, the system (4.45) has a unique solution $\psi^{(k)}(t) = (\psi_i^{(k)}(t); \; i \in D_m)^\top$ for $t \in [0, T]$. Moreover, $\psi^{(k)}(t) \gg 0$ for all $t \in [0, T]$. Let us set

$$\varepsilon^{(k)} := \min_{i=1,\ldots,m} \left\{ \inf_{t \in [0,T]} \psi_i^{(k)}(t) \right\}. \qquad (4.49)$$

The continuity of $\psi^{(k)}(t)$ in $t \in [0, T]$ and $\psi^{(k)}(t) \gg 0$ for all $t \in [0, T]$ lead to $\varepsilon^{(k)} > 0$[1].

Take $a < \varepsilon^{(k)}$, and we consider the following system described as

$$\begin{cases} \dfrac{d}{dt} \psi_a^{(k)}(t) = f_a^{(k)}(\psi_a^{(k)}(t)), & \text{in } (0, T], \\ \psi_a^{(k)}(0) = \mathbf{1}_m. \end{cases} \qquad (4.50)$$

Here, the vector-valued function $f_a^{(k)}(x) = (f_{a,i}^{(k)}(x); \; i \in D_m)^\top$ is specified by, for $i \in D_m$,

$$f_{a,i}^{(k)}(x) := \sum_{j=1}^n q_{ij} x_j - \left(\frac{\theta}{2} r(i) + \sum_{j \notin \{j_1,\ldots,j_k\}} h_j^{(k)}(i) \right) x_i$$
$$- \beta_i \{|x_i| \vee a\}. \qquad (4.51)$$

It is easy to verify that $\psi_a^{(k)}(t) = \psi^{(k)}(t)$ for all $t \in [0, T]$. This yields that

$$\psi_a^{(k)}(t) \geq \varepsilon^{(k)} \mathbf{1}_m \gg 0, \quad \forall t \in [0, T]. \qquad (4.52)$$

To apply Lemma 4.4, we rewrite (4.44) in the following form:

$$\begin{cases} \dfrac{d}{dt} \tilde{\varphi}_a^{(k)}(t) = f_a^{(k)}(\tilde{\varphi}_a^{(k)}(t)) + \tilde{f}_a^{(k)}(t, \tilde{\varphi}_a^{(k)}(t)), & \text{in } (0, T], \\ \tilde{\varphi}_a^{(k)}(0) = \mathbf{1}_m. \end{cases} \qquad (4.53)$$

Here, the Lipschitz continuous function $\tilde{f}_a^{(k)}(t, x) = (\tilde{f}_{a,i}^{(k)}(t, x); \; i \in D_m)^\top$ on $(t, x) \in [0, T] \times \mathbb{R}^m$ is given by

$$\tilde{f}_{a,i}^{(k)}(t, x) := G_a^{(k)}(T - t, x) + \beta_i \{|x_i| \vee a\}, \quad \forall i \in D_m. \qquad (4.54)$$

---

[1] If $m = +\infty$ in (4.49), it may be invalid.

By the recursive assumption that $\varphi^{(k+1),j}(t) \gg 0$ on $[0,T]$ for $j \notin \{j_1,\ldots,j_k\}$, we have, for each $i \in D_m$ and $(t,x) \in [0,T] \times \mathbb{R}^m$,

$$G_i^{(k)}(T-t, x \vee a\mathbf{1}_m)$$

$$= \inf_{\pi^{(k)} \in U^{(k)}} \left\{ \sum_{j \notin \{j_1,\ldots,j_k\}} \varphi^{(k+1),j}(T-t,i)(1-\pi_j^{(k)})^{-\frac{\theta}{2}} \lambda_j^{(k)}(i) \right.$$

$$\left. + H^{(k)}(\pi^{(k)};i)(x_i \vee a) \right\}$$

$$\geq \{x_i \vee a\} \inf_{\pi^{(k)} \in U^{(k)}} \mathcal{H}^{(k)}(\pi^{(k)};i) \geq -\beta_i \{|x_i| \vee a\}. \quad (4.55)$$

It follow from (4.54) that, for all $(t,x) \in [0,T] \times \mathbb{R}^m$,

$$\tilde{f}_{a,i}^{(k)}(t,x) = G_i^{(k)}(T-t, x \vee ae_m) + \beta_i\{|x_i| \vee a\} \geq 0, \quad \forall i \in D_m. \quad (4.56)$$

Similarly, it follows from (4.48) that $f_a^{(k)}(x)$ is also of type K. Then, we apply Lemma 4.4 to systems (4.53) and (4.50) to have

$$\tilde{\varphi}_a^{(k)}(t) \geq \psi_a^{(k)}(t) \geq \varepsilon^{(k)} \mathbf{1}_m, \quad \forall t \in [0,T], \quad (4.57)$$

since $\tilde{\varphi}_a^{(k)}(0) = \psi_a^{(k)}(0) = \mathbf{1}_m$. Note that $a \in (0, \varepsilon^{(k)})$. Then, $G_a^{(k)}(T - t, \tilde{\varphi}_a^{(k)}(t)) = G^{(k)}(T-t, \tilde{\varphi}_a^{(k)}(t) \vee a\mathbf{1}_m) = G^{(k)}(T-t, \tilde{\varphi}_a^{(k)}(t))$ on $[0,T]$. By using (4.44), it follows that $\tilde{\varphi}_a^{(k)}(t) \geq \varepsilon^{(k)} \mathbf{1}_m$ on $[0,T]$ is a solution to the system (4.28). The uniqueness can follow from the verification result in Proposition 4.2 below. Thus, the proof of the theorem is complete. $\square$

By virtue of Theorem 4.1, we present the following result as a characterization of the optimal strategy $\pi^{(k)} \in U^{(k)}$ at the default state $h = 0^{j_1,\ldots,j_k}$ where $k = 0, 1, \ldots, n-1$.

Proposition 4.1. *For each $k = 0, 1, \ldots, n-1$, assume that Eq. (4.23) has a positive unique solution $\varphi^{(k+1),j}(t)$ on $t \in [0,T]$ for $j \notin \{j_1,\ldots,j_k\}$. Let $\varphi^{(k)}(t) = (\varphi^{(k)}(t,i); i \in D_m)^\top$ be the solution to (4.28). Then, there exists a unique optimal feedback strategy $\pi^{(k,*)} = \pi^{(k,*)}(t,i)$ for $(t,i) \in [0,T] \times D_m$, which is given explicitly by*

$$\pi^{(k,*)} = \pi^{(k,*)}(t,i)$$

$$= \arg\min_{\pi^{(k)} \in U^{(k)}} \left\{ \sum_{j \notin \{j_1,\ldots,j_k\}} \varphi^{(k+1),j}(t,i)(1-\pi_j^{(k)})^{-\frac{\theta}{2}} \lambda_j^{(k)}(i) \right.$$

$$\left. + \mathcal{H}^{(k)}(\pi^{(k)};i)\varphi^{(k)}(t,i) \right\} \quad (4.58)$$

$$= \operatorname*{arg\,min}_{\substack{\pi^{(k)} \in \{\pi \in U^{(k)}: \\ |\pi| \leq C\}}} \left\{ \sum_{j \notin \{j_1,\ldots,j_k\}} \varphi^{(k+1),j}(t,i)(1-\pi_j^{(k)})^{-\frac{\theta}{2}} \lambda_j^{(k)}(i) \right.$$

$$\left. + \mathcal{H}^{(k)}(\pi^{(k)}; i)\varphi^{(k)}(t,i) \right\},$$

where $C > 0$ is some positive constant.

Proof. Recall Eq. (4.28), in other words:

$$\begin{cases} \dfrac{d}{dt}\varphi^{(k)}(t) = -A^{(k)}\varphi^{(k)}(t) - G^{(k)}(t,\varphi^{(k)}(t)), & \text{in } [0,T), \\ \varphi^{(k)}(T) = \mathbf{1}_m. \end{cases}$$

Theorem 4.1 shows that the above dynamical system admits a unique positive solution $\varphi^{(k)}(t)$ on $[0,T]$. Moreover, $\varphi^{(k)}(t) \geq \varepsilon^{(k)} \mathbf{1}_m^\top$ for all $t \in [0,T]$. Here, $\varepsilon^{(k)} > 0$ is given by (4.49). Thus, we have from (4.38) that, there exists a constant $C(\varepsilon^{(k)}) > 0$ which depends on $\varepsilon^{(k)} > 0$ such that, for all $i \in D_m$,

$$G_i^{(k)}(t, \varphi^{(k)}(t,i))$$

$$= \inf_{\substack{\pi^{(k)} \in \{\pi \in U^{(k)}; \\ |\pi| \leq C(\varepsilon^{(k)})\}}} \left\{ \sum_{j \notin \{j_1,\ldots,j_k\}} \varphi^{(k+1),j}(t,i)(1-\pi_j^{(k)})^{-\frac{\theta}{2}} \lambda_j^{(k)}(i) \right.$$

$$\left. + \mathcal{H}^{(k)}(\pi^{(k)}; i)\varphi^{(k)}(t,i) \right\}.$$

Here, for each $i = 1,\ldots,m$, the function $G_i^{(k)}(t,x)$ on $(t,x) \in [0,T] \times \mathbb{R}^m$ is given by (4.30). Also, for each $i = 1,\ldots,m$, $\varphi^{(k+1),j}(t,i)$ on $t \in [0,T]$ is the i-th element of the positive solution $\varphi^{(k+1),j}(t)$ of (4.23) at the default state $h = 0^{j_1,\ldots,j_k,j}$ for $j \notin \{j_1,\ldots,j_k\}$. Recall that $\mathcal{H}^{(k)}(\pi^{(k)}; i)$ for $(\pi^{(k)}, i) \in U^{(k)} \times D_m$ is given by (4.31). Then, it is not difficult to see that, with fixed $(t,i) \in [0,T] \times D_m$,

$$h^{(k)}(\pi^{(k)}, i) := \sum_{j \notin \{j_1,\ldots,j_k\}} \varphi^{(k+1),j}(t,i)(1-\pi_j^{(k)})^{-\frac{\theta}{2}} \lambda_j^{(k)}(i)$$

$$+ \mathcal{H}^{(k)}(\pi^{(k)}; i)\varphi^{(k)}(t,i)$$

is continuous and strictly convex in $\pi^{(k)} \in \bar{U}^{(k)}$. Additionally, note that the space $\{\pi^{(k)} \in \bar{U}^{(k)}; |\pi^{(k)}| \leq C(\varepsilon^{(k)})\} \subset \mathbb{R}^{n-k}$ is compact. Hence, there exist a unique optimum $\pi^{(k,*)} = \pi^{(k,*)}(t,i) \in \bar{U}^{(k)}$. Furthermore, it is noted that $h^{(k)}(\pi^{(k)}, i) = +\infty$ when $\pi^{(k)} \in \bar{U}^{(k)} \setminus U^{(k)}$ while $h^{(k)}(\pi^{(k)}, i) < +\infty$ for all $\pi^{(k)} \in U^{(k)}$. Consequently, we in fact obtain the optimum $\pi^{(k,*)} = \pi^{(k,*)}(t,i) \in \bar{U}^{(k)}$ satisfying the representation (4.58) by taking $C = C(\varepsilon^{(k)})$, which completes the proof of the proposition. □

As one of our main results, we next provide the verification theorem for the *finite state* space of the Markov chain:

Proposition 4.2 (Verification for Finite State Case). *Let $\varphi(t,h) = (\varphi(t,i,h); i \in D_m)^\top$ with $(t,h) \in [0,T] \times \mathcal{S}$ be the solution of Eq. (4.23). For $(t,i,h) \in [0,T] \times D_m \times \mathcal{S}$, define*

$$\pi^*(t,i,h) \qquad (4.59)$$
$$:= \mathrm{diag}((1-h_j)_{j=1}^m)\arg\min_{\pi \in U} \tilde{\mathcal{H}}\left(\pi; i, h, (\varphi(t,i,h^j); j=0,1,\ldots,m)\right),$$

where $\tilde{\mathcal{H}}(\pi; i, h, \bar{f}(h))$ is given by (4.20). Let $\tilde{\pi}^* = (\tilde{\pi}^*_t)_{t \in [0,T]}$ with $\tilde{\pi}^*_t := \pi^*(t, I_{t-}, H_{t-})$. Then $\tilde{\pi}^* \in \tilde{\mathcal{U}}$, and it is the optimal feedback strategy, i.e.,

$$-\frac{2}{\theta}\ln \mathbb{E}^{\tilde{\pi}^*}_{t,i,h}\left[\exp\left(\frac{\theta}{2}\int_t^T L(\tilde{\pi}^*_s; I_s, H_s)ds\right)\right] = \bar{V}(t,i,h)$$
$$= -\frac{2}{\theta}\ln \varphi(t,i,h). \qquad (4.60)$$

Proof. Using Proposition 4.1, $\tilde{\pi}^*$ is a bounded and predictable process taking values on U. We next prove that $\tilde{\pi}^*$ is uniformly away from 1. In fact, for fixed $(i,h,x) \in D_m \times \mathcal{S} \times (0,\infty)^{m+1}$, we have $\tilde{\mathcal{H}}(\pi; i, h, x)$ is strictly convex w.r.t. $\pi \in U$, thus $\Phi(i,h,x) := \arg\min_{\pi \in U}\tilde{\mathcal{H}}(\pi; i, h, x)$ is well-defined. Note that $\Phi(i,h,\cdot)$ maps $(0,\infty)^{n+1}$ to U and satisfies the first-order condition $\frac{\partial \tilde{\mathcal{H}}}{\partial \pi_j}(\Phi(i,h,x); i, h, x) = 0$ for $j = 1, \ldots, m$. Then, the implicit function theorem yields that $\Phi(i,h,x)$ is continuous in x. Further, for $j = 1, \ldots, m$, if $H^j_{t-} = 0$, the first-order condition gives that

$$(1-\tilde{\pi}^{*,j}_t)^{-\frac{\theta}{2}-1} = \left[(\mu_j(I_{t-}) - r(I_{t-})) - \frac{\theta}{2}\left(1+\frac{\theta}{2}\right)\sum_{i=1}^m (\sigma^\top(I_{t-})\sigma(I_{t-}))_{ji}\tilde{\pi}^{*,i}_t \right.$$
$$\left. + \frac{\theta}{2}\lambda_j(I_{t-}, H_{t-})\right]\frac{\varphi(t, I_{t-}, H_{t-})}{\lambda_j(I_{t-}, H_{t-})\varphi(t, I_{t-}, H^j_{t-})}. \qquad (4.61)$$

Because, for all $(i,h) \in D_m \times \mathcal{S}$, $\varphi(\cdot, i, h)$ has a strictly positive lower bound using (4.57). Together with Proposition 4.1, it follows that, there exists a constant $C > 0$ such that $\sup_{t \in [0,T]}(1-\tilde{\pi}^{*,j}_t)^{-\frac{\theta}{2}-1} \leq C$ for all $j = 1, \ldots, m$.

Hence, the estimate (4.61) yields that $\tilde{\pi}^*$ is uniformly bounded away from 1. Thus, the following generalized Novikov's condition holds:

$$\mathbb{E}\left[\exp\left(\frac{\theta^2}{8}\int_0^T |\sigma(I_t)^\top \tilde{\pi}_t^*|^2 dt + \sum_{j=1}^m \int_0^T \left|(1-\tilde{\pi}_t^{*,j})^{-\frac{\theta}{2}} - 1\right|^2 \lambda_j(I_t, H_t)dt\right)\right]$$
$$< +\infty. \tag{4.62}$$

The above Novikov's condition (4.62) implies that $\tilde{\pi}^*$ is admissible. We next prove (4.60). Note that $\varphi(t,h) = (\varphi(t,i,h); i \in D_m)^\top$ with $(t,h) \in [0,T] \times \mathcal{S}$ is the unique classical solution of (4.23). Also, Note that, there exists a constant $C_L = C_L(m,i,z) > 0$ such that $L(\pi;i,h) > -C_L$ for $(\pi,i,h) \in U \times D_m \times \mathcal{S}$. For $\ell \geq 1$, set $L_\ell(\pi;i,h) := L(\pi;,i,h) \wedge \ell$. Then L_ℓ is bounded and $L_\ell(\pi;i,h) \uparrow L(\pi;i,h)$ as $\ell \to \infty$. Therefore, for any admissible strategy $\tilde{\pi} \in \tilde{\mathcal{U}}$, Itô's formula gives that, for $0 \leq t < s \leq T$,

$$\mathbb{E}_{t,i,h}^{\tilde{\pi}}\left[\varphi(s,I_s,H_s)\exp\left(\frac{\theta}{2}\int_t^s L_\ell(\tilde{\pi}_u; I_u, H_u)du\right)\right]$$

$$= \varphi(t,i,h) + \mathbb{E}_{t,i,h}^{\tilde{\pi}}\left[\int_t^s \exp\left(\frac{\theta}{2}\int_t^u L_\ell(\tilde{\pi}_v; I_v, H_v)dv\right)\right.$$

$$\times \left\{\frac{\partial \varphi(u, I_u, H_u)}{\partial t} + \sum_{l \neq I_u} q_{I_u l}\left(\varphi(u,l,H_u) - \varphi(u,I_u,H_u)\right)\right.$$

$$\left.+ \tilde{\mathcal{H}}\left(\tilde{\pi}_u; I_u, H_u, (\varphi(t, I_u, H_u^j); j=0,1,\ldots,n)\right)\right\} du\right] \tag{4.63}$$

$$+ \mathbb{E}_{t,i,h}^{\tilde{\pi}}\left[\int_t^s \exp\left(\frac{\theta}{2}\int_t^u L_\ell(\tilde{\pi}_v; I_v, H_v)dv\right)\right.$$

$$\left.\times \varphi(u, I_u, H_u)(L_\ell - L)(\tilde{\pi}_u; I_u, H_u)du\right]$$

$$\geq \varphi(t,i,h) + \mathbb{E}_{t,i,h}^{\tilde{\pi}}\left[\int_t^s \exp\left(\frac{\theta}{2}\int_t^u L_\ell(\tilde{\pi}_v; I_v, H_v)dv\right)\right.$$

$$\left.\times \varphi(u, I_u, H_u)(L_\ell - L)(\tilde{\pi}_u; I_u, H_u)du\right].$$

In the last inequality above, the integral term in the expectation is negative. On the other hand, note that φ is bounded and positive, this integral also admits that, $\mathbb{P}_{t,i,z}^{\tilde{\pi},\theta}$-a.s., for some constant $C_\varphi > 0$,

$$\int_t^s \exp\left(\frac{\theta}{2}\int_t^u L_\ell(\tilde{\pi}_v; I_v, H_v)dv\right)\varphi(u, I_u, H_u)(L_\ell - L)(\tilde{\pi}_u; I_u, H_u)du$$

$$\geq -C_\varphi \int_t^s \exp\left(\frac{\theta}{2}\int_t^u [L(\tilde{\pi}_v; I_v, H_v) + C_L]dv\right)(L(\tilde{\pi}_u; I_u, H_u) + C_L)du$$

$$= \frac{2C_\varphi}{\theta}\left[1 - e^{\frac{\theta}{2}C_L(s-t)}\exp\left(\frac{\theta}{2}\int_t^s L(\tilde{\pi}_u; I_u, H_u)du\right)\right].$$

By taking $s = T$ above, it follows from DCT that

$$\mathbb{E}^{\tilde{\pi}}_{t,i,h}\left[\varphi(T, I_T, H_T)\exp\left(\frac{\theta}{2}\int_t^T L(\tilde{\pi}_u; I_u, H_u)du\right)\right] \geq \varphi(t, i, h). \quad (4.64)$$

Note that $\varphi(T, i, h) = 1$ in (4.64), we obtain

$$\inf_{\tilde{\pi}\in\tilde{\mathcal{U}}_t}\mathbb{E}^{\tilde{\pi}}_{t,i,h}\left[\exp\left(\frac{\theta}{2}\int_t^T L(\tilde{\pi}_u; I_u, H_u)du\right)\right] \geq \varphi(t, i, h). \quad (4.65)$$

From (4.63) and (4.59), it follows that, for $0 \leq t < s \leq T$,

$$\mathbb{E}^{\tilde{\pi}^*}_{t,i,h}\left[\exp\left(\frac{\theta}{2}\int_t^T L(\tilde{\pi}^*_u; I_u, H_u)du\right)\right] = \varphi(t, i, h). \quad (4.66)$$

Since $\tilde{\pi}^* \in \tilde{\mathcal{U}}$, we deduce from (4.66) that

$$\varphi(t, i, h) \geq \inf_{\tilde{\pi}\in\tilde{\mathcal{U}}_t}\mathbb{E}^{\tilde{\pi}}_{t,i,h}\left[\exp\left(\frac{\theta}{2}\int_t^T L(\tilde{\pi}_u; I_u, H_u)du\right)\right]. \quad (4.67)$$

We then have from (4.65) and (4.67) that

$$\varphi(t, i, h) = \inf_{\tilde{\pi}\in\tilde{\mathcal{U}}_t}\mathbb{E}^{\tilde{\pi},\theta}_{t,i,h}\left[\exp\left(\frac{\theta}{2}\int_t^T L(\tilde{\pi}_u; I_u, H_u)du\right)\right]. \quad (4.68)$$

The equality (4.68) is equivalent to $\varphi(t, i, h) = e^{-\frac{\theta}{2}\bar{V}(t,i,h)}$ due to (4.14). Hence, Eq. (4.66) together with (4.68) imply that (4.60) holds, which ends the proof of the proposition. □

4.4.2 Countably Infinite State Case of Markov Chain

This subsection studies the existence of classical solutions to the prime HJB equation (4.19) and the corresponding verification theorem when $S_I = \mathbb{N}$. In fact, the truncation argument for nonlinear coefficient of HJB equation used in the finite state case fails to work in the case $\mathbb{N}$ (c.f. the constant defined in (4.49) may be zero when $S_I = \mathbb{N}$).

We propose a truncation technique for the state space of the Markov chain. Then, we will consider a sequence of approximating risk sensitive control problems with finite state set $D^0_m := D_m \cup \{0\}$ for $m \in \mathbb{N}$. Based on the results for the finite state case in the previous section, and by establishing valid uniform estimates, we can arrive at the desired conclusion that the smooth value functions corresponding to the approximating control problems converge to the classical solution of (4.19) with countably infinite set $S_I = \mathbb{N}$, as $m \to \infty$.

We introduce the truncated counterpart of the Markov chain I by, for $t \in [0, T]$,

$$I_t^{(m)} := I_t \mathbf{1}_{\{\tau_m > t\}}, \quad \tau_m^t := \inf\{s \geq t; \ I_s \notin D_m\}, \qquad (4.69)$$

where $\tau_m := \tau_m^0$ for $m \in \mathbb{N}$. By convention, we set $\inf \emptyset = +\infty$. Then, the process $I^{(m)} = (I_t^{(m)})_{t \in [0,T]}$ is a continuous-time Markov chain with finite state space D_m^0. Here, "0" is viewed as an absorbing state. The generator of $I^{(m)}$ can therefore be given by the following $m+1$-dimensional square matrix:

$$A_m := \begin{bmatrix} 0 & 0 & \cdots & 0 \\ q_{10}^{(m)} & q_{11} & \cdots & q_{1m} \\ q_{20}^{(m)} & q_{21} & \cdots & q_{2m} \\ \vdots & \vdots & \vdots & \vdots \\ q_{m0}^{(m)} & q_{m1} & \cdots & q_{mm} \end{bmatrix}, \qquad (4.70)$$

where $q_{l0}^{(m)} = -\sum_{i=1}^{m} q_{li} = \sum_{i \neq l, i > m} q_{li}$ for all $l \in D_m$. Thus, $I^{(m)}$ is conservative. Here, q_{il} for $i, l = 1, \ldots, m$ are the same as given in Section 4.1. Since "0" is an absorbing state, we arrange values for the model coefficients at this state. More precisely, we set

$$r(0) = 0, \quad \mu(0) = 0, \quad \lambda(0, h) = \frac{\theta}{2} \mathbf{1}_m^\top, \quad \sigma(0)\sigma(0)^\top = \frac{4}{2+\theta} I_m.$$

It follows from (4.8) and Taylor's expansion that, for all $(\pi, h) \in U \times \mathcal{S}$,

$$L(\pi; 0, h) = |\pi|^2 + \sum_{j=1}^{n}(1 - h_j)\left\{(1 - \pi_j)^{-\frac{\theta}{2}} - 1 - \frac{\theta}{2}\pi_j\right\} \geq 0.$$

We next propose a sequence of approximating risk-sensitive control problems where the Markov chain I takes values on D_m^0. Toward this end, define $\tilde{\mathcal{U}}_t^m$ as the admissible control set $\tilde{\mathcal{U}}_t$ for $t \in [0, T]$, but the Markov chain I is replaced with $I^{(m)}$. We then consider the following objective functional given by, for $\tilde{\pi} \in \tilde{\mathcal{U}}^m$ and $(t, i, h) \in [0, T] \times D_m^0 \times \mathcal{S}$,

$$J_m(\tilde{\pi}; t, i, h) := \mathbb{E}_{t,i,h}^{\tilde{\pi}} \left[\exp\left(\frac{\theta}{2} \int_t^{T \wedge \tau_m^t} L(\tilde{\pi}_s; I_s, H_s) ds\right) \right]$$

$$=\mathbb{E}_{t,i,h}^{\tilde{\pi}}\left[\exp\left(\frac{\theta}{2}\int_t^{T\wedge\tau_m^t}L(\tilde{\pi}_s;I_s^{(m)},H_s)ds\right)\right]. \qquad (4.71)$$

Here, the risk-sensitive cost function $L(\pi;i,h)$ for $(\pi,i,h)\in U\times S_I\times S$ is given by (4.8). To apply the results in the finite state case obtained in Section 4.4.1, we also need to propose the following objective functional given by, for $\tilde{\pi}\in\tilde{\mathcal{U}}^m$ and $(t,i,h)\in[0,T]\times D_m^0\times S$,

$$\tilde{J}_m(\tilde{\pi};t,i,h):=\mathbb{E}_{t,i,h}^{\tilde{\pi}}\left[\exp\left(\frac{\theta}{2}\int_t^T L(\tilde{\pi}_s;I_s^{(m)},H_s)ds\right)\right]. \qquad (4.72)$$

We will consider the auxiliary value function defined by

$$V_m(t,i,h):=-\frac{2}{\theta}\inf_{\tilde{\pi}\in\tilde{\mathcal{U}}_t^m}\ln\tilde{J}_m(\tilde{\pi};t,i,h). \qquad (4.73)$$

The characterization of the value function V_m which will play an important role in the study of the convergence of V_m as $m\to\infty$.

Lemma 4.5. *Let $(t,i,h)\in[0,T]\times D_m^0\times S$. Then, it holds that*

$$V_m(t,i,h)=-\frac{2}{\theta}\inf_{\tilde{\pi}\in\tilde{\mathcal{U}}_t^m}\ln J_m(\tilde{\pi};t,i,h).$$

Proof. It follows from (4.71) and (4.72) that, for all $\tilde{\pi}\in\tilde{\mathcal{U}}^m$,

$$\ln\tilde{J}_m(\tilde{\pi};t,i,h)=\ln\mathbb{E}_{t,i,h}^{\tilde{\pi}}\left[\exp\left(\frac{\theta}{2}\int_t^T L(\tilde{\pi}_s;I_s^{(m)},H_s)ds\right)\right]$$

$$=\ln\mathbb{E}_{t,i,h}^{\tilde{\pi}}\left[\exp\left(\frac{\theta}{2}\int_t^{T\wedge\tau_m^t}L(\tilde{\pi}_s;I_s^{(m)},H_s)ds+\frac{\theta}{2}\int_{T\wedge\tau_m^t}^T L(\tilde{\pi}_s;I_s^{(m)},H_s)ds\right)\right]$$

$$=\ln\mathbb{E}_{t,i,h}^{\tilde{\pi}}\left[\exp\left(\frac{\theta}{2}\int_t^{T\wedge\tau_m^t}L(\tilde{\pi}_s;I_s^{(m)},H_s)ds+\frac{\theta}{2}\int_{T\wedge\tau_m^t}^T L(\tilde{\pi}_s;0,H_s)ds\right)\right]$$

$$\geq\ln\mathbb{E}_{t,i,h}^{\tilde{\pi}}\left[\exp\left(\frac{\theta}{2}\int_t^{T\wedge\tau_m^t}L(\tilde{\pi}_s;I_s^{(m)},H_s)ds\right)\right]$$

$$=\ln J_m(\tilde{\pi};t,i,h)\geq\inf_{\tilde{\pi}\in\tilde{\mathcal{U}}_t^m}\ln J_m(\tilde{\pi};t,i,h).$$

In the above display, we used the positivity of $L(\pi;0,z)$ for all $(\pi,h)\in U\times S$. Since $\theta>0$, it follows from (4.73) that

$$V_m(t,i,h)\leq-\frac{2}{\theta}\inf_{\tilde{\pi}\in\tilde{\mathcal{U}}_t^m}\ln J_m(\tilde{\pi};t,i,h). \qquad (4.74)$$

On the other hand, for any $\tilde{\pi} \in \tilde{\mathcal{U}}^m$, define $\hat{\pi}_t = \tilde{\pi}(t)\mathbf{1}_{\{t \leq \tau_m\}}$ for $t \in [0,T]$. It is clear that $\hat{\pi} \in \tilde{\mathcal{U}}^m$, and it holds that $\Gamma^{\hat{\pi}}(t,T) := \frac{\Gamma_T^{\hat{\pi}}}{\Gamma_t^{\hat{\pi}}} = \frac{\Gamma_{T \wedge \tau_m^t}^{\hat{\pi}}}{\Gamma_t^{\hat{\pi}}} =: \Gamma^{\hat{\pi}}(t, T \wedge \tau_m^t)$. Then, it holds that

$$\ln J_m(\tilde{\pi}; t, i, h) = \ln \mathbb{E}_{t,i,h}\left[\Gamma^{\tilde{\pi}}(t,T)\exp\left(\frac{\theta}{2}\int_t^{T \wedge \tau_m^t} L(\tilde{\pi}_s; I_s^{(m)}, H_s)ds\right)\right]$$

$$= \ln \mathbb{E}_{t,i,h}\left[\exp\left(\frac{\theta}{2}\int_t^{T \wedge \tau_m^t} L(\tilde{\pi}_s; I_s^{(m)}, H_s)ds\right)\mathbb{E}\left[\Gamma^{\tilde{\pi}}(t,T)|\mathcal{F}_{T \wedge \tau_m^t}\right]\right]$$

$$= \ln \mathbb{E}_{t,i,h}\left[\Gamma^{\tilde{\pi}}(t, T \wedge \tau_m^t)\exp\left(\frac{\theta}{2}\int_t^{T \wedge \tau_m^t} L(\tilde{\pi}_s; I_s^{(m)}, H_s)ds\right)\right]$$

$$= \ln \mathbb{E}_{t,i,h}^{\hat{\pi}}\left[\exp\left(\frac{\theta}{2}\int_t^{T \wedge \tau_m^t} L(\hat{\pi}_s; I_s^{(m)}, H_s)ds\right)\right]$$

$$= \ln \mathbb{E}_{t,i,h}^{\hat{\pi}}\left[\exp\left(\frac{\theta}{2}\int_t^{T \wedge \tau_m^t} L(\hat{\pi}_s; I_s^{(m)}, H_s)ds + \frac{\theta}{2}\int_{T \wedge \tau_m^t}^T L(0; 0, H_s)ds\right)\right]$$

$$= \ln \mathbb{E}_{t,i,h}^{\hat{\pi}}\left[\exp\left(\frac{\theta}{2}\int_t^T L(\hat{\pi}_s; I_s^{(m)}, H_s)ds\right)\right]$$

$$= \ln \tilde{J}_m(\hat{\pi}; t, i, h)$$

$$\geq \inf_{\tilde{\pi} \in \tilde{\mathcal{U}}_t^m} \ln \tilde{J}_m(\tilde{\pi}; t, i, h).$$

The above inequality and the arbitrariness of $\tilde{\pi}$ jointly give that

$$-\frac{2}{\theta}\inf_{\tilde{\pi} \in \tilde{\mathcal{U}}_t^m} \ln J_m(\tilde{\pi}; t, i, h) \leq V_m(t, i, h). \tag{4.75}$$

Thus, the desired result follows by combining (4.74) and (4.75) above. □

Using Lemma 4.5 together with Theorem 4.1 and Proposition 4.2 in Section 4.4.1 for the finite state space of the Markov chain, we have the following conclusion:

Proposition 4.3. *Recall the value function $V_m(t,i,h)$ for $m \geq 1$ defined by (4.73). We define $\varphi_m(t,i,h) := \exp(-\frac{\theta}{2}V_m(t,i,h))$. Then, $\varphi_m(t,i,h)$ is the unique solution of the recursive HJB system given by*

$$0 = \frac{\partial \varphi_m(t,i,h)}{\partial t} + \sum_{l \neq i, 1 \leq l \leq m} q_{il}\left(\varphi_m(t,l,h) - \varphi_m(t,i,h)\right)$$

$$+ q_{i0}^{(m)}(\varphi_m(t,0,h) - \varphi_m(t,i,h)) \tag{4.76}$$

$$+ \inf_{\pi \in U} \tilde{\mathcal{H}}\left(\pi; i, h, (\varphi_m(t, i, h^j);\ j = 0, 1, \ldots, n)\right),$$

where $(t, i, h) \in [0, T] \times D_m^0 \times \mathcal{S}$, and the terminal condition is given by $\varphi_m(T, i, h) = 1$ for all $(i, h) \in D_m^0 \times \mathcal{S}$. Moreover, $\varphi_m(t, i, h) \in [0, 1]$ and it is decreasing in m for all $(t, i, h) \in [0, T] \times D_m^0 \times \mathcal{S}$.

Proof. For fixed $m \geq 1$, the state space of $I^{(m)}$ is given by D_m^0 which is a finite set. We have from (4.73) that $\varphi_m(t, i, h)$ is the unique solution of the recursive system (4.76) by applying Theorem 4.1 and Proposition 4.2. To verify that $\varphi_m(t, i, h) \in [0, 1]$ and it is decreasing in m, it suffices to prove that $V_m(t, i, h) \geq 0$ and it is nondecreasing in m. Thanks to Lemma 4.5, and $L(0, i, h) = -r(i) \leq 0$ by (4.8), also note that $\tilde{\pi}_t^0 \equiv 0$ is admissible (i.e., $\tilde{\pi}^0 \in \tilde{\mathcal{U}}^m$), then

$$\inf_{\tilde{\pi} \in \tilde{\mathcal{U}}_t^m} \ln J_m(\tilde{\pi}; t, i, h) \leq \ln J_m(\tilde{\pi}^0; t, i, h)$$

$$= \ln \mathbb{E}_{t,i,h}^{\tilde{\pi}^0}\left[\exp\left(\frac{\theta}{2} \int_t^{T \wedge \tau_m^t} L(0; I_s, H_s) ds\right)\right]$$

$$= \ln \mathbb{E}_{t,i,h}^{\tilde{\pi}^0}\left[\exp\left(-\frac{\theta}{2} \int_t^{T \wedge \tau_m^t} r(I_s) ds\right)\right]$$

$$\leq 0.$$

This yields that $V_m(t, i, h) \geq 0$ for all $(t, i, h) \in [0, T] \times D_m^0 \times \mathcal{S}$. On the other hand, for any $\tilde{\pi} \in \tilde{\mathcal{U}}^m$, we define $\hat{\pi}_t := \tilde{\pi}_t \mathbf{1}_{\{\tau_m \geq t\}}$ for $t \in [0, T]$. It is clear that $\hat{\pi} \in \tilde{\mathcal{U}}^m \cap \tilde{\mathcal{U}}^{m+1}$. Recall the density process given by (4.9), and we have, for $\tilde{\pi}, \hat{\pi} \in \tilde{\mathcal{U}}^m$,

$$\Gamma^{\tilde{\pi}} = \mathcal{E}(\Pi^{\tilde{\pi}}), \quad \Pi^{\tilde{\pi}} = -\frac{\theta}{2} \int_0^{\cdot} \tilde{\pi}_s^{\top} \sigma(I_s^{(m)}) dW_s + \sum_{j=1}^n \int_0^{\cdot} \{(1 - \tilde{\pi}_s^j)^{-\frac{\theta}{2}} - 1\} dM_s^j,$$

$$\Gamma^{\hat{\pi}} = \mathcal{E}(\Pi^{\hat{\pi}}), \quad \Pi^{\hat{\pi}} = -\frac{\theta}{2} \int_0^{\cdot} \hat{\pi}_s^{\top} \sigma(I_s^{(m)}) dW_s + \sum_{j=1}^n \int_0^{\cdot} \{(1 - \hat{\pi}_s^j)^{-\frac{\theta}{2}} - 1\} dM_s^j.$$

This shows that $\Gamma_{t \wedge \tau_m}^{\tilde{\pi}} = \Gamma_t^{\hat{\pi}}$ for $t \in [0, T]$. Then, we deduce from (4.71) that

$$\ln J_m(\tilde{\pi}; t, i, h) = \ln \mathbb{E}_{t,i,h}^{\tilde{\pi}}\left[\exp\left(\frac{\theta}{2} \int_t^{T \wedge \tau_m^t} L(\tilde{\pi}_s; I_s, H_s) ds\right)\right]$$

$$\geq \ln \mathbb{E}_{t,i,h}^{\hat{\pi}}\left[\exp\left(\frac{\theta}{2} \int_t^{T \wedge \tau_n^t} L(\tilde{\pi}_s; I_s, H_s) ds + \frac{\theta}{2} \int_{T \wedge \tau_m^t}^{T \wedge \tau_{m+1}^t} L(0; I_s, H_s)\right)\right]$$

$$= \ln \mathbb{E}^{\hat{\pi}}_{t,i,h} \left[\exp\left(\frac{\theta}{2} \int_t^{T \wedge \tau^t_{m+1}} L(\hat{\pi}_s; I_s, H_s) ds \right) \right]$$

$$= \ln J_{m+1}(\hat{\pi}; t, i, h) \geq \inf_{\tilde{\pi} \in \tilde{\mathcal{U}}^{m+1}_t} \ln J_{m+1}(\tilde{\pi}; t, i, h). \tag{4.77}$$

In view of (4.73) and Lemma 4.5, $V_m(t, i, h)$ is nondecreasing in m for fixed $(t, i, h) \in [0, T] \times D^0_m \times \mathcal{S}$. This ends the proof. $\square$

In light of Proposition 4.3, for any $(t, i, h) \in [0, T] \times S_I \times \mathcal{S}$, we set $V^*(t, i, h) := \lim_{m \to \infty} V_m(t, i, h)$. Thus, we arrive at

$$\lim_{m \to \infty} \varphi_m(t, i, h) = \exp\left(-\frac{\theta}{2} V^*(t, i, h) \right) =: \varphi^*(t, i, h). \tag{4.78}$$

On the other hand, it is easy from Eq. (4.73) to see that $\varphi_m(t, 0, h) = 1$ for all $(t, z) \in [0, T] \times \mathcal{S}$. Then, Eq. (4.76) above can be rewritten as:

$$\frac{\partial \varphi_m(t, i, h)}{\partial t} = - q_{ii}\varphi_m(t, i, h) - \sum_{l \neq i, 1 \leq l \leq m} q_{il}\varphi_m(t, l, h) - \sum_{l > m} q_{il}$$

$$- \inf_{\pi \in U} \tilde{\mathcal{H}} \left(\pi; i, h, (\varphi_m(t, i, h^j); j = 0, 1, \ldots, n) \right). \tag{4.79}$$

We have from (4.20) that, for $(\pi; i, h) \in U \times S_I \times \mathcal{S}$, $\tilde{\mathcal{H}}(\pi; i, h, x)$ is concave in every component of $x \in [0, \infty)^{n+1}$, so is $\inf_{\pi \in U} \tilde{\mathcal{H}}(\pi; i, h, x)$.

We present the main result in this chapter for the case of the countably infinite state space.

Theorem 4.2 (Well-posedness of Primal HJB Equation). *Let $(t, i, h) \in [0, T] \times S_I \times \mathcal{S}$. Then, the limit function $\varphi^*(t, i, h)$ given in (4.78) is a classical solution to the prime HJB equation (4.19), i.e.,*

$$0 = \frac{\partial \varphi^*(t, i, h)}{\partial t} + \sum_{l \neq i} q_{il} \left(\varphi^*(t, l, h) - \varphi^*(t, i, h) \right)$$

$$+ \inf_{\pi \in U} \tilde{\mathcal{H}} \left(\pi; i, h, (\varphi^*(t, i, h^j); j = 0, 1, \ldots, n) \right)$$

with terminal condition $\varphi^*(T, i, h) = 1$ for all $(i, h) \in S_I \times \mathcal{S}$.

The proof of Theorem 4.2 will be split into proving a sequence of auxiliary lemmas. We first prove the following result.

Lemma 4.6. *Let $(i, h) \in S_I \times \mathcal{S}$. Then $\left(\frac{\partial \varphi_m(t,i,h)}{\partial t} \right)_{m \geq i}$ is uniformly bounded in $t \in [0, T]$.*

Proof. We rewrite Eq. (4.79) as in the following form:
$$\frac{\partial \varphi_m(t,i,h)}{\partial t} = -q_{ii}\varphi_m(t,i,h) - \sum_{l\neq i, 1\leq l\leq m} q_{il}\varphi_m(t,l,h) - \sum_{l>m} q_{il} \quad (4.80)$$
$$- \inf_{\pi\in U} \hat{\mathcal{H}}\left(\pi; i, h, (\varphi_m(t,i,h^j); j = 0, 1, \ldots, n)\right) + C(i,h)\varphi_m(t,i,h),$$

where, for $(i,h) \in S_I \times S$,
$$C(i,h) := \left| \inf_{\pi\in U} \left\{ -\frac{\theta}{2}r(i) - \frac{\theta}{2}\pi^\top(\mu(i) - r(i)\mathbf{1}_m) + \frac{\theta}{4}\left(1 + \frac{\theta}{2}\right)|\sigma(i)^\top\pi|^2 \right. \right.$$
$$\left. \left. + \sum_{j=1}^n \left(-1 - \frac{\theta}{2}\pi_j\right)(1-h_j)\lambda_j(i,h) \right\} \right|, \quad (4.81)$$

and the nonnegative function
$$\hat{\mathcal{H}}(\pi; i, h, \bar{f}(h)) := \tilde{\mathcal{H}}(\pi; i, h, \bar{f}(h)) + C(i,h)f(h). \quad (4.82)$$

Since $\hat{\mathcal{H}}(\pi; i, h, x)$ is concave in every component of $x \in [0,\infty)^{n+1}$, $\Phi(x) := \inf_{\pi\in U} \hat{\mathcal{H}}(\pi; i, h, x)$ is also concave in every component of $x \in [0,\infty)^{n+1}$. It follows from Proposition 4.3 that $x^{(m)} := (\varphi_m(t,i,h^j); j = 0, 1, \ldots, n) \in [0,1]^{n+1}$. Using Lemma 4.13, there exists a constant $C > 0$ which is independent of $x^{(m)}$ such that $0 \leq \Phi(x^{(m)}) \leq C$ for all $m \geq 1$. Moreover, for fixed $(i,h) \in S_I \times S$,

$$\left| -q_{ii}\varphi_m(t,i,h) - \sum_{l\neq i, 1\leq l\leq m} q_{il}\varphi_m(t,l,h) - \sum_{l>m} q_{il} + C(i,h)\varphi_m(t,i,h) \right|$$
$$\leq -2q_{ii} + C(i,h).$$

The desired result follows from Eq. (4.80). □

Lemma 4.7. *Let $(i,h) \in S_I \times S$, then $(\varphi_m(t,i,h))_{m\geq i}$ (decreasingly) converges to $\varphi^*(t,i,h)$ uniformly in $t \in [0,T]$ as $m \to +\infty$.*

Proof. From Proposition 4.3, Lemma 4.6, and Azelà-Ascoli's theorem, it follows that $(\varphi_m(\cdot, i, h))_{m\geq i}$ contains an uniformly convergent subsequence. Moreover, Proposition 4.3 together with (4.78) yields that $\varphi_m(t,i,h)$ (decreasingly) converges to $\varphi^*(t,i,h)$ uniformly in $t \in [0,T]$ as $m \to +\infty$. □

Lemma 4.8. *For $m \geq 1$, consider the following linear system, with $(t,i,h) \in (0,T] \times D_m^0 \times S$,*
$$\begin{cases} \partial_t \phi_m(t,i,h) = (q_{ii} - C(i,h))\phi_m(t,i,h) \\ \qquad\qquad + \sum_{l\neq i, 1\leq l\leq m} q_{il}\phi_m(t,l,h), \quad (4.83) \\ \phi_m(0,i,h) = 1, \end{cases}$$

where $C(i,h)$ is given by (4.81). Then, there exists a measurable function $\phi^*(t,i,h)$ such that $\phi_m(t,i,h) \nearrow \phi^*(t,i,h)$ as $m \to +\infty$ for each fixed (t,i,h). Moreover, we have $0 < \phi_m(T-t,i,h) \leq \varphi_n(t,i,h) \leq 1$ for all $(t,i,h) \in [0,T] \times D_m^0 \times \mathcal{S}$.

Proof. Define $g_m(t,i,h) := \varphi_m(T-t,i,h)$. It follows from (4.80) that $g_m(\cdot,i,h) \in C^1((0,T]) \cap C([0,T])$ for each (i,h), and satisfies that

$$\frac{\partial g_m(t,i,h)}{\partial t} = (q_{ii} - C(i,h))g_m(t,i,h) + \sum_{l \neq i, 1 \leq l \leq m} q_{il} g_m(t,l,h) + \sum_{l > m} q_{il}$$

$$+ Q(t,i,h, g_m(t,i,h)),$$

$$g_m(0,i,h) = 1, \qquad (4.84)$$

where $Q(t,i,h,x) := \inf_{\pi \in U} \hat{\mathcal{H}}(\pi; i, h, x, g_m(t,i,h^1), \ldots, g_m(t,i,h^n))$ for $x \geq 0$. We have from (4.82) that $Q(t,i,h,x) \geq 0$ for all $(t,x) \in [0,T] \times [0,\infty)$. Then $\sum_{l>m} q_{il} + Q(t,i,h,x) \geq 0$. Note that the linear part of Eq. (4.84) satisfies the type K condition. Then, using the comparison result of Lemma 4.4, it shows that $g_m(t,i,h) \geq \phi_m(t,i,h)$, and hence $\varphi_m(t,i,h) \geq \phi_m(T-t,i,h)$. Moreover, we deduce from Lemma 4.1 that $\phi_m(t,i,h) > 0$. In light of Eq. (4.83), $\phi_{m+1}(t,i,h)$ with $(t,i,h) \in [0,T] \times D_{m+1}^0 \times \mathcal{S}$ satisfies that

$$\frac{\partial \phi_{m+1}(t,i,h)}{\partial t} = (q_{ii} - C(i,h))\phi_{m+1}(t,i,h) + \sum_{l \neq i, 1 \leq l \leq m} q_{il} \phi_{m+1}(t,l,h)$$

$$+ q_{i,m+1} \phi_{m+1}(t, m+1, h),$$

$$\phi_{m+1}(0,i,h) = 1.$$

Since $q_{i,m+1} \phi_{m+1}(t, m+1, h) \geq 0$ for $i \in D_m^0$, Lemma 4.4 shows that $\phi_{m+1}(t,i,h) \geq \phi_m(t,i,h)$ for all $(t,i,h) \in [0,T] \times D_m^0 \times \mathcal{S}$. Hence, there exists a measurable function $\phi^*(t,i,h)$ such that $\phi_m(t,i,h) \nearrow \phi^*(t,i,h)$ as $m \to \infty$ for any $(t,i,h) \in [0,T] \times S_I \times \mathcal{S}$. □

Lemma 4.9. Let $(i,z) \in S_I \times \mathcal{S}$. Then, there exists $\delta = \delta(i,h) > 0$ such that $\varphi^*(t,i,h) \geq \delta$ for all $t \in [0,T]$.

Proof. We have from Lemma 4.8 that $\varphi_m(t,i,h) \geq \phi_m(T-t,i,h)$. Letting $m \to \infty$ and using Lemma 4.7, it follows that $\varphi^*(t,i,h) \geq \phi^*(T-t,i,h) \geq \phi_i(T-t,i,h)$. Since $\phi_i(t,i,h) > 0$ is continuous in $t \in [0,T]$, there exists a positive constant $\delta = \delta(i,h)$ such that $\inf_{t \in [0,T]} \phi_i(t,i,h) \geq \delta$. Hence $\varphi^*(t,i,h) \geq \delta$ for all $t \in [0,T]$. □

We next conclude the proof of Theorem 4.2 by using previous auxiliary results.

Proof of Theorem 4.2. We first prove that there exists a measurable function $\tilde{\varphi}(t,i,h)$ for $(t,i,h) \in [0,T] \times S_I \times S$ such that $\lim_{m\to\infty} \frac{\partial \varphi_m(t,i,h)}{\partial t} = \tilde{\varphi}(t,i,h)$ for $(t,i,h) \in [0,T] \times S_I \times S$. Note that, for $(t,i,h) \in [0,T] \times D_m^0 \times S$, $0 \leq \varphi_{m+1}(t,i,h) \leq \varphi_m(t,i,h) \leq 1$ for all $m \geq 1$. Then

$$\sum_{l\neq i, 1\leq l\leq m} q_{il}\varphi_m(t,l,h) + \sum_{l>m} q_{il} \geq \sum_{l\neq i, 1\leq l\leq m+1} q_{il}\varphi_{m+1}(t,l,h) + \sum_{l>m+1} q_{il}.$$

This shows from (4.78) that $q_{ii}\varphi_m(t,i,h) \nearrow q_{ii}\varphi^*(t,i,h)$ when $m \to \infty$, and

$$\sum_{l\neq i, 1\leq l\leq m} q_{il}\varphi_m(t,l,h) + \sum_{l>m} q_{il} \searrow \sum_{l\neq i, l\geq 1} q_{il}\varphi^*(t,l,h). \quad (4.85)$$

On the other hand, let $\Phi(x) := \inf_{\pi\in U} \tilde{\mathcal{H}}(\pi; i, h, x)$ for $x \in [0,\infty)^{n+1}$. Then, $\Phi(x) : [0,\infty)^{n+1} \to \mathbb{R}$ is concave in every component of x. Let $x_t^* := (\varphi^*(t, i, h^j); \ j = 0, 1, \ldots, n)$ and $x_t^{(m)} := (\varphi_m(t, i, h^j); \ j = 0, 1, \ldots, n)$ for $m \geq 1$. Thus, $0 \leq x_t^* \leq x_t^{(m)}$ for $m \geq 1$, and $\lim_{m\to\infty} x_t^{(m)} = x_t^*$ by using (4.78). Furthermore, Lemma 4.9 gives that $\delta \ll x^* \ll 2$. It follows from Lemma 4.12 that $\lim_{m\to\infty} \Phi(x_t^{(m)}) = x_t^*$. Thus, in light of (4.79), as $m \to \infty$,

$$\frac{\partial \varphi_m(t,i,h)}{\partial t} \to \tilde{\varphi}(t,i,h) := -q_{ii}\varphi^*(t,i,h) - \sum_{l\neq i, l\geq 1} q_{il}\varphi^*(t,l,h)$$

$$- \Phi(x_t^*). \quad (4.86)$$

We next prove that, for $(i,h) \in S_I \times S$, $\frac{\partial \varphi_m(t,i,h)}{\partial t} \rightrightarrows \tilde{\varphi}(t,i,h)$ in $t \in [0,T]$, as $m \to \infty$. Here, $\rightrightarrows$ denotes the uniform convergence. Eq. (4.80) together with (4.86) first give that, for $(t,i,h) \in [0,T] \times D_m^0 \times S$,

$$\frac{\partial \varphi_m(t,i,h)}{\partial t} - \tilde{\varphi}(t,i,h) = \sum_{i=1}^{3} B_i^{(m)}(t,i,h), \quad (4.87)$$

where we have defined that

$$B_1^{(m)}(t,i,h) := -q_{ii}(\varphi_m(t,i,h) - \varphi^*(t,i,h))$$
$$+ C(i,h)(\varphi_m(t,i,h) - \varphi^*(t,i,h)),$$
$$B_2^{(m)}(t,i,h) := \sum_{l\neq i, 1\leq l\leq m} q_{il}(\varphi_m(t,l,h) - \varphi^*(t,l,h)) \quad (4.88)$$
$$+ \sum_{l>m} q_{il}(1 - \varphi^*(t,i,h)),$$

$$B_3^{(m)}(t,i,h) := \Phi(x_t^{(m)}) - \Phi(x_t^*).$$

Here, $\Phi(x) := \inf_{\pi \in U} \tilde{\mathcal{H}}(\pi; i, h, x)$ for $x \in [0,\infty)^{n+1}$, $x_t^{(m)} := (\varphi_m(t,i,h^j); \; j = 0, 1, \ldots, n)$, and $x_t^* := (\varphi^*(t,i,h^j); \; j = 0, 1, \ldots, n)$.

Lemma 4.7 guarantees that $\varphi_m(t,i,h) \rightrightarrows \varphi^*(t,i,h)$ in $t \in [0,T]$ as $m \to \infty$, and hence $B_1^{(m)}(t,i,h) \rightrightarrows 0$ in $t \in [0,T]$ as $m \to \infty$. On the other hand, for any small $\varepsilon > 0$, since $\sum_{l \neq i} q_{il} < \infty$, there exists $n_1 \geq 1$ such that $\sum_{l > n_1, l \neq i} q_{il} < \frac{\varepsilon}{2}$. Note that, for all $1 \leq l \leq n_1$, $\varphi_m(t,l,h) \rightrightarrows \varphi^*(t,l,h)$ in $t \in [0,T]$ as $m \to \infty$, there exists $n_2 \geq 1$ such that $\sup_{t \in [0,T]} \sum_{l \neq i, 1 \leq l \leq n_1} q_{il}(\varphi_m(t,l,h) - \varphi^*(t,l,h)) \leq \frac{\varepsilon}{2}$ for $m > n_2$. Hence, for all $m > n_1 \vee n_2$, noting that $0 \leq \varphi^*(t,i,h) \leq \varphi_m(t,i,h) \leq 1$, it holds that

$$|B_2^{(m)}(t,i,h)| = \sum_{l \neq i, 1 \leq l \leq n_1} q_{il}(\varphi_m(t,l,h) - \varphi^*(t,l,h))$$
$$+ \sum_{l \neq i, n_1 < l < m} q_{il}(\varphi_m(t,l,h) - \varphi^*(t,l,h)) + \sum_{l > m} q_{il}(1 - \varphi^*(t,i,h))$$
$$\leq \frac{\varepsilon}{2} + \sum_{l > n_1} q_{il} \leq \frac{\varepsilon}{2} + \frac{\varepsilon}{2} = \varepsilon. \tag{4.89}$$

Thus, we deduce that $B_2^{(m)}(t,i,h) \rightrightarrows 0$ in $t \in [0,T]$ as $m \to \infty$. We can have from Lemma 4.13 that, for all $x \in \mathbb{R}^{n+1}$ satisfying $0 \leq x \leq 2$, $0 \leq \Phi(x) \leq C$ for some constant $C > 0$. Since, for $j = 0, 1, \ldots, n$, $\varphi_m(t,i,h^j) \rightrightarrows \varphi^*(t,i,h^j)$ in $t \in [0,T]$ as $m \to \infty$, Lemma 4.9 yields that there exists a constant $\delta > 0$ such that $1 \geq \varphi_m(t,i,h^j) \geq \varphi^*(t,i,h^j) \geq \delta > 0$ for all $t \in [0,T]$. Moreover, there exists $\lambda_m^j(t) \in [0,1]$ such that $\varphi_m(t,i,h^j) = (1 - \lambda_m^j(t))\varphi^*(t,i,h^j) + 2\lambda_m^j(t)$. In turn, $\lambda_m^j(t) = \frac{\varphi_m(t,i,h^j) - \varphi^*(t,i,h^j)}{2 - \varphi^*(t,i,h^j)}$, and hence for all $j = 0, 1, \ldots, n$, $\lambda_m^j(t) \rightrightarrows 0$ in $t \in [0,T]$ as $m \to \infty$. Similarly to that in (4.102), we can derive that

$$\Phi(x_t^{(m)}) \geq \Phi(x_t^*) \prod_{j=0}^{n}(1 - \lambda_m^j(t)) + \Lambda_1^{(m)}(t). \tag{4.90}$$

Similar to the first term in RHS of the inequality (4.90), every term in $\Lambda_1^{(m)}(t)$ above has $n+1$ multipliers and at least one of these multipliers is of the form $\lambda_m^j(t)$, while other multipliers are non-negative and bounded by $1 \vee C$. Due to the fact that $\lambda_m^j(t) \rightrightarrows 0$ in $t \in [0,T]$ as $m \to \infty$, we have that $\Lambda_1^{(m)}(t) \rightrightarrows 0$ in $t \in [0.T]$ as $m \to \infty$. Moreover, it follows from (4.90) that

$$\left(1 - \prod_{j=0}^{n}(1 - \lambda_m^j(t))\right)\Phi(x_t^*) - \Lambda_1^{(m)}(t) \geq \Phi(x_t^*) - \Phi(x^{(m)}(t))$$

$$= -B_3^{(m)}(t,i,h). \tag{4.91}$$

It is not difficult to see that the l.h.s. of the inequality (4.91) tends to 0 uniformly in $t \in [0,T]$ as $m \to \infty$. On the other hand, there exists $\tilde{\lambda}_m^j(t) \in [0,1]$ such that $\varphi^*(t,i,h^j) = (1 - \tilde{\lambda}_m^j(t))\varphi_m(t,i,h^j) + 0 \cdot \tilde{\lambda}_m^j(t)$, and in turn $\tilde{\lambda}_m^j(t) = \frac{\varphi_m(t,i,h^j) - \varphi^*(t,i,h^j)}{\varphi_m(t,i,h^j)} \rightrightarrows 0$ in $t \in [0,T]$ as $m \to \infty$, since $\varphi_m(t,i,h^j) \geq \delta > 0$. So that

$$\left(1 - \prod_{j=0}^{n}(1 - \tilde{\lambda}_m^j(t))\right) \Phi(x_t^{(m)}) - \Lambda_2^{(m)}(t) \geq \Phi(x_t^{(m)}) - \Phi(x_t^*)$$

$$= B_3^{(m)}(t,i,h), \tag{4.92}$$

where, the form of $\Lambda_2^{(m)}(t)$ is similar to that of $\Lambda_1^{(m)}(t)$, but it is related to $\tilde{\lambda}_m^j(t)$ for $j = 0, 1, \ldots, n$. As in (4.91), the LHS of the inequality (4.92) tends to 0 uniformly in $t \in [0,T]$ as $m \to \infty$. Hence, it follows from (4.91) and (4.92) that $B_3^{(m)}(t,i,h) \rightrightarrows 0$ uniformly in $t \in [0,T]$ as $m \to \infty$. Thus, we proved that for $(i,h) \in S_I \times S$, $\frac{\partial \varphi_m(t,i,h)}{\partial t} \rightrightarrows \tilde{\varphi}(t,i,h)$ in $t \in [0,T]$ as $m \to \infty$.

Lastly, we show that, for $(i,h) \in S_I \times S$, $\varphi^*(T,i,h) - \varphi^*(t,i,h) = \int_t^T \tilde{\varphi}(s,i,h)ds$ for $t \in [0,T]$. For $m \geq 1$, it follows from Proposition 4.3 that $\varphi_m(\cdot,i,h) \in C^1([0,T)) \cap C([0,T])$ for $(i,h) \in D_m^0 \times S$. This implies that

$$\varphi^*(T,i,h) - \varphi^*(t,i,h) = \varphi^*(T,i,h) - \varphi^*(t,i,h) - (\varphi_m(T,i,h) - \varphi_m(t,i,h))$$
$$+ \int_t^T \frac{\partial \varphi_m(s,i,h)}{\partial t}(s,i,h)ds. \tag{4.93}$$

Lemma 4.7 ensures that $\varphi(T,i,h) - \varphi(t,i,h) - (\varphi_m(T,i,h) - \varphi_m(t,i,h)) \to 0$ as $m \to \infty$. From Lemma 4.6 and the uniform convergence of $\frac{\partial \varphi_m(t,i,h)}{\partial t}$ to $\tilde{\varphi}(t,i,h)$ in $t \in [0,T]$, it follows that $\tilde{\varphi}(t,i,h)$ is continuous in $t \in [0,T]$ and $\int_t^T \frac{\partial \varphi_m(s,i,h)}{\partial t}ds \to \int_t^T \tilde{\varphi}(s,i,h)ds$ as $m \to \infty$. Moreover, as $\varphi^*(T,i,h) - \varphi^*(t,i,h) = \int_t^T \tilde{\varphi}(s,i,h)ds$ for each $t \in [0,T]$, $\frac{\partial \varphi^*(t,i,h)}{\partial t} = \tilde{\varphi}(t,i,h)$ holds for all $t \in [0,T]$. Hence, $\varphi^*(t,i,h)$ is indeed a classical solution of (4.19). □

4.5 The Verification Result

This section proves the verification result for the case of countably infinite state space $S_I = \mathbb{N}$. Before doing it, we impose the following conditions on model coefficients throughout the section:

Assumption 4.1. there exist positive constants c_1, c_2, δ and K such that

$$c_1|\xi|^2 \leq \xi^\top \sigma(i)\sigma(i)^\top \xi \leq c_2|\xi|^2, \quad \forall \xi \in \mathbb{R}^n, \ i \in S_I,$$
$$\delta \leq \lambda(i,h) \leq K, \quad \forall(i,h) \in S_I \times S,$$
$$r(i) + |\mu(i)| \leq K, \quad \forall i \in S_I.$$

The first condition on $\sigma(i)$ is actually related to the uniformly elliptic property of the volatility matrix $\sigma(i)$ of stocks.

Proposition 4.4. Let $\varphi^*(t,i,h)$ for $(t,i,h) \in [0,T] \times S_I \times S$ be given by (4.78). Then, for all $(t,i,h) \in [0,T] \times S_I \times S$,

$$\varphi^*(t,i,h) = \inf_{\tilde{\pi} \in \tilde{\mathcal{U}}_t} \mathbb{E}^{\tilde{\pi}}_{t,i,h}\left[\exp\left(\frac{\theta}{2}\int_t^T L(\tilde{\pi}_s; I_s, H_s)ds\right)\right]. \quad (4.94)$$

Proof. It follows from Proposition 4.2 and Lemma 4.5 that, for $m \geq 1$,

$$\varphi_m(t,i,h) = \inf_{\tilde{\pi} \in \tilde{\mathcal{U}}_t^m} \mathbb{E}^{\tilde{\pi}}_{t,i,h}\left[\exp\left(\frac{\theta}{2}\int_t^T L(\tilde{\pi}_s; I_s^{(m)}, H_s)ds\right)\right]$$
$$= \inf_{\tilde{\pi} \in \tilde{\mathcal{U}}_t^m} \mathbb{E}^{\tilde{\pi}}_{t,i,h}\left[\exp\left(\frac{\theta}{2}\int_t^{T \wedge \tau_m^t} L(\tilde{\pi}_s; I_s, H_s)ds\right)\right].$$

Then, for any $\varepsilon > 0$, there exists $\tilde{\pi}^\varepsilon \in \tilde{\mathcal{U}}^m$ such that

$$\varphi_m(t,i,h) + \varepsilon > \mathbb{E}^{\tilde{\pi}^\varepsilon}_{t,i,h}\left[\exp\left(\frac{\theta}{2}\int_t^{T \wedge \tau_m^t} L(\tilde{\pi}_s^\varepsilon; I_s, H_s)ds\right)\right]. \quad (4.95)$$

We introduce $\hat{\pi}_t^\varepsilon := \tilde{\pi}_t^\varepsilon \mathbf{1}_{\{t \leq \tau_m\}}$ for $t \in [0,T]$. Then, $\hat{\pi}^\varepsilon \in \tilde{\mathcal{U}}$, and $\Gamma^{\hat{\pi}^\varepsilon}(t,T) = \Gamma^{\tilde{\pi}^\varepsilon}(t, T \wedge \tau_m^t)$ for $t \in [0,T]$. Note that $L(0,i,h) = -r(i) \leq 0$ for all $(i,h) \in S_I \times S$. Then, the inequality (4.95) continues that

$$\varphi_m(t,i,h) + \varepsilon > \mathbb{E}^{\tilde{\pi}^\varepsilon}_{t,i,h}\left[\exp\left(\frac{\theta}{2}\int_t^{T \wedge \tau_m^t} L(\tilde{\pi}_s^\varepsilon; I_s, H_s)ds\right)\right]$$
$$= \mathbb{E}^{\hat{\pi}^\varepsilon}_{t,i,h}\left[\exp\left(\frac{\theta}{2}\int_t^{T \wedge \tau_m^t} L(\hat{\pi}_s^\varepsilon; I_s, H_s)ds\right)\right]$$
$$\geq \mathbb{E}^{\hat{\pi}^\varepsilon}_{t,i,h}\left[\exp\left(\frac{\theta}{2}\int_t^T L(\hat{\pi}_s^\varepsilon; I_s, H_s)ds\right)\right]$$
$$\geq \inf_{\tilde{\pi} \in \tilde{\mathcal{U}}_t} \mathbb{E}^{\tilde{\pi}}_{t,i,h}\left[\exp\left(\frac{\theta}{2}\int_t^T L(\tilde{\pi}_s^\varepsilon; I_s, H_s)ds\right)\right].$$

By passing $n \to \infty$, and then $\varepsilon \to 0$, we get

$$\varphi^*(t,i,h) \geq \inf_{\tilde{\pi} \in \tilde{\mathcal{U}}_t} \mathbb{E}^{\tilde{\pi}}_{t,i,h} \left[\exp\left(\frac{\theta}{2} \int_t^T L(\tilde{\pi}_s; I_s, H_s) ds\right) \right]. \quad (4.96)$$

On the other hand, using Theorem 4.2 and Proposition 4.2, the mapping $\varphi^*(t,i,h)$ is strictly positive and $\varphi^*(t,i,h) \leq \varphi_m(t,i,h) \leq 1$ for all $m \geq 1$. Then, under Assumption 4.1, by applying a similar argument in the proof of (4.64), we have, for all $\tilde{\pi} \in \tilde{\mathcal{U}}$,

$$\mathbb{E}^{\tilde{\pi}}_{t,i,h}\left[\varphi^*(T, I_T, H_T) \exp\left(\frac{\theta}{2} \int_t^T L(\tilde{\pi}_u; I_u, H_u) du\right)\right] \geq \varphi^*(t,i,h).$$

Since $\varphi(T, i, h) = 1$, $\forall (i, h) \in S_I \times S$, we deduce that

$$\inf_{\tilde{\pi} \in \tilde{\mathcal{U}}_t} \mathbb{E}^{\tilde{\pi}}_{t,i,h}\left[\exp\left(\frac{\theta}{2} \int_t^T L(\tilde{\pi}_s; I_s, H_s) ds\right)\right] \geq \varphi^*(t,i,h). \quad (4.97)$$

Thus, the equality (4.94) follows by combining (4.96) and (4.97), and the validity of the proposition is verified. □

Similar to that in Proposition 4.2, we can construct a candidate optimal $\mathbb{G}$-predictable feedback strategy $\tilde{\pi}^*$ by, for $t \in [0, T]$,

$$\tilde{\pi}^*_t := \text{diag}((1 - H^j_{t-})^n_{j=1})$$
$$\times \arg\min_{\pi \in U} \tilde{\mathcal{H}}(\pi; I_{t-}, H_{t-}, (\varphi^*(t, I_{t-}, H^j_{t-}); j = 0, 1, \ldots, n)). \quad (4.98)$$

We next prove that $\tilde{\pi}^*$ can be characterized as an approximation limit by a sequence of well-defined admissible strategies.

Lemma 4.10. *There exists a sequence of strategies* $(\tilde{\pi}^{(m,*)})_{m \geq 1} \subset \tilde{\mathcal{U}}$ *such that* $\lim_{m \to \infty} \tilde{\pi}^{(m,*)}_t = \tilde{\pi}^*_t$ *for all* $t \in [0, T]$, $\mathbb{P}$*-a.s.. Moreover* $\lim_{m \to \infty} J(\tilde{\pi}^{(m,*)}; t, i, h) = \varphi^*(t, i, h)$ *for* $(t, i, h) \in [0, T] \times S_I \times S$, $\mathbb{P}$*-a.s..* *Here, the objective functional* J *is defined in* (4.14).

Proof. For fixed $(i, h, x) \in \mathbb{N} \times \mathcal{S} \times (0, \infty)^{m+1}$, we have $\tilde{\mathcal{H}}(\pi; i, h, x)$ is strictly concave w.r.t. $\pi \in U$, and hence $\Phi(i, h, x) := \arg\min_{\pi \in U} \tilde{\mathcal{H}}(\pi; i, h, x)$ is well defined. Note that $\Phi(i, h, \cdot)$ maps $(0, \infty)^{m+1}$ to U and satisfies the first-order condition $\frac{\partial \tilde{\mathcal{H}}}{\partial \pi_j}(\Phi(i, h, x); i, h, x) = 0$ for $j = 1, \ldots, m$. Then, the implicit function theorem yields that $\Phi(i, h, x)$ is continuous in x. Let $x^{(m)}_t := (\varphi_m(t, I^{(m)}_{t-}, H^j_{t-}); j = 0, 1, \ldots, n)$. It follows from Proposition 4.2 and Lemma 4.5 that, for $t \in [0, T]$,

$$\tilde{\pi}^{(m,*)}_t := \text{diag}((1 - H^j_{t-})^n_{j=1})\Phi(I_{t-}, H_{t-}, x^{(m)}_t)\mathbf{1}_{\{\tau_m \geq t\}}$$

belongs to $\tilde{\mathcal{U}}^m \cap \tilde{\mathcal{U}}$, and further it satisfies that

$$\varphi_m(t,i,h) = \mathbb{E}_{t,i,h}^{\tilde{\pi}^{(m,*)}}\left[\exp\left(\frac{\theta}{2}\int_t^{T\wedge\tau_m^t} L(\tilde{\pi}_s^{(m,*)};I_s,H_s)ds\right)\right]. \quad (4.99)$$

Lemma 4.7 implies that $\lim_{m\to\infty}|x_t^{(m)} - x_t^*| = 0$ for $t \in [0,T]$, $\mathbb{P}$-a.s., where $x_t^* := (\varphi^*(t,I_{t-},H_{t-}^j);\ j=0,1,\ldots,n)$. We define the predictable process $\tilde{\pi}_t^* := \mathrm{diag}((1-H_{t-}^j)_{j=1}^n)\Phi(I_{t-},H_{t-},x_t^*)$ for $t \in [0,T]$. Using Lemma 4.9 and the continuity of $\Phi(i,h,\cdot)$, we obtain $\lim_{m\to\infty}\tilde{\pi}_t^{(m,*)} = \tilde{\pi}_t^*$ for all $t \in [0,T]$, a.s. Moreover, we also have

$$J(\tilde{\pi}^{(m,*)};t,i,h) = \mathbb{E}_{t,i,h}^{\tilde{\pi}^{(m,*)}}\left[\exp\left(\frac{\theta}{2}\int_t^T L(\tilde{\pi}_s^{(m,*)};I_s,H_s)ds\right)\right]$$

$$= \mathbb{E}_{t,i,h}^{\tilde{\pi}^{(m,*)}}\left[\exp\left(\frac{\theta}{2}\int_t^{T\wedge\tau_m^t} L(\tilde{\pi}_s^{(m,*)};I_s,H_s)ds\right.\right.$$

$$\left.\left.+ \frac{\theta}{2}\int_{T\wedge\tau_m^t}^T L(0;I_s,H_s)ds\right)\right]$$

$$\leq \mathbb{E}_{t,i,h}^{\tilde{\pi}^{(m,*)}}\left[\exp\left(\frac{\theta}{2}\int_t^{T\wedge\tau_m^t} L(\tilde{\pi}_s^{(m,*)};I_s,H_s)ds\right)\right]$$

$$= \varphi_m(t,i,h).$$

It follows from Proposition 4.4 that $\varphi^*(t,i,h) \leq J(\tilde{\pi}^{(m,*)};t,i,h) \leq \varphi_m(t,i,h)$ for $m \geq 1$. This verifies that $\lim_{m\to\infty}J(\tilde{\pi}^{(m,*)};t,i,h) = \varphi^*(t,i,h)$ for all $(t,i,h) \in [0,T]\times S_I \times \mathcal{S}$, a.s. by applying Lemma 4.7. $\square$

Proposition 4.5. *The optimal feedback strategy $\tilde{\pi}^*$ given by (4.98) is admissible, i.e., $\tilde{\pi}^* \in \tilde{\mathcal{U}}$.*

Proof. From Assumption 4.1, it is not difficult to verify that, there exists a constant $C > 0$ such that $L(\pi;i,h) \geq -C$ for all $(\pi,i,h) \in U \times S_I \times \mathcal{S}$. Thanks to Proposition 4.4, for all $(t,i,h) \in [0,T]\times S_I \times \mathcal{S}$,

$$\varphi^*(t,i,h) = \inf_{\tilde{\pi}\in\tilde{\mathcal{U}}_t}\mathbb{E}_{t,i,h}^{\tilde{\pi}}\left[\exp\left(\frac{\theta}{2}\int_t^T L(\tilde{\pi}_s;I_s,H_s)ds\right)\right]$$

$$\geq \inf_{\tilde{\pi}\in\tilde{\mathcal{U}}_t}\mathbb{E}_{t,i,h}^{\tilde{\pi}}\left[\exp\left(-\frac{\theta}{2}\int_t^T Cds\right)\right] = \exp\left(-\frac{\theta}{2}C(T-t)\right).$$

Therefore, for $t \in [0,T]$,

$$x_t^* = (\varphi^*(t,I_{t-},H_{t-}^j);\ j=0,1,\ldots,n)) \in [e^{-\frac{\theta}{2}C(T-t)},1]^{n+1}. \quad (4.100)$$

The continuity of $\Phi(i,h,x) := \arg\min_{\pi \in U} \tilde{\mathcal{H}}(\pi;i,h,x)$ gives that $\tilde{\pi}_t^*$ for $t \in [0,T]$ is uniformly bounded by some constant $C_1 > 0$. Moreover, the first-order condition yields that, for all $j = 1,\ldots,n$, if $H_{t-}^j = 0$,

$$(1-\tilde{\pi}_t^{*,j})^{-\frac{\theta}{2}-1} = \left[(\mu_j(I_{t-}) - r(I_{t-})) - \frac{\theta}{2}\left(1+\frac{\theta}{2}\right)\sum_{i=1}^n (\sigma(I_{t-})^\top \sigma(I_{t-}))_{ji} \tilde{\pi}_t^{*,i}\right.$$
$$\left. + \frac{\theta}{2}\lambda_j(I_{t-}, H_{t-})\right] \frac{\varphi^*(t, I_{t-}, H_{t-})}{\lambda_j(I_{t-}, H_{t-})\varphi^*(t, I_{t-}, H_{t-}^j)}$$
$$\leq C_2, \qquad (4.101)$$

where we used Assumption 4.1 and (4.100). Note that $\tilde{\pi}_t^{*,j} = 0$ if $H_{t-}^j = 1$, then $\tilde{\pi}^*$ is also uniformly bounded away from 1. This implies that the generalized Novikov's condition holds in the countably infinite state case, and hence $\tilde{\pi}^*$ is admissible. $\square$

4.6 The Convergence Rate

The verification results provided in Proposition 4.4 and Proposition 4.5 in Section 4.5 can be seen as a uniqueness result for the HJB equation.

With Assumption 4.1, we can also establish an error estimate on the approximation of the sequence of strategies $\tilde{\pi}^{(m,*)}$ to the optimal strategy π^* in terms of the objective functional J (c.f. (4.14)), which is given by

Lemma 4.11 (Convergence Rate). *Let $m \geq 1$ and Assumption 4.1 hold. Then, for any $(t,i,h) \in [0,T] \times D_m \times \mathcal{S}$, there exists a constant $C > 0$ independent of m such that*

$$\left|J(\tilde{\pi}^{(m,*)}; t,i,h) - J(\tilde{\pi}^{(*)}; t,i,h)\right| \leq C\left[1 - \sum_{j=1}^m a_{ij}^{(m)}(T-t)\right].$$

Here, the coefficient

$$a_{ij}^{(m)}(T-t) = \delta_{ij} + (T-t)q_{ij} + \sum_{k=1}^{+\infty}\sum_{1\leq l_1,\ldots,l_k \leq m} \frac{(T-t)^{k+1}}{(k+1)!} q_{il_1}q_{l_1l_2}\cdots q_{l_kj}.$$

Proof. It follows from Proposition 4.5 that $J(\tilde{\pi}^{(m,*)}; t,i,h) \to \varphi^*(t,i,h) = J(\tilde{\pi}^*; t,i,h)$ as $m \to \infty$. On the other hand, it can be verified that there exists constants $\gamma \in (0,1)$ and $C_1 > 0$ such that $\tilde{\pi}_t^* \in [-C_1, 1-\gamma]^n$ for all

$t \in [0,T]$, a.s. Then, using (4.8), it follows that $L(\tilde{\pi}_t^*; I_t, H_t) \leq C_2$, a.s. for $t \in [0,T]$. Here, C_2 is a positive constant. By noting $\tilde{\pi}^* \in \tilde{\mathcal{U}}^m$, we have

$$\varphi^*(t,i,h) = \mathbb{E}_{t,i,h}^{\tilde{\pi}^*}\left[\exp\left(\frac{\theta}{2}\int_t^T L(\tilde{\pi}_s^*; I_s, H_s)ds\right)\right]$$

$$\geq \mathbb{E}_{t,i,h}^{\tilde{\pi}^*}\left[\exp\left(\frac{\theta}{2}\int_t^T L(\tilde{\pi}_s^*; I_s, H_s)ds\right) \mathbf{1}_{\{\tau_m^t > T\}}\right]$$

$$= \mathbb{E}_{t,i,h}^{\tilde{\pi}^*}\left[\exp\left(\frac{\theta}{2}\int_t^{T\wedge\tau_m^t} L(\tilde{\pi}_s^*; I_s, H_s)ds\right) \mathbf{1}_{\{\tau_m^t > T\}}\right]$$

$$= \mathbb{E}_{t,i,h}^{\tilde{\pi}^*}\left[\exp\left(\frac{\theta}{2}\int_t^{T\wedge\tau_m^t} L(\tilde{\pi}_s^*; I_s, H_s)ds\right)\right]$$

$$\quad - \mathbb{E}_{t,i,h}^{\tilde{\pi}^*}\left[\exp\left(\frac{\theta}{2}\int_t^{T\wedge\tau_m^t} L(\tilde{\pi}_s^*; I_s, H_s)ds\right) \mathbf{1}_{\{\tau_m^t \leq T\}}\right]$$

$$\geq \varphi_m(t,i,h) - \mathbb{E}_{t,i,h}^{\tilde{\pi}^*}\left[e^{\frac{\theta C_2}{2}(T\wedge\tau_m^t - t)} \mathbf{1}_{\{\tau_m^t \leq T\}}\right]$$

$$\geq \varphi_m(t,i,h) - C_3 \mathbb{P}_{t,i,h}^{\tilde{\pi}^*}(\tau_m^t \leq T),$$

where $C_3 := e^{\frac{\theta C_2 T}{2}}$ and $\varphi_m(t,i,h)$ is defined in Proposition 4.3. Using the inequality $\varphi^*(t,i,h) \leq J(\tilde{\pi}^{(m,*)}; t,i,h) \leq \varphi_m(t,i,h)$ in the proof of Lemma 4.10, under Assumption 4.1, we arrive at

$$\left|J(\tilde{\pi}^{(m,*)}; t,i,h) - J(\tilde{\pi}^{(*)}; t,i,h)\right| = J(\tilde{\pi}^{(m,*)}; t,i,h) - \varphi^*(t,i,h)$$

$$\leq \varphi_m(t,i,h) - \varphi^*(t,i,h)$$

$$\leq C_3 \mathbb{P}_{t,i,h}^{\tilde{\pi}^*}(\tau_m^t \leq T).$$

Note that, by Proposition 4.5, the process I is also a Markov chain with the generator $Q = (q_{ij})$ under $\mathbb{P}_{t,i,z}^{\tilde{\pi}^*}$. Then $\mathbb{P}_{t,i,z}^{\tilde{\pi}^*}(\tau_m^t \leq T) \to 0$ as $m \to \infty$. On the other hand, τ_m^t is the absorption time of $(I_s^{(m)})_{s \in [t,T]}$ whose generator is given as A_m given by (4.70). Hence, using Bielecki and Rutkowski (2002, Section 11.2.3 in Chapter 11), we also have $\mathbb{P}_{t,i,h}^{\tilde{\pi}^*}(\tau_m^t \leq T) = 1 - \sum_{j=1}^m a_{ij}^{(m)}(T-t)$. This completes the proof of the lemma. $\square$

We next provide an example in which the error estimate in Lemma 4.11 has a closed-form representation.

Example 4.1 (Explicit Convergence Rate). Consider the following specific generator of the Markvo chain I, which is given by

$$Q = \begin{bmatrix} -1 & \frac{1}{2} & \frac{1}{4} & \cdots & \frac{1}{2^{m-1}} & \frac{1}{2^m} & \cdots \\ \frac{1}{2} & -1 & \frac{1}{4} & \cdots & \frac{1}{2^{m-1}} & \frac{1}{2^m} & \cdots \\ \frac{1}{2} & \frac{1}{4} & -1 & \cdots & \frac{1}{2^{m-1}} & \frac{1}{2^m} & \cdots \\ \vdots & \vdots & \vdots & & \vdots & & \\ \frac{1}{2} & \frac{1}{4} & \frac{1}{8} & \cdots & \frac{1}{2^{m-1}} & -1 & \cdots \\ \vdots & \vdots & \vdots & & \vdots & & \end{bmatrix}.$$

Then, for any $l \leq m$, $\sum_{j=1}^{m} q_{lj} = \sum_{j=1}^{m-1} \frac{1}{2^j} - 1 = \frac{-1}{2^{m-1}}$. Therefore, for all $i \leq m$,

$$\sum_{j=1}^{m} a_{ij}^{(m)}(T-t) = \sum_{k=0}^{\infty} \frac{(T-t)^k}{k!} \left(\frac{-1}{2^{m-1}}\right)^k = e^{-\frac{T-t}{2^{m-1}}}.$$

Thus, for all $(t, i, h) \in [0, T] \times D_m \times \mathcal{S}$, we arrive at

$$\left| J(\tilde{\pi}^{(m,*)}; t, i, h) - J(\tilde{\pi}^{(*)}; t, i, h) \right| \leq C \left(1 - e^{-\frac{T-t}{2^{m-1}}} \right),$$

where the constant $C > 0$ is independent of m.

It is also worth mentioning here that our method used in the chapter can be applied to treat the case where the process I is a time-inhomogeneous Markov chain with a time-dependent generator given by $Q(t) = (q_{ij}(t))_{i,j \in S_I}$ for $t \in [0, T]$. Here, for $t \in [0, T]$, $q_{ii}(t) \leq 0$ for $i \in S_I$, $q_{ij}(t) \geq 0$ for $i \neq j$, and $\sum_{j=1}^{\infty} q_{ij}(t) = 0$ for $i \in S_I$ (i.e., $\sum_{j \neq i} q_{ij}(t) = -q_{ii}(t)$ for $i \in S_I$). Also, for $i, j \in S_I$, $t \to q_{ij}(t)$ is continuous on $[0, T]$, and the infinite summation $\sum_{j \in S_I} q_{ij}(t)$ is uniformly convergent in $t \in [0, T]$.

At the end of this chapter, we provide two auxiliary lemmas which serve to prove the previous results.

Lemma 4.12. *Let the function $\Phi(x) : [0, \infty)^{n+1} \to \mathbb{R}$ be concave in every component of x. Assume that there exists $\bar{x}$, x^*, $\underline{x} \in [0, \infty)^{n+1}$ such that $\underline{x} \ll x^* \ll \bar{x}$. Let $\{x^{(m)}\}_{m \geq 1} \subset [0, \infty)^{n+1}$ satisfy $x^* \leq x^{(m)}$ for $m \geq 1$ and $\lim_{m \to \infty} x^{(m)} = x^*$. Then $\lim_{m \to \infty} \Phi(x^{(m)}) = \Phi(x^*)$.*

Proof. By the given conditions in the lemma, there exists $n_0 \geq 1$ such that $x^* \leq x^{(m)} \leq \bar{x}$ for all $m \geq n_0$. For each $m \geq n_0$, there exists a vector $\lambda^{(m)} \in [0, 1]^{n+1}$ satisfying $\lim_{m \to \infty} \lambda^{(m)} = 0$ such that $x_k^{(m)} = \lambda_k^{(m)} \bar{x}_k + (1 - \lambda_k^{(m)}) x_k^*$ for $k = 1, \ldots, n+1$. Therefore

$$\Phi(x^{(m)}) = \Phi\left(\lambda_1^{(m)} \bar{x}_1 + (1 - \lambda_1^{(m)}) x_1^*, \lambda_2^{(m)} \bar{x}_2 + (1 - \lambda_2^{(m)}) x_2^*, \ldots,\right.$$

$$\ldots, \lambda_{n+1}^{(k)} \overline{x}_{n+1} + (1 - \lambda_{n+1}^{(n)}) x_{n+1}^*\Big)$$

$$\geq \lambda_1^{(m)} \Phi\Big(\overline{x}_1, \lambda_2^{(m)} \overline{x}_2 + (1 - \lambda_2^{(m)}) x_2^*, \ldots, \lambda_{n+1}^{(k)} \overline{x}_{n+1} + (1 - \lambda_{m+1}^{(m)}) x_{n+1}^*\Big)$$
$$+ (1 - \lambda_1^{(m)}) \Phi\Big(x_1^*, \lambda_2^{(m)} \overline{x}_2 + (1 - \lambda_2^{(m)}) x_2^*, \ldots,$$
$$\ldots, \lambda_{n+1}^{(k)} \overline{x}_{n+1} + (1 - \lambda_{n+1}^{(m)}) x_{n+1}^*\Big)$$

$$\geq \lambda_1^{(m)} \lambda_2^{(m)} \Phi\Big(\overline{x}_1, \overline{x}_2, \ldots, \lambda_{n+1}^{(m)} \overline{x}_{n+1} + (1 - \lambda_{N+1}^{(m)}) x_{n+1}^*\Big)$$
$$+ \lambda_1^{(m)} (1 - \lambda_2^{(m)}) \Phi\Big(\overline{x}_1, x_2^*, \ldots, \lambda_{n+1}^{(k)} \overline{x}_{m+1} + (1 - \lambda_{n+1}^{(m)}) x_{n+1}^*\Big)$$
$$+ (1 - \lambda_1^{(m)}) \lambda_2^{(m)} \Phi\Big(\overline{x}_1, x_2^*, \ldots, \lambda_{n+1}^{(k)} \overline{x}_{n+1} + (1 - \lambda_{n+1}^{(m)}) x_{n+1}^*\Big)$$
$$+ (1 - \lambda_1^{(m)})(1 - \lambda_2^{(m)}) \Phi\Big(x_1^*, x_2^*, \ldots, \lambda_{n+1}^{(k)} \overline{x}_{n+1} + (1 - \lambda_{n+1}^{(m)}) x_{n+1}^*\Big)$$

$$\geq \Phi(x^*) \prod_{k=1}^{n+1} (1 - \lambda_k^{(m)}) + \Sigma_1^{(m)}. \tag{4.102}$$

We observe that every term in $\Sigma_1^{(m)}$ above has one or more multipliers which is of the form $\lambda_k^{(m)}$ for $k = 1, \ldots, n+1$. Since $\lim_{m \to \infty} \lambda_k^{(m)} = 0$ for $k = 1, \ldots, n+1$, we have $\Sigma_1^{(m)} \to 0$ as $m \to \infty$. It follows from (4.102) that $\liminf_{m \to \infty} \Phi(x^{(m)}) \geq \Phi(x^*)$. Similarly, as $x^{(m)} \geq x^* \gg \underline{x}$ for all $m \in \mathbb{N}$, there exists a vector $\tilde{\lambda}^{(m)} \in [0,1]^{n+1}$ satisfying $\lim_{m \to \infty} \tilde{\lambda}^{(m)} = 0$ such that $x_k^* = \tilde{\lambda}_k^{(m)} \underline{x}_k + (1 - \tilde{\lambda}_k^{(m)}) x_k^{(m)}$ for $k = 1, \ldots, n+1$. Using the similar argument in the proof of (4.102), we deduce that

$$\Phi(x^*) \geq \Phi(x^{(m)}) \prod_{k=1}^{n+1} (1 - \tilde{\lambda}_k^{(m)}) + \Sigma_2^{(m)}, \tag{4.103}$$

where every term in $\Sigma_2^{(m)}$ above has one or more multipliers which is of the form $\lambda_k^{(m)}$, $k = 1, \ldots, n+1$. The inequality (4.103) gives that $\Phi(x^*) \geq \limsup_{m \to \infty} \Phi(x^{(m)})$. Putting the above two inequalities together, we obtain $\lim_{m \to \infty} \Phi(x^{(m)}) = \Phi(x^*)$, which completes the proof. □

Lemma 4.13. *Let the function* $\Phi(x) : [0, \infty)^{n+1} \to [0, \infty)$ *be concave in every component of* x. *Then, for any* $\alpha, \beta \in [0, \infty)^{n+1}$ *satisfying* $\alpha \leq \beta$, *there exists a constant* $C = C(\alpha, \beta) > 0$ *such that* $0 \leq \Phi(x) \leq C$ *for all* $\alpha \leq x \leq \beta$.

Proof. For any $\alpha \leq x \leq \beta$ where $\alpha, \beta \in [0, \infty)^{n+1}$, there exists a vector $\nu \in (0, \infty)^{n+1}$ such that $\beta \ll \nu$. This implies that there exists $\lambda \in [0,1]^{n+1}$

such that $\beta_k = \lambda_k x_k + (1-\lambda_k)\nu_k$, $k = 1,\ldots,n+1$. Since $\alpha \le x \le \beta \ll \nu$, there exists $\delta > 0$ such that $1 \ge \lambda_k = \frac{\nu_k - \beta_k}{\nu_k - x_k} \ge \frac{\nu_k - \beta_k}{\nu_k - \alpha_k} \ge \delta$. Using the concave property of $x \to \Phi(x)$, we have

$$\Phi(\beta) = \Phi(\lambda_1 x_1 + (1-\lambda_1)\nu_1, \lambda_2 x_2 + (1-\lambda_2)\nu_2, \ldots,$$
$$\ldots, \lambda_{n+1} x_{n+1} + (1-\lambda_{n+1})\nu_{n+1})$$
$$\ge \Phi(x) \prod_{k=1}^{n+1} \lambda_k + \sum_{\substack{1 \le j_1 < j_2 < \ldots < j_k < n+1 \\ 1 \le k \le n+1}} (1-\lambda_{j_1}) \times \cdots \times (1-\lambda_{j_k}) \lambda_{j_{k+1}} \times \cdots$$
$$\times \cdots \times \lambda_{j_n} \Phi(C_{j_1 \ldots j_k})$$
$$\ge \delta^{n+1} \Phi(x), \tag{4.104}$$

for some $C_{j_1\ldots j_k} \in [0,\infty)^{n+1}$, and $\{j_{k+1},\ldots,j_{n+1}\} = \{1,\ldots,n+1\} \setminus \{j_1,\ldots,j_k\}$. Thus, the lemma follows. $\square$

Chapter 5

Risk Sensitive Credit Portfolio Optimization: Partial Information*

This chapter continues studying the risk-sensitive portfolio optimization among multiple credit risky assets. Similar to Chapter 4, the default contagion is considered in the sense that the default intensities of surviving names depend on the default events of all other assets as well as regime states. In particular, the regime switching process is described by a continuous time Markov chain with countable states and the default events of risky assets are depicted via some pure jump indicators. The joint impacts on the optimal portfolio by contagion risk and changes of market and credit regimes can be analyzed in an integrated fashion.

As opposed to Chapter 4, we recast the problem into a more practical setting when the regime-switching process is not observable, in which the filtering procedure becomes necessary. Consequently, the contagion risk comes from two distinct sources: the *physical* contagion that is from our way to model default intensity as a function depending on all other default indicators and the *information-induced* contagion that is generated by our estimation of the regime transition probability of the incoming default using observations of past default events.

Portfolio optimization under partial information. The existing work on portfolio optimization under a hidden Markov chain include Pham and Quenez (2001), Sass and Haussmann (2004), Bäuerle and Rieder (2007), Branger et al. (2014), Lim and Quenez (2015), Bo and Capponi (2017), Xiong and Zhou (2007) and many others. This chapter focuses on the risk-sensitive portfolio optimization with both default contagion and partial observations based on countable regimes states. Comparing with Chapter 4, the countable regime states results in an infinite-dimensional filter process

---
*This chapter is based on the work Bo et al. (2022).

and we confront a more complicated infinite-dimensional system of coupled nonlinear PDEs due to default contagion and the infinite-dimensional filter process in Proposition 5.1. We are lack of adequate tools to tackle this infinite-dimensional system by means of standard PDE theories such as operator method or fixed point method (c.f. Cerrai (2001) and Delong and Klüppelberg (2008)). On the other hand, BSDE approach has become a powerful tool in financial applications with default risk or incomplete information; c.f. Jiao et al. (2013) in the context of utility maximization under contagion risk and complete information, and Papanicolaou (2019) on stochastic control under partial observations without default jumps. Herein, we choose to employ the BSDE method to tackle the risk-sensitive control problem and it is interesting to see that the associated BSDE in (5.56) has a non-standard driver term that deserves some careful investigations.

Contributions of this chapter. In this chapter, we establish a martingale representation theorem under partial and phasing-out information. We also extend the study of quadratic BSDE with jumps by considering a random driver induced from our control problem. More detailed explanations are summarized as **(i)** regarding the aspect of partial observations, we are interested in the incomplete information filtration that possesses a phasing out feature due to sequential defaults of multiple assets. That is, the information of the Brownian motion will be terminated after the associated risky asset defaults. This assumption can better match with the real life situation that the investor can no longer perceive any information from the asset once it exits the market. We therefore focus on the filtration $\mathbb{F}^M$ defined in (5.5) that is generated by stopped Brownian motions and the default indicator processes, and a new martingale representation theorem under $\mathbb{F}^M$, i.e., Theorem 5.1, is needed. By applying the changing of measure and technical modifications of some arguments in Frey and Schmidt (2012) together with the approximation scheme and MCT, we can conclude Theorem 5.1, which is an interesting new result. **(ii)** there are many existing works on quadratic BSDE with jumps. Morlais (2009) studies the existence of solution to the BSDE with jumps arising from an exponential utility maximization problem with a bounded terminal condition. Morlais (2010) extends the work when the jump measure satisfies the infinite-mass. Kazi-Tani et al. (2015) apply a fixed point method to study the quadratic BSDE with jumps given a small L^∞-terminal condition. Antonelli and Mancini (2016) further refines the results of the previous work by considering a generator depending on all components and unbounded terminal conditions.

All aforementioned work crucially rely on the same quadratic-exponential structure of the driver term, namely quadratic growth in the

Brownian component and exponential growth with respect to the jump term, which entails *a priori* estimates of the solution. On the contrary, the random driver in our quadratic BSDE (5.56) does not satisfy this property, which results from the risk sensitive preference engaging contagion dependence and the filtering process. Consequently, the existence of solution can not follow from the same analysis in the literature. This is the main motivation for us to conduct this research, which not only can contribute to the risk sensitive portfolio optimization under default contagion, but will also enrich the study of quadratic BSDE with jumps by allowing some non-standard random drivers. To overcome some new difficulties caused by the random driver, we follow a two-step procedure. In the first step, we propose some tailor-made truncations on the driver term to make it Lipschitz uniformly in time and in sample path such that the existence and uniqueness of the solution can easily follow. The challenging part is to derive a uniform *a priori* estimates for all truncated solutions, in which the bounded estimate of the jump solution of the truncated quadratic BSDE will become helpful when the random driver does not exhibit the standard structure. In the second step, we adopt and modify some approximation arguments in Kobylanski (2000) to fit into our setting with jumps and verify that the limiting process from step one solves the original BSDE in an appropriate space. We believe that the analysis of BSDE (5.56) can be further extended to tackle more general random drivers that stem from other default contagion models.

The rest of the chapter is organized as follows. Section 5.1 formulates the model of credit risky assets with regime-switching. Section 5.2 introduces the partial information and filter processes. Section 5.3 focuses on the filter process and establishes a martingale representation theorem. Section 5.4 discusses a general correlation volatility matrix case for multiple credit assets. Section 5.5 relates the risk-sensitive portfolio optimization problem under partial information to a quadratic BSDE with jumps. Section 5.6 is devoted to the proof of the existence of solution to the BSDE problem. In Section 5.7, the verification theorem is concluded by using our BSDE results. Section 5.8 concludes the uniqueness of solutions to the BSDE.

5.1 The Market Model

The market model consisting of credit risky assets with default contagion and regime-switching which are the same to that in Chapter 4, however, the regime-switching process $I = (I_t)_{t \in [0,T]}$ as a continuous time Markov chain is assume to be *unobservable*.

To the completion of description for the model under partial information, let $(\Omega, \mathcal{F}, \mathbb{F}, \mathbb{P})$ be a filtered probability space with the filtration $\mathbb{F} = (\mathcal{F}_t)_{t\in[0,T]}$ satisfying the usual conditions. We consider n defaultable risky assets and one riskless bond, whose dynamics are $\mathbb{F}$-adapted processes and are defined via the following components:

(i) A hidden Markov chain process $I = (I_t)_{t\in[0,T]}$ with state space $S_I := \{1, \ldots, m\}$, which is a continuous-time Markov chain with the generator matrix $Q = (q_{ij})_{1\le i,j\le m}$, where $2 \le m \le +\infty$.

(ii) A default indicator process $H = (H_t^i;\ i = 1,\ldots,n)_{t\in[0,T]}$, which denotes the default indicator process with the state space $\mathcal{S} = \{0,1\}^n$. The default mechanism is the same to that introduced in Eq. (4.1) of Section 4.1 in Chapter 4.

(iii) As in Eq. (4.2) of Section 4.1, the *pre-default* price process of the n risky assets is given by

$$dP_t = \mathrm{diag}(P_t)\{(\mu(I_t) + \lambda(I_t, H_t))dt + \sigma dW_t\}, \qquad (5.1)$$

where $W = (W_t^i;\ i = 1,\ldots,n)_{t\in[0,T]}^\top$ is an n-dimensional Brownian motion, which is independent of Markov chain I. The volatility $\sigma = \mathrm{diag}((\sigma_i)_{i=1,\ldots,n})$ with $\sigma_i > 0$ for all $i = 1, \ldots, n$.

(iv) In the setup of jump-to-default, the price process $\tilde{P}_t^i := (1 - H_t^i)P_t^i$ of the i-th defaultable asset is

$$d\tilde{P}_t = \mathrm{diag}(\tilde{P}_{t-})\{\mu(I_t)dt + \sigma dW_t - dM_t\}. \qquad (5.2)$$

Here, $M = (M_t^j;\ j = 1,\ldots,n)_{t\in[0,T]}^\top$ is the pure jump $\mathbb{F}$-martingale defined by (4.4) in Chapter 4.

5.2 Partial Information and Filter Processes

This section formulates the partial information and introduces the corresponding filter processes.

The information of the hidden Markov chain I is not accessible by the investor, who can only observe public prices of risky assets continuously and the default events of assets (i.e., the information generated by $\tilde{P}$ and H).

Partial information. We formulate the partial information filtration. For an adapted process $X = (X_t)_{t\in[0,T]}$, let $\mathbb{F}^X = (\mathcal{F}_t^X)_{t\in[0,T]} = (\sigma(X_s; s \le t))_{t\in[0,T]}$ be the natural filtration generated by X. We then introduce the auxiliary process $W^o = (W_t^{o,1}, \ldots, W_t^{o,n})_{t \in [0,T]}^\top$ given by, for $i = 1, \ldots, n$,

$$W_t^{o,i} := \sigma_i^{-1} \int_0^t (\mu_i(I_s) + \lambda_i(I_s, H_s)) ds + W_t^i. \quad (5.3)$$

We next consider the stopped process $W^{o,\tau} = (W_t^{o,1,\tau}, \ldots, W_t^{o,n,\tau})_{t \in [0,T]}^\top$ by the default times $(\tau_1, \ldots \tau_n)$ in the sense that

$$W_t^{o,i,\tau} := W_{t \wedge \tau_i}^{o,i}, \quad \forall i = 1, \ldots, n. \quad (5.4)$$

In view of (5.1) and (2.55), the available *market information* filtration $\mathbb{F}^M := (\mathcal{F}_t^M)_{t \in [0,T]}$ is then described as:

$$\mathcal{F}_t^M := \mathcal{F}_t^{\tilde{P}} \vee \mathcal{F}_t^H = \mathcal{F}_t^{W^{o,\tau}} \vee \mathcal{F}_t^H. \quad (5.5)$$

We here recall that $(\mathcal{F}_t^{W^{o,\tau}})_{t \in [0,T]}$ and $(\mathcal{F}_t^H)_{t \in [0,T]}$ are the filtration generated by $W^{o,\tau}$ and H respectively, i.e., $\mathcal{F}_t^{W^{o,\tau}} = \bigvee_{i=1}^n \mathcal{F}_t^{W^{o,i,\tau}}$ and $\mathcal{F}_t^H = \bigvee_{i=1}^n \mathcal{F}_t^{H_i}$ for $t \in [0,T]$.

From now on, we impose the following assumption for handling the case with the infinite number of regime states (i.e., $m = +\infty$).

Assumption 5.1. For $(i, k, h) \in \{1, \ldots, n\} \times S_I \times S$, there exist positive constants ε and C independent of k such that $\varepsilon \leq |\lambda_i(k,h)| + |\mu_i(k)| \leq C$.

Filter processes. We introduce the filter process of the hidden Markov chain I under the partial information $\mathbb{F}^M$ by, for $k \in S_I$,

$$p_t^{M,k} := \mathbb{P}(I_t = k | \mathcal{F}_t^M), \quad \forall t \in [0,T]. \quad (5.6)$$

The state space of $p^M = (p_t^{M,k}; k \in S_I)_{t \in [0,T]}^\top$ is denoted by S_{p^M}. When $m < +\infty$, it is shown in Lemma B.1 in Capponi et al. (2015) that $S_{p^M} = \{p \in (0,1)^m; \sum_{i=1}^m p_i = 1\}$. Consider $W^M = (W_t^{M,1}, \ldots, W_t^{M,n})_{t \geq 0}^\top$ defined by, for $i = 1, \ldots, n$,

$$W_t^{M,i} := W_t^{o,i,\tau} - \sigma_i^{-1} \int_0^{t \wedge \tau_i} (\mu_i^M(p_s^M) + \lambda_i^M(p_s^M, H_s)) ds, \quad (5.7)$$

in which we define, for all $(p, h) \in S_{p^M} \times S_H$,

$$\mu^M(p) := \sum_{k \in S_I} \mu(k) p_k, \quad \lambda^M(p, z) := \sum_{k \in S_I} \lambda(k, h) p_k. \quad (5.8)$$

Note that $\mu^M(p_t^M)$ and $\lambda^M(p_t^M, h)$ are conditional expectations of $\mu(I_t)$ and $\lambda(I_t, h)$ given the filtration $\mathcal{F}_t^M$. Assumption 5.1 guarantees that $\mu^M(p)$ and $\lambda^M(p, h)$ defined in (5.8) are finite. Therefore, it is not difficult to verify that, under Assumption 5.1, the process $W^M = (W_t^{M,i}; i = 1, \ldots, n)_{t \in [0,T]}^\top$

is a continuous $(\mathbb{P}, \mathbb{F}^{\mathrm{M}})$-martingale. Moreover, we can show that, for $i = 1, \ldots, n$, the pure jump process defined by

$$M_t^{\mathrm{M},i} := H_t^i - \int_0^t \lambda_i^{\mathrm{M}}(p_s^{\mathrm{M}}, H_s) ds, \quad \forall t \in [0, T] \tag{5.9}$$

is a $(\mathbb{P}, \mathbb{F}^{\mathrm{M}})$-martingale.

We first have the following auxiliary result.

Lemma 5.1. *Let* $t \in [0, T]$ *and* $i = 1, \ldots, n$. *Denote by* $\check{\mathcal{F}}_t^i := \mathcal{F}_t^{W^{o,i}} \vee \mathcal{F}_t^{H_i}$ *and* $\mathcal{F}_t^{\mathrm{M}i} := \mathcal{F}_t^{W^{o,i,\tau}} \vee \mathcal{F}_t^{H_i}$. *Then, for any bounded $\mathbb{R}$-valued r.v.* $\xi \in \check{\mathcal{F}}_t^i$, *we have* $\xi \mathbf{1}_{\{\tau_i \geq t\}} \in \mathcal{F}_t^{\mathrm{M}i}$.

Proof. Let $\mathcal{L}$ be the family of all bounded $\mathbb{R}$-valued r.v.s in the sense that

$$\mathcal{L} := \{\xi \in \check{B}_t^i; \; \xi \mathbf{1}_{\{\tau_i \geq t\}} \in \mathcal{F}_t^{\mathrm{M}i}\},$$

where $\check{B}_t^i$ stands for all bounded $\mathbb{R}$-valued r.v.s that are $\check{\mathcal{F}}_t^i$-measurable. The class $\mathcal{L}$ is nonempty as all constants are in $\mathcal{L}$. Also, it holds that

(i) Let $\xi_k \in \mathcal{L}$ for $k \geq 1$ such that $\lim_{k \to \infty} \xi_k = \xi$, then $\xi \mathbf{1}_{\{\tau_i \geq t\}} = \lim_{k \to \infty} \xi_k \mathbf{1}_{\{\tau_i \geq t\}} \in \mathcal{F}_t^{\mathrm{M}i}$.
(ii) Let $\xi_i \in \mathcal{L}$ with $i = 1, 2$. Then, for all $a, b \in \mathbb{R}$, $\{a\xi_1 + b\xi_2\} \mathbf{1}_{\{\tau_i \geq t\}} = a\xi_1 \mathbf{1}_{\{\tau_i \geq t\}} + b\xi_2 \mathbf{1}_{\{\tau_i \geq t\}} \in \mathcal{F}_t^{\mathrm{M}i}$.

We define another class of r.v.s by

$$\mathcal{M} := \left\{ \prod_{\ell=1}^k \mathbf{1}_{\{[W_{t_\ell}^{o,i}]^{-1}(A_\ell)\}}; \; 0 \leq t_1 < \ldots < t_k \leq t, \; A_\ell \in \mathcal{B}(\mathbb{R}), \right.$$
$$\left. \ell = 1, \ldots, k \in \mathbb{N} \right\}. \tag{5.10}$$

It is not difficult to see that $\mathcal{M}$ is a multiplicative class, and it holds that $\mathcal{F}_t^{W^{o,i}} = \sigma(\mathcal{M})$. Furthermore, each $\xi \in \mathcal{M}$ admits the form that

$$\xi = \prod_{\ell=1}^k \mathbf{1}_{\{[W_{t_\ell}^{o,i}]^{-1}(A_\ell)\}}, \text{ with } 0 \leq t_1 < \ldots < t_k \leq t, \; A_\ell \in \mathcal{B}(\mathbb{R}), \; \ell = 1, \ldots, k.$$

Therefore, we obtain

$$\xi \mathbf{1}_{\{\tau_i \geq t\}} = \prod_{\ell=1}^k \mathbf{1}_{\{[W_{t_\ell}^{o,i}]^{-1}(A_\ell)\}} \mathbf{1}_{\{\tau_i \geq t\}} = \prod_{\ell=1}^k \mathbf{1}_{\{[W_{t_\ell}^{o,i,\tau}]^{-1}(A_\ell)\}} \mathbf{1}_{\{\tau_i \geq t\}} \in \mathcal{F}_t^{\mathrm{M}i}.$$

This implies that $\mathcal{M} \subset \mathcal{L}$. The monotone class theorem entails that $\mathcal{L}$ contains all bounded $\sigma(\mathcal{M})$-measurable r.v.s. On the other hand, we have $\mathcal{F}_t^{H_i} \subset \mathcal{L}$ by definition. We next consider

$$\check{\mathcal{M}} := \left\{ \mathbf{1}_A(\omega)\mathbf{1}_B(\omega);\ A \in \mathcal{F}_t^{W^{o,i}},\ B \in \mathcal{F}_t^{H_i} \right\}.$$

It holds that $\check{\mathcal{M}}$ is a multiplicative class and $\check{\mathcal{F}}_t^i = \sigma(\check{\mathcal{M}})$. Then, for any $\eta \in \check{\mathcal{M}}$, we have $\eta = \mathbf{1}_A \mathbf{1}_B$ with $A \in \mathcal{F}_t^{W^{o,i}}$ and $B \in \mathcal{F}_t^{H_i}$. It has been proved that both $\mathbf{1}_A$ and $\mathbf{1}_B$ are in $\mathcal{L}$, and hence

$$\eta \mathbf{1}_{\{\tau_i \geq t\}} = \mathbf{1}_A \mathbf{1}_B \mathbf{1}_{\{\tau_i \geq t\}} = (\mathbf{1}_A \mathbf{1}_{\{\tau_i \geq t\}})(\mathbf{1}_B \mathbf{1}_{\{\tau_i \geq t\}}) \in \mathcal{F}_t^{M_i},$$

which shows that $\eta \in \mathcal{L}$. By the monotone class theorem again, it holds that $\mathcal{L}$ contains all bounded $\check{\mathcal{F}}_t^i$-measurable r.v.s. □

5.3 Martingale Representation Theorem

This section establishes a martingale representation theorem with respect to the $(\mathbb{P}, \mathbb{F}^M)$-martingales W^M and M^M which are defined respectively by (5.7) and (5.9) in the previous section.

The martingale representation theorem is stated as follows:

Theorem 5.1 (Martingale Representation Theorem). *Let $L = (L_t)_{t \in [0,T]}$ be a real-valued $(\mathbb{P}, \mathbb{F}^M)$-square integrable martingale with bounded jumps. Then, there exist two $\mathbb{F}^M$-predictable and square integrable $\alpha^M = (\alpha_t^{M,1}, \ldots, \alpha_t^{M,n})_{t \in [0,T]}^\top$ and $\beta^M = (\beta_t^{M,1}, \ldots, \beta_t^{M,n})_{t \in [0,T]}^\top$ such that, for all $t \in [0, T]$,*

$$L_t = L_0 + \sum_{i=1}^n \int_0^t \alpha_s^{M,i} dW_s^{M,i} + \sum_{i=1}^n \int_0^t \beta_s^{M,i} dM_s^{M,i}. \quad (5.11)$$

Note that, the observable information $\mathbb{F}^M$ is generated by $W^{o,\tau}$ and H, where $W^{o,\tau}$ is a stopped Brownian motion under $\mathbb{P}$. The roadmap for showing Theorem 5.1 is as follows:

- Firstly, we prove a martingale representation with respect to $\mathbb{F}^M$ using an auxiliary probability measure $\mathbb{P}^*$, under which the observed $W^{o,\tau}$ has zero drift and H has the unit default intensity.
- Secondly, we change the measure and establish the martingale representation under the original measure $\mathbb{P}$.

To implement it, fix $t \in [0, T]$, we define

$$\Gamma_u^t := \sum_{i=1}^n \int_t^u (\lambda_i^{-1}(s-) - 1) dM_s^i$$

$$-\sum_{i=1}^{n}\sigma_{i}^{-1}\int_{t}^{u\wedge\tau_{i}^{t}}(\mu_{i}(s)+\lambda_{i}(s))dW_{s}^{i}, \quad \forall t\in[0,T], \qquad (5.12)$$

where we have used the simplified notations $\mu_i(t) := \mu_i(I_t)$ and $\lambda_i(t) := \lambda_i(I_t, H_t)$. We then introduce the following change of measure:

$$\frac{d\mathbb{P}^*}{d\mathbb{P}}\Big|_{\mathcal{F}_T} = \mathcal{E}_T(\Gamma^0), \qquad (5.13)$$

with $\Gamma^0 = (\Gamma_t^0)_{t \in [0,T]}$. Assumption 5.1 guarantees that $\mathbb{P}^* \sim \mathbb{P}$ is a probability measure. Furthermore, W^o is an $\mathbb{F}$-Brownian motion under $\mathbb{P}^*$, while the observed process $W^{o,\tau}$ is a stopped $\mathbb{F}$-Brownian motion. The $\mathbb{F}$-intensity of H is 1, that is, for $i = 1, \ldots, n$,

$$M_t^{*,i} := H_t^i - \int_0^t (1 - H_s^i) ds, \quad t \in [0,T] \qquad (5.14)$$

is an $\mathbb{F}$-martingale of pure jumps (it is in fact also an $\mathbb{F}^M$-martingale). The following result serves as the first step to prove Theorem 5.1.

Lemma 5.2. *Let $L = (L_t)_{t\in[0,T]}$ be a $(\mathbb{P}^*, \mathbb{F}^M)$-square integrable martingale with bounded jumps. Then, there exist two $\mathbb{F}^M$-predictable processes $\alpha^M = (\alpha_t^{M,1}, \ldots, \alpha_t^{M,n})_{t\in[0,T]}^\top$ and $\beta^M = (\beta_t^{M,1}, \ldots, \beta_t^{M,n})_{t\in[0,T]}^\top$ such that, for all $t \in [0,T]$,*

$$L_t = L_0 + \sum_{i=1}^n \int_0^t \alpha_s^{M,i} dW_s^{o,i,\tau} + \sum_{i=1}^n \int_0^t \beta_s^{M,i} dM_s^{*,i}. \qquad (5.15)$$

Proof. Let $\mathcal{L}$ be the family of all bounded $\mathcal{F}_T^M$-measurable r.v.s that can be represented by stochastic integrals w.r.t. $W^{o,\tau}$ and M^*, i.e., $\xi \in \mathcal{L}$ if and only if there exist $\mathbb{F}^M$-predictable processes (α, β) such that

$$\xi = \mathbb{E}^*[\xi] + \sum_{i=1}^n \int_0^T \alpha_s^i dW_s^{o,i,\tau} + \sum_{i=1}^n \int_0^T \beta_s^i dM_s^{*,i}, \qquad (5.16)$$

where $\mathbb{E}^*$ denotes the expectation under $\mathbb{P}^*$. It is easy to see that all constants are in $\mathcal{L}$, and $\mathcal{L}$ is a vector space. Moreover, let us consider non-negative increasing r.v.s $(\xi_k)_{k\geq 1} \subset \mathcal{L}$ such that $\lim_{k\to\infty} \xi_k = \xi$ a.s. and ξ is bounded. Then, the BCT yields that $\xi_k \to \xi$, in $L^2(\Omega)$, as $k \to \infty$. Therefore, for each $k \geq 1$, there exist $\mathbb{F}^M$-predictable processes $(\alpha^{(k)}, \beta^{(k)})$ such that ξ_k admits (5.16). It follows that, for all distinct $k, l \geq 1$,

$$\xi_k - \xi_l = \mathbb{E}^*[\xi_k - \xi_l] + \sum_{i=1}^n \int_0^T (\alpha_s^{(k),i} - \alpha_s^{(l),i}) dW_s^{o,i,\tau}$$

$$+ \sum_{i=1}^{n} \int_{0}^{T} (\beta_s^{(k),i} - \beta_s^{(l),i}) dM_s^{*,i}.$$

Hence, $4\mathbb{E}^*[|\xi_k - \xi_l|^2] \geq \int_0^T \mathbb{E}^*[|\alpha_s^{(k)} - \alpha_s^{(l)}|^2 + |\beta_s^{(k)} - \beta_s^{(l)}|^2]ds$. This gives that $(\alpha^{(k)}, \beta^{(k)})_{k \geq 1}$ is a Cauchy sequence in $L^2(\Omega \times [0,T])$, and there exist $\mathbb{F}^{\mathrm{M}}$-predictable processes (α^*, β^*) such that $(\alpha^{(k)}, \beta^{(k)}) \to (\alpha^*, \beta^*)$ in $L^2(\Omega \times [0,T])$, as $k \to \infty$. Let us define

$$\tilde{\xi} := \mathbb{E}^*[\xi] + \sum_{i=1}^{n} \int_0^T \alpha_s^{*,i} dW_s^{o,i,\tau} + \sum_{i=1}^{n} \int_0^T \beta_s^{*,i} dM_s^{*,i}.$$

It follows that $\xi_k \to \tilde{\xi}$ in $L^2(\Omega)$, as $k \to \infty$. The uniqueness of L^2-limit gives that $\xi = \tilde{\xi}$, and hence $\xi \in \mathcal{L}$.

We next define a multiplicative class of r.v.s by

$$\mathcal{M} := \left\{ \prod_{i=1}^{n} \xi_i;\ \xi_i \in \mathcal{F}_T^{\mathrm{M}i} \text{ is bounded } \forall i = 1, \ldots, n \right\}, \quad (5.17)$$

where we recall that $\mathcal{F}_T^{\mathrm{M}i}$ is defined in Lemma 5.1. It is obvious to have that $\mathcal{F}_T^{\mathrm{M}} = \sigma(\mathcal{M})$. Consider bounded r.v.s $\xi_i \in \mathcal{F}_T^{\mathrm{M}i}$ for $i = 1, \ldots, n$. Since $\mathcal{F}_T^{\mathrm{M}i} \subset \check{\mathcal{F}}_T^i$ for $i = 1, \ldots, n$, the classical martingale representation under $\check{\mathcal{F}}_T^i$ (c.f. Proposition 7.1.3 of Bielecki and Rutkowski (2002)) gives the existence of $\check{\mathbb{F}}^i$-predictable processes $\check{\alpha}_i = (\check{\alpha}_t^i)_{t \in [0,T]}$ and $\check{\beta}_i = (\check{\beta}_t^i)_{t \in [0,T]}$ such that $\xi_i = \mathbb{E}^*[\xi_i] + \int_0^T \check{\alpha}_s^i dW_s^{o,i} + \int_0^T \check{\beta}_i(s) dM_s^{*,i}$. For $i = 1, \ldots, n$, and $t \in [0,T]$, it holds that $W_t^{o,i,\tau}, H_t^i \in \check{\mathcal{F}}_{t \wedge \tau_i}^i$, and hence $\mathcal{F}_T^{\mathrm{M}i} \subset \check{\mathcal{F}}_{T \wedge \tau_i}^i$. Thus

$$\xi_i = \mathbb{E}^*[\xi_i | \check{\mathcal{F}}_{T \wedge \tau_i}^i] = \mathbb{E}^*[\xi_i] + \int_0^{T \wedge \tau_i} \check{\alpha}_s^i dW_s^{o,i} + \int_0^{T \wedge \tau_i} \check{\beta}_s^i dM_s^{*,i}$$

$$= \mathbb{E}^*[\xi_i] + \int_0^{T \wedge \tau_i} \check{\alpha}_s^i dW_s^{o,i,\tau} + \int_0^{T \wedge \tau_i} \check{\beta}_s^i dM_s^{*,i}.$$

By using Lemma 5.1, we have $\alpha_t^i := \check{\alpha}_t^i \mathbf{1}_{\{\tau_i \geq t\}}$ and $\beta_t^i := \check{\beta}_t^i \mathbf{1}_{\{\tau_i \geq t\}}$ are $\mathcal{F}_t^{\mathrm{M}i}$-predictable for $t \in [0,T]$ since $\mathbf{1}_{\{\tau_i \geq t\}}$ is $\mathcal{F}_t^{\mathrm{M}i}$-predictable. Thus, each $\xi_i \in \mathcal{F}_T^{\mathrm{M}i}$ has the representation given by

$$\xi_i = \mathbb{E}^*[\xi_i] + \int_0^T \alpha_s^i dW_s^{o,i,\tau} + \int_0^T \beta_s^i dM_s^{*,i}, \quad \forall i = 1, \ldots, n.$$

We define $\mathbb{F}^{\mathrm{M}}$-predictable processes by $\alpha_t^{\mathrm{M},i} := \prod_{k \neq i} \xi_{t-}^{\mathrm{M},k} \alpha_t^i$ and $\beta_t^{\mathrm{M},i} := \prod_{k \neq i} \xi_{t-}^{\mathrm{M},k} \beta_t^i$ with $\xi_t^{\mathrm{M},i} := \mathbb{E}^*[\xi_i] + \int_0^t \alpha_s^i dW_s^{o,i,\tau} + \int_0^t \beta_s^i dM_s^{*,i}$. The Itô's rule yields that

$$\prod_{i=1}^{n} \xi_i = \mathbb{E}^*\left[\prod_{i=1}^{n} \xi_i\right] + \sum_{i=1}^{n} \int_0^T \alpha_s^{\mathrm{M},i} dW_s^{o,i,\tau} + \sum_{i=1}^{n} \int_0^T \beta_s^{\mathrm{M},i} dM_s^{*,i}. \quad (5.18)$$

The representation (5.18) then implies that $\mathcal{M} \subset \mathcal{L}$ and the monotone class theorem yields that $\mathcal{L}$ contains all bounded $\mathcal{F}_T^M$-measurable r.v.s. Note that the jumps of M^* are bounded. We can hence apply the localization techniques to L and obtain the desired martingale representation under $\mathbb{P}^*$ as stated in (5.15). □

We are now at position to complete the proof of Theorem 5.1.

Proof of Theorem 5.1. Fix $t \in [0,T]$, and we define

$$\Gamma_u^{M,t} := \sum_{i=1}^n \int_t^u (\lambda_i^M(s-)^{-1} - 1) dM_s^{M,i}$$
$$- \sum_{i=1}^n \sigma_i^{-1} \int_t^u (\mu_i^M(s) + \lambda_i^M(s)) dW_s^{M,i}, \quad \forall u \in [t,T]. \quad (5.19)$$

Using Assumption 5.1, the process $\psi(u) := \mathcal{E}_u(\Gamma^{M,t})$ for $u \in [t,T]$, is an $\mathbb{F}^M$-martingale satisfying

$$d\psi(u) = \psi(u-)\left\{\sum_{i=1}^n (\lambda_i^M(u-)^{-1} - 1) dM_u^{M,i}\right.$$
$$\left.- \sum_{i=1}^n \sigma_i^{-1}(\mu_i^M(u) + \lambda_i^M(u)) dW_u^{M,i}\right\}.$$

Consider an arbitrary bounded r.v. $\xi \in \mathcal{F}_T^M$. The process $\zeta_t^{M,*} := \mathbb{E}^*[\psi(T)^{-1}\xi|\mathcal{F}_t^M]$ for $t \in [0,T]$ is a square integrable $(\mathbb{P}^*, \mathbb{F}^M)$-martingale by Assumption 5.1. In light of Lemma 5.2, there exist $\mathbb{F}^M$-predictable processes $\alpha^M = (\alpha_t^{M,1}, \ldots, \alpha_t^{M,n})_{t \in [0,T]}^\top$ and $\beta^M = (\beta_t^{M,1}, \ldots, \beta_t^{M,n})_{t \in [0,T]}^\top$ such that

$$\zeta^{M,*}(T) = \psi(T)^{-1}\xi = \mathbb{E}^*[\psi(T)^{-1}\xi] + \sum_{i=1}^n \int_0^T \alpha_s^{M,i} dW_s^{o,i,\tau}$$
$$+ \sum_{i=1}^n \int_0^T \beta_s^{M,i} dM_s^{*,i}.$$

Hence, we deduce that

$$\xi = \psi(T)\mathbb{E}^*[\psi(T)^{-1}\xi] + \psi(T)\sum_{i=1}^n \int_0^T \alpha_s^{M,i} dW_s^{o,i,\tau}$$
$$+ \psi(T)\sum_{i=1}^n \int_0^T \beta_s^{M,i} dM_s^{*,i}. \quad (5.20)$$

We have additionally that

$$\psi(T)\mathbb{E}^*[\psi(T)^{-1}\xi] = \mathbb{E}^*[\psi(T)^{-1}\xi]$$
$$+ \mathbb{E}^*[\psi(T)^{-1}\xi]\sum_{i=1}^n \int_0^T \psi(s-)(\lambda_i^M(s-)^{-1} - 1)dM_s^{M,i}$$
$$- \mathbb{E}^*[\psi(T)^{-1}\xi]\sum_{i=1}^n \int_0^T \psi(s)\sigma_i^{-1}(\mu_i^M(s) + \lambda_i^M(s))dW_s^{M,i}. \quad (5.21)$$

Integration by parts yields that

$$\psi(T)\sum_{i=1}^n \int_0^T \alpha_s^{M,i}dW_s^{o,i,\tau} = \sum_{i=1}^n \int_0^T \psi(s)\alpha_s^{M,i}dW_s^{M,i} \quad (5.22)$$
$$+ \sum_{j=1}^n \int_0^T \psi(s-)\left(\sum_{i=1}^n \int_0^s \alpha_u^M(u)dW_u^{o,i,\tau}\right)(\lambda_j^M(s-)^{-1} - 1)dM_s^{M,j}$$
$$- \sum_{j=1}^n \int_0^T \psi(s)\left(\sum_{i=1}^n \int_0^s \alpha_u^{M,i}dW_u^{o,i,\tau}\right)\sigma_j^{-1}(\mu_j^M(s) + \lambda_j^M(s))dW_s^{M,j},$$

and

$$\psi(T)\sum_{i=1}^n \int_0^T \beta_s^{M,i}dM_s^{*,i} = \sum_{i=1}^n \int_0^T \psi(s-)\beta_s^{M,i}\lambda_j^M(s-)^{-1}dM_s^{M,i} \quad (5.23)$$
$$+ \sum_{j=1}^n \int_0^T \psi(s-)\left(\sum_{i=1}^n \int_0^{s-} \beta_u^{M,i}dM_u^{*,i}\right)(\lambda_j^M(s-)^{-1} - 1)dM_s^{M,j}$$
$$- \sum_{j=1}^n \int_0^T \psi(s)\left(\sum_{i=1}^n \int_0^s \beta_u^{M,i}dM_u^{*,i}\right)\sigma_j^{-1}(\mu_j^M(s) + \lambda_j^M(s))dW_s^{M,j}.$$

We deduce from (5.20)–(5.23) that, any bounded r.v. $\xi \in \mathcal{F}_T^M$ admits the representation as a stochastic integrals w.r.t $\mathbb{P}$-martingales W^M and M^M. Since the jumps of M^* are bounded, the localization technique can be applied to L and the desired martingale representation under $\mathbb{P}$ in (5.31) follows. □

An implication of Theorem 5.1 is to derive the dynamics of the filter process $p^M = (p^{M,k})_{k \in S_I}$ explicitly.

Proposition 5.1. *The filter process $p^{M,k}$ for $k \in S_I$ defined in (5.6) has the following dynamics, for $t \in [0, T]$,*

$$dp_t^{M,k} = \sum_{j \in S_I} q_{jk}p_t^{M,j}dt + p_{t-}^{M,k}\sum_{i=1}^n \left\{\frac{\lambda_i(k, H_{t-})}{\sum_{l \in S_I}\lambda_i(l, H_{t-})p_{t-}^{M,l}} - 1\right\}dM_t^{M,i} \quad (5.24)$$

$$+ p_t^{M,k} \sum_{i=1}^{n} \left\{ \sigma_i^{-1}(\mu_i(k) + \lambda_i(k, H_t)) - \sum_{l \in S_I} p_t^{M,l} \sigma_i^{-1}(\mu_i(l) + \lambda_i(l, H_t)) \right\} dW_t^{M,i}.$$

Here, we recall that $(\mathbb{P}, \mathbb{F}^M)$-martingales W^M and M^M are defined by (5.7) and (5.9), respectively.

Proof. For $t \in [0,T]$, let us define $\zeta_t^k := \mathbf{1}_{\{I_t = k\}}$ for $k \in S_I$. It is clear that $J_t^k := \zeta_t^k - \zeta_0^k - \int_0^t \sum_{i \in S_I} q_{ik} \zeta_s^i ds$, $t \in [0,T]$, is a $(\mathbb{P}, \mathbb{F})$-martingale with bounded jumps. Taking the $\mathbb{P}$-conditional expectation under $\mathcal{F}_t^M$ on both sides, we obtain $J_t^{M,k} = p_t^{M,k} - p_0^{M,k} - \sum_{i \in S_I} \int_0^t q_{ik} p_s^{M,i} ds$ for $t \in [0,T]$ is a square-integrable $(\mathbb{P}, \mathbb{F}^M)$-martingale with bounded jumps. Theorem 5.1 gives the existence of $\mathbb{F}^M$-predictable processes $\alpha^M = (\alpha_t^{M,1}, \ldots, \alpha_t^{M,n})^\top_{t \in [0,T]}$ and $\beta^M = (\beta_t^{M,1}, \ldots, \beta_t^{M,n})^\top_{t \in [0,T]}$ such that, for all $t \in [0,T]$,

$$J_t^{M,k} = J_0^{M,k} + \sum_{i=1}^{n} \int_0^t \alpha_s^{M,i} dW_s^{M,i} + \sum_{i=1}^{n} \int_0^t \beta_s^{M,i} dM_s^{M,i}.$$

Hence, we deduce that

$$p_t^{M,k} = p_0^{M,k} + \sum_{j \in S_I} \int_0^t q_{jk} p_s^{M,j} ds + \sum_{i=1}^{n} \int_0^t \alpha_s^{M,i} dW_s^{M,i}$$
$$+ \sum_{i=1}^{n} \int_0^t \beta_s^{M,i} dM_s^{M,i}. \tag{5.25}$$

We next identify α^M and β^M by taking $W^{o,\tau}$ defined by (5.4) as a test process. We have by using (5.7) that $W_t^{o,i,\tau} = W_t^{M,i} + \sigma_i^{-1} \int_0^{t \wedge \tau_i} (\mu_i^M(p_s^M) + \lambda_i^M(p_s^M, H_s)) ds$ for $t \in [0,T]$ which is $\mathbb{F}^M$-adapted. Then, for $i = 1, \ldots, n$,

$$(\zeta_t^k W_t^{o,i,\tau})^M = \zeta_t^{M,k} W_t^{o,i,\tau} = p_t^{M,k} W_t^{o,i,\tau}, \quad \forall k \in S_I. \tag{5.26}$$

Note that J_k is a semimartingale of pure jumps while $W_i^{o,i,\tau}$ is continuous. It is clear that $[\zeta^k, W^{o,i,\tau}] = [J^k, W^{o,i,\tau}] \equiv 0$. Using integration by parts, we arrive at

$$\zeta_t^k W_t^{o,i,\tau} = \int_0^t W_s^{o,i,\tau} \sum_{j \in S_I} q_{jk} \zeta_s^j ds + \int_0^t W_s^{o,i,\tau} dJ_s^k + \int_0^{t \wedge \tau_i} \zeta_s^k dW_s^i$$
$$+ \sigma_i^{-1} \int_0^{t \wedge \tau_i} (\mu_i(k) + \lambda_i(k, H_s)) \zeta_s^k ds. \tag{5.27}$$

Since both $W^{o,i,\tau}$ and J_k are square-integrable semimartingales under $\mathbb{P}$, the 2nd and 3rd terms on RHS of (5.27) are true $\mathbb{F}$-martingales. Taking the

$\mathbb{P}$-conditional expectation under $\mathbb{F}^{\mathrm{M}}$ on both sides of (5.27), we can write the $\mathbb{F}^{\mathrm{M}}$-semimartingale $(\zeta^k W^{o,i,\tau})^{\mathrm{M}} := (\mathbb{E}[\zeta_t^k W_t^{o,i,\tau} | \mathcal{F}_t^{\mathrm{M}}])_{t \in [0,T]}$ as follows:

$$(\zeta_t^k W_t^{o,i,\tau})^{\mathrm{M}} = \mathbb{E}\left[\int_0^t W_s^{o,i,\tau} dJ_s^k + \int_0^{t \wedge \tau_i} \zeta_s^k dW_s^i \Big| \mathcal{F}_t^{\mathrm{M}}\right] \quad (5.28)$$
$$+ \int_0^t W_s^{o,i,\tau} \sum_{j \in S_I} q_{jk} p_s^{\mathrm{M},j} ds + \sigma_i^{-1} \int_0^{t \wedge \tau_i} (\mu_i(k) + \lambda_i(k, H_s)) p_s^{\mathrm{M},k} ds,$$

where the 1st term on RHS of (5.28) is a $(\mathbb{P}, \mathbb{F}^{\mathrm{M}})$-martingale, and the remaining terms are finite variation processes in the canonical decomposition of $(\zeta_k W^{o,i,\tau})^{\mathrm{M}}$. Additionally, we also have

$$p_t^{\mathrm{M},k} W_t^{o,i,\tau}$$
$$= \int_0^t W_s^{o,i,\tau} \sum_{j \in S_I} q_{jk} p_s^{\mathrm{M},j} ds + \int_0^t W_s^{o,i,\tau} dJ_s^{\mathrm{M},k} + \int_0^{t \wedge \tau_i} p_s^{\mathrm{M},k} dW_s^{\mathrm{M},i}$$
$$+ \sigma_i^{-1} \int_0^{t \wedge \tau_i} p_s^{\mathrm{M},k}(\mu_i^{\mathrm{M}}(p_s^{\mathrm{M}}) + \lambda_i^{\mathrm{M}}(p_s^{\mathrm{M}}, H_s)) ds + \int_0^{t \wedge \tau_i} \alpha_s^{\mathrm{M},i} ds,$$

where the 2nd and 3rd terms of RHS of the above display are true $\mathbb{F}^{\mathrm{M}}$-martingale due to the square integrability of $W^{o,i,\tau}$ and $p^{\mathrm{M},k}$. In light of (5.26), we can compare the finite variation parts of $(\zeta_t^k \phi_t^i)^{\mathrm{M}}$ and $p_t^{\mathrm{M},k} W_t^{o,i,\tau}$ to obtain that, on $\{0 < t \leq \tau_i\}$,

$$\alpha_t^{\mathrm{M},i} = \sigma_i^{-1} p_t^{\mathrm{M},k} \left\{ \mu_i(k) + \lambda_i(k) - \mu_i^{\mathrm{M}}(p_t^{\mathrm{M}}) - \lambda_i^{\mathrm{M}}(p_t^{\mathrm{M}}, H_t) \right\}$$
$$= \sigma_i^{-1} p_t^{\mathrm{M},k} \left\{ (\mu_i(k) + \lambda_i(k, H_t) - \sum_{j \in S_I} \mu_i(j) p_t^{\mathrm{M},j} - \sum_{j \in S_I} \lambda_i(j, H_s) p_t^{\mathrm{M},j} \right\}.$$

Finally, we replace the test process $W^{o,i,\tau}$ by the test process H_t^i. Note that the Markov chain I do not jump simultaneously with the default indicator process H. It holds that $[\zeta^k, H^i] \equiv 0$. By applying a similar argument to identify α^{M}, one can show that, on $\{0 < t \leq \tau_i\}$,

$$\beta_t^{\mathrm{M},i} = \lambda_i^{\mathrm{M}}(p_{t-}^{\mathrm{M}}, H_{t-})^{-1} p_{t-}^{\mathrm{M},k} \lambda_i(k, H_{t-}) - p_{t-}^{\mathrm{M},k}$$
$$= p_{t-}^{\mathrm{M},k} \left\{ \frac{\lambda_i(k, H_{t-})}{\sum_{l \in S_I} \lambda_i(l, H_{t-}) p_{t-}^{\mathrm{M},l}} - 1 \right\}.$$

By substituting $(\alpha^{\mathrm{M}}, \beta^{\mathrm{M}})$ in (5.25), we obtain the desired dynamics given in (5.24). □

5.4 General Correlation Volatility Matrix

In the price dynamics (5.1), the volatility matrix σ is assumed to be diagonal, i.e., all defaultable assets are driven by *independent* Brownian motions. This setting can actually be relaxed as shown in this section.

Consider the following price dynamics of defaultable asset i given by

$$d\tilde{P}_t^i = \tilde{P}_{t-}^i \left\{ \mu_i(I_t)dt + \sum_{j=1}^n \sigma_{ij} dW_t^j - dM_t^i \right\}, \quad i = 1, \ldots, n. \quad (5.29)$$

Here, the volatility matrix $\sigma = (\sigma_{ij}) \in \mathbb{R}^{n \times n}$ is non-diagonal. We next transfer (5.29) into the one with a diagonal volatility matrix, but noises are no longer independent. To do it, we define

$$\tilde{W}_t^i := \tilde{\sigma}_i^{-1} \sum_{k=1}^n \sigma_{ik} W_t^k, \quad \forall t \in [0, T],$$

where $\tilde{\sigma}_i := \sqrt{\sum_{k=1}^n \sigma_{ik}^2}$ for $i = 1, \ldots, n$. Then, $\tilde{W}_i = (\tilde{W}_t^i)_{t \in [0,T]}$ is a Brownian motion with correlation $\langle \tilde{W}^i, \tilde{W}^j \rangle_t = \tilde{\sigma}_i^{-1} \tilde{\sigma}_j^{-1} \sum_{k=1}^n \sigma_{ik} \sigma_{jk} t$ for $i \neq j$. Then, the price process (5.29) can be rewritten as:

$$d\tilde{P}_t = \text{diag}(\tilde{P}_{t-})\{\mu(I_t)dt + \tilde{\sigma} d\tilde{W}_t - dM_t\}. \quad (5.30)$$

Here, $\tilde{\sigma} := \text{diag}(\tilde{\sigma}_1, \ldots, \tilde{\sigma}_n)$ is still diagonal and $\tilde{W} = (\tilde{W}^1, \ldots, \tilde{W}^n)^\top$ is an n-dimensional correlated Brownian motion. Thus, we can still consider the price dynamics (2.55) with correlated Brownian motions $(W^1, \ldots, W^n)$. Using the approximation argument and the monotone class theorem, Lemma 5.1 still holds. However, it will be difficult to prove Lemma 5.2 and Theorem 5.1 when $(W^1, \ldots, W^n)$ are *not* independent.

Recall that the proof of Lemma 5.1 is based on the filtration generated by the price process and the default event of every asset i (i.e., the sub-filtration $\mathcal{F}_t^{Mi} := \mathcal{F}_t^{W^{o,i,\tau}} \vee \mathcal{F}_t^{H^i}$ for $t \in [0, T]$). When $(W^1, \ldots, W^n)$ are independent, we first establish the martingale representation result under each sub-filtration $\mathcal{F}_T^{Mi}$. That is, for $i = 1, \ldots, n$, any bounded r.v. $\xi_i \in \mathcal{F}_T^{Mi}$ admits the following representation:

$$\xi_i = \mathbb{E}^*[\xi_i] + \int_0^T \alpha_s^i dW_s^{o,i,\tau} + \int_0^T \beta_s^i dM_s^{*,i},$$

where α_i and β_i are $(\mathcal{F}_t^{Mi})_{t \in [0,T]}$-predictable processes. Thus, integration by parts can be applied to yield a general representation result under the filtration $\mathcal{F}_T^M$, as the underlying driving martingales $(W^{o,i,\tau}, M^{*,i})$ are orthogonal for $i = 1, \ldots, n$, and hence Lemma 5.2 can be proved by the approximation scheme and the monotone class theorem. On the other hand, if $(W^1, \ldots, W^n)$ are *not* independent, the orthogonality of these martingales does not hold.

However, we can still make the same conclusion using an alternative argument. For $i = 1, \ldots, n$, under each $\mathbb{F}^{Mi} = (\mathcal{F}_t^{Mi})_{t \in [0,T]}$, it first follows from the same techniques used in Lemma 5.1, Theorem 5.1, and Lemma 5.2 with independent $(W^1, \ldots, W^n)$ that for any real-valued $\mathbb{F}^{Mi}$-square integrable $(\mathbb{P}, \mathbb{F}^{Mi})$-martingale $L = (L_t)_{t \in [0,T]}$ with bounded jumps, there exist $\mathbb{F}^{Mi}$-predictable and square integrable processes α_i^M and β_i^M such that

$$L_t = L_0 + \int_0^t \alpha_s^{M,i} dW_s^{M,i} + \int_0^t \beta_s^{M,i} dM_s^{M,i}, \quad \forall t \in [0, T]. \tag{5.31}$$

We next prove Theorem 5.1 using Jacod-Yor Theorem (see, e.g. Theorem IV.57 in Protter (2005) or Theorem III.4.29 in Jacod and Shiryaev (2003)). To this purpose, let us consider a filtered probability space $(\Omega, \mathcal{G}, \mathbb{G}, P)$. Let $\mathcal{H}^2$ be the space of $(P, \mathbb{G})$-special semimartingales with finite $\mathcal{H}^2$-norm. The $\mathcal{H}^2$-norm for a special semimartingale with canonical decomposition $X = N + A$ with N (resp. A) being a local P-martingale (resp. a predictable process of finite variation under P) is defined by

$$\|X\|_{\mathcal{H}^2} := \left\| [N, N]_T^{1/2} \right\|_{L^2(P)} + \left\| \int_0^T |dA_s| \right\|_{L^2(P)}.$$

Let $\mathcal{A} \subset \mathcal{H}^2$, which contains constant martingales. Denote by $\mathcal{S}(\mathcal{A})$ the stable subspace of stochastic integrals generated by $\mathcal{A}$, and $\mathcal{M}(\mathcal{A})$ the space of probability measures making all elements of $\mathcal{A}$ square integrable martingales. We consider the space

$$\mathcal{A} = \left\{ W^{M,1}, \ldots, W^{M,n}, M^{M,1}, \ldots, M^{M,n} \right\}, \quad \mathcal{G} = \mathcal{F}_T^M.$$

It is easy to see that $\mathbb{P} \in \mathcal{M}(\mathcal{A})$. In lieu of Theorem IV.57 in Protter (2005), establishing the martingale representation property is equivalent to show that $\mathbb{P}$ is an extremal point of $\mathcal{M}(\mathcal{A})$, i.e., for any given probability measures $\mathbb{Q}, \mathbb{K} \in \mathcal{M}(\mathcal{A})$ satisfying

$$\lambda \mathbb{Q} + (1 - \lambda) \mathbb{K} = \mathbb{P}, \quad \exists \lambda \in [0, 1], \tag{5.32}$$

it holds that $\mathbb{Q} = \mathbb{K} = \mathbb{P}$. For $i = 1, \ldots, n$, let us consider

$$\mathcal{G}_i = \mathcal{F}_T^{Mi}, \quad \mathcal{A}_i = \left\{ W^{M,i}, M^{M,i} \right\}.$$

Let $\mathbb{P}_i$, $\mathbb{Q}_i$ and $\mathbb{K}_i$ be the restriction of $\mathbb{P}$, $\mathbb{Q}$ and $\mathbb{K}$ on $\mathcal{G}_i$, respectively. Consequently, $\mathbb{P}_i$, $\mathbb{Q}_i$ and $\mathbb{K}_i \in \mathcal{M}(\mathcal{A}_i)$ for $i = 1, \ldots, n$ and $\mathbb{P}_i$ is an extremal point of $\mathcal{M}(\mathcal{A}_i)$. On the other hand, it follows from (5.32) that

$$\lambda \mathbb{Q}_i + (1 - \lambda) \mathbb{K}_i = \mathbb{P}_i, \quad \exists \lambda \in [0, 1].$$

Since $\mathbb{P}_i$, $\mathbb{Q}_i$ and $\mathbb{K}_i$ are the restriction of $\mathbb{P}$, $\mathbb{Q}$ and $\mathbb{K}$ on $\mathcal{G}_i$, it holds that $\mathbb{Q}_i = \mathbb{K}_i = \mathbb{P}_i$ for $i = 1, \ldots, n$. Recall that $\mathcal{F}_T^M = \bigvee_{i=1}^n \mathcal{F}_T^{Mi}$ and $\mathbb{Q} = \mathbb{K} = \mathbb{P}$ on $\mathcal{F}_T^{Mi}$ for $i = 1, \ldots, n$, we have that $\mathbb{Q} = \mathbb{K} = \mathbb{P}$ on $\mathcal{G}$, which shows Theorem 5.1 when $(W^1, \ldots, W^n)$ are *not* independent.

5.5 Risk Sensitive Control under Partial Information

This section formulates the risk-sensitive portfolio optimization under the partial information $\mathbb{F}^M$. Let us first introduce the prime value function and transform it into an equivalent objective functional using the martingale representation result in Section 5.3 and change of measure. This formulation, together with the appropriate set of admissible trading strategies, can connect the control problem to a non-standard quadratic BSDE with jumps.

Let $\pi = (\pi_t^i; i = 1, \ldots, n)_{t \in [0,T]}^\top$ be an $\mathbb{F}^M$-predictable process, which represents the vector of proportions of wealth invested in n defaultable assets with price $\tilde{P}$ under partial observations. The resulting wealth process $X^\pi = (X_t^\pi)_{t \in [0,T]}$ evolves as follows:

$$dX_t^\pi = rX_t^\pi dt + X_{t-}^\pi \pi_t^\top \{(\mu(I_t) - r\mathbf{1}_n)dt + \sigma dW_t - dM_t\}, \qquad (5.33)$$

where $r \geq 0$ is the constant interest rate. Since the price of the i-th asset jumps to zero when it defaults by (2.55), the corresponding fraction of wealth held by the investor in this asset stays at zero after it defaults. It consequently follows that $\pi_t^i = (1 - H_{t-}^i)\pi_t^i$ for $i = 1, \ldots, n$.

We next introduce the admissible set of all candidate dynamic investment strategies.

Definition 5.1 (Admissible Control Set). For $t \in [0, T]$, $\mathcal{U}_t^{ad}$ denotes the set of admissible controls $\pi_s = (\pi_s^i; i = 1, \ldots, n)^\top$, $s \in [t, T]$, which are $\mathbb{F}^M$-predictable processes such that SDE (5.33) has a unique positive strong solution with $X_t^\pi = x \in \mathbb{R}_+$ and $(\mathcal{E}_u(\Lambda^{\pi,t}))_{u \in [t,T]}$ is a true $(\mathbb{P}^*, \mathbb{F}^M)$-martingale, where $\mathbb{P}^*$ is given by (5.13) and $\Lambda^{\pi,t}$ is defined later by (5.53). It also follows that the process π should take values in $U := (-\infty, 1)^n$.

The constraint on admissible investment strategies with the martingale property is by no means restrictive. It will be shown in Section 5.7 that the first-order condition leads to the optimal solution $\pi^* \in \mathcal{U}_t^{ad}$ since $(\mathcal{E}(\Lambda_u^{\pi^*,t}))_{u \in [t,T]}$ can be verified to be a $(\mathbb{P}^*, \mathbb{F}^M)$-martingale. This additional constraint on admissibility can facilitate our future transformation of the original control problem into a simplified form. For $\pi \in \mathcal{U}_t^{ad}$, the

Risk Sensitive Credit Portfolio Optimization: Partial Information 127

wealth process can be rewritten equivalently by

$$X_T^\pi = X_t^\pi \exp\left\{\int_t^T [r + \pi_s^\top(\mu(I_s) - r\mathbf{1}_n)]ds + \int_t^T \pi_s^\top \sigma dW_s \right.$$
$$\left. - \frac{1}{2}\int_t^T \pi_s^\top \sigma\sigma^\top \pi_s ds + \sum_{i=1}^n \int_t^T \ln(1-\pi_s^i)dM_s^i \right. \quad (5.34)$$
$$\left. + \sum_{i=1}^n \int_t^T \lambda_i(I_s, H_s)(1-H_s^i)[\pi_s^i + \ln(1-\pi_s^i)]ds \right\}.$$

For $\pi \in \mathcal{U}_0^{ad}$ and $(X_0^\pi, H_0) = (x, h) \in \mathbb{R}_+ \times S_H$, the risk-sensitive objective functional is given by

$$\tilde{J}(\pi; x, h) := -\frac{2}{\theta}\ln\mathbb{E}\left[\exp\left(-\frac{\theta}{2}\ln X_T^\pi\right)\right], \quad \theta > 0. \quad (5.35)$$

The investor seeks to maximize $\tilde{J}$ over all admissible strategies $\pi \in \mathcal{U}_0^{ad}$. Then, the value function of the control problem is

$$\tilde{V}(x, h) := \sup_{\pi \in \mathcal{U}_0^{ad}} \left\{-\frac{2}{\theta}\ln\mathbb{E}\left[\exp\left(-\frac{\theta}{2}\ln X_T^\pi\right)\right]\right\}$$
$$= \sup_{\pi \in \mathcal{U}_0^{ad}} \left\{-\frac{2}{\theta}\ln\mathbb{E}\left[(X_0^\pi)^{-\frac{\theta}{2}}\left(\frac{X_T^\pi}{X_0^\pi}\right)^{-\frac{\theta}{2}}\right]\right\}$$
$$= \ln x - \frac{2}{\theta}\inf_{\pi \in \mathcal{U}_0^{ad}}\left\{\ln\mathbb{E}\left[\left(\frac{X_T^\pi}{X_0^\pi}\right)^{-\frac{\theta}{2}}\right]\right\}$$
$$= \ln x - \frac{2}{\theta}\ln\left\{\inf_{\pi \in \mathcal{U}_0^{ad}}\mathbb{E}\left[\left(\frac{X_T^\pi}{X_0^\pi}\right)^{-\frac{\theta}{2}}\right]\right\}. \quad (5.36)$$

The control problem is then transformed to $\inf_{\pi \in \mathcal{U}_0^{ad}} \mathbb{E}[(X_T^\pi/X_0^\pi)^{-\frac{\theta}{2}}]$. Hence, for $(t, p, h) \in [0, T] \times S_{p^M} \times S_H$, it is equivalent to study the dynamic minimization problem:

$$V(t, p, h) := \inf_{\pi \in \mathcal{U}_t^{ad}} J(\pi; t, p, h) := \inf_{\pi \in \mathcal{U}_t^{ad}} \mathbb{E}_{t,p,h}\left[\left(\frac{X_T^\pi}{X_t^\pi}\right)^{-\frac{\theta}{2}}\right], \quad (5.37)$$

where $\mathbb{E}_{t,p,h}[\cdot] := \mathbb{E}[\cdot | p_t^M = p, H_t = h]$ and $\frac{X_T^\pi}{X_t^\pi}$ can be expressed by (5.34).

We next rewrite the objective functional J in (5.37) under $\mathbb{P}^*$. First, it is not difficult to verify that (5.34) is equivalent to

$$\left(\frac{X_T^\pi}{X_t^\pi}\right)^{-\frac{\theta}{2}} = \exp\left\{-\frac{\theta}{2}\int_t^T r(1-\pi_s^\top \mathbf{1}_n)ds - \frac{\theta}{2}\int_t^T \pi_s^\top \sigma dW_s^{o,\tau}\right.$$

$$+\frac{\theta}{4}\int_t^T \pi_s^\top \sigma\sigma^\top \pi_s ds - \frac{\theta}{2}\sum_{i=1}^n \int_t^T \ln(1-\pi_s^i)dH_s^i\Bigg\}, \quad (5.38)$$

where the last equality holds using $\pi_t^i = (1-H_{t-}^i)\pi_t^i$. We note that all terms in (5.38) are $\mathbb{F}^M$-adapted. In view of (5.37), the objective functional is reformulated to

$$J(\pi;t,q,h) = \mathbb{E}_{t,p,h}\left[\left(\frac{X_T^\pi}{X_t^\pi}\right)^{-\frac{\theta}{2}}\right] = \mathbb{E}_{t,p,h}^*\left[\eta^{-1}(t,T)\left(\frac{X_T^\pi}{X_t^\pi}\right)^{-\frac{\theta}{2}}\right]. \quad (5.39)$$

Here, the density process is defined by $\eta(t,u) := \mathcal{E}_u(\Gamma^t)$ with Γ^t given in (5.12) and $u \geq t$, and $\mathbb{E}^*$ denotes the expectation operator under $\mathbb{P}^*$ given in (5.13). Note that $\eta(t,T)$ is not necessarily $\mathbb{F}^M$-adapted due to the presence of I in $\eta(t,T)$. In order to transform J into a fully observable form, let us introduce

$$\eta^M(t,u) := \mathbb{E}[\eta(t,u)|\mathcal{F}_u^M], \quad \forall u \in [t,T]. \quad (5.40)$$

Lemma 5.3. *We have $\eta^M(t,u) = \mathcal{E}_u(\phi^t)$ for $u \in [t,T]$. Here, we defined that*

$$\phi^t(\cdot) := \sum_{i=1}^n \int_t^\cdot (\lambda_i^M(p_{s-}^M, H_{s-}))^{-1} - 1)dM_s^{M,i}$$

$$-\sum_{i=1}^n \int_t^\cdot \sigma_i^{-1}(1-H_s^i)(\mu_i^M(p_s^M) + \lambda_i^M(p_s^M, H_s))dW_s^{M,i}.$$

Proof. It follows by definition that, for $u \in [t,T]$,

$$\frac{d\eta(t,u)}{\eta(t,u-)} = \sum_{i=1}^n (\lambda_i(I_{u-}, H_{u-})^{-1} - 1)dM_u^i$$

$$-\sum_{i=1}^n \sigma_i^{-1}(1-H_u^i)(\mu_i(I_u) + \lambda_i(I_u, H_u))dW_u^i.$$

As in the proof of Proposition 5.1, we still choose $W^{o,i,\tau}$ to be the test process for $i = 1,\ldots,n$. Note that $W^{o,i,\tau}$ is a stopped $\mathbb{F}$-Brownian motion under $\mathbb{P}^*$, we obtain that $\eta^M = (\eta^M(t,u))_{u \in [t,T]}$ and $(\eta W^{o,i,\tau})^M = (\mathbb{E}[\eta(t,u)W_u^{o,i,\tau}|\mathcal{F}_u^M])_{u \in [t,T]}$ are both square-integrable $\mathbb{F}^M$-martingales under $\mathbb{P}$. In light of Theorem 5.1, there exist $\mathbb{F}^M$-predictable processes $\alpha^M = (\alpha_t^{M,1},\ldots,\alpha_t^{M,n})_{t \in [0,T]}^\top$ and $\beta^M = (\beta_t^{M,1},\ldots,\beta_t^{M,n})_{t \in [0,T]}^\top$ such that, for $u \in [t,T]$,

$$\eta^M(t,u) = 1 + \sum_{i=1}^n \int_t^u \alpha_s^{M,i}dW_s^{M,i} + \sum_{i=1}^n \int_t^u \beta_s^{M,i}dM_s^{M,i}. \quad (5.41)$$

Thus, integration by parts gives that

$$\eta^{\mathrm{M}}(t,u)W_u^{o,i,\tau} = W_t^{o,i,\tau} + \int_t^u W_s^{o,i,\tau} d\eta^{\mathrm{M}}(t,s) + \int_t^u \eta^{\mathrm{M}}(t,s) dW_s^{\mathrm{M},i}$$
$$+ \sigma_i^{-1} \int_t^u \eta^{\mathrm{M}}(t,s)(1-H_s^i)(\mu_i^{\mathrm{M}}(s) + \lambda_i^{\mathrm{M}}(s)) ds$$
$$+ \int_t^u (1-H_s^i)\alpha_s^{\mathrm{M},i} ds.$$

Note that the $\mathbb{F}^{\mathrm{M}}$-adapted finite variation part in the canonical decomposition of $(\eta W^{o,i,\tau})^{\mathrm{M}}$ vanishes. Using the equality $(\eta W^{o,i,\tau})^{\mathrm{M}} = \eta^{\mathrm{M}} W^{o,i,\tau}$ and comparing their finite variation parts, we deduce that

$$\alpha_s^{\mathrm{M},i} = -\sigma_i^{-1}\eta^{\mathrm{M}}(t,s)(\mu_i^{\mathrm{M}}(s) + \lambda_i^{\mathrm{M}}(s)), \quad t \leq s \leq \tau_i^t. \quad (5.42)$$

We next choose a test process $\phi_i(t) := H_t^i - t \wedge \tau_i$ for $t \in [0,T]$ to identify β^{M} in (5.41). By Girsanov's theorem, $\eta\phi_i$ is a $(\mathbb{P},\mathbb{F})$-martingale. Then, the $\mathbb{F}^{\mathrm{M}}$-adapted finite variation part of $(\eta\phi_i)^{\mathrm{M}}$ vanishes. Moreover, integration by parts also yields that

$$\eta^{\mathrm{M}}(t,u)\phi_i(u) = \phi_i(t) + \int_t^u \phi_i(s-) d\eta^{\mathrm{M}}(t,s) + \int_t^u (\eta^{\mathrm{M}}(t,s-) + \beta_s^{\mathrm{M},i}) dM_s^{\mathrm{M},i}$$
$$+ \sigma_i^{-1} \int_t^u \eta^{\mathrm{M}}(t,s)(1-H_s^i)(\lambda_i^{\mathrm{M}}(s) - 1) ds$$
$$+ \int_t^u (1-H_s^i)\lambda_i^{\mathrm{M}}(s)\beta_s^{\mathrm{M},i} ds.$$

Comparing the finite variation parts of stochastic processes $(\eta\phi_i)^{\mathrm{M}} = (\mathbb{E}[\eta(t,u)\phi_i(u)|\mathcal{F}_u^{\mathrm{M}}])_{u \in [t,T]}$ and $\eta^{\mathrm{M}}\phi_i = (\eta^{\mathrm{M}}(t,u)\phi_i(u))_{u \in [t,T]}$, we have

$$\beta_s^{\mathrm{M},i} = \eta^{\mathrm{M}}(t,s-)(\lambda_i^{\mathrm{M}}(s-)^{-1} - 1), \quad t \leq s \leq \tau_i^t. \quad (5.43)$$

The proof is finalized by plugging α^{M} into (5.42) and β^{M} in (5.43) back into (5.41). □

We next give the reformulation of the objective functional J in (5.39) under partial information $\mathbb{F}^{\mathrm{M}}$. The proof is straightforward, and hence we omit it.

Lemma 5.4. *Let $\mathbb{P}^*$ be the probability measure defined in (5.13). Then, for all $(\pi;t,p,h) \in \mathcal{U}_t^{ad} \times [0,T] \times S_{p^{\mathrm{M}}} \times S_H$, it holds that*

$$J(\pi;t,p,h) = \mathbb{E}_{t,p,h}\left[\left(\frac{X_T^\pi}{X_t^\pi}\right)^{-\frac{\theta}{2}}\right] = \mathbb{E}_{t,p,h}^*\left[e^{Q_T^{\tau,t}}\right]. \quad (5.44)$$

Here, the $\mathbb{F}^M$-adapted process $Q_u^{\pi,t}$ for $u \in [t,T]$ is defined by

$$Q_u^{\pi,t} := -\frac{r\theta}{2}(u-t) + \sum_{i=1}^n \int_t^u \left\{\sigma_i^{-1}(\mu_i^M(s) + \lambda_i^M(s)) - \frac{\theta\sigma_i}{2}\pi_s^i\right\} dW_s^{o,i,\tau}$$

$$- \sum_{i=1}^n \int_t^u \left\{\frac{\theta}{2}\ln(1-\pi_s^i) - \ln(\lambda_i^M(s-))\right\} dM_s^{*,i}$$

$$+ \sum_{i=1}^n \int_t^{u \wedge \tau_i^t} \left\{1 - \lambda_i^M(s) + \ln(\lambda_i^M(s)) - \frac{1}{2}\sigma_i^{-2}(\mu_i^M(s) + \lambda_i^M(s))^2\right\} ds$$

$$+ \sum_{i=1}^n \int_t^{u \wedge \tau_i^t} \left\{\frac{r\theta}{2}\pi_s^i + \frac{\theta\sigma_i^2}{4}(\pi_s^i)^2 - \frac{\theta}{2}\ln(1-\pi_s^i)\right\} ds, \quad (5.45)$$

where the process $M^* = (M_t^{*,1}, \ldots, M_t^{*,n})_{t \in [0,T]}^\top$ is defined by (5.14).

5.6 Quadratic BSDE with Jumps

This section introduces a quadratic BSDE with jumps associated to the control problem (5.37).

To do it, let $(t, p, h) \in [0, T] \times S_{p^M} \times S_H$, and $(p_t^M, H_t) = (p, h)$. Consider the following BSDE defined on the filtered probability space $(\Omega, \mathcal{F}, \mathbb{F}^M, \mathbb{P}^*)$ with $\mathbb{P}^*$ being defined in (5.13) that

$$\begin{cases} dY_u = f(p_u^M, H_u, Z_u, V_u) du + Z_u^\top dW_u^{o,\tau} + V_u^\top dM_u^*, & u \in [t,T), \\ Y_T = 0, \end{cases} \quad (5.46)$$

where, for $(p, h, \xi, v) \in S_{p^M} \times S_H \times \mathbb{R}^n \times \mathbb{R}^n$, the driver of the BSDE (5.46) is given by

$$f(p, h, \xi, v) := \sup_{\pi \in (-\infty, 1)^n} \mathcal{H}(\pi; p, h, \xi, v), \quad (5.47)$$

in which $h(\pi; p, h, \xi, v)$ is specified as:

$$\mathcal{H}(\pi; p, h, \xi, v) := \mathcal{H}_L(p, h, \xi, v) + \sum_{i=1}^n \mathcal{H}_i(\pi_i; p, h, \xi_i, v_i). \quad (5.48)$$

Here, $\mathcal{H}_L(p, h, \xi, v)$ is a linear strategy-independent function in (ξ, v), which is defined by

$$\mathcal{H}_L(p, h, \xi, v) := -\sum_{i=1}^n (1 - h_i)\xi_i \sigma_i^{-1}(\mu_i^M(p) + \lambda_i^M(p, h))$$

$$+ \sum_{i=1}^n (1 - h_i)v_i + \frac{r\theta}{2}. \quad (5.49)$$

For $i = 1, \ldots, n$, the mapping

$$\mathcal{H}_i(\pi_i; p, h, \xi_i, v_i) := (1 - h_i) \left\{ -\frac{\theta}{4} \sigma_i^2 \pi_i^2 + \frac{\theta}{2} \left(\mu_i^{\mathrm{M}}(p) + \lambda_i^{\mathrm{M}}(p, h) - r \right) \pi_i \right.$$

$$\left. - \frac{1}{2} \left| \frac{\theta}{2} \sigma_i \pi_i - \xi_i \right|^2 + \lambda_i^{\mathrm{M}}(p, h) - \lambda_i^{\mathrm{M}}(p, h)(1 - \pi_i)^{-\frac{\theta}{2}} e^{v_i} \right\}. \quad (5.50)$$

The functions $\mu^{\mathrm{M}}(p)$ and $\lambda^{\mathrm{M}}(p, h)$ are given in (5.8). From this point onwards, we will write the first component Y_u of the solution of the BSDE (5.46) as $Y_u(t, p, h)$ to emphasize its dependence on the initial data (p, h) at time t.

The prime relationship between the value function and the solution of BSDE (5.46) is built in the first verification result on the optimality as below.

Lemma 5.5. *Let (Y, Z, V) be a solution to BSDE (5.46) with initial data $(p_t^{\mathrm{M}}, H_t) = (p, h) \in S_{p^{\mathrm{M}}} \times S_H$ at time t. Then, for any $\pi \in \mathcal{U}_t^{ad}$, it holds that $J(\pi; t, p, h) \geq e^{Y_t(t,p,h)}$. Furthermore, if there exists a process $\pi^* \in \mathcal{U}_t^{ad}$ such that, for $u \in [t, T]$,*

$$\mathcal{H}(\pi_u^*; p_{u-}^{\mathrm{M}}, H_{u-}, Z_u, V_u) = f(p_{u-}^{\mathrm{M}}, H_{u-}, Z_u, V_u), \quad a.s., \quad (5.51)$$

then π^* is an optimal strategy for the risk sensitive control problem (5.36).

Proof. Lemma 5.4 yields that, for $\pi \in \mathcal{U}_t^{ad}$,

$$J(\pi; t, p, h) = \mathbb{E}_{t,p,h} \left[\left(\frac{X_T^\pi}{X_t^\pi} \right)^{-\frac{\theta}{2}} \right] = \mathbb{E}_{t,p,h}^* \left[e^{Q_T^{\pi,t}} \right], \quad (5.52)$$

where $Q_T^{\pi,t}$ is given by (5.45). For $u \in [t, T]$, we define

$$\Lambda_u^{\pi,t} := \sum_{i=1}^n \int_t^u \left\{ \sigma_i^{-1}(\mu_i^{\mathrm{M}}(s) + \lambda_i^{\mathrm{M}}(s)) - \frac{\theta \sigma_i}{2} \pi_s^i + Z_s^i \right\} dW_s^{o,i,\tau}$$

$$+ \sum_{i=1}^n \int_t^u \left\{ (1 - \pi_s)^{-\frac{\theta}{2}} \lambda_i^{\mathrm{M}}(s-) e^{V_s^i} - 1 \right\} dM_s^{*,i}. \quad (5.53)$$

Since (Y, Z, V) solves the BSDE (5.46), a direct calculation yields that

$$J(\pi; t, p, h) e^{-Y_t(t,p,h)} = \mathbb{E}_{t,p,h}^* \left[e^{Q_T^{\pi,t} - Y_t(t,p,h)} \right]$$

$$= \mathbb{E}_{t,p,h}^* \left[\mathcal{E}_T(\Lambda^{\pi,t}) \exp \left(\int_t^T (f(u) - \mathcal{H}(\pi_u; u)) du \right) \right].$$

Here, we used notations $f(u) := f(p_{u-}^M, H_{u-}, Z_u, V_u)$ and $\mathcal{H}(\pi_u; u) := \mathcal{H}(\pi_u; p_{u-}^M, H_{u-}, Z_u, V_u)$. By the definition of f in (5.47), it is easy to see that $f(u) - \mathcal{H}(\pi_u; u) \geq 0$ for all $u \in [t, T]$. Therefore, for all $s \in [t, T]$,

$$e^{Q_s^{\pi,t}} e^{Y_s(t,p,h) - Y_t(t,p,h)} = \mathcal{E}_s(\Lambda^{\pi,t}) \exp\left(\int_t^s (f(u) - \mathcal{H}(\pi_u; u)) du\right)$$
$$\geq \mathcal{E}_s(\Lambda^{\pi,t}). \tag{5.54}$$

Note that, for all $\pi \in \mathcal{U}_t^{ad}$, the process $(\mathcal{E}_s(\Lambda^{\pi,t}))_{s \in [t,T]}$ is a $(\mathbb{P}^*, \mathbb{F}^M)$-martingale by Definition 5.1. This implies that, for any $\pi \in \mathcal{U}_t^{ad}$,

$$J(\pi; t, p, h) e^{-Y_t(t,p,h)} = \mathbb{E}_{t,p,h}^* \left[e^{Q_T^{\pi,t} - Y_t(t,p,h)} \right]$$
$$= \mathbb{E}_{t,p,h}^* \left[\mathcal{E}_T(\Lambda^{\pi,t}) \exp\left(\int_t^T (f(u) - \mathcal{H}(\pi_u; u)) du\right) \right]$$
$$\geq \mathbb{E}_{t,p,h}^* \left[\mathcal{E}_T(\Lambda^{\pi,t}) \right] = 1. \tag{5.55}$$

On the other hand, if (5.51) holds, then $f(u) = h(\pi_u^*; u) = 0$ for $u \in [t, T]$, a.s.. This further entails that the inequality (5.55) holds as an equality. Hence, for all $\pi \in \mathcal{U}_t^{ad}$, we get that $J(\pi; t, p, h) \geq e^{Y_t(t,p,h)} = J(\pi^*; t, p, h)$, which confirms that $\pi^* \in \mathcal{U}_t^{ad}$ is an optimal strategy. $\square$

We next focuses on the existence of solutions to the BSDE (5.46) under the partial information probability space $(\Omega, \mathcal{F}, \mathbb{F}^M, \mathbb{P}^*)$ with $\mathbb{P}^*$ given by (5.13). Toward this end, we introduce the following regularized form of the BSDE (5.46) that

$$\begin{cases} d\tilde{Y}_u = \tilde{f}(p_u^M, H_u, \tilde{Z}_u, \tilde{V}_u) du + \tilde{Z}_u^\top dW_u^{o,\tau} + \tilde{V}_u^\top dM_u^*, \ u \in [t, T), \\ \tilde{Y}_T = \int_t^T f(p_u^M, H_u, 0, 0) du, \end{cases} \tag{5.56}$$

where $\tilde{f}(p, h, \xi, v) := f(p, h, \xi, v) - f(p, h, 0, 0)$, and hence $\tilde{f}(p, h, 0, 0) = 0$ for all $(p, h) \in S_{p^M} \times S_H$. Note that (Y, Z, V) solves (5.46) on $[t, T]$ if and only if $(Y - \int_t^\cdot f(p_u^M, H_u, 0, 0) du, Z, V)$ solves (5.56) on $[t, T]$. Therefore, it suffices to prove the existence of $\mathbb{F}^M$-solutions of the BSDE (5.56) with the random terminal condition.

5.6.1 Formulation of Truncated BSDEs

In this section, we introduce the following truncated BSDE under $(\Omega, \mathbb{F}, \mathbb{F}^M, \mathbb{P}^*)$ as follows, for $N \geq 1$,

$$d\tilde{Y}_u^N = \tilde{f}^N(u, \tilde{Z}_u^N, \tilde{V}_u^N) du + (\tilde{Z}_u^N)^\top dW_u^{o,\tau} + (\tilde{V}_u^N)^\top dM_u^*, \ u \in [t, T),$$

$$\tilde{Y}_T^N = \int_t^T f^N(u,0,0)du. \tag{5.57}$$

For $(\omega, u, \xi, v) \in \Omega \times [t,T] \times \mathbb{R}^n \times \mathbb{R}^n$, the truncated *random* driver $\tilde{f}^N$ is defined by

$$\tilde{f}^N(\omega, u, \xi, v) := f^N(\omega, u, \xi, v) - f^N(\omega, u, 0, 0), \tag{5.58}$$

where, the truncating mapping

$$f^N(\omega, u, \xi, v) := \mathcal{H}_L(p_u^{\mathrm{M}}(\omega), H_u(\omega), \xi)$$
$$+ \sum_{i=1}^n (1 - H_u^i(\omega)) \sup_{\pi_i \in (-\infty, 1)} \mathcal{H}_i^N(\pi_i; p_u^{\mathrm{M}}(\omega), H_u(\omega), \xi, v),$$

$$\mathcal{H}_i^N(\pi_i; p, h, \xi_i, v_i) := -\frac{\theta}{4}\sigma_i^2 \pi_i^2 + \frac{\theta}{2}\left(\mu_i^{\mathrm{M}}(p) + \lambda_i^{\mathrm{M}}(p,h) - r\right)\pi_i \tag{5.59}$$
$$- \frac{1}{2}\left|\frac{\theta}{2}\sigma_i \pi_i - \xi_i\right|^2 \rho_N(\xi_i) + \lambda_i^{\mathrm{M}}(p,h) - \lambda_i^{\mathrm{M}}(p,h)(1-\pi_i)^{-\frac{\theta}{2}}\hat{\rho}_N(e^{v_i}).$$

Here, for $N \geq 1$,

- $\rho_N : \mathbb{R} \to \mathbb{R}_+$ is a chosen truncation function whose first-order derivative is bounded by 1, such that $\rho_N(x) = 1$ if $|x| \leq N$, $\rho_N(x) = 0$ if $|x| \geq N+2$, and $0 \leq \rho_N(x) \leq 1$ if $N \leq |x| \leq N+2$.
- $\hat{\rho}_N : \mathbb{R}_+ \to \mathbb{R}_+$ is chosen as an increasing C^1-function whose first-order derivative is bounded by 1, such that $\hat{\rho}_N(x) = x$, if $0 \leq x \leq N$, $\hat{\rho}_N(x) = N+1$, if $x \geq N+2$, and $N \leq \hat{\rho}(x) \leq N+1$, if $N \leq x \leq N+2$.

We next show that, for each $N \geq 1$, the truncated random driver $\tilde{f}^N(\omega, u, \xi, v)$ is Lipschtiz in $(\xi, v) \in \mathbb{R}^n \times \mathbb{R}^n$ uniformly in $(\omega, u) \in \Omega \times [t,T]$. To this end, we first present the following auxiliary result:

Lemma 5.6. *Let* $(p, h, \xi_i, v_i) \in S_{p^{\mathrm{M}}} \times S_H \times \mathbb{R} \times \mathbb{R}$ *with* $i = 1, \ldots, n$. *For each* $N \geq 1$, *there exists a constant* $R_N > 0$ *depending on* N *only such that*

$$\sup_{\pi_i \in (-\infty, 1)} \mathcal{H}_i^N(\pi_i; p, h, \xi_i, v_i) = \sup_{\pi_i \in [-R_N, 1)} \mathcal{H}_i^N(\pi_i; p, h, \xi_i, v_i). \tag{5.60}$$

Proof. With the aid of (5.59) and Assumption 5.1, we have, for $i = 1, \ldots, n$,

$$\mathcal{H}_i^N(0; p, h, \xi_i, v_i) \geq -\frac{1}{2}|\xi_i|^2 \rho_N(\xi_i)\mathbf{1}_{\{|\xi_i| \leq N+2\}} - \lambda_i^{\mathrm{M}}(p,h)e^{v_i}\hat{\rho}_N(e^{v_i})$$
$$\geq -\left\{\frac{(N+2)^2}{2} + C(N+1)\right\}.$$

Further, for $\pi_i \in (-\infty, 1)$,

$$\mathcal{H}_i^N(\pi; p, h, \xi, v) \leq -\frac{\theta}{4}\sigma_i^2\pi_i^2 + \frac{\theta}{2}\left(\mu_i^M(p) + \lambda_i^M(p,h) - r\right)\pi_i + \lambda_i^M(p,h)$$

$$\leq -\frac{\theta}{4}\sigma_i^2\pi_i^2 + \frac{\theta}{2}(2C+r)|\pi_i| + C.$$

For $i = 1, \ldots, n$, we can take a constant $R_N > 0$ only depending on N such that, for all $\pi_i \in (-\infty, 1)$ satisfying $|\pi_i| > R_N$, we have

$$-\frac{\theta}{4}\sigma_i^2\pi_i^2 + \frac{\theta}{2}(2C+r)|\pi_i| + C < -\left\{\frac{(N+2)^2}{2} + C(N+1)\right\}.$$

Thus, it holds that $\mathcal{H}_i^N(\pi_i; p, h, \xi_i, v_i) < \mathcal{H}_i^N(0; p, h, \xi_i, v_i)$ for all $\pi_i \in (-\infty, -R_N)$. This concludes the validity of the equality (5.60). □

The following result helps to derive *a priori* estimate for the solution of the truncated BSDE (5.57).

Lemma 5.7. *For each $N \geq 1$, the (random) driver $\tilde{f}^N(\omega, u, \xi, v)$ defined by (5.58) is Lipschtiz continuous in $(\xi, v) \in \mathbb{R}^n \times \mathbb{R}^n$ uniformly on $(\omega, u) \in \Omega \times [t, T]$.*

Proof. In light of (5.58), (5.59) and Lemma 5.6, it suffices to prove that, for each $i = 1, \ldots, n$, $\bar{\mathcal{H}}_i^N(p, h, \xi_i, v_i) := \sup_{\pi_i \in [-R_N, 1)} \mathcal{H}_i^N(\pi_i; p, h, \xi_i, v_i)$ is Lipschtizian continuous in $(\xi_i, v_i) \in \mathbb{R} \times \mathbb{R}$ uniformly on $(p, h) \in S_{p^M} \times S_H$. For each $(p, h, \xi_i, v_i) \in S_{p^M} \times S_H \times \mathbb{R} \times \mathbb{R}$, using the first-order condition, the critical point $\pi_i^* = \pi_i^*(p, h, \xi_i, v_i)$ satisfies that

$$\lambda_i^M(p,h)(1-\pi_i^*)^{-\frac{\theta}{2}-1}\hat{\rho}_N(e^{v_i}) \quad (5.61)$$

$$= -\left(1 + \frac{\theta}{2}\rho_N(\xi_i)\right)\sigma_i^2\pi_i^* + \mu_i^M(p) + \lambda_i^M(p,h) - r + \sigma_i\xi_i\rho_N(\xi_i).$$

With the aid of Lemma 5.6 and the strict convexity of $\pi_i \to \mathcal{H}_i^N(\pi_i; p, h, \xi_i, v_i)$, we get $\pi_i^* \in [-R_N, 1)$. Moreover, in view of (5.61), it follows that, the positive term

$$(1-\pi_i^*)^{-\frac{\theta}{2}}\hat{\rho}_N(e^{v_i}) = \frac{1-\pi_i^*}{\lambda_i^M(p,h)}\left[-\left(1+\frac{\theta}{2}\rho_N(\xi_i)\right)\sigma_i^2\pi_i^* + \mu_i^M(p)\right.$$

$$\left. + \lambda_i^M(p,h) - r + \sigma_i\xi_i\rho_N(\xi_i)\right] \leq R_{N,1}, \quad (5.62)$$

where the constant $R_{N,1} > 0$ satisfies that

$$R_{N,1} \geq \frac{1+R_N}{\varepsilon}\max_{i=1,\ldots,n}\left[\left(1+\frac{\theta}{2}\right)\sigma_i^2 R_N + 2C + r + \sigma_i(N+2)\right].$$

The implicit function theorem yields that

$$\frac{\partial}{\partial v_i}\bar{\mathcal{H}}_i^N(p,h,\xi_i,v_i) = \frac{\partial}{\partial v_i}h_i^N(\pi_i^*(p,h,\xi_i,v_i);p,h,\xi_i,v_i)$$

$$= \frac{\partial}{\partial v_i}\mathcal{H}_i^N(\pi_i;p,h,\xi_i,v_i)\Big|_{\pi_i=\pi_i^*(p,h,\xi_i,v_i)}$$

$$+ \frac{\partial \pi_i^*}{\partial v_i}(p,h,\xi_i,v_i)\frac{\partial}{\partial \pi_i}\mathcal{H}_i^N(\pi_i;p,h,\xi_i,v_i)\Big|_{\pi_i=\pi_i^*(p,h,\xi_i,v_i)}$$

$$= \frac{\partial}{\partial v_i}\mathcal{H}_i^N(\pi_i;p,h,\xi_i,v_i)\Big|_{\pi_i=\pi_i^*(p,h,\xi_i,v_i)}$$

$$= -\lambda_i^{\mathrm{M}}(p,h)(1-\pi_i^*)^{-\frac{\theta}{2}}e^{v_i}\hat{\rho}_N'(e^{v_i}).$$

Above, we applied the first-order condition (5.61) for π_i^* in the last equality. Note that the increasing function $\hat{\rho}_N$ enjoys the property that

$$\frac{x\hat{\rho}_N'(x)}{\hat{\rho}_N(x)} = \begin{cases} 1, & \text{if } x \in (0,N], \\ \in [0,\frac{N+2}{N}], & \text{if } x \in [N,N+2], \\ 0, & \text{if } x \geq N+2. \end{cases} \quad (5.63)$$

Taking into account Assumption 5.1 and (5.62), we arrive at

$$\left|\frac{\partial}{\partial v_i}\bar{\mathcal{H}}_i^N(p,h,\xi_i,v_i)\right| = \lambda_i^{\mathrm{M}}(p,h)(1-\pi_i^*)^{-\frac{\theta}{2}}\hat{\rho}_N(e^{v_i})\frac{e^{v_i}\hat{\rho}_N'(e^{v_i})}{\hat{\rho}_N(e^{v_i})}$$

$$\leq R_{N,2}, \quad (5.64)$$

where $R_{N,2} := C\frac{N+2}{N}R_{N,1}$ is a positive constant that only depends on N. Additionally, we have

$$\frac{\partial}{\partial \xi_i}\bar{\mathcal{H}}_i^N(p,h,\xi_i,v_i) = \frac{\partial}{\partial \xi_i}\mathcal{H}_i^N(\pi_i;p,h,\xi_i,v_i)\Big|_{\pi_i=\pi_i^*(p,h,\xi_i,v_i)}$$

$$= \left(\frac{\theta}{2}\sigma_i\pi_i^* - \xi_i\right)\rho_N(\xi_i) - \frac{1}{2}\left|\frac{\theta}{2}\sigma_i\pi_i^* - \xi_i\right|^2 \rho_N'(\xi_i).$$

It then holds that

$$\left|\frac{\partial}{\partial \xi_i}\bar{\mathcal{H}}_i^N(p,h,\xi_i,v_i)\right| \leq \frac{\theta}{2}\sigma_i(R_N \vee 1) + |\xi_i|\rho_N(\xi_i)\mathbf{1}_{|\xi_i|\leq N+2}$$

$$+ \frac{\theta^2}{4}\sigma_i^2(R_N \vee 1)^2 + |\xi_i|^2|\rho_N'(\xi_i)|\mathbf{1}_{|\xi_i|\leq N+2}$$

$$\leq R_{N,3}, \quad (5.65)$$

where the following constant is positive, which only depends on N:

$$R_{N,3} := \max_{i=1,\ldots,n}\left\{\frac{\theta}{2}\sigma_i(R_N \vee 1) + \frac{\theta^2}{4}\sigma_i^2(R_N \vee 1)^2 + N + 2 + (N+2)^2\right\}.$$

By combining (5.64) and (5.65), we obtain the desired result. □

It follows from (5.59) that $f^N(u,0,0) = f(p_u^M, H_u, 0, 0)$ for $u \in [t,T]$. Therefore, the terminal condition of the truncated BSDE (5.57) coincides with the one of the regular BSDE (5.56), i.e.,

$$\zeta := \tilde{Y}_T^N = \tilde{Y}_T, \quad \forall N \geq 1. \tag{5.66}$$

The next auxiliary result claims that this random terminal condition is in fact bounded.

Lemma 5.8. *For fixed $t \in [0,T]$, the random terminal value $\zeta = \int_t^T f(p_s^M, H_s, 0, 0) ds$ is bounded.*

Proof. By virtue of (5.50), we have, for $(p, h, \pi_i) \in S_{p^M} \times S_H \times (-\infty, 1)$,

$$\mathcal{H}_i(\pi_i; p, h, 0, 0) = -\left(\frac{\theta}{4} + \frac{\theta^2}{8}\right)\sigma_i^2 \pi_i^2 + \frac{\theta}{2}(\mu_i^M(p) + \lambda_i^M(p,h) - r)\pi_i$$
$$+ \lambda_i^M(p,h) - \lambda_i^M(p,h)(1-\pi_i)^{-\frac{\theta}{2}}.$$

Using Assumption 5.1, it follows that, $\left|\frac{\theta}{2}(\mu_i^M(p) + \lambda_i^M(p,h) - r)\pi_i\right| \leq \frac{\theta}{4}\{\pi_i^2 + (2C+r)^2\}$. On the other hand, for $\pi_i \in (-\infty, 1)$, it holds that $R_2(\pi_i) \leq \mathcal{H}_i(\pi_i; p, h, 0, 0) \leq R_1$ with $R_1 := \frac{\theta}{4}(2C+r)^2 + \frac{\theta}{4} + C$, and for $\pi_i \in (-\infty, 1)$,

$$R_2(\pi_i) := -\left(\frac{\theta}{4} + \frac{\theta^2}{8} + \frac{\theta}{4\sigma_i^2}\right)\sigma_i^2\pi_i^2 - C(1-\pi_i)^{-\frac{\theta}{2}} - \frac{\theta}{4}(2C+r)^2 + \varepsilon.$$

Note that $R_3 := |\sup_{\pi_i \in (-\infty, 1)} R_2(\pi_i)| < +\infty$. Then, for all $(p,h) \in S_{p^M} \times S_H$,

$$\left|\sup_{\pi_i \in (-\infty, 1)} h_i(\pi_i; p, h, 0, 0)\right| \leq R_1 \vee R_3, \quad \forall i = 1, \ldots, n.$$

Thanks to (5.48), we deduce that $h_L(p, h, 0, 0) = \frac{r\theta}{2}$ for all $(p,h) \in S_{p^M} \times S_H$. This verifies that ζ is a bounded r.v.. □

Using the martingale representation result in Theorem 5.1, Lemma 5.7 and Lemma 5.8, we next prove that there exists a unique solution of the truncated BSDE (5.57) under Assumption 5.1. In accordance with conventional notations, let us first introduce the following spaces of processes, for fixed $t \in [0,T]$ and $p \in [1, +\infty)$,

- $\mathcal{S}_t^p$: the space of $\mathbb{F}^M$-adapted r.c.l.l. real-valued processes $Y = (Y_u)_{u \in [t,T]}$ such that $\mathbb{E}^*[\sup_{u \in [t,T]} |Y_u|^p] < +\infty$.
- $\mathcal{S}_t^\infty$: the space of $\mathbb{F}^M$-adapted r.c.l.l. real-valued processes $Y = (Y_u)_{u \in [t,T]}$ such that $\|Y\|_{t,\infty} := \operatorname{esssup}_{(u,\omega) \in [t,T] \times \Omega} |Y_u(\omega)| < +\infty$.

- L_t^2: the space of $\mathbb{F}^M$-predictable $\mathbb{R}^n$-valued processes $X = (X_u)_{u \in [t,T]}$ such that $\sum_{i=1}^n \mathbb{E}^*[\int_t^{T \wedge \tau_i^t} |X_u^i|^2 du] < +\infty$.

- $\mathbb{H}_{t,\text{BMO}}^2$: the space of $\mathbb{F}^M$-predictable $\mathbb{R}^n$-valued processes $Z = (Z_u)_{u \in [t,T]}$ such that

$$\|Z\|_{t,\text{BMO}}^2 := \sup_{\zeta \in \mathcal{T}_{[t,T]}} \sum_{i=1}^n \mathbb{E}^*\left[\int_\zeta^T (1 - H_u^i)|Z_u^i|^2 du \Big| \mathcal{F}_\zeta^M\right] < +\infty,$$

where $\mathcal{T}_{[t,T]}$ denotes the set of all $\mathbb{F}^M$-stopping times taking values on $[t,T]$.

Then, we have

Lemma 5.9. *For each $N \geq 1$, the truncated BSDE (5.57) has a unique solution $(\tilde{Y}^N, \tilde{Z}^N, \tilde{V}^N) \in \mathcal{S}_t^2 \times L_t^2 \times L_t^2$.*

Proof. We here modify the argument used in Carbone et al. (2008) to fit into our framework. By Lemma 5.7, the driver $\tilde{f}^N$ of the BSDE (5.57) is uniformly Lipschitz. Moreover, the predictable quadratic variation process of $K_s := (W_s^{o,\tau}, M_s^*)$ for $s \in [t,T]$ is given by $\langle K, K \rangle_s = \int_0^s k_u k_u^\top du$ with $k_u := \text{diag}(1 - H_u, 1 - H_u) \in \mathbb{R}^{2n \times 2n}$. Theorem 3.1 in Carbone et al. (2008) implies that, there is a unique $(\tilde{Y}^N, \tilde{Z}^N, \tilde{V}^N) \in \mathcal{S}_t^2 \times L_t^2 \times L_t^2$ and a square integrable $(\mathbb{P}^*, \mathbb{F}^M)$-martingale $U = (U_u)_{u \in [t,T]}$ satisfying $[U, W^{o,i,\tau}]_u = [U, M^{*,i}]_u = 0$ for $u \in [t,T]$ such that

$$\tilde{Y}_T^N - \tilde{Y}_s^N = \int_s^T \tilde{f}^N(u, \tilde{Z}_u^N, \tilde{V}_u^N) du + \int_s^T (\tilde{Z}_u^N)^\top dW_u^{o,\tau} + \int_s^T (\tilde{V}_u^N)^\top dM_u^*$$
$$+ U_T - U_s, \quad \forall s \in [t,T] \tag{5.67}$$

with $\tilde{Y}_T^N = \int_t^T f^N(u, 0, 0) du$. Using the martingale representation in Lemma 5.2, there exist $\alpha, \beta \in L_t^2$ such that

$$U_s = U_t + \sum_{i=1}^n \int_t^s \alpha_u^i dW_u^{o,i,\tau} + \sum_{i=1}^n \int_t^s \beta_u^i dM_u^{*,i}, \quad \forall s \in [t,T].$$

A direct calculation yields that, for all $s \in [t,T]$,

$$[U, U]_s = \sum_{i=1}^n \int_t^s \alpha_u^i d[U, W^{o,i,\tau}]_u + \sum_{i=1}^n \int_t^s \beta_u^i d[U, M^{*,i}]_u = 0.$$

This gives that $U_T - U_s = 0$ for all $s \in [t,T]$, and it follows from (5.67) that $(\tilde{Y}^N, \tilde{Z}^N, \tilde{V}^N) \in \mathcal{S}_t^2 \times L_t^2 \times L_t^2$ is the unique solution of the BSDE (5.57). □

5.6.2 A Priori Estimates of Truncated Solutions

This section establishes *a priori* estimates and a comparison result of the solution to the truncated BSDE (5.57) under Assumption 5.1.

We start with the following simple estimate on the solution of the BSDE depending on N.

Lemma 5.10. *For any $N \geq 1$, let $(\tilde{Y}^N, \tilde{Z}^N, \tilde{V}^N) \in \mathcal{S}_t^2 \times L_t^2 \times L_t^2$ be the solution of (5.57). There exists a constant $R_{T,N} > 0$ depending on N and the bound of $|\zeta|$ such that $d\mathbb{P}^* \times du$-a.e.,*

$$\|\tilde{Y}^N\|_{t,\infty} \leq R_{T,N}, \quad \tilde{V}_u^N \leq R_{T,N}. \tag{5.68}$$

Proof. By applying Itô's rule to $e^{\beta u}|\tilde{Y}_u^N|^2$ with a constant $\beta \in \mathbb{R}$ to be determined, we get that, for all $u \in [t,T]$,

$$e^{\beta T}\zeta - e^{\beta u}\left|\tilde{Y}_u^N\right|^2 = \int_u^T \beta e^{\beta s}\left|\tilde{Y}_s^N\right|^2 ds + 2\int_u^T e^{\beta s}\tilde{Y}_s^N \tilde{f}^N(s, \tilde{Z}_s^N, \tilde{V}_s^N) ds$$
$$+ 2\int_u^T e^{\beta s}\tilde{Y}_s^N(\tilde{Z}_s^N)^\top dW_s^{o,\tau} - 2\sum_{i=1}^n \int_u^{T\wedge\tau_i^u} e^{\beta u}\tilde{Y}_s^N \tilde{V}_s^{N,i} ds$$
$$+ \sum_{i=1}^n \int_u^T e^{\beta s}\{|\tilde{Y}_{s-}^N + \tilde{V}_s^{N,i}|^2 - |\tilde{Y}_{s-}^N|^2\}dH_s^i$$
$$+ \sum_{i=1}^n \int_u^{T\wedge\tau_i^u} e^{\beta s}\left|\tilde{Z}_s^{N,i}\right|^2 ds. \tag{5.69}$$

By rearranging terms on both sides of (5.69), one has

$$e^{\beta u}\left|\tilde{Y}_u^N\right|^2 + \int_u^T \beta e^{\beta s}\left|\tilde{Y}_s^N\right|^2 ds + \sum_{i=1}^n \int_u^{T\wedge\tau_i^u} e^{\beta s}\left|\tilde{Z}_s^{N,i}\right|^2 ds \tag{5.70}$$
$$= e^{\beta T}\zeta - 2\int_u^T e^{\beta s}\tilde{Y}_s^N \tilde{f}^N(s, \tilde{Z}_s^N, \tilde{V}_s^N) ds - \sum_{i=1}^n \int_u^{T\wedge\tau_i^u} e^{\beta s}\left|\tilde{V}_s^{N,i}\right|^2 ds$$
$$- 2\int_u^T e^{\beta s}\tilde{Y}_s^N(\tilde{Z}_s^N)^\top dW_s^{o,\tau} - \sum_{i=1}^n \int_u^T e^{\beta s}\{2\tilde{Y}_{s-}^N \tilde{V}_s^{N,i} + |\tilde{V}_s^{N,i}|^2\}dM_s^{*,i}.$$

By taking into account (5.48) and (5.59), the random driver $\tilde{f}^N(u,\xi,v)$ satisfies that $\tilde{f}^N(u,\xi,v) = \tilde{f}^N(u,(1-H_u)\xi,(1-H_u)v)$. Using Lemma 5.7, there is a constant $L_N > 0$ depending only on N such that, for all $\epsilon > 0$,

$$\left|2\int_u^T e^{\beta s}\tilde{Y}_s^N \tilde{f}^N(s, \tilde{Z}_s^N, \tilde{V}_s^N) ds\right|$$

$$\leq 2L_N \sum_{i=1}^{n} \int_{u}^{T\wedge\tau_i^u} e^{\beta s} \left|\tilde{Y}_s^N\right| (|\tilde{Z}_s^{N,i}| + |\tilde{V}_s^{N,i}|) ds \tag{5.71}$$

$$\leq n\epsilon^{-1} L_N \int_{u}^{T} e^{\beta s} \left|\tilde{Y}_s^N\right|^2 ds + 2\epsilon L_N \sum_{i=1}^{n} \int_{u}^{T\wedge\tau_i^u} e^{\beta s}(|\tilde{Z}_s^{N,i}|^2 + |\tilde{V}_s^{N,i}|^2) ds.$$

Taking $\epsilon = (4L_N)^{-1}$ and $\beta = n\epsilon^{-1} L_N$, we obtain from (5.70) and (5.71) that $e^{\beta u}|\tilde{Y}_u^N|^2 \leq \mathbb{E}^*[e^{\beta T}|\zeta|^2|\mathcal{F}_u^M]$, a.s. for $u \in [t,T]$. Thanks to Lemma 5.8, it follows that $\|\tilde{Y}^N\|_{t,\infty} \leq e^{\beta T}\|\zeta\|_{0,\infty}$, which proves the 1st term in (5.68).

In light of $\Delta \tilde{Y}_u^N = (\tilde{V}_u^N)^\top \Delta M_u^*$, we obtain $|\tilde{V}^N(u)^\top \Delta M^*(u)| \leq 2\|\tilde{Y}^N\|_{t,\infty}$. The fact that $\Delta M_u^{*,i} \in \{0,1\}$ for all $i = 1,\ldots,n$ leads to that $(\tilde{V}_u^N)^\top \Delta M_u^* = (\tilde{V}_u^N)^\top \Delta M^*(u)$. For $i = 1,\ldots,n$, we define

$$\hat{V}_u^{N,i} := \tilde{V}_u^{N,i} \wedge (2\|\tilde{Y}^N\|_{t,\infty}) \vee (-2\|\tilde{Y}^N\|_{t,\infty}). \tag{5.72}$$

Then, the stochastic integral $(\tilde{V}^N - \hat{V}^N) \cdot M^*$ is a continuous martingale of finite variation, which implies that $(\tilde{V}^N - \hat{V}^N) \cdot M^* \equiv 0$. Hence, it follows from $[(\tilde{V}^N - \hat{V}^N) \cdot M^*] \equiv 0$ that

$$(1 - H_u)\tilde{V}_u^N = (1 - H_u)\hat{V}_u^N, \quad d\mathbb{P}^* \times du\text{-a.e.}. \tag{5.73}$$

Here, for any $\alpha \in \mathbb{R}^n$, $(1 - H_u)\alpha := ((1 - H_u^1)\alpha_1, \ldots, (1 - H_u^n)\alpha_n)^\top$. Therefore, $(\tilde{Y}^N, \tilde{Z}^N, \hat{V}^N)$ also solves the BSDE (5.57) in view of (5.73). Since $\hat{V}^N \in L_t^2$, the uniqueness of solution in Lemma 5.9 entails that $\tilde{V}_u^N = \hat{V}_u^N$, $d\mathbb{P}^* \otimes du$-a.e., which completes the proof of the estimate (5.68). □

The following result improves the estimation by establishing a uniform bound of $(\tilde{Y}^N, \tilde{Z}^N, \tilde{V}^N)_{N\geq 1}$, which is independent of N. In particular, the BMO property plays an important role in the proof of the verification theorem.

Lemma 5.11. *For any $N \geq 1$, let $(\tilde{Y}^N, \tilde{Z}^N, \tilde{V}^N) \in \mathcal{S}_t^2 \times L_t^2 \times L_t^2$ be the solution of (5.57). There is some constant $C_T > 0$ depending on the bound of $|\zeta|$ such that, $d\mathbb{P}^* \otimes du$-a.e.,*

$$\|\tilde{Z}^N\|_{t,\mathrm{BMO}} \vee \|\tilde{Y}^N\|_{t,\infty} \leq C_T, \quad \tilde{V}_u^N \leq C_T. \tag{5.74}$$

Proof. The key step of the proof is to construct an equivalent probability measure under which $\tilde{Y}^N = (\tilde{Y}_u^N)_{u\in[t,T]}$ is an $(\mathbb{P}^*, \mathbb{F}^M)$-martingale. By applying Lemma 5.8, the boundedness of $\tilde{Y}^N$ follows by the martingale property of $\tilde{Y}^N = (\tilde{Y}_u^N)_{u\in[t,T]}$ under the new probability measure and the fact that $\hat{Y}_T^N = \zeta$ is bounded. It follows from Lemma 5.10 that, there exists an $\mathbb{F}^M$-predictable $\mathbb{R}^n$-valued (bounded) process $\hat{V}^N$ defined in (5.72) such that $\mathbb{P}^* \otimes du$-a.e., $(1 - H_u)\tilde{V}_u^N = (1 - H_u)\hat{V}_u^N$.

To establish the aforementioned equivalent probability measure, we define, for $i = 1, \ldots, n$,
$$\bar{Z}_u^{N,i} := (\tilde{Z}_u^{N,1}, \ldots, \tilde{Z}_u^{N,i}, 0, \ldots, 0), \quad \bar{V}_u^{N,i} = (\hat{V}_u^{N,1}, \ldots, \hat{V}_u^{N,i}, 0, \ldots, 0).$$
We also set $\bar{Z}_u^{N,0} = \bar{V}_u^{N,0} = 0$. Consider the following processes:
$$\gamma_i(u) := \begin{cases} \dfrac{\tilde{f}^N(u, \bar{Z}_u^{N,i}, \tilde{V}_u^N) - \tilde{f}^N(u, \bar{Z}_u^{N,i-1}, \tilde{V}_u^N)}{\tilde{Z}_u^{N,i}}, & \text{if } (1 - H_u^i)\tilde{Z}_u^{N,i} \neq 0, \\ 0, & \text{if } (1 - H_u^i)\tilde{Z}_u^{N,i} = 0, \end{cases}$$
and
$$\eta_i(u) := \begin{cases} \dfrac{\tilde{f}^N(u, 0, \bar{V}_u^{N,i}) - \tilde{f}^N(u, 0, \bar{V}_u^{N,i-1})}{\hat{V}_u^{N,i}}, & \text{if } (1 - H_u^i)\hat{V}_u^{N,i} \neq 0, \\ 0, & \text{if } (1 - H_u^i)\hat{V}_u^{N,i} = 0. \end{cases}$$

Note that $\tilde{f}^N(u, 0, 0) = 0$. Then, for all $t \in [0, T]$,
$$\int_t^T (\tilde{Z}_u^N)^\top \gamma(u) du + \int_t^T (\hat{V}_u^N)^\top \eta(u) du = \int_t^T \tilde{f}^N(u, \tilde{Z}_u^N, \hat{V}_u^N) du. \quad (5.75)$$

On the other hand, Lemma 5.8 yields that the $\mathbb{R}^n$-valued process $\gamma = (\gamma(u))_{u \in [0,T]}$ is bounded. Furthermore, Lemma 5.10 states that the $\mathbb{F}^M$-predictable $\mathbb{R}^n$-valued process $\hat{V}^N$ is bounded by some constant $C_{T,N} > 0$ depending on T and N.

We next prove that, there exists some positive constant $\delta_{T,N}$ depending on N such that, for $i = 1, \ldots, n$,
$$-1 + \delta_{T,N} \leq -\eta_i(u) \leq L_N, \quad \text{a.e.}, \quad (5.76)$$
where $L_N > 0$ is the Lipchitz coefficient of the driver $\tilde{f}^N$ (c.f. Lemma 5.7). In fact, if $H_u^i = 1$, then $\eta_i(u) = 0$. It suffices to assume that $H_u^i = 0$. For $\tilde{V}_u^{N,i} \neq 0$, we have from (5.63) that
$$\frac{\tilde{f}^N(u, 0, \bar{V}_u^{N,i}) - \tilde{f}^N(u, 0, \bar{V}_u^{N,i-1})}{\hat{V}_u^{N,i}}$$
$$= \int_0^1 \frac{\partial}{\partial v_i} \tilde{f}^N(u, 0, s\bar{V}_u^{N,i} + (1-s)\bar{V}_u^{N,i-1}) ds$$
$$= 1 - \int_0^1 (1 - \pi_u^{*,i})^{-\frac{\theta}{2}} \hat{\rho}_N(e^{s\hat{V}_u^{N,i}}) \frac{e^{s\hat{V}_u^{N,i}} \hat{\rho}_N'(e^{u\hat{V}_u^{N,i}})}{\hat{\rho}_N(e^{s\hat{V}_u^{N,i}})} ds$$
$$\leq 1 - (1 + R_N)^{-\frac{\theta}{2}} \int_0^{1 \wedge R_{T,N}^{-1} \ln N} \hat{\rho}_N(e^{s\hat{V}_u^{N,i}}) \frac{e^{s\hat{V}_u^{N,i}} \hat{\rho}_N'(e^{s\hat{V}_u^{N,i}})}{\hat{\rho}_N(e^{s\hat{V}_u^{N,i}})} ds$$

$$= 1 - (1+R_N)^{-\frac{\theta}{2}} \int_0^{1 \wedge R_{T,N}^{-1} \ln N} \hat{\rho}_N(e^{s\hat{V}_u^{N,i}}) ds$$

$$\leq 1 - \frac{(1+R_N)^{-\frac{\theta}{2}}}{R_{T,N}} \left\{ 1 - e^{-(R_{T,N} \wedge \ln N)} \right\}$$

$$=: 1 - \delta_{T,N}.$$

Here, the positive constants R_N and $R_{T,N}$ are given in Lemma 5.6 and Lemma 5.10 respectively. We next define the probability measure $\mathbb{Q} \sim \mathbb{P}^*$ by

$$\left.\frac{d\mathbb{Q}}{d\mathbb{P}^*}\right|_{\mathcal{F}_s^{\mathrm{M}}} = \mathcal{E}_s\left(-\int_0^\cdot \gamma(u)^\top dW_u^{o,\tau} - \int_0^\cdot \eta(u)^\top dM_u^*\right). \quad (5.77)$$

In view of (5.76) and the boundedness of $\gamma = (\gamma(s))_{s \in [0,T]}$, $\hat{W}^{o,\tau} = (\hat{W}_s^{o,\tau})_{s \in [0,T]}$ and $\hat{M}^* = (\hat{M}_s^*)_{s \in [0,T]}$ are both $(\mathbb{Q}, \mathbb{F}^{\mathrm{M}})$-martingales, where for $s \in [0,T]$,

$$\hat{W}_s^{o,\tau} := W_s^{o,\tau} + \int_0^s \gamma(u)du, \quad \hat{M}_s^*(s) := M_s^* + \int_0^s \eta(u)du. \quad (5.78)$$

It follows from (5.57) and (5.75) that, for all $u \in [t,T]$,

$$\tilde{Y}_u^N - \tilde{Y}_T^N = -\int_u^T (\tilde{Z}_s^N)^\top d\hat{W}_s^{o,\tau} - \int_u^T (\tilde{V}_s^N)^\top d\hat{M}_s^*. \quad (5.79)$$

Let $\theta_k^t \geq t$ be a localizing sequence as $\mathbb{F}^{\mathrm{M}}$ stopping times satisfying $\lim_{k \to \infty} \theta_k^t = T$, a.s.. Using (5.79), it holds that $\tilde{Y}_u^N = \mathbb{E}^{\mathbb{Q}}[\tilde{Y}_{T \wedge \tau_k}^N | \mathcal{F}_u^{\mathrm{M}}]$ for all $k \geq 1$. Lemma 5.10 and BCT result in $\tilde{Y}_u^N = \mathbb{E}^{\mathbb{Q}}[\zeta | \mathcal{F}_u^{\mathrm{M}}]$ for all $u \in [t,T]$. This, together with Lemma 5.8, implies the uniform bound of $\tilde{Y}^N$, i.e., $\|\tilde{Y}^N\|_{t,\infty} \leq \|\zeta\|_{0,\infty}$.

We again construct $\hat{V}_u^N$ as in (5.72), which gives that $|\hat{V}_u^N| \leq 2\|\tilde{Y}^N\|_{t,\infty}$. We then have $\|\hat{V}^N\|_{t,\infty} \leq 2\|\zeta\|_{0,\infty}$ by the using argument above. Following the same proof of Lemma 5.10, the uniqueness of the solution to the BSDE (5.57) entails the 2nd estimate in (5.74).

We next apply Itô's rule to $e^{\beta \tilde{Y}_u^N}$ on $u \in [t,T]$, where $\beta \in \mathbb{R}$ is a constant to be determined, and get that

$$e^{\beta \zeta} - e^{\beta \tilde{Y}_u^N}$$

$$= \sum_{i=1}^n \int_u^T \{e^{\beta(\tilde{Y}_{s-}^N + \hat{V}_s^{N,i})} - e^{\beta \tilde{Y}_{s-}^N}\} dH_s^i - \sum_{i=1}^n \int_u^{T \wedge \tau_i^u} \beta e^{\beta \tilde{Y}_s^N} \hat{V}_s^{N,i} ds$$

$$+ \int_u^T \beta e^{\beta \tilde{Y}_s^N} \tilde{f}^N(s, \tilde{Z}_s^N, \hat{V}_s^N) ds + \int_u^T \beta e^{\beta \tilde{Y}_s^N} (\tilde{Z}_s^N)^\top dW_s^{o,\tau}$$

$$+ \frac{\beta^2}{2} \sum_{i=1}^{n} \int_{u}^{T \wedge \tau_i^u} e^{\beta \tilde{Y}_s^N} \left| \tilde{Z}_s^{N,i} \right|^2 ds. \tag{5.80}$$

Note that $\|(1-H)\hat{V}^N\|_{t,\infty} \leq 2\|\zeta\|_{0,\infty}$. Then, for all $N \geq 1$ and $s \in [0,T]$, we claim here that, there exist positive constants R_4 and R_5 independent of (N,s) such that

$$\left| \tilde{f}^N(s, Z_s^N, \hat{V}_s^N) \right| \leq R_4 + R_5 \sum_{i=1}^{n} (1-H_s^i) \left| \tilde{Z}_s^{N,i} \right|^2. \tag{5.81}$$

To see this, note that the following estimates are independent of N:

$$-\left| \frac{\theta}{2} \sigma_i \pi_i \right|^2 - \xi_i^2 \leq -\frac{1}{2} \left| \frac{\theta}{2} \sigma_i \pi_i - \xi_i \right|^2 \rho_N(\xi_i) \leq 0,$$

and

$$0 \geq -\lambda_i^{\mathrm{M}}(p,h)(1-\pi_i)^{-\frac{\theta}{2}} \hat{\rho}_N(e^{v_i}) \geq -\lambda_i^{\mathrm{M}}(p,h)(1-\pi_i)^{-\frac{\theta}{2}} e^{v_i}$$
$$\geq -\frac{|\lambda_i^{\mathrm{M}}(p,h)|^2 (1-\pi_i)^{-\theta} + e^{2v_i}}{2}.$$

It then follows that

$$-\xi_i^2 - \frac{1}{2} e^{2v_i} + \mathcal{H}_i^{(1)}(\pi_i; p, h) \leq \mathcal{H}_i^N(\pi_i; p, h, \xi_i, v_i) \leq \mathcal{H}_i^{(2)}(\pi_i; p, h), \tag{5.82}$$

where the lower and upper bound functions are respectively given by

$$\mathcal{H}_i^{(1)}(\pi_i; p, h) := -\frac{\theta}{2} \sigma_i^2 \pi_i^2 + \frac{\theta}{2} \left(\mu_i^{\mathrm{M}}(p) + \lambda_i^{\mathrm{M}}(p,h) - r \right) \pi_i + \lambda_i^{\mathrm{M}}(p,h)$$
$$- \frac{1}{2} |\lambda_i^{\mathrm{M}}(p,h)|^2 (1-\pi_i)^{-\theta},$$

$$\mathcal{H}_i^{(2)}(\pi_i; p, h) := -\frac{\theta}{4} \sigma_i^2 \pi_i^2 + \frac{\theta}{2} \left(\mu_i^{\mathrm{M}}(p) + \lambda_i^{\mathrm{M}}(p,h) - r \right) \pi_i + \lambda_i^{\mathrm{M}}(p,h).$$

Note that $\mathcal{H}_i^{(1)}(\pi_i; p, h)$ and $\mathcal{H}_i^{(2)}(\pi_i; p, h)$ are independent of (N, ξ_i, v_i). As a consequence, under Assumption 5.1, there exists a constant C independent of N such that

$$\sup_{\pi_i \in (-\infty, 1)} |\mathcal{H}_i^{(1)}(\pi_i; p, h)| + \sup_{\pi_i \in (-\infty, 1)} |\mathcal{H}_i^{(2)}(\pi_i; p, h)| \leq C. \tag{5.83}$$

By combining (5.82) and (5.83), we have

$$\left| \sum_{i=1}^{n} (1 - H_u^i(\omega)) \sup_{\pi_i \in (-\infty, 1)} \mathcal{H}_i^N(\pi_i; p_u^{\mathrm{M}}(\omega), H_u(\omega), \xi, v) \right|$$
$$\leq C_1 \sum_{i=1}^{n} (1 - H_u^i(\omega)) \left(\xi_i^2 + \sum_{i=1}^{n} e^{v_i} + 1 \right).$$

Analogously, we have the estimate of $\mathcal{H}_L$ that

$$|\mathcal{H}_L(p,h,\xi,v)| \leq C_2 \sum_{i=1}^{n}(1 - H_u^i(\omega))(\xi_i^2 + |v_i| + 1), \qquad (5.84)$$

where C_2 is independent of N. Plugging (5.83) and (5.84) into (5.59), we obtain

$$|f^N(\omega,u,\xi,v)| \leq C_3 \sum_{i=1}^{n}(1 - H_u^i(\omega))\left(\xi_i^2 + |v_i| + \sum_{i=1}^{n} e^{v_i} + 1\right),$$

in which C_3 is hence independent of N. As a result, we get

$$\left|\tilde{f}^N(s, Z_s^N, \hat{V}_s^N)\right| = \left|f^N(\omega, s, Z_s^N, \hat{V}_s^N) + f^N(\omega, s, 0, 0)\right|$$

$$\leq C_3 \sum_{i=1}^{n}(1 - H_u^i(\omega))\left(|Z_s^{N,i}|^2 + |\hat{V}_s^{N,i}| + \sum_{i=1}^{n} e^{\hat{V}_s^{N,i}} + 1\right)$$

$$+ C_3(n+1)\sum_{i=1}^{n}(1 - H_u^i(\omega)). \qquad (5.85)$$

Therefore, the existence of R_4 and R_5 in the claim (5.81) follows from (5.85) and the fact that $\|(1-H)\hat{V}^N\|_{t,\infty} \leq 2\|\zeta\|_{0,\infty}$. Plugging (5.81) into (5.80) and taking the conditional expectation under $\mathcal{F}_u^M$, we attain that, for all $u \in [t,T]$,

$$\left(\frac{\beta^2}{2} - R_5\beta\right)\sum_{i=1}^{n}\mathbb{E}^*\left[\int_u^{T\wedge\tau_i^u} e^{\beta\tilde{Y}_s^N}\left|\tilde{Z}_s^{N,i}\right|^2 ds\Big|\mathcal{F}_u^M\right] \leq \mathbb{E}^*\left[e^{\beta\zeta}\big|\mathcal{F}_u^M\right] - e^{\beta\tilde{Y}_u^N}$$

$$+ R_4\beta\mathbb{E}^*\left[\int_u^T e^{\beta\tilde{Y}_s^N} ds\Big|\mathcal{F}_u^M\right] + \sum_{i=1}^{n}\mathbb{E}^*\left[\int_u^{T\wedge\tau_i^u} \beta e^{\beta\tilde{Y}_s^N}\hat{V}_s^{N,i} ds\Big|\mathcal{F}_u^M\right]$$

$$- \sum_{i=1}^{n}\mathbb{E}^*\left[\int_u^{T\wedge\tau_i^u} \{e^{\beta(\tilde{Y}_s^N + \hat{V}_s^{N,i})} - e^{\beta\tilde{Y}_s^N}\} ds\Big|\mathcal{F}_u^M\right]. \qquad (5.86)$$

For any constant $R_0 > 0$ independent of N, there exists a constant $\beta_0 > 0$ such that $\frac{\beta_0^2}{2} - R_5\beta_0 = R_0$. Note that each term in RHS of (5.86) is bounded by a positive constant, uniformly in N, say R_6. We then arrive at, a.s.

$$\sum_{i=1}^{n}\mathbb{E}^*\left[\int_u^{T\wedge\tau_i^u} e^{-\beta_0\|\zeta\|_{0,\infty}}\left|\tilde{Z}_s^{N,i}\right|^2 ds\Big|\mathcal{F}_u^M\right]$$

$$\leq \sum_{i=1}^{n}\mathbb{E}^*\left[\int_u^{T\wedge\tau_i^u} e^{\beta_0\tilde{Y}_s^N}\left|\tilde{Z}_s^{N,i}\right|^2 ds\Big|\mathcal{F}_u^M\right] \leq R_0^{-1}R_6.$$

This implies that

$$\sum_{i=1}^{n} \mathbb{E}^* \left[\int_u^T \left| \tilde{Z}_s^{N,i} \right|^2 ds \bigg| \mathcal{F}_u^M \right] \leq e^{\beta_0 \|\zeta\|_{0,\infty}} R_0^{-1} R_6.$$

This concludes the desired estimate (5.74). □

We also state here a comparison result for the truncated BSDE that will be used in forthcoming sections.

Lemma 5.12. *For any $N \geq 1$, let $(\tilde{Y}^N, \tilde{Z}^N, \tilde{V}^N) \in \mathcal{S}_t^2 \times L_t^2 \times L_t^2$ be the solution of (5.57). There exists a constant $N_0 > 0$ such that, for all $u \in [t,T]$, $\tilde{Y}_u^N$ is increasing for all $N \geq N_0$, $\mathbb{P}^*$-a.s..*

Proof. Let $u \in [t,T]$, and we define, for $i = 1, \ldots, n$,

$$\tilde{Z}_u^{N+1,N,i} := (\tilde{Z}_u^{N+1,1}, \ldots, \tilde{Z}_u^{N+1,i}, \tilde{Z}_u^{N,i+1}, \ldots, \tilde{Z}_u^{N,n}),$$
$$\tilde{V}_u^{N+1,N,i} := (\tilde{V}_u^{N+1,1}, \ldots, \tilde{V}_u^{N+1,i}, \tilde{V}_u^{N,i+1}, \ldots, \tilde{V}_u^{N,n}).$$

Here, $\tilde{V}^N$ is the $\mathbb{F}^M$-predictable $\mathbb{R}^n$-valued bounded process satisfying (5.74) in Lemma 5.11. We also set $\tilde{Z}_u^{N+1,N,0} = \tilde{Z}_u^N$, $\tilde{Z}_u^{N+1,N,n} = \tilde{Z}_u^{N+1}$, $\tilde{V}_u^{N+1,N,0} = \tilde{V}_u^N$ and $\tilde{V}_u^{N+1,N,n} = \tilde{V}_u^{N+1}$. For $i = 1, \ldots, n$, let us define that

$$\gamma_i(u) := \frac{\tilde{f}^{N+1}(u, \tilde{Z}_u^{N+1,N,i}, \tilde{V}_u^{N+1}) - \tilde{f}^{N+1}(u, \tilde{Z}_u^{N+1,N,i-1}, \tilde{V}_u^{N+1})}{\tilde{Z}_u^{N+1,i} - \tilde{Z}_u^{N,i}},$$

provided $(1 - H_u^i)\tilde{Z}_u^{N+1,i} \neq (1 - H_u^i)\tilde{Z}_u^{N,i}$, and it is 0 otherwise. We also define

$$\eta_i(u) := \frac{\tilde{f}^{N+1}(u, \tilde{Z}_u^N, \tilde{V}_u^{N+1,N,i}) - \tilde{f}^{N+1}(u, \tilde{Z}_u^N, \tilde{V}_u^{N+1,N,i-1})}{\tilde{V}_u^{N+1,i} - \tilde{V}_u^{N,i}},$$

provided $(1 - H_u^i)\tilde{V}_u^{N+1,i} \neq (1 - H_u^i)\tilde{V}_u^{N,i}$, and it is 0 otherwise. Moreover, let us consider the probability measure $\mathbb{Q} \sim \mathbb{P}^*$ defined in (5.77) with $(\gamma_i(u), \eta_i(u))$ given above. By using Lemma 5.11, for all $s \in [0,1]$ and $u \in [t,T]$, a.s.

$$s\tilde{V}_u^{N+1,i} + (1-s)\tilde{V}_u^{N,i} \leq C_T, \tag{5.87}$$

for some constant $C_T > 0$ depending on $T > 0$ only. By taking constant $N_0 > C_T$, we have, for all $N \geq N_0$,

$$\frac{\tilde{f}^{N+1}(u, \tilde{Z}_u^N, \tilde{V}_u^{N+1,N,i}) - \tilde{f}^{N+1}(u, \tilde{Z}_u^N, \tilde{V}_u^{N+1,N,i-1})}{\tilde{V}_u^{N+1,i} - \tilde{V}_u^{N,i}}$$
$$\leq 1 - (1 + R_{N+1})^{-\frac{\theta}{2}} e^{-C_T}.$$

This yields that $\hat{W}^{o,\tau} = (\hat{W}^{o,\tau}_s)_{s\in[0,T]}$ and $\hat{M}^* = (\hat{M}^*_s)_{s\in[0,T]}$ defined in (5.78) are $(\mathbb{Q}, \mathbb{F}^M)$-martingales. It follows from (5.58) that $\tilde{f}^N(\omega, u, \xi, v) \geq \tilde{f}^{N+1}(\omega, u, \xi, v)$ for all (ω, u, ξ, v). By putting all the pieces together, the BSDE (5.57) implies that, for all $u \in [t, T]$,

$$\tilde{Y}^{N+1}_u - \tilde{Y}^N_u \geq -\int_u^T (\tilde{Z}^{N+1}_s - \tilde{Z}^N_s)^\top d\hat{W}^{o,\tau}_s - \int_u^T (\tilde{V}^{N+1}_s - \tilde{V}^N_s)^\top d\hat{M}^*_s.$$

This confirms the desired comparison result that $\tilde{Y}^{N+1}_u \geq \tilde{Y}^N_u$, $\mathbb{P}^*$-a.s., since we have $\mathbb{Q} \sim \mathbb{P}^*$. □

5.6.3 Convergence of Solutions to Truncated BSDEs

To prove the existence of solutions of the prime BSDE (5.56), we show that the sequence of solutions to the truncated BSDE (5.57) converge as $N \to \infty$ and the resulting limit process is the desired solution of the BSDE (5.56) in an appropriate function space.

For any compact set $\mathcal{C} \subset \mathbb{R}^n$, we choose N large enough such that $e^{|y|} \leq N$ for all $y \in \mathcal{C}$. By virtue of (5.59), we have, $\mathbb{P}$-a.s., $f^N(u, \xi, v) = f(u, \xi, v)$ for all $u \in [t, T]$ and $\xi, v \in \mathcal{C}$. This implies the locally uniform (almost surely) convergence of f^N to f, i.e., $\sup_{(u,\xi,v)\in[t,T]\times\mathcal{C}^2} |f^N(u, \xi, v) - f(u, \xi, v)| \to 0$, $N \to \infty$, a.s. We first have the next convergence result of the truncated solutions $(\tilde{Y}^N, \tilde{Z}^N, \tilde{V}^N)$ given in Lemma 5.9. Thanks to Lemma 5.11, it is known that $\tilde{V}^N$ is $d\mathbb{P}^* \times du$-a.e. bounded by a constant C_T for all $N \geq 1$.

Lemma 5.13. *There exist $\mathbb{F}^M$-adapted processes $\tilde{Y} = (\tilde{Y}_u)_{u\in[t,T]}$ and $(\tilde{Z}, \tilde{V}) \in L^2_t \times L^2_t$ such that, for $u \in [t, T]$, $\tilde{Y}^N_u \to \tilde{Y}_u$, $\mathbb{P}^*$-a.e., $\tilde{Z}^N \to \tilde{Z}$ weakly in L^2_t, and $\tilde{V}^N \to \tilde{V}$ weakly in L^2_t, as $N \to \infty$.*

Proof. In view of Lemma 5.12, $N \to \tilde{Y}^N_u$ is increasing, $\mathbb{P}^*$-a.e. for $u \in [t, T]$. Lemma 5.11 gives that $\tilde{Y}^N = (\tilde{Y}^N_u)_{u\in[t,T]}$ is uniformly bounded in $\mathcal{S}^\infty_t$. Then, there exists an $\mathbb{F}^M$-adapted process $\tilde{Y} = (\tilde{Y}_u)_{u\in[t,T]}$ such that, for $u \in [t, T]$, $\tilde{Y}^N_u \to \tilde{Y}_u$, as $N \to \infty$, $\mathbb{P}^*$-a.e.. It follows from Lemma 5.11 that the sequence of $\mathbb{F}^M$-predictable solutions $\tilde{Z}^N = (\tilde{Z}^N_u)_{u\in[t,T]}$ for $N \geq 1$ is bounded in L^2_t. Hence, there exists a process $\tilde{Z} = (\tilde{Z}_u)_{u\in[t,T]} \in L^2_t$ such that $\tilde{Z}^N \to \tilde{Z}$ weakly in L^2_t. Moreover, by Lemma 5.9, the sequence of $\int_t^\cdot (\tilde{V}^N_u)^\top dM^*_u$ for $N \geq 1$ is bounded in L^2_t. Thanks to the martingale representation theorem in Protter (2005) and the weak compactness of L^2, there exists a process $\tilde{V} = (\tilde{V}_u)_{u\in[t,T]} \in L^2_t$ such that $\tilde{V}^N \to \tilde{V}$ (up to a subsequence) weakly in L^2_t as $N \to \infty$. We claim that $\tilde{V}$ is predictable.

Indeed, by using Mazur's lemma, we deduce the existence of a sequence of convex combinations of $\tilde{V}^N$ for $N \geq 1$, which converges to $\tilde{V}$ pointwise. Since every convex combination of $\tilde{V}^N$ is predictable, $\tilde{V}$ is also predictable. Thus, we complete the proof of the lemma. □

We continue to prove the strong convergence result of the truncated solutions $(\tilde{Y}^N, \tilde{Z}^N, \tilde{V}^N)$ for $N \geq 1$ given in Lemma 5.9 to the limit process $(\tilde{Y}, \tilde{Z}, \tilde{V})$ given in Lemma 5.13.

Lemma 5.14. *The sequence $(\tilde{Z}^N)_{N \geq 1}$ converges to $\tilde{Z}$ in L_t^2, as $N \to \infty$.*

Proof. To ease the notation, set $\tilde{f}^N(u) := \tilde{f}^N(u, \tilde{Z}_u^N, \tilde{V}_u^N)$ for $u \in [t,T]$. Let $N_2 \geq N_1 \geq 1$ be two integers and $\phi : \mathbb{R} \to \mathbb{R}_+$ be a smooth function that will be determined later. For $Y_u^{\mathrm{d}} := \tilde{Y}_u^{N_2} - \tilde{Y}_u^{N_1} \geq 0$, a.s., we have from Lemma 5.9 and Itô's rule that

$$\phi(0) - \phi(Y_t^{\mathrm{d}}) = \int_t^T \phi'(Y_u^{\mathrm{d}})(\tilde{f}^{N_2}(u) - \tilde{f}^{N_1}(u))du$$

$$+ \int_t^T \phi'(Y_u^{\mathrm{d}})(\tilde{Z}_u^{N_2} - \tilde{Z}_u^{N_1})^\top dW_u^{o,\tau} - \sum_{i=1}^n \int_t^{T \wedge \tau_i^t} \phi'(Y_u^{\mathrm{d}})(\tilde{V}_u^{N_2,i} - \tilde{V}_u^{N_1,i})du$$

$$+ \frac{1}{2} \sum_{i=1}^n \int_t^{T \wedge \tau_i^t} \phi''(Y_u^{\mathrm{d}}) \left| \tilde{Z}_u^{N_2,i} - \tilde{Z}_u^{N_1,i} \right|^2 du \quad (5.88)$$

$$+ \sum_{i=1}^n \int_t^T \{\phi(Y_{u-}^{\mathrm{d}} + \tilde{V}_u^{N_2,i} - \tilde{V}_u^{N_1,i}) - \phi(Y_{u-}^{\mathrm{d}})\}dH_u^i.$$

It follows from (5.58) and Lemma 5.11 that, for all $u \in [t,T]$, there exist positive constants R_i with $i = 1,2,3$ which are independent of N and u such that, a.s.

$$\left| \tilde{f}^{N_2}(u) - \tilde{f}^{N_1}(u) \right| \leq R_1 + R_2 \sum_{i=1}^n (1 - H_u^i) \left\{ \left| \tilde{Z}_u^{N_1,i} \right|^2 + \left| \tilde{Z}_u^{N_2,i} \right|^2 \right\} \quad (5.89)$$

$$\leq R_1 + R_3 \sum_{i=1}^n (1 - H_u^i) \left\{ \left| \tilde{Z}_u^{N_1,i} - \tilde{Z}_u^{N_2,i} \right|^2 + \left| \tilde{Z}_u^{N_1,i} - \tilde{Z}_u^i \right|^2 + \left| \tilde{Z}_u^i \right|^2 \right\}.$$

We choose $\phi(x) = e^{\beta x} - \beta x - 1$ for $x \in \mathbb{R}$, where β is a positive constant satisfying $\beta > 4R_3$. Then ϕ enjoys the properties that $\phi(x) \geq 0$ for all $x \in \mathbb{R}$, $\phi(0) = \phi'(0) = 0$, $\phi'(x) \geq 0$ for $x \in \mathbb{R}_+$, and $\phi''(x) - 4R_3\phi'(x) = (\beta^2 - 4R_3\beta)e^{\beta x} + 4R_3\beta > 0$ for all $x \in \mathbb{R}$. Plugging (5.89) into (5.88) and manipulating terms on both sides, we obtain

$$\frac{1}{2} \sum_{i=1}^n \int_t^{T \wedge \tau_i^t} \phi''(Y_u^{\mathrm{d}}) \left| \tilde{Z}_u^{N_1,i} - \tilde{Z}_u^{N_2,i} \right|^2 du$$

$$-R_3 \sum_{i=1}^{n} \int_0^{T\wedge\tau_i^t} \phi'(Y_u^{\mathrm{d}}) \left| \tilde{Z}_u^{N_1,i} - \tilde{Z}_u^{N_2,i} \right|^2 du$$

$$\leq \phi(0) - \phi(Y_t^{\mathrm{d}}) + R_3 \sum_{i=1}^{n} \int_t^{T\wedge\tau_i^t} \phi'(Y_u^{\mathrm{d}}) \left| \tilde{Z}_u^{N_1,i} - \tilde{Z}_u^i \right|^2 du$$

$$+ R_1 \int_t^T \phi'(Y_u^{\mathrm{d}}) du + R_3 \sum_{i=1}^{n} \int_t^{T\wedge\tau_i^t} \phi'(Y_u^{\mathrm{d}}) \left| \tilde{Z}_u^i \right|^2 du$$

$$- \int_t^T \phi'(Y_u^{\mathrm{d}})(\tilde{Z}_u^{N_2} - \tilde{Z}_u^{N_1})^\top dW_u^{o,\tau} + \sum_{i=1}^{n} \int_t^{T\wedge\tau_i^t} \phi'(Y_u^{\mathrm{d}})(\tilde{V}_u^{N_2,i} - \tilde{V}_u^{N_1,i}) du$$

$$- \sum_{i=1}^{n} \int_t^T \{\phi(Y_{u-}^{\mathrm{d}} + \tilde{V}_u^{N_2,i} - \tilde{V}_u^{N_1,i}) - \phi(Y_{u-}^{\mathrm{d}})\} dH_u^i. \tag{5.90}$$

Using Lemma 5.13, one has that $\tilde{Z}^{N_2}$ converges weakly to $\tilde{Z}$ in L_t^2 as $N_2 \to \infty$.

We next prove that, as $N_2 \to \infty$,

$$\sqrt{\left(\frac{1}{2}\phi'' - R_3\phi'\right)(Y_u^{\mathrm{d}})}(1 - H^i)(\tilde{Z}^{N_1,i} - \tilde{Z}^{N_2,i}) \tag{5.91}$$

$$\to \sqrt{\left(\frac{1}{2}\phi'' - R_3\phi'\right)}(\tilde{Y} - \tilde{Y}^{N_1})(1 - H^i)(\tilde{Z}^{N_1,i} - \tilde{Z}^i),$$

weakly in $L^2([t,T] \times \Omega; \mathbb{P}^*)$. Using the fact that $(\tilde{Y}^N)_{N\geq 1}$ and $\tilde{Y}$ are bounded, we have

$$\delta Y_u^{N_2} := \left(\frac{1}{2}\phi'' - R_3\phi'\right)^{\frac{1}{2}}(\tilde{Y}_u^{N_2} - \tilde{Y}_u^{N_1})$$

$$- \left(\frac{1}{2}\phi'' - R_3\phi'\right)^{\frac{1}{2}}(\tilde{Y}_u - \tilde{Y}_u^{N_1}), \quad u \in [t,T]$$

is also bounded and tends to 0 as $N_2 \to \infty$. Furthermore, the weak convergence of $(\tilde{Z}^N)_{N\geq 1}$ in L_t^2 implies that they are uniformly bounded in L_t^2 by the resonance theorem, which can also be deduced from $(\tilde{Z}^N)_{N\geq 1} \subset \mathbb{H}_{t,\mathrm{BMO}}^2$ by Lemma 5.11. The Cauchy–Schwartz inequality then gives that, for all $X \in L^2([t,T] \times \Omega; \mathbb{P}^*)$,

$$\lim_{N_2 \to \infty} \mathbb{E}^* \left[\int_t^{T\wedge\tau_i^t} \delta Y_u^{N_2}(\tilde{Z}_u^{N_1,i} - \tilde{Z}_u^{N_2,i}) X_u du \right] = 0.$$

Thus, we obtain

$$\lim_{N_2 \to \infty} \mathbb{E}^* \left[\int_t^{T\wedge\tau_i^t} \left(\frac{1}{2}\phi'' - R_3\phi'\right)^{\frac{1}{2}}(Y_u^{\mathrm{d}})(\tilde{Z}_u^{N_1,i} - \tilde{Z}_u^{N_2,i}) X_u du \right]$$

$$= \lim_{N_2 \to \infty} \mathbb{E}^* \left[\int_t^{T \wedge \tau_i^t} \left(\frac{1}{2} \phi'' - R_3 \phi' \right)^{\frac{1}{2}} (Y_u - Y_u^{N_1})(\tilde{Z}_u^{N_1,i} - \tilde{Z}_u^{N_2,i}) X_u du \right]$$

$$+ \lim_{N_2 \to \infty} \mathbb{E}^* \left[\int_t^{T \wedge \tau_i^t} \delta Y_u^{N_2} (\tilde{Z}_u^{N_1,i} - \tilde{Z}_u^{N_2,i}) X_u du \right]$$

$$= \mathbb{E}^* \left[\int_t^{T \wedge \tau_i^t} \left(\frac{1}{2} \phi'' - R_3 \phi' \right)^{\frac{1}{2}} (Y_u - Y_u^{N_1})(\tilde{Z}_u^{N_1,i} - \tilde{Z}_u^i) X_u du \right],$$

which yields (5.91). By using the property of convex functional and weak convergence (c.f. Theorem 1.4 in De Figueiredo (1991)), as $N_2 \to \infty$, we deduce that, the LHS of (5.90) satisfies that

$$\liminf_{N_2 \to \infty} \sum_{i=1}^n \mathbb{E}^* \left[\int_t^{T \wedge \tau_i^t} \left(\frac{1}{2} \phi'' - R_3 \phi' \right) (Y_u^{\mathrm{d}}) \left| \tilde{Z}_u^{N_1,i} - \tilde{Z}_u^{N_2,i} \right|^2 du \right] \quad (5.92)$$

$$\geq \sum_{i=1}^n \mathbb{E}^* \left[\int_t^{T \wedge \tau_i^t} \left(\frac{1}{2} \phi'' - R_3 \phi' \right) (\tilde{Y}_u - \tilde{Y}_u^{N_1}) \left| \tilde{Z}_u^{N_1,i} - \tilde{Z}_u^i \right|^2 du \right].$$

For the jump term in the RHS of (5.90), as $\phi(x) \geq 0$ for all $x \in \mathbb{R}$, we get

$$\sum_{i=1}^n \mathbb{E}^* \left[\int_t^{T \wedge \tau_i^t} \phi'(Y_u^{\mathrm{d}})(\tilde{V}_u^{N_2,i} - \tilde{V}_u^{N_1,i}) du \right]$$

$$- \sum_{i=1}^n \mathbb{E}^* \left[\int_t^T \left(\phi(Y_{u-}^{\mathrm{d}} + \tilde{V}_u^{N_2,i} - \tilde{V}_u^{N_1,i}) - \phi(Y_{u-}^{\mathrm{d}}) \right) dH_u^i \right]$$

$$= - \sum_{i=1}^n \mathbb{E}^* \left[\int_t^{T \wedge \tau_i^t} e^{\beta Y_u^{\mathrm{d}}} \phi(\tilde{V}_u^{N_2,i} - \tilde{V}_u^{N_1,i}) du \right] \leq 0. \quad (5.93)$$

Thanks to (5.92), (5.93) and DCT, it follows from (5.90) that

$$\sum_{i=1}^n \mathbb{E}^* \left[\int_t^{T \wedge \tau_i^t} \left(\frac{1}{2} \phi'' - R_3 \phi' \right) (\tilde{Y}_u - \tilde{Y}_u^{N_1}) \left| \tilde{Z}_u^{N_1,i} - \tilde{Z}_u^i \right|^2 du \right]$$

$$\leq R_3 \sum_{i=1}^n \mathbb{E}^* \left[\int_t^{T \wedge \tau_i^t} \phi'(\tilde{Y}_u - \tilde{Y}_u^{N_1}) \left| \tilde{Z}_u^{N_1,i} - \tilde{Z}_u^i \right|^2 du \right]$$

$$+ R_3 \sum_{i=1}^n \mathbb{E}^* \left[\int_t^T \phi'(\tilde{Y}_u - \tilde{Y}_u^{N_1}) \left| \tilde{Z}_u^i \right|^2 du \right]$$

$$+ R_1 \mathbb{E}^* \left[\int_t^T \phi'(\tilde{Y}_u - \tilde{Y}_u^{N_1}) du \right].$$

We have from Lemma 5.11 and Lemma 5.13 that $\|\tilde{Y}\|_{t,\infty} \leq \|\varsigma\|_{0,\infty}$. By choosing $R_4 := \frac{1}{2}(\beta^2 - 4R_3\beta)e^{-2\beta|\varsigma|_\infty} > 0$, we arrive at

$$R_4 \sum_{i=1}^{n} \mathbb{E}^* \left[\int_t^{T \wedge \tau_i^t} \left| \tilde{Z}_u^{N_1,i} - \tilde{Z}_u^i \right|^2 du \right]$$

$$\leq \frac{1}{2} \sum_{i=1}^{n} \mathbb{E}^* \left[\int_t^{T \wedge \tau_i^t} \{\phi'' - 4R_3\phi'\}(\tilde{Y}_u - \tilde{Y}_u^{N_1}) \left| \tilde{Z}_u^{N_1,i} - \tilde{Z}_u^i \right|^2 du \right]$$

$$\leq R_3 \sum_{i=1}^{n} \mathbb{E}^* \left[\int_t^{T \wedge \tau_i^t} \phi'(\tilde{Y}_u - \tilde{Y}_u^{N_1}) \left| \tilde{Z}_u^i \right|^2 du \right]$$

$$+ R_1 \mathbb{E}^* \left[\int_t^{T \wedge \tau_i^t} \phi'(\tilde{Y}_u - \tilde{Y}_u^{N_1}) du \right]. \tag{5.94}$$

Note that $\phi'(0) = 0$ and that for each $u \in [t,T]$, $\tilde{Y}_u^N \uparrow \tilde{Y}_u$ as $N \to \infty$. The DCT gives that the RHS of (5.94) tends to zero as $N_1 \to \infty$. Then, the estimate (5.94) yields that

$$\lim_{N_1 \to \infty} \sum_{i=1}^{n} \mathbb{E}^* \left[\int_t^{T \wedge \tau_i^t} \left| \tilde{Z}_u^{N_1,i} - \tilde{Z}_u^i \right|^2 du \right] = 0,$$

which completes the proof of the lemma. □

We also have

Lemma 5.15. *The sequence $(\tilde{V}^N)_{N \geq 1}$ converges to $\tilde{V}$ in L_t^2 as $N \to \infty$. Therefore, $\tilde{V}$ is also $d\mathbb{P}^* \times du$-a.s. bounded by some constant C_T.*

Proof. Let us take $\phi(x) = x^2$ for $x \in \mathbb{R}$ in (5.88). Then, Eq. (5.88) can be reduced to

$$-\mathbb{E}\left[|Y_t^d|^2\right] = 2\mathbb{E}^*\left[\int_t^T Y_u^d(\tilde{f}^{N_2}(u) - \tilde{f}^{N_1}(u))du\right]$$

$$- 2\sum_{i=1}^{n} \mathbb{E}^*\left[\int_t^{T \wedge \tau_i^t} Y_u^d(\tilde{V}_u^{N_2,i} - \tilde{V}_u^{N_1,i})du\right]$$

$$+ \sum_{i=1}^{n} \mathbb{E}^*\left[\int_t^{T \wedge \tau_i^t} \left|\tilde{Z}_u^{N_2,i} - \tilde{Z}_u^{N_1,i}\right|^2 du\right]$$

$$+ \sum_{i=1}^{n} \mathbb{E}^*\left[\int_t^{T \wedge \tau_i^t} \left(|Y_{u-}^d + \tilde{V}_u^{N_2,i} - \tilde{V}_u^{N_1,i}|^2 - |Y_{u-}^d|^2\right) du\right].$$

It follows from (5.89) that

$$\sum_{i=1}^{n} \mathbb{E}^* \left[\int_{t}^{T \wedge \tau_i^t} \left| \tilde{V}_u^{N_2,i} - \tilde{V}_u^{N_1,i} \right|^2 du \right]$$
$$\leq 2R_2 \sum_{i=1}^{n} \mathbb{E}^* \left[\int_{t}^{T \wedge \tau_i^t} |Y_u^{\mathrm{d}}| \left(|\tilde{Z}_u^{N_1,i}|^2 + |\tilde{Z}_u^{N_2,i}|^2 \right) du \right]$$
$$- \mathbb{E}^* \left[|Y_t^{\mathrm{d}}|^2 \right] + 2R_1 \mathbb{E}^* \left[\int_{t}^{T} |Y_u^{\mathrm{d}}| \, du \right]$$
$$- \sum_{i=1}^{n} \mathbb{E}^* \left[\int_{t}^{T \wedge \tau_i^t} \left| \tilde{Z}_u^{N_2,i} - \tilde{Z}_u^{N_1,i} \right|^2 du \right]. \tag{5.95}$$

We also have that, for $i = 1, \ldots, n$,

$$\mathbb{E}^* \left[\int_{t}^{T \wedge \tau_i^t} |Y_u^{\mathrm{d}}| \left| \tilde{Z}_u^{N_2,i} \right|^2 du \right] \leq 2\mathbb{E}^* \left[\int_{t}^{T \wedge \tau_i^t} |Y_u^{\mathrm{d}}| \left| \tilde{Z}_u^{N_2,i} - \tilde{Z}_u^i \right|^2 du \right]$$
$$+ 2\mathbb{E}^* \left[\int_{t}^{T \wedge \tau_i^t} |Y_u^{\mathrm{d}}| \left| \tilde{Z}_u^i \right|^2 du \right] \leq 4\|\zeta\|_{0,\infty} \mathbb{E}^* \left[\int_{t}^{T \wedge \tau_i^t} \left| \tilde{Z}_u^{N_2,i} - \tilde{Z}_u^i \right|^2 du \right]$$
$$+ 2\mathbb{E}^* \left[\int_{t}^{T \wedge \tau_i^t} |Y_u^{\mathrm{d}}| \left| \tilde{Z}_u^i \right|^2 du \right]. \tag{5.96}$$

We can derive from (5.95) and (5.96) that

$$\sum_{i=1}^{n} \mathbb{E}^* \left[\int_{t}^{T \wedge \tau_i^t} \left| \tilde{V}_u^{N_2,i} - \tilde{V}_u^{N_1,i} \right|^2 du \right] \leq 2R_1 \mathbb{E}^* \left[\int_{t}^{T} |Y_u^{\mathrm{d}}| \, du \right]$$
$$+ 2R_2 \sum_{i=1}^{n} \mathbb{E} \left[\int_{t}^{T \wedge \tau_i^t} |Y_u^{\mathrm{d}}| \left| \tilde{Z}_u^{N_1,i} \right|^2 du \right]$$
$$+ 4R_2 \sum_{i=1}^{n} \mathbb{E}^* \left[\int_{t}^{T \wedge \tau_i^t} |Y_u^{\mathrm{d}}| \left| \tilde{Z}_u^i \right|^2 du \right]$$
$$+ 8R_2 \|\zeta\|_{0,\infty} \mathbb{E}^* \left[\int_{t}^{T \wedge \tau_i^t} \left| \tilde{Z}_u^{N_2,i} - \tilde{Z}_u^i \right|^2 du \right].$$

Letting $N_2 \to \infty$ and using DCT and Lemma 5.14, we deduce that

$$\liminf_{N_2 \to \infty} \sum_{i=1}^{n} \mathbb{E}^* \left[\int_{t}^{T \wedge \tau_i^t} \left| \tilde{V}_u^{N_2,i} - \tilde{V}_u^{N_1,i} \right|^2 du \right] \leq 2R_1 \mathbb{E}^* \left[\int_{t}^{T} \left| \tilde{Y}_u - \tilde{Y}_u^{N_1} \right| du \right]$$
$$+ 2R_2 \sum_{i=1}^{n} \mathbb{E}^* \left[\int_{t}^{T \wedge \tau_i^t} \left| \tilde{Y}_u - \tilde{Y}_u^{N_1} \right| \left| \tilde{Z}_u^{N_1,i} \right|^2 du \right]$$

$$+ 4R_2 \sum_{i=1}^{n} \mathbb{E}^* \left[\int_{t}^{T \wedge \tau_i^t} \left| \tilde{Y}_u - \tilde{Y}_u^{N_1} \right| \left| \tilde{Z}_u^i \right|^2 du \right]$$

$$\leq 2R_1 \mathbb{E}^* \left[\int_{t}^{T} \left| \tilde{Y}_u - \tilde{Y}_u^{N_1} \right| du \right] + 8R_2 \sum_{i=1}^{n} \mathbb{E}^* \left[\int_{t}^{T \wedge \tau_i^t} \left| \tilde{Y}_u - \tilde{Y}_u^{N_1} \right| \left| \tilde{Z}_u^i \right|^2 du \right]$$

$$+ 8R_2 \|\zeta\|_{0,\infty} \sum_{i=1}^{n} \mathbb{E}^* \left[\int_{t}^{T \wedge \tau_i^t} \left| \tilde{Z}_u^{N_1,i} - \tilde{Z}_u^i \right|^2 du \right].$$

With the property of convex functional and weak convergence (Theorem 1.4 in De Figueiredo (1991)), one can get

$$\sum_{i=1}^{n} \mathbb{E}^* \left[\int_{t}^{T \wedge \tau_i^t} \left| \tilde{V}_u^i - \tilde{V}_u^{N_1,i} \right|^2 du \right] \leq 2R_1 \mathbb{E}^* \left[\int_{t}^{T} \left| \tilde{Y}_u - \tilde{Y}_u^{N_1} \right| du \right] \quad (5.97)$$

$$+ 8R_2 \sum_{i=1}^{n} \mathbb{E}^* \left[\int_{t}^{T \wedge \tau_i^t} \left| \tilde{Y}_u - \tilde{Y}_u^{N_1} \right| \left| \tilde{Z}_u^i \right|^2 du \right]$$

$$+ 8R_2 \|\zeta\|_{0,\infty} \sum_{i=1}^{n} \mathbb{E}^* \left[\int_{t}^{T \wedge \tau_i^t} \left| \tilde{Z}_u^{N_1,i} - \tilde{Z}_u^i \right|^2 du \right].$$

The desired convergence that $\tilde{V}^N \to \tilde{V}$ in L_t^2 can be derived by DCT and Lemma 5.14 as $N_1 \to \infty$. The boundedness of $\tilde{V}$ is consequent on the uniform boundedness of $(\tilde{V}^N)_{N \geq 1}$. □

Lastly, we present the main result of this section on the existence of solutions to the prime BSDE (5.56).

Theorem 5.2 (Well-posedness of Prime BSDE). *Let $(\tilde{Y}, \tilde{Z}, \tilde{V})$ be the limiting process given in Lemma 5.13. Then, $(\tilde{Y}, \tilde{Z}, \tilde{V}) \in \mathcal{S}_t^\infty \times \mathbb{H}_{t,\mathrm{BMO}}^2 \times L_t^2$ is a solution of the BSDE (5.56).*

Proof. We first prove that $\tilde{Y}^N$ converges to $\tilde{Y}$ in the uniform norm as $N \to \infty$, a.s. In fact, for fixed $t \in [0, T]$ and any $u \in [t, T]$, we have

$$\sup_{u \in [t,T]} \left| \tilde{Y}_u^{N_1} - \tilde{Y}_u^{N_2} \right| \leq \int_{t}^{T} \left| \tilde{f}^{N_1}(s) - \tilde{f}^{N_2}(s) \right| ds \quad (5.98)$$

$$+ \sup_{u \in [t,T]} \left| \int_{u}^{T} (\tilde{Z}_s^{N_1} - \tilde{Z}_s^{N_2})^\top dW_s^{o,\tau} \right| + \sup_{u \in [t,T]} \left| \int_{u}^{T} (\tilde{V}_s^{N_1} - \tilde{V}_s^{N_2})^\top dM_s^* \right|.$$

Taking into account Lemma 5.14 and Lemma 2.5 in Kobylanski (2000), we obtain that, there exists a subsequence (N_l) such that, as $l \to \infty$,

$$(1 - H)\tilde{Z}^{N_l} \to (1 - H)\tilde{Z}, \quad d\mathbb{P}^* \times du\text{-a.e.},$$

$$\hat{Z} = (\hat{Z}_1, \ldots, \hat{Z}_n) \in L_t^2, \tag{5.99}$$

where $\hat{Z}_u^i := \sup_{l \geq 1} |(1-H_u^i)\tilde{Z}_u^{N_l,i}|$ for $u \in [t,T]$. Moreover, Lemma 5.15 implies that, for some subsequence $(N_{l_k}) \subset (N_l)$, it holds that $(1-H)\tilde{V}^{N_{l_k}} \to (1-H)\tilde{V}$, as $k \to \infty$, $d\mathbb{P}^* \times du$-a.e.. To ease the notation, the subsequence is still denoted by (N). By the definition of $\tilde{f}^N$ and the fact that the random function $\tilde{f}$ is a.s. continuous in its domain, we have

$$\lim_{N \to \infty} \tilde{f}^N(u, \tilde{Z}_u^N, \tilde{V}_u^N) du = \tilde{f}(u, \tilde{Z}_u, \tilde{V}_u), \quad d\mathbb{P}^* \times du\text{-a.e.} \tag{5.100}$$

In light of (5.58) and Lemma 5.11, for all $u \in [t,T]$, there exist constants $R_1, R_2 > 0$ independent of (N, u) such that

$$\left|\tilde{f}^N(u, \tilde{Z}_u^N, \tilde{V}_u^N)\right| \leq R_1 + R_2 \sum_{i=1}^n (1 - H_u^i)\left|\tilde{Z}_u^{N,i}\right|^2$$

$$\leq R_1 + R_2 \sum_{i=1}^n (1 - H_u^i)\left|\hat{Z}_u^i\right|^2.$$

Note that $\hat{Z} \in L_t^2$. Together with above inequality and (5.100), the DCT implies that

$$\lim_{N \to \infty} \mathbb{E}\left[\int_t^T \left|\tilde{f}^N(u, \tilde{Z}_u^N, \tilde{V}_u^N) - \tilde{f}(u, \tilde{Z}_u, \tilde{V}_u)\right| du\right] = 0. \tag{5.101}$$

The BDG inequality then implies the existence of constants $R_3, R_4 > 0$ independent of N such that

$$\mathbb{E}^*\left[\sup_{u \in [t,T]}\left|\int_u^T (\tilde{Z}_s^N - \tilde{Z}_s)^\top dW_s^{o,\tau}\right|^2\right] \leq 2\mathbb{E}^*\left[\left|\int_t^T (\tilde{Z}_s^N - \tilde{Z}_s)^\top dW_s^{o,\tau}\right|^2\right]$$

$$+ 2\mathbb{E}^*\left[\sup_{u \in [t,T]}\left|\int_t^u (\tilde{Z}_s^N - \tilde{Z}_s)^\top dW_s^{o,\tau}\right|^2\right]$$

$$\leq R_3 \sum_{i=1}^n \mathbb{E}^*\left[\int_t^{T \wedge \tau_i^t} \left|\tilde{Z}_s^{N,i} - \tilde{Z}_s^i\right|^2 ds\right].$$

In a similar fashion, we also attain that

$$\mathbb{E}^*\left[\sup_{u \in [t,T]}\left|\int_u^T (\tilde{V}_s^N - \tilde{V}_s)^\top dM_s^*\right|^2\right] \leq R_4 \sum_{i=1}^n \mathbb{E}^*\left[\int_t^{T \wedge \tau_i^t} \left|\tilde{V}_s^{N,i} - \tilde{V}_s^i\right|^2 ds\right].$$

It follows from Lemmas 5.14 and 5.15 that

$$\lim_{N \to \infty} \mathbb{E}^*\left[\sup_{u \in [t,T]}\left|\int_u^T (\tilde{Z}_s^N - \tilde{Z}_s)^\top dW_s^{o,\tau}\right|^2\right]$$

$$= \lim_{N\to\infty} \mathbb{E}^* \left[\sup_{u\in[t,T]} \left| \int_u^T (\tilde{V}_s^N - \tilde{V}_s)^\top dM_s^* \right|^2 \right] = 0.$$

As a consequence, there exists a subsequence (still denoted by N) such that (5.101) holds, and a.e.

$$\lim_{N\to\infty} \sup_{u\in[t,T]} \left| \int_t^T (\tilde{Z}_s^N - \tilde{Z}_s)^\top dW_s^{o,\tau} \right| = 0,$$

$$\lim_{N\to\infty} \sup_{u\in[t,T]} \left| \int_t^T (\tilde{V}_s^N - \tilde{V}_s)^\top dM_s^* \right| = 0. \tag{5.102}$$

Deduce from (5.102) that $(\tilde{Y}^N)_{N\geq 1}$ is a Cauchy sequence a.e. under the uniform norm, and its limiting process coincides with $\tilde{Y}$ by Lemma 5.13. Thus, it holds that $\lim_{N\to\infty} \sup_{u\in[t,T]} |\tilde{Y}_u^N - \tilde{Y}_u| = 0$, a.e.. By taking the limit on both sides of the equation, it holds that

$$\zeta - Y_t^N = \int_t^T \tilde{f}^N(u, \tilde{Z}_u^N, \tilde{V}_u^N) du + \int_t^T (\tilde{Z}_u^N)^\top dW_u^{o,\tau} + \int_t^T (\tilde{V}_u^N)^\top dM_u^*.$$

By applying the established convergence results in (5.101) and (5.102), we can conclude that $(\tilde{Y}, \tilde{Z}, \tilde{V}) \in \mathcal{S}_t^\infty \times \mathbb{H}_{t,\text{BMO}}^2 \times L_t^2$ is indeed a solution of the BSDE (5.56). □

5.7 The Optimal Admissible Strategy

In this section, we characterize the optimal (admissible) investment strategy using the verification result provided in Lemma 5.5.

The following theorem provides the existence of an optimal investment strategy (which is in essence unique proved in Section 5.8) for the prime risk sensitive portfolio optimization problem.

Theorem 5.3 (Optimal Admissible Strategy). *Let $(\tilde{Y}, \tilde{Z}, \tilde{V}) \in \mathcal{S}_t^\infty \times \mathbb{H}_{t,\text{BMO}}^2 \times L_t^2$ be a solution of the BSDE (5.56) in Theorem 5.2. We define*

$$\pi_u^* := \arg\max_{\pi\in U} \mathcal{H}(\pi; p_{u-}^M, H_{u-}, \tilde{Z}_u, \tilde{V}_u), \quad u \in [t, T], \tag{5.103}$$

where $\mathcal{H}(\pi; p, h, \xi, v)$ is given by (5.48). Then, $\pi^* \in \mathcal{U}_t^{ad}$, and it is an optimal investment strategy for problem (5.36).

Proof. The key step of the proof is to verify that $\pi^* \in \mathcal{U}_t^{ad}$. More precisely, by Definition 5.1, it is enough to show that $(\mathcal{E}_u(\Lambda^{\pi^*,t}))_{u\in[t,T]}$ is a true $(\mathbb{P}^*, \mathbb{F}^M)$-martingale. In view of (5.103), it holds that, a.e.

$$\mathcal{H}(\pi_u^*; p_{u-}^M, H_{u-}, \tilde{Z}_u, \tilde{V}_u) \geq \mathcal{H}(0; p_{u-}^M, H_{u-}, \tilde{Z}_u, \tilde{V}_u), \quad u \in [0, T].$$

Similar to the proof of Lemma 5.6, we can manipulate the RHS of the above inequality and attain the existence of constants $R_1, R_2 > 0$ depending on the essential upper bound of $\tilde{V}$ such that, a.e.

$$|\pi_u^*|^2 \leq R_1 |(1 - H_{u-})\tilde{Z}_u|^2 + R_2, \quad u \in [t, T]. \quad (5.104)$$

For $u \in [t, T]$, we define

$$\Lambda_u^{\pi^*, t, 1} := \sum_{i=1}^{n} \int_t^u \left\{ \sigma_i^{-1}(\mu_i^M(s) + \lambda_i^M(s)) - \frac{\theta \sigma_i}{2} \pi_s^{*,i} + \tilde{Z}_s^i \right\} dW_s^{o,i,\tau}. \quad (5.105)$$

Thanks to the fact that $\tilde{Z} \in \mathbb{H}^2_{t,\text{BMO}}$ and (5.104), it follows that $\Lambda^{\pi^*, t, 1} = (\Lambda_u^{\pi^*, t, 1}(u))_{u \in [t,T]}$ is a continuous BMO $(\mathbb{P}^*, \mathbb{F}^M)$-martingale. By using Theorem 3.4 in Kazamaki (1994), there exists $\rho > 1$ such that

$$\mathbb{E}^*_{t,p,h}\left[\mathcal{E}_T(\Lambda^{\pi^*, t, 1})^\rho\right] < +\infty. \quad (5.106)$$

Moreover, the first-order condition yields that, for $i = 1, \ldots, n$, a.s.

$$\mu_i^M(u-) + \lambda_i^M(u-) - r + \sigma_i(1 - H_{u-}^i)\tilde{Z}_u^i = \left(1 + \frac{\theta}{2}\right)\sigma_i^2 \pi_u^{*,i} \quad (5.107)$$
$$+ \lambda_i^M(u-)(1 - \pi_u^{*,i})^{-\frac{\theta}{2}-1} e^{\tilde{V}_u^i}.$$

We next prove the existence of constants $R_3, R_4 > 0$ depending on the essential upper bound of $\tilde{V}$ such that, for $i = 1, \ldots, n$, a.e.

$$\lambda_i^M(u-)(1 - \pi_u^{*,i})^{-\frac{\theta}{2}-1} e^{\tilde{V}_u^i} \leq R_3 \left|(1 - H_{u-}^i)\tilde{Z}_u^i\right| + R_4. \quad (5.108)$$

In fact, if $\pi_u^{*,i} \leq 0$, then the LHS of (5.108) is bounded by the constant $R_\lambda e^{|\tilde{V}^i|_{t,\infty}}$ with $R_\lambda := \max_{(i,k,h) \in \{1,\ldots,n\} \times S_I \times S_H} \lambda_i(k, h) > 0$ being finite under Assumption 5.1. If $\pi_u^{*,i} \in (0, 1)$, it follows from (5.107) that

$$\lambda_i^M(u-)(1 - \pi_u^{*,i})^{-\frac{\theta}{2}-1} e^{\tilde{V}_u^i} \leq \left(1 + \frac{\theta}{2}\right)\sigma_i^2 \pi_u^{*,i} + \lambda_i^M(u-)(1 - \pi_u^{*,i})^{-\frac{\theta}{2}-1} e^{\tilde{V}_u^i}$$
$$= \mu_i^M(u-) + \lambda_i^M(u-) - r + \sigma_i(1 - H_{u-}^i)\tilde{Z}_u^i.$$

This shows (5.108) again using Assumption 5.1. To continue, the estimate (5.108) in turn entails the existence of constants $R_5, R_6 > 0$ such that, for $i = 1, \ldots, n$, a.e.

$$|\lambda_i^M(u-)|^2 (1 - \pi_u^{*,i})^{-\theta} e^{2\tilde{V}_u^i} \leq R_5 (1 - H_{u-}^i)|\tilde{Z}_u^i|^2 + R_6. \quad (5.109)$$

For $u \in [t, T]$, we define

$$\Lambda_u^{\pi^*, t, 2} := \sum_{i=1}^n \Lambda_u^{\pi^*, t, 2, i}$$

$$:= \sum_{i=1}^{n} \int_{t}^{u} \{(1-\pi_s^{*,i})^{-\frac{\theta}{2}} \lambda_i^{M}(s-) e^{\tilde{V}_s^i} - 1\} dM_s^{*,i}. \tag{5.110}$$

We next define a probability measure $\mathbb{P}^{(0)} \sim \mathbb{P}^*$ via $\frac{d\mathbb{P}^{(0)}}{d\mathbb{P}^*}|_{\mathcal{F}_T^M} = \mathcal{E}_T(\Lambda^{\pi^*,0,1})$. Then, for $i = 1, \ldots, n$, H^i admits the $\mathbb{P}^{(0)}$-intensity given by 1. It holds that

$$\mathcal{E}_u(\Lambda^{\pi^*,t,2,1}) = \exp\left(\int_t^u \{1-(1-\pi_s^{*,1})^{-\frac{\theta}{2}} \lambda_1^M(s) e^{\tilde{V}_s^1}\} ds\right) \prod_{s \le u}(1+\Delta\Lambda_s^{\pi^*,t,2,1})$$

$$\le e^{T-t}\left\{1+\int_t^T (1-\pi_s^{*,1})^{-\frac{\theta}{2}} \lambda_1^M(s-) e^{\tilde{V}_s^1} dH_s^1\right\}. \tag{5.111}$$

Let $R_T > 0$ be a constant depending on T that may refer to different values from line to line. Then, it follows from (5.106) and (5.109) that, for $(t,p,h) \in [0,T] \times S_{p^M} \times S_H$,

$$\mathbb{E}_{t,p,h}^{(0)}\left[\mathcal{E}_u(\Lambda^{\pi^*,t,2,1})^2\right] \le R_T \mathbb{E}_{t,p,h}^{(0)}\left[1+\int_t^T (1-\pi_s^{*,1})^{-\theta} \left|\lambda_1^M(u-)\right|^2 e^{2\tilde{V}_s^1} dH_s^1\right]$$

$$\le R_T\left\{1+\mathbb{E}_{t,p,h}^*\left[\mathcal{E}_T(\Lambda^{\pi^*,t,1}) \int_t^{T \wedge \tau_1^t} |\tilde{Z}_u^1|^2 du\right]\right\}$$

$$\le R_T\left\{\mathbb{E}_{t,p,h}^*\left[\mathcal{E}_T(\Lambda^{\pi^*,t,1})^p\right]\right\}^{\frac{1}{p}} \left\{\mathbb{E}_{t,p,h}^*\left[\left(\int_t^{T \wedge \tau_1^t} |\tilde{Z}_u^1|^2 du\right)^q\right]\right\}^{\frac{1}{q}} + R_T$$

$$\le R_T, \tag{5.112}$$

where $q > 1$ satisfies that $\frac{1}{p} + \frac{1}{q} = 1$, and we used Corollary 2.1 in Kazamaki (1994) for BMO $(\mathbb{P}^*, \mathbb{F}^M)$-martingales in the last inequality. This yields that $(\mathcal{E}_u(\Lambda^{\pi^*,t,2,1}))_{u \in [t,T]}$ is uniformly integrable (U.I.) under $\mathbb{P}^{(0)}$. By using the orthogonality of $\mathbb{P}^*$-martingales $\Lambda^{\pi^*,t,1}$ and $\Lambda^{\pi^*,t,2,1}$, it holds that

$$\mathbb{E}_{t,p,h}^{(0)}\left[\mathcal{E}_T(\Lambda^{\pi^*,t,2,1})\right] = \mathbb{E}_{t,p,h}^*\left[\mathcal{E}_T(\Lambda^{\pi^*,t,1})\mathcal{E}_T(\Lambda^{\pi^*,t,2,1})\right] = 1. \tag{5.113}$$

We next define a probability measure $\mathbb{P}^{(1)} \sim \mathbb{P}^*$ via $\frac{d\mathbb{P}^{(1)}}{d\mathbb{P}^*}|_{\mathcal{F}_T^M} = \mathcal{E}_T(\Lambda^{\pi^*,t,1})\mathcal{E}_T(\Lambda^{\pi^*,t,2,1})$. Note that H^1 and H^2 do not jump simultaneously. Then, H^2 admits the unit intensity under $\mathbb{P}^{(1)}$. Therefore, in the light of (5.109) and (5.112), we can derive

$$\mathbb{E}_{t,p,h}^{(1)}\left[\mathcal{E}_u(\Lambda^{\pi^*,t,2,2})^2\right] \le R_T \mathbb{E}_{t,p,h}^{(1)}\left[1+\int_t^T (1-\pi_s^{*,2})^{-\theta} \left|\lambda_2^M(u-)\right|^2 e^{2\tilde{V}_s^2} dH_s^2\right]$$

$$\leq R_T \left\{ 1 + \mathbb{E}_{t,p,h}^{(0)} \left[\mathcal{E}_T(\Lambda^{\pi^*,t,2,1})_T \int_t^{T\wedge\tau_2^t} |\tilde{Z}_u^2|^2 du \right] \right\}$$

$$\leq R_T \left\{ \mathbb{E}_{t,p,h}^{(0)} \left[\mathcal{E}_T(\Lambda^{\pi^*,t,2,1})^2 \right] \right\}^{\frac{1}{2}} \left\{ \mathbb{E}_{t,p,h}^{(0)} \left[\left(\int_t^{T\wedge\tau_2^t} |\tilde{Z}_u^2|^2 du \right)^2 \right] \right\}^{\frac{1}{2}} + R_T$$

$$\leq R_T \left\{ \mathbb{E}_{t,p,h}^{(0)} \left[\left(\int_t^{T\wedge\tau_2^t} |\tilde{Z}_u^2|^2 du \right)^2 \right] \right\}^{\frac{1}{2}} + R_T. \tag{5.114}$$

The term $\mathbb{E}_{t,p,h}^{(0)}[(\int_t^{T\wedge\tau_2^t} |\tilde{Z}_u^2|^2 du)^2]$ can be estimated by

$$\mathbb{E}_{t,p,h}^{(0)} \left[\left(\int_t^{T\wedge\tau_2^t} \left|\tilde{Z}_u^2\right|^2 du \right)^2 \right]$$

$$\leq \left\{ \mathbb{E}_{t,p,h}^* \left[\mathcal{E}_T(\Lambda^{\pi^*,t,1})^p \right] \right\}^{\frac{1}{p}} \left\{ \mathbb{E}_{t,p,h}^* \left[\left(\int_t^{T\wedge\tau_2^t} \left|\tilde{Z}_u^2\right|^2 du \right)^{2q} \right] \right\}^{\frac{1}{q}}.$$

Thus, there exists a constant $R_T^{(1)} > 0$ depending on T such that, for all $u \in [t, T]$,

$$\mathbb{E}_{t,p,h}^{(1)} \left[\mathcal{E}_u(\Lambda^{\pi^*,t,2,2})^2 \right] = \mathbb{E}_{t,p,h}^* \left[\mathcal{E}_u(\Lambda^{\pi^*,t,1}) \mathcal{E}_u(\Lambda^{\pi^*,t,2,1}) \mathcal{E}_u(\Lambda^{\pi^*,t,2,2})^2 \right]$$

$$\leq R_T^{(1)}. \tag{5.115}$$

Up to now, we have proved the following estimate with $l = 2$: there exists a constant $R_T^{(l-1)} > 0$ depending on T such that, for all $u \in [t, T]$,

$$\mathbb{E}_{t,p,h}^* \left[\mathcal{E}_u(\Lambda^{\pi^*,t,1}) \mathcal{E}_u \left(\sum_{i=1}^{l-1} \Lambda^{\pi^*,t,2,i} \right) \mathcal{E}_u(\Lambda^{\pi^*,t,2,l})^2 \right] \leq R_T^{(l-1)}. \tag{5.116}$$

We next verify the estimate (5.116) for all $l \leq n$ using the mathematical induction argument. To this end, assume the estimate (5.116) holds for all $l \leq k$ (where $2 \leq k \leq n$). The aim is to validate the estimate (5.116) for $l = k+1$. First, following similar lines of argument to prove (5.113), we can obtain inductively that, for all $2 \leq l \leq k$,

$$\mathbb{E}_{t,p,h}^* \left[\mathcal{E}_T(\Lambda^{\pi^*,t,1}) \prod_{i=1}^{l} \mathcal{E}_T(\Lambda^{\pi^*,t,2,i}) \right] = 1. \tag{5.117}$$

Let us define a probability measure $\mathbb{P}^{(l)} \sim \mathbb{P}^*$ by

$$\left. \frac{d\mathbb{P}^{(l)}}{d\mathbb{P}^*} \right|_{\mathcal{F}_T^M} := \mathcal{E}_T(\Lambda^{\pi^*,t,1}) \prod_{i=1}^{l} \mathcal{E}_T(\Lambda^{\pi^*,t,2,i}), \quad 2 \leq l \leq k. \tag{5.118}$$

Note again that $H^1, \ldots, H^k, H^{k+1}$ do not jump simultaneously and hence H^{k+1} admits the unit intensity under $\mathbb{P}^{(k)}$. We deduce from (5.109) and (5.116) with $l \leq k$ that

$$\mathbb{E}^{(k)}_{t,p,h}\left[\mathcal{E}_u(\Lambda^{\pi^*,t,2,k+1})^2\right]$$

$$\leq R_T\left\{1 + \mathbb{E}^{(k-1)}_{t,p,h}\left[\mathcal{E}_T(\Lambda^{\pi^*,t,2,k})\int_t^{T\wedge\tau^t_{k+1}}|\tilde{Z}^{k+1}_u|^2 du\right]\right\}$$

$$\leq R_T\left\{\mathbb{E}^{(k-1)}_{t,p,h}\left[\mathcal{E}_T(\Lambda^{\pi^*,t,2,k})^2\right]\right\}^{\frac{1}{2}}\left\{\mathbb{E}^{(k-1)}_{t,p,h}\left[\left(\int_t^{T\wedge\tau^t_{k+1}}|\tilde{Z}^{k+1}_u|^2 du\right)^2\right]\right\}^{\frac{1}{2}}$$

$$+ R_T$$

$$\leq R_T\left\{\mathbb{E}^{(k-1)}_{t,p,h}\left[\left(\int_t^{T\wedge\tau^t_{k+1}}|\tilde{Z}^{k+1}_u|^2 du\right)^2\right]\right\}^{\frac{1}{2}} + R_T$$

$$= R_T\left\{\mathbb{E}^{(k-2)}_{t,p,h}\left[\mathcal{E}_T(\Lambda^{\pi^*,t,2,k-1})\left(\int_t^{T\wedge\tau^t_{k+1}}|\tilde{Z}^{k+1}_u|^2 du\right)^2\right]\right\}^{\frac{1}{2}} + R_T$$

$$\leq R_T\left\{\mathbb{E}^{(k-2)}_{t,p,h}\left[\left(\int_t^{T\wedge\tau^t_{k+1}}|\tilde{Z}^{k+1}_u|^2 du\right)^{2^2}\right]\right\}^{\frac{1}{2^2}} + R_T$$

$$\leq \quad \cdots\cdots$$

$$\leq R_T\left\{\mathbb{E}^{(0)}_{t,p,h}\left[\left(\int_t^{T\wedge\tau^t_{k+1}}|\tilde{Z}^{k+1}_u|^2 du\right)^{2^k}\right]\right\}^{\frac{1}{2^k}} + R_T \quad (5.119)$$

$$\leq R_T\left\{\mathbb{E}^*_{t,p,h}\left[\mathcal{E}_T(\Lambda^{\pi^*,t,1})^\rho\right]\right\}^{\frac{1}{\rho 2^k}}\left\{\mathbb{E}^*_{t,p,h}\left[\left(\int_t^{T\wedge\tau^t_{k+1}}|\tilde{Z}^{k+1}_u|^2 du\right)^{q 2^k}\right]\right\}^{\frac{1}{q 2^k}}$$

$$+ R_T$$

$$\leq R_T.$$

Above, the constant $R_T > 0$ is a generic constant which may be different from line to line. This confirms the estimate (5.116) with $l = k+1$. As a result of the previous induction and the orthogonality of $\Lambda^{\pi^*,t,1}$, $\Lambda^{\pi^*,t,2,1}, \ldots, \Lambda^{\pi^*,t,2,n}$, we have

$$\mathbb{E}^*_{t,p,h}\left[\mathcal{E}_T(\Lambda^{\pi^*,t})\right] = \mathbb{E}^*_{t,p,h}\left[\mathcal{E}_T(\Lambda^{\pi^*,t,1})\prod_{i=1}^n \mathcal{E}_T(\Lambda^{\pi^*,t,2,i})\right] = 1. \quad (5.120)$$

This shows that $(\mathcal{E}_s(\Lambda^{\pi^*,t}))_{s\in[t,T]}$ is a U.I. $(\mathbb{P}^*,\mathbb{F}^M)$-martingale, which verifies the first claim that $\pi^* \in \mathcal{U}_t^{ad}$.

Toward this end, the first order condition in the definition of π^* and Theorem 5.2 can entail that (5.51) in Lemma 5.5 holds. We can readily conclude the second assertion that π^* is indeed an optimal strategy using Lemma 5.5. □

5.8 Uniqueness of Solutions to BSDE

Theorem 5.2 provides the existence of a solution $(\tilde{Y},\tilde{Z},\tilde{V}) \in \mathcal{S}_t^\infty \times \mathbb{H}_{t,\text{BMO}}^2 \times L_t^2$ of BSDE (5.56), while the uniqueness of the solution remains open.

The following result confirms that our constructed solution in Theorem 5.2 is unique that is a consequence of Lemma 5.5 and Theorem 5.3, which in turn implies that π^* constructed in (5.103) is the unique optimal portfolio.

Proposition 5.2 (Uniqueness of Solutions to BSDE). *The limiting process $(\tilde{Y},\tilde{Z},\tilde{V})$ in Lemma 5.13 is the unique (in the sense of $d\mathbb{P}^* \times du$-a.e.) solution of the BSDE (5.56) in the space $\mathcal{S}_t^\infty \times \mathbb{H}_{t,\text{BMO}}^2 \times L_t^2$. Moreover, the portfolio process π^* defined in (5.103) by $(\tilde{Y},\tilde{Z},\tilde{V})$ is the unique (in the sense of $d\mathbb{P}^* \times du$-a.e.) optimal investment strategy for problem (5.36).*

Proof. In Theorem 5.2, we have proved that there exists one solution $(\tilde{Y},\tilde{Z},\tilde{V}) \in \mathcal{S}_t^\infty \times \mathbb{H}_{t,\text{BMO}}^2 \times L_t^2$ to the BSDE (5.56) such that $(\tilde{Y} + \int_t^\cdot f(p_s^M, H_s, 0, 0)ds, \tilde{Z}, \tilde{V})$ solves the original BSDE (5.46). Recall $U = (-\infty,1)^n$, and we next define the set, for $t \in [0,T]$,

$$\hat{\mathcal{U}}_t^{ad} := \bigg\{ \pi = (\pi_u^i;\, i=1,\ldots,n)_{u\in[t,T]}^\top \in U;\, \pi \text{ is } \mathbb{F}^M\text{-predictable such that}$$

$$\sum_{i=1}^n \int_t^u \pi_s^i dW_s^{o,i,\tau} \text{ and } \sum_{i=1}^n \int_t^u (1-\pi_s^i)^{-\frac{\theta}{2}} dW_s^{o,i,\tau},\, u \in [t,T],$$

$$\text{are } (\mathbb{P}^*,\mathbb{F}^M)\text{-BMO martingales} \bigg\}.$$

Let $(\tilde{Y},\tilde{Z},\tilde{V}) \in \mathcal{S}_t^\infty \times \mathbb{H}_{t,\text{BMO}}^2 \times L_t^2$ be a solution of the BSDE (5.56) and let $\pi^* = (\pi_u^*)_{u\in[t,T]}$ be defined by (5.103) using $(\tilde{Z},\tilde{V})$ from this solution. Then, it follows from (5.104), (5.109) and $\tilde{Z} \in \mathbb{H}_{t,\text{BMO}}^2$ that $\pi^* \in \hat{\mathcal{U}}_t^{ad}$.

Now, for any $\pi \in \hat{\mathcal{U}}_t^{ad}$, we define, for $i=1,\ldots,n$,

$$\hat{Z}_u^i := |\pi_u^i| + (1-\pi_u^i)^{-\frac{\theta}{2}},\quad \forall u \in [t,T].$$

Then $\hat{Z} = (\hat{Z}_u^i; i = 1, \ldots, n)_{u \in [t,T]}^\top \in \mathbb{H}_{t,\text{BMO}}^2$, and we can obtain the same estimates (5.104) and (5.109) with $(\pi^*, \tilde{Z})$ replaced by $(\pi, \hat{Z})$. Moreover, by applying a similar induction to prove (5.120), we deduce that $\hat{\mathcal{U}}_t^{ad} \subset \mathcal{U}_t^{ad}$. This implies that π^* constructed by $(\tilde{Z}, \tilde{V})$ satisfies that

$$\inf_{\pi \in \hat{\mathcal{U}}_t^{ad}} J(\pi; t, p, h) = e^{Y_t(t,p,h)} = J(\pi^*; t, p, h), \quad (5.121)$$

where $J(\pi; t, p, h)$ is given by (5.44), and $Y_t(t, p, h) = \tilde{Y}_t$ as $Y := \tilde{Y} + \int_t^\cdot f(p_s^M, H_s, 0, 0) ds$ in the proof of Lemma 5.5. That is, we have constructed an admissible control subset $\hat{\mathcal{U}}_t^{ad} \subset \mathcal{U}_t^{ad}$ independent of $(\tilde{Y}, \tilde{Z}, \tilde{V})$ such that the optimal strategy π^* given by (5.103) is still in $\hat{\mathcal{U}}_t^{ad}$.

We next apply this subset $\hat{\mathcal{U}}_t^{ad}$ to conclude the uniqueness of solutions to the BSDE (5.56). To this end, let $(\tilde{Y}^\ell, \tilde{Z}^\ell, \tilde{V}^\ell) \in \mathcal{S}_t^\infty \times \mathbb{H}_{t,\text{BMO}}^2 \times L_t^2$, $\ell = 1, 2$ be two solutions of the BSDE (5.56) with the same terminal condition. We can then define $\pi^{\ell,*} \in \hat{\mathcal{U}}_t^{ad}$ as in (5.103) by using $(\tilde{Y}^\ell, \tilde{Z}^\ell, \tilde{V}^\ell)$ respectively for $\ell = 1, 2$. The verification of optimality in Lemma 5.5, together with (5.121), yields that

$$e^{\tilde{Y}_t^1} = e^{\tilde{Y}_t^2} = \inf_{\pi \in \hat{\mathcal{U}}_t^{ad}} J(\pi; t, p, h).$$

This implies that

$$J(\pi^{1,*}; t, p, h) e^{-\tilde{Y}_t^2} = \mathbb{E}_{t,p,h}^* \left[\mathcal{E}_T(\Lambda^{\pi^{1,*},t}) \exp\left(\int_t^T \left(f(p_{u-}^M, H_{u-}, \tilde{Z}_u^2, \tilde{V}_u^2) \right.\right.\right.$$
$$\left.\left.\left. - \mathcal{H}(\pi_u^{1,*}; p_{u-}^M, H_{u-}, \tilde{Z}_u^2, \tilde{V}_u^2) \right) du \right) \right] = 1,$$

where $\Lambda^{\pi,t} = (\Lambda_u^{\pi,t})_{u \in [t,T]}$ for $\pi \in \mathcal{U}_t^{ad}$ is defined by (5.53). Therefore, $d\mathbb{P}^* \otimes du$-a.e.

$$f(p_{u-}^M, H_{u-}, \tilde{Z}_u^2, \tilde{V}_u^2) = \mathcal{H}(\pi_u^{1,*}; p_{u-}^M, H_{u-}, \tilde{Z}_u^2, \tilde{V}_u^2).$$

Let $J(\pi; u) := \mathbb{E}[(\frac{X_T^\pi}{X_u^\pi})^{-\frac{\theta}{2}} | \mathcal{F}_u^M]$ for $u \in [t, T]$. Then, for all $u \in [t, T]$, we have

$$J(\pi^{1,*}; u) e^{-\tilde{Y}_u^2 + \int_t^u f(p_{s-}^M, H_{s-}, 0, 0) ds} \quad (5.122)$$
$$= \mathbb{E}^* \left[\mathcal{E}_T(\Lambda^{\pi^{1,*},u}) \exp\left(\int_u^T (f(p_{s-}^M, H_{s-}, \tilde{Z}_s^2, \tilde{V}_s^2) \right.\right.$$
$$\left.\left. - \mathcal{H}(\pi_s^{1,*}; p_{s-}^M, H_{s-}, \tilde{Z}_s^2, \tilde{V}_s^2)) ds \right) \bigg| \mathcal{F}_u^M \right] = 1.$$

On the other hand, by Lemma 5.5, we have, for all $u \in [t, T]$,

$$J(\pi^{1,*}; u)e^{-\tilde{Y}_u^1 + \int_t^u f(p_{s-}^M, H_{s-}, 0, 0)ds} \tag{5.123}$$

$$= \mathbb{E}^*\left[\mathcal{E}_T(\Lambda^{\pi^{1,*},u}) \exp\left(\int_u^T (f(p_{s-}^M, H_{s-}, \tilde{Z}_s^1, \tilde{V}_s^1) \right.\right.$$

$$\left.\left. - \mathcal{H}(\pi_s^{1,*}; p_{s-}^M, H_{s-}, \tilde{Z}_s^1, \tilde{V}_s^1))ds\right) \bigg| \mathcal{F}_u^M\right] = 1.$$

It follows from (5.122) and (5.123) that, for $u \in [t, T]$, $\tilde{Y}_u^{(1)} = \tilde{Y}_u^{(2)}$, $\mathbb{P}^*$-a.s.. Note that $(\tilde{Y}^\ell, \tilde{Z}^\ell, \tilde{V}^\ell) \in \mathcal{S}_t^\infty \times \mathbb{H}_{t,\text{BMO}}^2 \times L_t^2$, $\ell = 1, 2$ satisfy the BSDE (5.56). Together with Theorem 5.1, the unique canonical decomposition of the semimartingale $\tilde{Y} = (\tilde{Y}_u)_{u \in [t,T]} \in \mathcal{S}_t^\infty$ under $\mathbb{P}^*$ (Theorem 34 in Chapter III in Protter (2005)) implies that, for $u \in [t, T]$, $\mathbb{P}^*$-a.e.

$$\int_t^u (\tilde{Z}_s^1)^\top dW_s^{o,\tau} = \int_t^u (\tilde{Z}_s^2)^\top dW_s^{o,\tau}, \quad \int_t^u (\tilde{V}_s^1)^\top dM_s^* = \int_t^u (\tilde{V}_s^2)^\top dM_s^*,$$

which proves the uniqueness of solutions to the BSDE (5.56) in the sense of $d\mathbb{P}^* \times du$-a.e..

For the unique solution $(\tilde{Y}, \tilde{Z}, \tilde{V}) \in \mathcal{S}_t^\infty \times \mathbb{H}_{t,\text{BMO}}^2 \times L_t^2$ of the BSDE (5.56), we then claim that the constructed strategy π^* in (5.103) is the unique optimal portfolio for the original control problem. In fact, for an arbitrary optimal strategy $\hat{\pi} \in \mathcal{U}_t^{ad}$, from the proof of Lemma 5.5, it follows that

$$J(\hat{\pi}; t, p, h)e^{-\tilde{Y}_t} = \mathbb{E}_{t,p,h}^*\left[\mathcal{E}_T(\Lambda^{\hat{\pi},t}) \exp\left(\int_t^T (f(p_{u-}^M, H_{u-}, \tilde{Z}_u, \tilde{V}_u) \right.\right.$$

$$\left.\left. - \mathcal{H}(\hat{\pi}_u; p_{u-}^M, H_{u-}, \tilde{Z}_u, \tilde{V}_u))du\right)\right] = 1.$$

Thus, it holds that, $d\mathbb{P}^* \times du$-a.s.

$$\mathcal{H}(\hat{\pi}_u; p_{u-}^M, H_{u-}, \tilde{Z}_u, \tilde{V}_u) = f(p_{u-}^M, H_{u-}, \tilde{Z}_u, \tilde{V}_u)$$

$$= \max_{\pi \in U} \mathcal{H}(\pi; p_{u-}^M, H_{u-}, \tilde{Z}_u, \tilde{V}_u).$$

It then follows from the strict convexity of $U \ni \pi \to h(\pi; p, h, \xi, v)$ that $\hat{\pi} = \pi^*$, $d\mathbb{P}^* \times du$-a.e. This verifies the uniqueness of the admissible optimal strategy π^*, which completes the proof of the proposition. □

Chapter 6

Portfolio Optimization with Consumption Habit in Incomplete Market*

This chapter focuses on the utility maximization problem with consumption habit in an incomplete market. It is well-known that the time separable "von Neumann–Morgenstern" preferences on consumption have been observed to be inconsistent with some empirical evidences. For instant, the magnitude of the equity premium (Mehra and Prescott (1985)) can not be reconciled with the preference $\mathbb{E}[\int_0^T U(t, c_t)dt]$, where the instantaneous utility function $U(t, x)$ is only derived from the consumption rate process (c.f. the classical Merton's optimal consumption problem documented in Section 1.1.1 of Chapter 1).

Linear additive habit formulation. As an extension to this preference, the "linear addictive" habit formation has been introduced, which is defined by $\mathbb{E}[\int_0^T U(t, c_t - F_t^c)dt]$, where $U : [0,T] \times \mathbb{R}_+ \to \mathbb{R}$ and the additional accumulative process $F^c = (F_t^c)_{t\in[0,T]}$, called the "habit formation" or the "standard of living" process, describes the consumption history impact. More precisely, $F^c = (F_t^c)_{t\in[0,T]}$ satisfies the following ODE:

$$dF_t^c = (\delta_t c_t - \alpha_t F_t^c)dt, \quad F_0^c = z \geq 0, \tag{6.1}$$

where the discounting factors α_t and δ_t are assumed to be nonnegative optional processes and $z \geq 0$ is called the initial habit or the initial standard of living. Here, the consumption habits are assumed to be "addictive" in the sense that, a.s.

$$c_t \geq F_t^c, \quad \forall t \in [0,T]. \tag{6.2}$$

Compared with the time separable case, a small drop in consumption may cause large fluctuation in consumption net of the subsistence level due to the standard of living constraint. The habit formation preference can possibly explain sizable excess returns on risky assets in equilibrium models

---
*This chapter is based on the work Yu (2015).

even for moderate values of the degree of risk aversion. Based on this, a vast literature recommends this time non-separable preference as the new economic paradigm. We refer readers to, for instance, Otrok et al. (2002), Samuelson (1969) and Campbell and Cochrane (1999).

The earlier study of habit formation in modern economics dates back to Hicks (1965) and Ryder and Heal (1973). More recently, there are some important contributions in complete markets (described as Itô processes), see among Detemple and Zapatero (1991), Detemple and Zapatero (1992), Schroder and Skiadas (2002), Munk (2008), Detemple and Karatzas (2003) and Egglezos and Karatzas (2009). Several pioneering work have derived the explicit feedback form of the optimal policies under different assumptions and market models. However, in the words by Egglezos and Karatzas (2009), *The existence of an optimal portfolio/consumption pair in an incomplete market is an open question....*, and new methodologies are needed to handle the problem. Therefore, we are interested in the general (incomplete) semimartingale market and aim to prove the existence and uniqueness of the optimal solution to this path dependent optimization problem. This depends on an appropriate application of the convex duality approach to this path dependent optimization problem. In fact, the convex duality plays an important role in the treatment of utility maximization problems in incomplete markets. To list a very small subset of the existing literature, we refer to Karatzas et al. (1991), Kramkov and Schachermayer (1999, 2003), Cvitanić et al. (2001), Karatzas and Žitković (2003), Hugonnier and Kramkov (2004), and Žitković (2002, 2005).

The critical step to build conjugate duality is to define the dual space as a proper extension of space $\mathcal{M}$, which is the set of equivalent local martingale measure density processes. The first natural choice is the bipolar set of the space $\mathcal{M}$, which is the smallest convex, closed and solid set containing the set $\mathcal{M}$. Kramkov and Schachermayer (1999, 2003) and Žitković (2002), proved that this bipolar set can be characterized as the solid hull of the set $\mathcal{Y}(y)$, which is defined as the set of supermartingale deflators:

$$\mathcal{Y}(y) := \{Y = (Y_t)_{t \in [0,T]} | Y_0 = y, \ Y_t > 0, \forall t \in [0,T] \text{ and } XY = (X_t Y_t)_{t \in [0,T]}$$
$$\text{is a supermartingale for each } X \in \mathcal{X}(x)\},$$

where $\mathcal{X}(x)$ denotes the set of accumulated gains/losses processes under some admissible portfolios with initial endowment less than or equal to x. However, according to the definition of habit formation process Z^c, if we derive the naive dual problem using the Fenchel–Legendre transform and the first-order condition, we have

$$\inf_{y>0, Y \in \mathcal{Y}(y)} \mathbb{E}\left[\int_0^T V\left(Y_t + \delta_t \mathbb{E}_t\left[\int_t^T e^{\int_t^s (\delta_v - \alpha_v) dv} Y_s ds\right]\right) dt\right]$$
$$- z\mathbb{E}\left[\int_0^T e^{\int_0^t (\delta_v - \alpha_v) dv} Y_t dt\right],$$

where $\mathbb{E}_t[\cdot] := \mathbb{E}[\cdot | \mathcal{F}_t]$. The first difficulty is the additional integral $\mathbb{E}[\int_0^T e^{\int_0^t (\delta_v - \alpha_v) dv} Y_t dt]$, from which we can see that $\mathcal{Y}(y)$ is not the appropriate space to show the existence of an optimal dual solution. However, it still reminds us to invoke the general treatment of random endowment developed by Cvitanić et al. (2001), Karatzas and Žitković (2003) and Žitković (2005). Therein, they proposed another extension of the set $\mathcal{M}$, which is now considered as the set of equivalent local martingale measures, to the set $\mathcal{D}$ of bounded finitely additive measures. Nevertheless, their approach is inadequate to deal with the first term of the dual problem, when the conditional integral part $\mathbb{E}_t[\int_t^T e^{\int_t^s (\delta_v - \alpha_v) dv} Y_s ds]$ in the conjugate function V is taken into account.

Reducing the complexity of path-dependence. To avoid the complexity of the path-dependence, we propose the transform from the consumption rate process c_t to the auxiliary process $\tilde{c}_t := c_t - F_t^c$, so that the primal utility maximization problem becomes time separable w.r.t. the process $\tilde{c}_t$. This substitution idea from c_t to $\tilde{c}_t$ appeared first in the market isomorphism result for complete markets by Schroder and Skiadas (2002). Meanwhile, for any equivalent local martingale measure density process $Y \in \mathcal{M}$, we define the auxiliary dual process $\Gamma = (\Gamma_t)_{t \in [0,T]}$ exactly as:

$$\Gamma_t := Y_t + \delta_t \mathbb{E}_t\left[\int_t^T e^{\int_t^s (\delta_v - \alpha_v) dv} Y_s ds\right], \quad \forall t \in [0,T].$$

The dual problem can therefore be formulated in terms of auxiliary process Γ instead of Y so that the path dependence of Y can also be hidden in the definition of process Γ. By introducing the process given by

$$\tilde{w}_t := e^{-\int_0^t \alpha_v dv}, \quad \forall t \in [0,T], \tag{6.3}$$

one can shift the integral $\mathbb{E}[\int_0^T e^{\int_0^t (\delta_v - \alpha_v) dv} Y_t dt]$ to the integral $\mathbb{E}[\int_0^T \tilde{w}_t \Gamma_t dt]$ with respect to its auxiliary process Γ. With the aid of this equality, we can treat the extra exogenous random term $\tilde{w}_t$ as the shadow random endowment density process and define the dual functional on the properly modified space of Γ instead of Y. By enlarging the effective domain of values for x and z, the original utility maximization problem with habit formation can be embedded into the framework of Hugonnier and Kramkov (2004) as an abstract time separable utility maximization problem on the product space. On the other hand, we are facing some troubles in applying the classical duality results since the auxiliary process Γ is not integrable. For instance, to show the existence of the dual optimizer, the trick of applying de la Vallée–Poussin theorem in the proof of Lemma 3.2 in Kramkov and Schachermayer (1999) does *not* work. The argument of contradiction in the proof of Lemma 1 in Kramkov and Schachermayer (2003) using the subsequence splitting lemma will also fail by observing that constants may not be contained in the corresponding space. Therefore, we impose the additional sufficient conditions on habit formation discounting factors α_t and δ_t, see Assumption 6.2, to guarantee the well-posedness of the primal optimization problem. We also ask for *reasonable asymptotic elasticity* conditions on utility functions U both at $x \to 0$ and $x \to \infty$ for the validity of several key assertions of our main results to hold true.

The rest of this chapter is organized as follows: Section 6.1 introduces the financial market and consumption habit formation process. Section 6.2 defines the auxiliary process space $\bar{\mathcal{A}}(x,z)$, the enlarged space $\widetilde{\mathcal{A}}(x,z)$ and the auxiliary dual space $\mathcal{M}$. The original path-dependent utility maximization problem is embedded into an abstract time separable optimization problem with the shadow random endowment. Section 6.3 is devoted to the formulation of the two dimensional dual problem over the properly enlarged dual space $\widetilde{\mathcal{Y}}(y,r)$ such that the shadow random endowment part can be hidden and our main results are stated at the end. Section 6.4 contains detail proofs of our main theorems.

6.1 The Market Model

Consider a financial market with d risky assets whose price is modeled by a d-dimensional semimartingale process $S = (S_t^1, \ldots, S_t^d)_{t \in [0,T]}$ on a given filtered probability space $(\Omega, \mathcal{F}, \mathbb{F}, \mathbb{P})$ with the filtration $\mathbb{F} = (\mathcal{F}_t)_{t \in [0,T]}$ satisfying the usual conditions. For the notational convenience, we take $\mathcal{F} = \mathcal{F}_T$ with $T > 0$ being the finite maturity date. We make the standard assumption that there exists one riskless bond $S_t^0 \equiv 1$ for all $t \in [0,T]$, which amounts to consider S_t^0 as the numéraire asset. Let $H = (H_t^1, \ldots, H_t^d)_{t \in [0,T]}$ be the portfolio process which is a predictable S-integrable process representing the number of shares of each risky asset held by the investor. The accumulated gains/losses process of the investor under his trading strategy

H by time t is given by

$$X_t^H = (H \cdot S)_t := \sum_{k=1}^{d} \int_0^t H_u^k dS_u^k, \quad \forall t \in [0, T].$$

The portfolio process H is called *admissible* if there exists a constant bound $a \in \mathbb{R}$ such that $X_t^H \geq a$, a.s. for all $t \in [0, T]$.

For a given initial wealth level $x > 0$, the investor will also choose an intermediate consumption plan during the whole investment horizon. We denote the consumption rate process by $c = (c_t)_{t \in [0,T]}$. Then, the resulting self-financing wealth process $(W_t^{x,H,c})_{t \in [0,T]}$ becomes that

$$W_t^{x,H,c} := x + (H \cdot S)_t - \int_0^t c_s ds.$$

Apart from the wealth process, the *consumption habit formation* process F^c defined by (6.1) is given equivalently by the following exponentially weighted average of the investor's past consumption integral and the initial habit

$$F_t^c = z e^{-\int_0^t \alpha_v dv} + \int_0^t \delta_s c_s e^{-\int_s^t \alpha_v dv} ds, \quad (6.4)$$

where discounting factors $\alpha = (\alpha_t)_{t \in [0,T]}$ and $\delta = (\delta_t)_{t \in [0,T]}$ measure, respectively, the persistence of the initial habits level and the intensity of consumption history. Herein, we will be mostly interested in the general case when discounting factors α and δ are stochastic processes which are allowed to be unbounded. However, for technical reasons, we will assume that $\int_0^t (\delta_s - \alpha_s) ds < \infty$ a.s. for all $t \in [0, T]$.

Throughout this chapter, we make the assumption (6.2) that the consumption habit is addictive, i.e., $c_t \geq F_t^c$, a.s., for all $t \in [0, T]$, which is to say, the investor's current consumption rate will never fall below "the standard of living" process. A consumption process $c = (c_t)_{t \in [0,T]}$ is defined to be (x, z)-*financeable* if there exists an admissible portfolio process H such that $W_t^{x,H,c} \geq 0$, a.s. $\forall t \in [0, T]$ and the addictive habit formation constraint $c_t \geq F_t^c$, a.s. with $t \in [0, T]$ holds. The class of all (x, z)-financeable consumption rate processes will be denoted by $\mathcal{A}(x, z)$, for $x > 0$ and $z \geq 0$.

6.1.1 Absence of Arbitrage

Using the definition in Delbaen and Schachermayer (1994, 1998), we call a probability measure $\mathbb{Q}$ the *equivalent local martingale measure* if (i) $\mathbb{Q} \sim \mathbb{P}$; and (ii) $X^H = (X_t^H)_{t \in [0,T]}$ is a local martingale under $\mathbb{Q}$. Here, $\mathbb{P}$ is the physical probability measure introduced in the previous section. Denote by $\mathcal{M}$ the family of equivalent local martingale measures. To rule out the arbitrage opportunities in the market, we also assume that

$$\mathcal{M} \neq \emptyset. \tag{6.5}$$

For $\mathbb{Q} \in \mathcal{M}$, we define the following r.c.l.l. (right-continuous with left limits) process $Y^{\mathbb{Q}} = (Y_t^{\mathbb{Q}})_{t \in [0,T]}$ by

$$Y_t^{\mathbb{Q}} := \mathbb{E}_t\left[\frac{d\mathbb{Q}}{d\mathbb{P}}\right], \quad \forall t \in [0,T],$$

which is called an equivalent local martingale measure density. With a slight abuse of notation, we denote $\mathcal{M}$ also as the set of all equivalent local martingale density processes.

The celebrated *optional decomposition theorem* (c.f. Kramkov (1996)) enables us to characterize the (x, z)-financeable consumption process by the following budget constraint condition:

Lemma 6.1. *The consumption process $c = (c_t)_{t \in [0,T]}$ is (x, z)-financeable if and only if $c_t \geq F_t^c$ for all $t \in [0,T]$, and it holds that*

$$\mathbb{E}\left[\int_0^T c_t Y_t dt\right] \leq x, \quad \forall Y \in \mathcal{M}. \tag{6.6}$$

In the forthcoming section, we will introduce the utility function used in this chapter.

6.1.2 The Utility Function

The individual investor's preference is represented by a utility function $U : [0,T] \times \mathbb{R}_+ \to \mathbb{R}$, such that, for all $x > 0$, $U(\cdot, x)$ is continuous on $[0,T]$, and for every $t \in [0,T]$, the function $U(t, \cdot)$ is strictly concave, strictly increasing, continuously differentiable and satisfies the Inada conditions:

$$U'(t,0) := \lim_{x \to 0} U'(t,x) = \infty, \quad U'(t,\infty) := \lim_{x \to \infty} U'(t,x) = 0 \tag{6.7}$$

with $U'(t,x) := \frac{\partial}{\partial x} U(t,x)$. For $t \in [0,T]$, we extend the definition of the utility function by $U(t,x) = -\infty$ for all $x < 0$, which is equivalent to the addictive habit formation constraint $c_t \geq F_t^c$. The convex conjugate of the utility function is defined by, for all $(t,y) \in [0,T] \times \mathbb{R}_+$,

$$V(t,y) := \sup_{x>0}\{U(t,x) - xy\}.$$

We make assumptions on the asymptotic behavior of U at both $x = 0$ and $x = \infty$ for future purposes:

Assumption 6.1. The utility function U satisfies the reasonable asymptotic elasticity condition both at $x = \infty$ and $x = 0$, i.e.,

$$AE_\infty[U] = \limsup_{x \to \infty} \left(\sup_{t \in [0,T]} \frac{x U'(t,x)}{U(t,x)} \right) < 1, \tag{6.8}$$

and

$$AE_0[U] = \limsup_{x \to 0} \left(\sup_{t \in [0,T]} \frac{x U'(t,x)}{|U(t,x)|} \right) < \infty. \tag{6.9}$$

Moreover, in order to get some inequalities uniformly in time t, we assume that

$$\lim_{x \to \infty} \left(\inf_{t \in [0,T]} U(t,x) \right) > 0, \tag{6.10}$$

and

$$\lim_{x \to 0} \left(\sup_{t \in [0,T]} U(t,x) \right) < 0. \tag{6.11}$$

The well known utility functions satisfy *reasonable asymptotic elasticity* conditions (6.8) and (6.9), for instant:

(i) the discounted logarithm utility function $U(t,x) = e^{-\beta t} \ln x$;
(ii) the discounted power utility function $U(t,x) = e^{-\beta t} \frac{x^p}{p}$ with $p < 1$ and $p \neq 0$, and $\beta > 0$.

However, it is also easy to check that the utility function $U(t,x) = -e^{\frac{1}{x}}$ does not satisfy the condition (6.9) and the utility function $U(t,x) = \frac{x}{\ln x}$ does not satisfy the condition (6.8). If the utility function satisfies the lower bound assumption $\inf_{t \in [0,T]} U(t,0) > -\infty$, then the condition (6.9) is automatically verified. Moreover, if the utility function satisfies the upper bound assumption $\sup_{t \in [0,T]} U(t,\infty) < \infty$, the condition (6.8) holds true.

In the sequel, the technical results give the equivalent characterizations of the *reasonable asymptotic elasticity* conditions (6.8) and (6.9). The proof is based on the fact that $-V$ is a concave function and similar arguments in Lemma 6.3 of Kramkov and Schachermayer (1999), see also Proposition 3.7 in Karatzas and Žitković (2003).

Lemma 6.2. *Let $U(t,x)$ be a utility function satisfying Assumption 6.1. Then, we have*

$$AE_0[U] < +\infty \iff AE_\infty[V] < 1.$$

Here, we have defined that

$$AE_\infty[V] := \limsup_{y \to \infty} \left(\sup_{t \in [0,T]} \frac{yV'(t,y)}{V(t,y)} \right) < 1. \tag{6.12}$$

In each of the subsequent assertions, the infimum of $\gamma_1 > 0$ for which these assertions hold true equals the reasonable asymptotic elasticity $AE_\infty[U]$, and the infimum of $\gamma_2 > 0$ equals the reasonable asymptotic elasticity $AE_\infty[V]$.

(i) *there exist $x_0 > 0$ and $y_0 > 0$ such that, for all $t \in [0,T]$,*

$$\begin{cases} U(t, \lambda x) < \lambda^{\gamma_1} U(t,x) & \text{for } \lambda > 1, \ x \geq x_0; \\ V(t, \lambda y) > \lambda^{\gamma_2} V(t,y) & \text{for } \lambda > 1, \ y \geq y_0. \end{cases}$$

(ii) *there exist $x_0 > 0$ and $y_0 > 0$ such that, for all $t \in [0,T]$,*

$$\begin{cases} U'(t,x) < \gamma_1 \dfrac{U(t,x)}{x} & \text{for } x \geq x_0; \\ V'(t,y) > \gamma_2 \dfrac{V(t,y)}{y} & \text{for } y \geq y_0. \end{cases}$$

(iii) *there exist $x_0 > 0$ and $y_0 > 0$ such that, for all $t \in [0,T]$,*

$$\begin{cases} V(t, \mu y) < \mu^{-\frac{\gamma_1}{1-\gamma_1}} V(t,y) & \text{for } 0 < \mu < 1, \ 0 < y \leq y_0; \\ U(t, \mu x) > \mu^{-\frac{\gamma_2}{1-\gamma_2}} U(t,x) & \text{for } 0 < \mu < 1, \ 0 < x \leq x_0. \end{cases}$$

(iv) *there exist $x_0 > 0$ and $y_0 > 0$ such that, for all $t \in [0,T]$,*

$$\begin{cases} -V'(t,y) < \left(\dfrac{\gamma_1}{1-\gamma_1} \right) \dfrac{V(t,y)}{y} & \text{for } 0 < y \leq y_0; \\ -U'(t,x) > \left(\dfrac{\gamma_2}{1-\gamma_2} \right) \dfrac{U(t,x)}{x} & \text{for } 0 < x \leq x_0. \end{cases}$$

6.2 The Characterization of Financeable Consumption

In the spirit of Bouchard and Pham (2004) which treats the wealth dependent problem (see also Žitković (2005) on consumption and endowment with stochastic clock), denote by $\mathcal{O}$ the σ-algebra of optional sets relative to the filtration $\mathbb{F}$. Let $d\bar{\mathbb{P}} = dt \times d\mathbb{P}$ be the measure on the product space $(\Omega \times [0,T], \mathcal{O})$ defined as:

$$\bar{\mathbb{P}}[A] = \mathbb{E}\left[\int_0^T 1_A(t,\omega)dt\right], \quad \forall A \in \mathcal{O},$$

where we recall that $\mathbb{E}$ is the expectation operator under $\mathbb{P}$. We denote by $\mathbb{L}^0(\Omega \times [0,T], \mathcal{O}, \bar{\mathbb{P}})$ ($\mathbb{L}^0$ for short) the set of all random variables on the product space $\Omega \times [0,T]$ w.r.t. the optional σ-algebra $\mathcal{O}$ endowed with the topology of convergence in measure $\bar{\mathbb{P}}$. From now on, we will identify the optional stochastic process $Y = (Y_t)_{t \in [0,T]}$ with the random variable $Y \in \mathbb{L}^0(\Omega \times [0,T], \mathcal{O}, \bar{\mathbb{P}})$. We also define the positive orthant $\mathbb{L}^0_+(\Omega \times [0,T], \mathcal{O}, \bar{\mathbb{P}})$ ($\mathbb{L}^0_+$ for short) as the set of $Y = Y(t,\omega) \in \mathbb{L}^0$ such that $Y \geq 0$, $\bar{\mathbb{P}}$ a.s.. Endow $\mathbb{L}^0_+$ with the bilinear form valued in $[0, \infty]$ as $\langle X, Y \rangle := \mathbb{E}[\int_0^T X_t Y_t dt]$ for all $X, Y \in \mathbb{L}^0_+$.

6.2.1 Path-dependence Reduction

We define the set of all (x,z)-financeable consumption rate processes as a set of random variables on the product space $(\Omega \times [0,T], \mathcal{O}, \bar{\mathbb{P}})$ and Lemma 6.1 states that

$$\mathcal{A}(x,z) = \left\{c \in \mathbb{L}^0_+; \ c_t \geq F_t^c, \ \forall t \in [0,T], \text{ and } \langle c, Y \rangle \leq x, \forall Y \in \mathcal{M}\right\}.$$

However, the family $\mathcal{A}(x,z)$ maybe empty for some values $x > 0$ and $z \geq 0$. We will restrict ourselves to the *effective domain* $\bar{\mathcal{H}}$ which is defined as the union of the *interior* of set such that $\mathcal{A}(x,z)$ is not empty and the boundary $\{x > 0, z = 0\}$:

$$\bar{\mathcal{H}} := \text{int}\left\{(x,z) \in (0,\infty) \times [0,\infty); \ \mathcal{A}(x,z) \neq \emptyset\right\} \cup \mathbb{R}_+ \times \{0\}.$$

From the definition, $\bar{\mathcal{H}}$ includes the special case of zero initial habit, i.e., $z = 0$.

Before stating the next result, we first impose some additional conditions on the discounting factors $\alpha = (\alpha_t)_{t \in [0,T]}$ and $\delta = (\delta_t)_{t \in [0,T]}$, which are essential for the well-posedness of the primal utility optimization problem:

Assumption 6.2. The nonnegative optional processes $\alpha = (\alpha_t)_{t \in [0,T]}$ and $\delta = (\delta_t)_{t \in [0,T]}$ satisfy

$$\sup_{Y \in \mathcal{M}} \mathbb{E}\left[\int_0^T e^{\int_0^t (\delta_v - \alpha_v)dv} Y_t dt\right] < +\infty. \tag{6.13}$$

Furthermore, there exists a constant $\bar{x} > 0$ such that

$$\mathbb{E}\left[\int_0^T U^-\left(t, \bar{x}e^{-\int_0^t \alpha_v dv}\right) dt\right] < +\infty. \tag{6.14}$$

If stochastic discounting processes $\alpha = (\alpha_t)_{t\in[0,T]}$ and $\delta = (\delta_t)_{t\in[0,T]}$ are assumed to be bounded, conditions (6.13) and (6.14) will be satisfied. The condition (6.13) is the well-known super-hedging property of the random variable $\int_0^T e^{\int_0^t (\delta_v - \alpha_v) dv} dt$ in the original market. Then, we have

Lemma 6.3. *Let the condition (6.13) hold. Then, the effective domain $\bar{\mathcal{H}}$ can be rewritten explicitly as follows:*

$$\bar{\mathcal{H}} = \left\{(x, z) \in \mathbb{R}_+ \times [0, \infty); \; x > z \sup_{Y \in \mathcal{M}} \mathbb{E}\left[\int_0^T e^{\int_0^t (\delta_v - \alpha_v) dv} Y_t dt\right]\right\}.$$

The proof of the lemma is straightforward, and the reader may refer to Yu (2012) for details.

By choosing $(x, z) \in \bar{\mathcal{H}}$, we can now define the preliminary version of the "primal utility maximization problem" as follows:

$$u(x, z) := \sup_{c \in \mathcal{A}(x,z)} \mathbb{E}\left[\int_0^T U(t, c_t - F_t^c) dt\right]. \tag{6.15}$$

To reduce the path dependence structure, for each (x, z)-financeable consumption rate process, we want to generalize the "market isomorphism" idea by Schroder and Skiadas (2002). We then introduce the auxiliary process $\tilde{c}_t := c_t - F_t^c$ with the auxiliary set of $\mathcal{A}(x, z)$ given by

$$\bar{\mathcal{A}}(x, z) := \left\{\tilde{c} \in \mathbb{L}_+^0; \; \tilde{c}_t = c_t - F_t^c, \; \forall t \in [0, T], \; c \in \mathcal{A}(x, z)\right\}. \tag{6.16}$$

It is straightforward to verify that the following lemma holds:

Lemma 6.4. *For each fixed $(x, z) \in \bar{\mathcal{H}}$, there is one to one correspondence between sets $\mathcal{A}(x, z)$ and $\bar{\mathcal{A}}(x, z)$, and hence $\bar{\mathcal{A}}(x, z) \neq \emptyset$ for all $(x, z) \in \bar{\mathcal{H}}$.*

We move to the set $\mathcal{M}$ of equivalent local martingale measure densities. For any $Y \in \mathcal{M}$, the auxiliary optional process w.r.t. $Y = (Y_t)_{t\in[0,T]}$ is defined as:

$$\Gamma_t := Y_t + \delta_t \mathbb{E}_t\left[\int_t^T e^{\int_t^s (\delta_v - \alpha_v) dv} Y_s ds\right], \quad \forall t \in [0, T]. \tag{6.17}$$

Denote the set of all these auxiliary optional processes by

$$\widetilde{\mathcal{M}} := \left\{ \Gamma \in \mathbb{L}_+^0; \; \Gamma_t = Y_t + \delta_t \mathbb{E}_t \left[\int_t^T e^{\int_t^s (\delta_v - \alpha_v) dv} Y_s ds \right], \right.$$
$$\left. \forall t \in [0, T], \; Y \in \mathcal{M} \right\}. \tag{6.18}$$

Note that, although stochastic discounting processes α and δ are unbounded, under the condition (6.13), the auxiliary dual process Γ is well defined in $\mathbb{L}_+^0$, but, it is not necessarily in $\mathbb{L}^1$. A direct application of Fubini–Tonelli Theorem induces the key equalities below, for the detail proof, we refer to Proposition 2.3.3 in Yu (2012):

Proposition 6.1. *Let the condition (6.13) hold. For each nonnegative optional process $c = (c_t)_{t \in [0,T]}$ such that $c_t \geq F_t^c$ with $F^c = (F_t^c)_{t \in [0,T]}$ defined by (6.4) for fixed initial standard of living $z \geq 0$ and the nonnegative optional process $Y = (Y_t)_{t \in [0,T]}$, we have the following equalities w.r.t. their corresponding auxiliary processes $\tilde{c}_t = c_t - F_t^c$ and Γ_t which is defined by (6.17) that*

$$\langle c, Y \rangle = \langle \tilde{c}, \Gamma \rangle + z \langle w, Y \rangle = \langle \tilde{c}, \Gamma \rangle + z \langle \tilde{w}, \Gamma \rangle. \tag{6.19}$$

Here, the processes $w, \tilde{w} \in \mathbb{L}_+^0$ are given by

$$w_t := e^{\int_0^t (\delta_v - \alpha_v) dv} \quad \text{and} \quad \tilde{w}_t := e^{-\int_0^t \alpha_v dv}, \quad \forall t \in [0, T]. \tag{6.20}$$

In light of Lemma 6.1 and Proposition 6.1, under conditions (6.13) and (6.14), we would have the alternative budget constraint characterization of the consumption rate process $c = (c_t)_{t \in [0,T]}$ as:

Proposition 6.2. *For any given pair $(x, z) \in \bar{\mathcal{H}}$, the consumption rate process c is (x, z)-financeable if and only if $c_t \geq F_t^c$ for all $t \in [0, T]$, and it holds that*

$$\langle c - F^c, \Gamma \rangle \leq x - z \langle \tilde{w}, \Gamma \rangle, \quad \forall \Gamma \in \widetilde{\mathcal{M}}.$$

The above proposition provides us the alternative definition of set $\bar{\mathcal{A}}(x, z)$ for $(x, z) \in \bar{\mathcal{H}}$ by

$$\bar{\mathcal{A}}(x, z) = \left\{ \tilde{c} \in \mathbb{L}_+^0; \; \langle \tilde{c}, \Gamma \rangle \leq x - z \langle \tilde{w}, \Gamma \rangle, \; \forall \Gamma \in \widetilde{\mathcal{M}} \right\}. \tag{6.21}$$

6.2.2 Embedding into Maximization Problem with Shadow Endowment

To apply the convex duality approach for the random endowment, we need to enlarge the domain of the set $\bar{\mathcal{H}}$ to $\mathcal{H}$ and also enlarge the corresponding auxiliary set $\bar{\mathcal{A}}(x, z)$ to $\widetilde{\mathcal{A}}(x, z)$:

$$\widetilde{\mathcal{A}}(x, z) := \left\{ \tilde{c} \in \mathbb{L}_+^0; \; \langle \tilde{c}, \Gamma \rangle \leq x - z \langle \tilde{w}, \Gamma \rangle, \; \forall \Gamma \in \widetilde{\mathcal{M}} \right\}, \tag{6.22}$$

where $(x, z) \in \mathbb{R}^2$, and is restricted in the enlarged domain $\mathcal{H} := \text{int}\{(x, z) \in \mathbb{R}^2;\ \widetilde{\mathcal{A}}(x, z) \neq \emptyset\}$. Under the condition (6.13) and Proposition 6.1, it is easy to verify the following equivalent characterization of $\mathcal{H}$:

Lemma 6.5. *It holds that*

$$\mathcal{H} = \left\{(x, z) \in \mathbb{R}^2;\ x > z\langle \tilde{w}, \Gamma\rangle,\ \forall \Gamma \in \widetilde{\mathcal{M}}\right\} \qquad (6.23)$$
$$= \left\{(x, z) \in \mathbb{R}^2;\ x > \bar{p}z,\ z \geq 0\right\} \cup \left\{(x, z) \in \mathbb{R}^2;\ x > \underline{p}z,\ z < 0\right\}.$$

Here $\bar{p} := \sup_{Y \in \mathcal{M}} \langle w, Y\rangle = \sup_{\Gamma \in \widetilde{\mathcal{M}}} \langle \tilde{w}, \Gamma\rangle$ and $\underline{p} := \inf_{Y \in \mathcal{M}} \langle w, Y\rangle = \inf_{\Gamma \in \widetilde{\mathcal{M}}} \langle \tilde{w}, \Gamma\rangle$, where $\bar{p}, \underline{p} < \infty$ and $\mathcal{H}$ is a well defined convex cone in $\mathbb{R}^2$. Moreover, it also holds that

$$\text{cl}\mathcal{H} = \left\{(x, z) \in \mathbb{R}^2;\ \widetilde{\mathcal{A}}(x, z) \neq \emptyset\right\}$$
$$= \left\{(x, z) \in \mathbb{R}^2;\ x \geq z\langle \tilde{w}, \Gamma\rangle,\ \forall\ \Gamma \in \widetilde{\mathcal{M}}\right\}, \qquad (6.24)$$

where we recall that $\text{cl}\mathcal{H}$ denotes the closure of the set $\mathcal{H}$ in $\mathbb{R}^2$.

We now define the "auxiliary primal utility maximization problem" based on the auxiliary domain $\widetilde{\mathcal{A}}(x, z)$ as follows:

$$\tilde{u}(x, z) := \sup_{\tilde{c} \in \widetilde{\mathcal{A}}(x, z)} \mathbb{E}\left[\int_0^T U(t, \tilde{c}_t)dt\right], \quad \forall (x, z) \in \mathcal{H}. \qquad (6.25)$$

By respective definitions of $\bar{\mathcal{A}}(x, z)$ for $(x, z) \in \bar{\mathcal{H}}$ and $\widetilde{\mathcal{A}}(x, z)$ for $(x, z) \in \mathcal{H}$, we embedded our original utility maximization problem (6.15) with consumption habit formation into the auxiliary abstract utility maximization problem (6.25) without habit formation, however, with the shadow random endowment. More precisely, the following equivalence can be guaranteed that, for any $(x, z) \in \bar{\mathcal{H}} \subset \mathcal{H}$, $\bar{\mathcal{A}}(x, z) = \widetilde{\mathcal{A}}(x, z)$ and the two value functions coincide $u(x, z) = \tilde{u}(x, z)$. In addition, we have that c_t^* is the optimal solution for $u(x, z)$ if and only if $\tilde{c}_t^* = c_t^* - F_t^{c^*} \geq 0$ for all $t \in [0, T]$ is the optimal solution for $\tilde{u}(x, z)$, when $(x, z) \in \bar{\mathcal{H}}$.

6.3 The Dual Optimization Problem and Its Solvability

Inspired by the idea in Hugonnier and Kramkov (2004) for optimal investment with random endowment, we now focus on the construction of the dual problem by introducing the following set:

$$\mathcal{R} := \text{ri}\left\{(y, r) \in \mathbb{R}^2;\ xy - zr \geq 0,\ \forall (x, z) \in \mathcal{H}\right\}. \qquad (6.26)$$

Here, "ri" represents the "relative interior".

Let us make the following assumption on stochastic discounting processes $\alpha = (\alpha_t)_{t\in[0,T]}$ and $\delta = (\delta_t)_{t\in[0,T]}$:

Assumption 6.3. The random variable defined by $E := \int_0^T w_t dt = \int_0^T e^{\int_0^t (\delta_v - \alpha_v) dv} dt$ is not replicable under our original financial market, i.e., there is no constant K such that $\mathbb{E}^{\mathbb{Q}}[E] = K$ for any $\mathbb{Q} \in \mathcal{M}$.

Under Assumption 6.3 above, the existence of "market isomorphism" by Schroder and Skiadas (2002) may no longer hold. Our work can generally extend their conclusions and provide the existence and uniqueness of optimal solutions in incomplete markets using convex analysis. Then, we have

Lemma 6.6. Let Assumption 6.3 hold. Then, the set $\mathcal{R}$ is an open convex cone in $\mathbb{R}^2$, which can be rewritten as:

$$\mathcal{R} = \{(y,r) \in \mathbb{R}^2;\ y > 0,\ \underline{p}y < r < \bar{p}y\} \quad (6.27)$$

with $\bar{p}$ and $\underline{p}$ being defined in Lemma 6.5 with $\bar{p} < \underline{p}$.

For an arbitrary pair $(y,r) \in \mathcal{R}$, we denote by $\widetilde{\mathcal{Y}}(y,r)$ the set of nonnegative processes as a proper extension of the auxiliary set $\widetilde{\mathcal{M}}$ in the way that

$$\widetilde{\mathcal{Y}}(y,r) \quad (6.28)$$
$$:= \left\{\Gamma \in \mathbb{L}_+^0;\ \langle \tilde{c}, \Gamma \rangle \leq xy - zr,\ \forall \tilde{c} \in \widetilde{\mathcal{A}}(x,z) \text{ and } (x,z) \in \mathcal{H}\right\}.$$

The "auxiliary dual utility maximization problem" to (6.25) can be given by

$$\tilde{v}(y,r) := \inf_{\Gamma \in \widetilde{\mathcal{Y}}(y,r)} \mathbb{E}\left[\int_0^T V(t,\Gamma_t) dt\right], \quad \forall (y,r) \in \mathcal{R}. \quad (6.29)$$

The following theorems constitute our main results. We will provide their proofs via a number of auxiliary results in the next section.

Theorem 6.1. Let Assumptions 6.2 and 6.3 hold. Suppose also that conditions (6.5), (6.7), (6.9) (i.e., $AE_0[U] < +\infty$), (6.10) and (6.11) are satisfied. Moreover, we assume that

$$\tilde{u}(x,z) < \infty,\quad \exists\, (x,z) \in \mathcal{H}. \quad (6.30)$$

Then, we have

(i) the function $\tilde{u}$ is $(-\infty,\infty)$-valued on $\mathcal{H}$ and $\tilde{v}$ is $(-\infty,\infty]$-valued on $\mathcal{R}$. For each $(y,r) \in \mathcal{R}$, there exists a constant $s = s(y,r) > 0$ such that $\tilde{v}(sy, sr) < +\infty$. The conjugate duality of value functions $\tilde{u}$ and $\tilde{v}$ holds:

$$\tilde{u}(x,z) = \inf_{(y,r)\in\mathcal{R}} \{\tilde{v}(y,r) + xy - zr\}, \quad \forall (x,z) \in \mathcal{H},$$

$$\tilde{v}(y,r) = \sup_{(x,z)\in\mathcal{H}} \{\tilde{u}(x,z) - xy + zr\}, \quad \forall (y,r) \in \mathcal{R}.$$

(ii) the solution $\Gamma^*(y,r)$ to the optimization problem (6.29) exists and is unique (in the sense of "=" under $\bar{\mathbb{P}}$ in $\mathbb{L}_+^0$) for all $(y,r) \in \mathcal{R}$ such that $\tilde{v}(y,r) < +\infty$.

The second main result is stated as follows:

Theorem 6.2. *In addition to Assumptions of Theorem 6.1, assume also that the condition (6.8) holds (i.e., $AE_\infty[U] < 1$). Then, we have*

(i) *the value function $\tilde{v}(y,r)$ is $(-\infty,\infty)$-valued on $(y,r) \in \mathcal{R}$ and $\tilde{v}$ is continuously differentiable on $\mathcal{L}$.*
(ii) *the solution $\tilde{c}^*(x,z)$ to the optimization problem (6.25) exists and is unique (in the sense of "=" under $\bar{\mathbb{P}}$ in $\mathbb{L}_+^0$) for any $(x,z) \in \mathcal{H}$. There exists a representation of the optimal solution such that $\tilde{c}_t^*(x,z) > 0$, $\mathbb{P}$-a.s. for all $t \in [0,T]$.*
(iii) *the superdifferential of $\tilde{u}$ maps $\mathcal{H}$ into $\mathcal{R}$, i.e., $\partial \tilde{u}(x,z) \subset \mathcal{R}$ for all $(x,z) \in \mathcal{H}$. Moreover, if $(y,r) \in \partial \tilde{u}(x,z)$, then there exists a representation of the optimal solution such that $\Gamma_t^*(y,r) > 0$, $\mathbb{P}$-a.s. for all $t \in [0,T]$, and $\tilde{c}^*(x,z)$ and $\Gamma^*(y,r)$ are related by*

$$\Gamma_t^*(y,r) = U'(t, \tilde{c}_t^*(x,z)) \text{ or } \tilde{c}^*(x,z) = I(t, \Gamma_t^*(y,r)), \quad \forall t \in [0,T],$$
$$\langle \Gamma^*(y,r), \tilde{c}^*(x,z)\rangle = xy - zr. \tag{6.31}$$

(iv) *if we restrict the choice of initial wealth x and initial standard of living z such that $(x,z) \in \bar{\mathcal{H}} \subset \mathcal{H}$, the solution $c_t^*(x,z)$ to our primal utility optimization problem (6.15) exists and is unique, moreover, it holds that, for all $t \in [0,T]$,*

$$\tilde{c}_t^*(x,z) = c_t^*(x,z) - F_t^{c^*}(x,z).$$

The forthcoming section will prove main results (Theorems 6.1 and 6.2) of the chapter.

6.4 Proofs of Main Results

This section provides the detailed proofs for Theorems 6.1 and 6.2 by providing a number of auxiliary results.

6.4.1 The Proof of First Main Result

The following proposition will serve as the key step to build bipolar relationships:

Proposition 6.3. *Let Assumptions of Theorem 6.1 hold. Then, the families $(\widetilde{\mathcal{A}}(x,z))_{(x,z)\in\mathcal{H}}$ and $(\widetilde{\mathcal{Y}}(y,r))_{(y,r)\in\mathcal{R}}$ satisfy the following properties:*

(i) *For any $(x,z) \in \mathcal{H}$, the set $\widetilde{\mathcal{A}}(x,z)$ contains a strictly positive random variable on the product space. A nonnegative random variable $\tilde{c}$ belongs to $\widetilde{\mathcal{A}}(x,z)$ if and only if*

$$\langle \tilde{c}, \Gamma \rangle \leq xy - zr, \quad \forall (y,r) \in \mathcal{R} \text{ and } \Gamma \in \widetilde{\mathcal{Y}}(y,r). \tag{6.32}$$

(ii) *For any $(y,r) \in \mathcal{R}$, the set $\widetilde{\mathcal{Y}}(y,r)$ contains a strictly positive random variable on the product space. A nonnegative random variable Γ belongs to $\widetilde{\mathcal{Y}}(y,r)$ if and only if*

$$\langle \tilde{c}, \Gamma \rangle \leq xy - zr, \quad \forall (x,z) \in \mathcal{H} \text{ and } \tilde{c} \in \widetilde{\mathcal{A}}(x,z). \tag{6.33}$$

To prove Proposition 6.3, for any $p > 0$, denote by $\mathcal{M}(p)$ the subset of $\mathcal{M}$ that consists of densities $Y \in \mathcal{M}$ such that $\langle w, Y \rangle = p$. For any density process $Y \in \mathcal{M}(p)$, define the auxiliary set as follows:

$$\widetilde{\mathcal{M}}(p) := \left\{ \Gamma \in \mathbb{L}_+^0; \ \Gamma_t = Y_t + \delta_t \mathbb{E}_t \left[\int_t^T e^{\int_t^s (\delta_v - \alpha_v) dv} Y_s ds \right], \right.$$

$$\left. \forall t \in [0,T], \ Y \in \mathcal{M}(p) \right\}. \tag{6.34}$$

We then have $\langle \tilde{w}, \Gamma \rangle = \langle w, Y \rangle = p$. Define $\mathcal{P}$ as the open interval $\mathcal{P} = (\underline{p}, \bar{p})$, where $\underline{p}$ and $\bar{p}$ are given in Lemma 6.5. We have the following result:

Lemma 6.7. *Let Assumptions of Proposition 6.3 hold and $p > 0$. Then, the set $\widetilde{\mathcal{M}}(p)$ is not empty if and only if $p \in \mathcal{P} = (\underline{p}, \bar{p})$. In particular, it holds that $\bigcup_{p \in \mathcal{P}} \widetilde{\mathcal{M}}(p) = \widetilde{\mathcal{M}}$, where the set $\widetilde{\mathcal{M}}$ is defined by (6.18).*

Proof. The proof reduces to verifying that $\mathcal{P} = \mathcal{P}'$ with $\mathcal{P}' := \{p > 0; \widetilde{\mathcal{M}}(p) \neq \emptyset\}$. Similar to the proof of Lemma 8 in Hugonnier and Kramkov (2004), one direction inclusion that $\mathcal{P} \subseteq \mathcal{P}'$ is obvious. For the inverse direction, let $p \in \mathcal{P}'$, $(x,z) \in \mathrm{cl}\mathcal{H}$, $\Gamma \in \widetilde{\mathcal{M}}(p)$, and we conclude that there exists a $\tilde{c} \in \widetilde{\mathcal{A}}(x,z)$ such that $\bar{\mathbb{P}}[\tilde{c} > 0] > 0$, which implies that $0 < \langle \tilde{c}, \Gamma \rangle \leq x - zp$. Since (x,z) is an arbitrary element of $\mathrm{cl}\mathcal{H}$, it follows that $p \in \mathcal{P}$. As for the above claim, according to Theorem 2.11 in Schachermayer (2004), the condition (6.3) guarantees that, for all $Y \in \mathcal{M}$, $\underline{p} < \langle w, Y \rangle < \bar{p}$, which is $\underline{p} < \langle \tilde{w}, \Gamma \rangle < \bar{p}$, for all $\Gamma \in \widetilde{\mathcal{M}}$. By the definition of $\mathrm{cl}\mathcal{H}$ in Lemma 6.5, for any $(x,z) \in \mathrm{cl}\mathcal{H}$, we have $x - z\langle \tilde{w}, \Gamma \rangle > 0$ for all $\Gamma \in \widetilde{\mathcal{M}}$, and the claim holds by the definition of $\widetilde{\mathcal{A}}(x,z)$. $\square$

By definitions of sets $\widetilde{\mathcal{A}}(x,z)$ and $\widetilde{\mathcal{Y}}(1,p)$, we immediately obtain

Lemma 6.8. *Let Assumptions in Proposition 6.3 hold and $p \in \mathcal{P} = (\underline{p}, \bar{p})$. Then, $\widetilde{\mathcal{M}}(p) \subseteq \widetilde{\mathcal{Y}}(1,p)$.*

In light of the definition of $\widetilde{\mathcal{A}}(x,z)$, similar to the proof of Lemma 10 in Hugonnier and Kramkov (2004), it is straightforward to show the following result:

Lemma 6.9. *Let Assumptions in Proposition 6.3 hold. For any $(x,z) \in \mathcal{H}$, a nonnegative random variable $\tilde{c}$ belongs to $\widetilde{\mathcal{A}}(x,z)$ if and only if $\langle \tilde{c}, \Gamma \rangle \leq x - zp$, for all $p \in \mathcal{P}$ and $\Gamma \in \widetilde{\mathcal{M}}(p)$.*

Proof of Proposition 6.3. For the validity of the item (i), given any $(x,z) \in \mathcal{H}$, there exists a $\lambda > 0$ such that $(x - \lambda, z) \in \mathcal{H}$, since $\mathcal{H}$ is an open set. Let $\tilde{c} \in \widetilde{\mathcal{A}}(x - \lambda, z)$. For any $\Gamma \in \widetilde{\mathcal{M}}$, and $\tilde{w}_t = e^{-\int_0^t \alpha_v dv} > 0$ for all $t \in [0,T]$, we have $\langle \tilde{c}, \Gamma \rangle \leq x - \lambda - z\langle \tilde{w}, \Gamma \rangle$. Using the condition (6.13) and Proposition 6.1, we define $\rho_t := \frac{\lambda}{\bar{p}}\tilde{w}_t > 0$ for all $t \in [0,T]$. Then, for all $\Gamma \in \widetilde{\mathcal{M}}$, we obtain

$$\langle \rho, \Gamma \rangle \leq \langle \tilde{c} + \rho, \Gamma \rangle \leq x - \lambda - z\langle \tilde{w}, \Gamma \rangle + \frac{\lambda}{\bar{p}}\langle \tilde{w}, \Gamma \rangle \leq x - \lambda - z\langle \tilde{w}, \Gamma \rangle + \lambda$$
$$\leq x - z\langle \tilde{w}, \Gamma \rangle.$$

Consequently, the existence of a strictly positive element $\rho_t \in \widetilde{\mathcal{A}}(x,z)$ follows by the definition of $\widetilde{\mathcal{A}}(x,z)$. If the estimate (6.32) holds for some $\tilde{c} \in \mathbb{L}_+^0$. The density process $\Gamma \in \widetilde{\mathcal{M}}(p)$ belongs to $\widetilde{\mathcal{Y}}(1,p)$ for all $p \in \mathcal{P}$ by Lemma 6.8, and hence $\langle \tilde{c}, \Gamma \rangle \leq x - zp$ for all $p \in \mathcal{P}$ and $\Gamma \in \widetilde{\mathcal{M}}(p)$. Lemma 6.9 then implies that $\tilde{c} \in \widetilde{\mathcal{A}}(x,z)$. Conversely, suppose $\tilde{c} \in \widetilde{\mathcal{A}}(x,z)$,

the definition of set $\widetilde{\mathcal{Y}}(y,r)$, $(y,r) \in \mathcal{R}$ implies (6.32) and we complete the proof of the item (i). For the item (ii), we first have

$$k\widetilde{\mathcal{Y}}(y,r) = \widetilde{\mathcal{Y}}(ky, kr), \quad \forall k > 0, \ (y,r) \in \mathcal{R}.$$

Hence, it is enough to consider $(y,r) = (1,p)$ for some $p \in \mathcal{P}$. Lemma 6.8 implies $\Gamma \in \widetilde{\mathcal{M}}(p) \subseteq \widetilde{\mathcal{Y}}(1,p)$, and the existence of strictly positive $Y \in \mathcal{M}(p)$ implies the existence $\Gamma \in \widetilde{\mathcal{M}}(p)$ and $\Gamma > 0$, $\bar{\mathbb{P}}$-a.s.. The second part is a direct consequence of the definition of $\widetilde{\mathcal{Y}}(y,r)$. □

For the proof of Theorem 6.1, we will also need the following results:

Lemma 6.10. *Let Assumptions of Theorem 6.1 hold. Then, the value function $\tilde{u}$ is $(-\infty, \infty)$-valued on $\mathcal{H}$.*

Proof. By applying Lemma 6.2, the condition $AE_0[U] < +\infty$ implies that, for any constant $s > 0$, there exist $s_1 > 0$ and $s_2 > 0$ such that, for all $t \in [0,T]$, $U(t, x/s) \geq s_1 U(t,x) + s_2$ with $x > 0$. Using the condition (6.14) and the proof of Proposition 6.3, for each fixed pair $(x,z) \in \mathcal{H}$, there exists $\lambda = \lambda(x,z) > 0$ such that $\frac{\lambda}{\bar{p}}\tilde{w}_t \in \widetilde{\mathcal{A}}(x,z)$. W then deduce that $\bar{x}\tilde{w}_t \in \widetilde{\mathcal{A}}(\frac{\bar{x}\bar{p}}{\lambda}x, \frac{\bar{x}\bar{p}}{\lambda}z)$, and

$$\tilde{u}\left(\frac{\bar{x}\bar{p}}{\lambda}x, \frac{\bar{x}\bar{p}}{\lambda}z\right) = \sup_{\tilde{c} \in \widetilde{\mathcal{A}}(\frac{\bar{x}\bar{p}}{\lambda}x, \frac{\bar{x}\bar{p}}{\lambda}z)} \mathbb{E}\left[\int_0^T U(t, \tilde{c}_t)dt\right]$$

$$\geq \mathbb{E}\left[\int_0^T U(t, \bar{x}\tilde{w}_t)dt\right] > -\infty.$$

Thus, for any $(x,z) \in \mathcal{H}$, there exists a constant $s(x,z) > 0$ such that $\tilde{u}(sx, sz) > -\infty$, with $s(x,z) = \frac{\bar{x}\bar{p}}{\lambda}$. For any constant $s > 0$, $\widetilde{\mathcal{A}}(x,z) = \widetilde{\mathcal{A}}(sx, sz)/s$ which implies that $\tilde{u}(x,z) > -\infty$ if $\tilde{u}(sx, sz) > -\infty$ holds for a constant $s = s(x,z) > 0$. With the above result, we can conclude that $\tilde{u}(x,z) > -\infty$ in the whole domain $\mathcal{H}$. Now, since the set $\mathcal{H}$ is open and $\tilde{u}(x,z) < +\infty$ for some $(x,z) \in \mathcal{H}$ under the condition (6.30), we deduce that $\tilde{u}$ is finitely valued on $\mathcal{H}$ by the concavity of $\tilde{u}$ on $\mathcal{H}$. Thus, the proof is complete. □

Before stating the following lemma, let us introduce the definition on "convexly compact set" given by Žitković (2009).

Definition 6.1 (Convexly Compact Set). A convex subset C of a topological vector space X is called *convexly compact* if for any non-empty set

A and any family $(F_a)_{a\in A}$ of closed, convex subsets of C, the condition

$$\forall D \in Fin(A), \quad \bigcap_{a\in D} F_a \neq \emptyset \implies \bigcap_{a\in A} F_a \neq \emptyset,$$

where the set $Fin(A)$ consists of all non-empty finite subsets of A.

An easy characterization on the space of non-negative and measurable functions was derived in Theorem 3.1 of Žitković (2009). Here, we will modify his result to fit into our framework.

Lemma 6.11. *A closed and convex subset C of $\mathbb{L}_+^0(\Omega \times [0,T], \mathcal{O}, \bar{\mathbb{P}})$ is convexly compact if and only if it is bounded in the finite measure $\bar{\mathbb{P}}$.*

Based on the above lemma, we have the following result on the convexly compactness of sets $\widetilde{\mathcal{A}}(x,z)$ and $\widetilde{\mathcal{Y}}(y,r)$:

Lemma 6.12. *For each pair $(x,z) \in \mathcal{H}$ and $(y,r) \in \mathcal{R}$, the sets $\widetilde{\mathcal{A}}(x,z)$ and $\widetilde{\mathcal{Y}}(y,r)$ are convex, solid and closed in the topology of convergence in measure $\bar{\mathbb{P}}$. Moreover, they are both bounded in $\mathbb{L}_+^0(\Omega \times [0,T], \mathcal{O}, \bar{\mathbb{P}})$, and hence they are both convexly compact.*

Proof. For $(y,r) \in \mathcal{R}$, we define auxiliary sets as follows:

$$\begin{cases} \mathfrak{H}(y,r) := \{(x,z) \in \mathcal{H}; \ xy - zr \leq 1\}, \\ \mathfrak{A}(k) := \bigcup_{(x,z) \in k\mathfrak{H}(y,r)} \widetilde{\mathcal{A}}(x,z). \end{cases} \quad (6.35)$$

Denote by $\widetilde{\mathfrak{A}}(k)$ the closure of $\mathfrak{A}(k)$ w.r.t. convergence in measure $\bar{\mathbb{P}}$. It follows from Proposition 6.3 that

$$\Gamma \in \widetilde{\mathcal{Y}}(y,r) \Leftrightarrow \langle \tilde{c}, \Gamma \rangle \leq 1, \quad \forall \tilde{c} \in \widetilde{\mathfrak{A}}(1).$$

Therefore, sets $\widetilde{\mathcal{Y}}(y,r)$ and $\widetilde{\mathfrak{A}}(1)$ satisfy $\widetilde{\mathcal{Y}}(y,r) = \widetilde{\mathfrak{A}}(1)^\circ$. Meanwhile, by its definition, $\widetilde{\mathfrak{A}}(1)$ itself is closed, convex and solid. The bipolar theorem in Brannath and Schachermayer (1999) asserts that $\widetilde{\mathfrak{A}}(1) = \widetilde{\mathfrak{A}}(1)^{\circ\circ}$, and hence the following bipolar relationship holds:

$$\widetilde{\mathfrak{A}}(1) = \widetilde{\mathcal{Y}}(y,r)^\circ, \quad \widetilde{\mathcal{Y}}(y,r) = \widetilde{\mathfrak{A}}(1)^\circ. \quad (6.36)$$

The bipolar theorem on $\mathbb{L}_+^0$ implies that $\mathcal{Y}(y,r)$ is convex, solid and closed under the convergence in measure $\bar{\mathbb{P}}$. Similarly, for $(x,z) \in \mathcal{H}$, we define the set by

$$\mathfrak{R}(x,z) := \{(y,r) \in \mathcal{R}; \ xy-zr \leq 1\}, \quad \mathfrak{Y}(k) := \bigcup_{(y,r) \in k\mathfrak{R}(x,z)} \widetilde{\mathcal{Y}}(y,r). \quad (6.37)$$

Denote by $\widetilde{\mathfrak{Y}}(k)$ the closure of $\mathfrak{Y}(k)$ w.r.t. convergence in measure $\bar{\mathbb{P}}$. Again, Proposition 6.3 implies that

$$\tilde{c} \in \widetilde{\mathcal{A}}(x,z) \Leftrightarrow \langle \tilde{c}, \Gamma \rangle \leq 1, \ \forall \Gamma \in \widetilde{\mathfrak{Y}}(1),$$

and the following bipolar relationship:

$$\widetilde{\mathfrak{Y}}(1) = \widetilde{\mathcal{A}}(x,z)^\circ, \quad \widetilde{\mathcal{A}}(x,z) = \widetilde{\mathfrak{Y}}(1)^\circ. \tag{6.38}$$

Hence, $\widetilde{\mathcal{A}}(x,z)$ is also convex, solid and closed under convergence in measure $\bar{\mathbb{P}}$. Thanks to the existence of strictly positive $\Gamma \in \widetilde{\mathcal{M}}(p)$ which is also in $\widetilde{\mathcal{Y}}(1,p)$, the set $\widetilde{\mathcal{A}}(x,z)$ is therefore bounded in measure $\bar{\mathbb{P}}$ by Proposition 6.3-(i). Similarly, as in the proof of Proposition 6.3, there exists $\lambda = \lambda(x,z)$ such that $\rho_t > 0$ for all $t \in [0,T]$ and $\rho_t = \frac{\lambda}{p}\tilde{w}_t \in \widetilde{\mathcal{A}}(x,z)$. Thanks to Proposition 6.3-(ii), the set $\widetilde{\mathcal{Y}}(y,r)$ is also bounded in measure $\bar{\mathbb{P}}$. Hence, both of them are convexly compact in $\mathbb{L}^0_+$. $\square$

Contrary to the existing literature, we can not mimic the classical proof of the existence of the dual optimizer due to the lack of integrability of the dual process $\Gamma \in \widetilde{\mathcal{Y}}(y,r)$ for $(y,r) \in \mathcal{R}$. Conditions $AE_0[U] < \infty$ and $\mathbb{E}[\int_0^T U(t, \bar{x}\tilde{w}_t)dt] > -\infty$ are critical to prove lemmas below.

Lemma 6.13. *Let Assumptions of Theorem 6.1 hold. Then, for any $(y,r) \in \mathcal{R}$ fixed, we have*

$$\sup_{\Gamma \in \widetilde{\mathcal{Y}}(y,r)} \mathbb{E}\left[\int_0^T V^-(t,\Gamma_t)dt\right] < +\infty.$$

Proof. The condition (6.14) yields the existence of $\bar{x}\tilde{w}_t \in \mathbb{L}^0_+$ such that $\mathbb{E}[\int_0^T U(t, \bar{x}\tilde{w}_t)dt] > -\infty$. Furthermore, by the proof of Proposition 6.3, for each fixed $(y,r) \in \mathcal{R}$, we may find a pair $(x,z) \in \mathfrak{H}(y,r)$ and there exists a constant $\lambda(x,z) > 0$ such that $\tilde{w} \in \widetilde{\mathfrak{A}}(\frac{\bar{x}}{\lambda})$. Taking into account the inequality $U(t,x) \leq V(t,y) + xy$, for any $\Gamma \in \widetilde{\mathcal{Y}}(y,r)$ and $y_0(t) := \inf\{y > 0; V(t,y) < 0\}$, we have

$$\mathbb{E}\left[\int_0^T V^-(t,\Gamma_t)dt\right] \leq -\mathbb{E}\left[\int_0^T V(t,\Gamma_t \mathbf{1}_{\{\Gamma_t \geq y_0(t)\}} + y_0(t)\mathbf{1}_{\{\Gamma_t < y_0(t)\}})dt\right]$$

$$\leq -\mathbb{E}\left[\int_0^T U(t,\bar{x}\tilde{w}_t)dt\right] + \bar{x}\mathbb{E}\left[\int_0^T \tilde{w}_t\Gamma_t dt\right]$$

$$+ \bar{x}\mathbb{E}\left[\int_0^T \tilde{w}_t(y_0(t) - \Gamma_t)\mathbf{1}_{\{\Gamma_t < y_0(t)\}}dt\right]$$

$$\leq -\mathbb{E}\left[\int_0^T U(t,\bar{x}\tilde{w}_t)dt\right] + \bar{x}\frac{\bar{p}}{\lambda} + \bar{x}\int_0^T y_0(t)dt.$$

The last term is finitely valued and independent of the initial choice of Γ, since $\tilde{w}_t = e^{\int_0^t(-\alpha_v)dv} \leq 1$ for all $t \in [0,T]$, and $\sup_{t\in[0,T]} y_0(t) < +\infty$ by the condition (6.11). Thus, the conclusion is valid. □

Lemma 6.14. *Let Assumptions of Theorem 6.1 hold. Then, for any $(y,r) \in \mathcal{R}$, the family $(V^-(\cdot,\Gamma))_{\Gamma \in \tilde{\mathcal{Y}}(y,r)}$ is uniformly integrable (U.I.).*

Proof. By Lemma 6.2, the condition $AE_0[U] < +\infty$ is equivalent to the following assertion:

$$\exists\, y_0 > 0 \text{ and } \mu \in (1,2), \text{ s.t. } V(t,2y) \geq \mu V(t,y), \forall y \geq y_0. \tag{6.39}$$

Let $y_0 > 0$ and $\mu \in (1,2)$ be constants in (6.39). Taking $\gamma = \log_2 \mu \in (0,1)$, define the auxiliary function $\tilde{V}(t,y) : [0,\infty) \times \mathbb{R}_+ \to \mathbb{R}$ by

$$\tilde{V}(t,y) := \begin{cases} -\frac{2y_0}{\gamma}V'(t,2y_0) - V(t,y), & y \geq 2y_0, \\ -V(t,2y_0) - \frac{2y_0}{\gamma}V'(t,2y_0)(\frac{y}{2y_0})^\gamma, & y < 2y_0. \end{cases} \tag{6.40}$$

For each fixed $t > 0$, $\tilde{V}(t,y)$ is a nonnegative, concave, and nondecreasing function which agrees with $-V(t,y)$ up to a constant for large enough values of y and satisfies that

$$\tilde{V}(t,2y) \leq \mu \tilde{V}(t,y), \quad \forall y > 0. \tag{6.41}$$

It follows from Lemma 6.13 that

$$\sup_{\Gamma \in \tilde{\mathcal{Y}}(y,r)} \mathbb{E}\left[\int_0^T V^-(t,\Gamma_t)dt\right] < \infty,$$

and hence in light of the fact that V^- and $\tilde{V}$ differ only by a constant in a neighborhood of $+\infty$, we will get

$$\sup_{\Gamma \in \tilde{\mathcal{Y}}(y,r)} \mathbb{E}\left[\int_0^T \tilde{V}(t,\Gamma_t)dt\right] < +\infty. \tag{6.42}$$

The validity of uniform integrability of the sequence $(V^-(\cdot,\Gamma^n))_{n\geq 1}$ for $\Gamma^n \in \tilde{\mathcal{Y}}(y,r)$, is therefore equivalent to the uniform integrability of $(\tilde{V}(\cdot,\Gamma^n))_{n\geq 1}$. To this end, we argue by contradiction. Suppose this sequence is not U.I., then by Rosenthal's subsequence splitting lemma, we can find a subsequence $(f^n)_{n\geq 1}$, a constant $\varepsilon > 0$ and a disjoint sequence $(A^n)_{n\geq 1}$ of $(\Omega \times [0,T], \mathcal{O})$ with $A^n \in \mathcal{O}$ and $A^i \cap A^j = \emptyset$ if $i \neq j$ such that

$$\mathbb{E}\left[\int_0^T \widetilde{V}(t, f_t^n) \mathbf{1}_{A^n} dt\right] \geq \varepsilon, \quad \forall n \geq 1.$$

Define the sequence of random variables $(h^n)_{n\geq 1}$ as $h_t^n := \sum_{k=1}^n f_t^k \mathbf{1}_{A^k}$. For any $\tilde{c} \in \widetilde{\mathfrak{A}}(1)$, we arrive at $\langle \tilde{c}, h^n \rangle \leq \sum_{k=1}^n \langle \tilde{c}, f^k \rangle \leq n$. Therefore, $\frac{h^n}{n} \in \widetilde{\mathcal{Y}}(y,r)$. Additionally, we have

$$\mathbb{E}\left[\int_0^T \widetilde{V}(t, h_t^n) dt\right] \geq \sum_{k=1}^n \mathbb{E}\left[\int_0^T \widetilde{V}(t, f_t^k) \mathbf{1}_{A^k} dt\right] \geq \varepsilon n.$$

Then, by taking $n = 2^m$, via iteration, it produces that

$$\mu^m \sup_{\Gamma_t \in \widetilde{\mathcal{Y}}(y,r)} \mathbb{E}\left[\int_0^T \widetilde{V}(t, \Gamma_t) dt\right] \geq \mu^m \mathbb{E}\left[\int_0^T \widetilde{V}(t, \frac{h_t^{2^m}}{2^m}) dt\right]$$

$$\geq \mathbb{E}\left[\int_0^T \widetilde{V}(t, h_t^{2^m}) dt\right] \geq 2^m \varepsilon.$$

Since $\mu \in (1,2)$, this contradicts (6.42) for m large enough, therefore the conclusion follows. $\square$

Following the classical analysis, Lemma 6.14 together with Fatou's lemma can deduce the existence of the dual optimizer.

Lemma 6.15. *For any pair $(y,r) \in \mathcal{R}$ such that $\tilde{v}(y,r) < +\infty$, the optimal solution Γ^* to the optimization problem (6.29) exists and is unique.*

For the proof of conjugate duality between value functions $\tilde{u}(x,z)$ and $\tilde{v}(y,r)$, similar to Lemma 11 in Hugonnier and Kramkov (2004), we have the following result:

Lemma 6.16. *Let $\mathcal{G} \subseteq \mathbb{L}_+^0$ be convex and contain a strictly positive random variable. Then, it holds that*

$$\sup_{g \in \mathcal{G}} \mathbb{E}\left[\int_0^T U(t, xg_t) dt\right] = \sup_{g \in \mathrm{cl}\mathcal{G}} \mathbb{E}\left[\int_0^T U(t, xg_t) dt\right], \quad \forall x > 0,$$

where $\mathrm{cl}\mathcal{G}$ denotes the closure of $\mathcal{G}$ w.r.t. convergence in the measure $\bar{\mathbb{P}}$.

It also holds that

Lemma 6.17. *For $\tilde{w}_t = e^{-\int_0^t \alpha_v dv}$, we have*

$$\mathbb{E}\left[\int_0^T V^-(t, U'(t, \tilde{w}_t)) dt\right] < +\infty. \quad (6.43)$$

Proof. Similar to the proof of Lemma 6.13, we have $\mathbb{E}[\int_0^T U(t, \bar{x}\tilde{w}_t)dt] > -\infty$. Using the inequality $U(t, x) < V(t, y) + xy$, for any $y_0(t) := \inf\{y > 0; V(t, y) < 0\}$, we have

$$\mathbb{E}\left[\int_0^T V^-(t, U'(t, \tilde{w}_t))dt\right]$$

$$\leq -\mathbb{E}\left[\int_0^T V(t, U'(t, \tilde{w}_t))\mathbf{1}_{\{U'(t,\tilde{w}_t)\geq y_0(t)\}} + y_0(t)\mathbf{1}_{\{U'(t,\tilde{w}_t)<y_0(t)\}})dt\right]$$

$$\leq -\mathbb{E}\left[\int_0^T U(t, \bar{x}\tilde{w}_t)dt\right] + \bar{x}\mathbb{E}\left[\int_0^T \tilde{w}_t U'(t, \tilde{w}_t)dt\right]$$

$$+ \bar{x}\mathbb{E}\left[\int_0^T \tilde{w}_t(y_0(t) - U'(t, \tilde{w}_t))\mathbf{1}_{\{U'(t,\tilde{w}_t)<y_0(t)\}}dt\right] \quad (6.44)$$

$$\leq -\mathbb{E}\left[\int_0^T U(t, \bar{x}\tilde{w}_t)dt\right] + \bar{x}\mathbb{E}\left[\int_0^T \tilde{w}_t U'(t, \tilde{w}_t)dt\right] + \bar{x}\int_0^T y_0(t)dt.$$

We already know the 1st term and the 3rd term are bounded, as for the 2nd term, we have two different cases:

Case 1. If $\bar{x} \leq 1$, the 2nd term can be rewritten as:

$$\mathbb{E}\left[\int_0^T \tilde{w}_t U'(t, \tilde{w}_t)dt\right] = \mathbb{E}\left[\int_0^T \tilde{w}_t U'(t, \tilde{w}_t)\mathbf{1}_{\{\tilde{w}_t \leq x_0\}}dt\right]$$

$$+ \mathbb{E}\left[\int_0^T \tilde{w}_t U'(t, \tilde{w}_t)\mathbf{1}_{\{\tilde{w}_t > x_0\}}dt\right],$$

where x_0 is the uniform constant in Lemma 6.2 such that, for all $t \in [0, T]$,

$$xU'(t, x) < \left(\frac{\gamma}{1-\gamma}\right)(-U(t, x)), \quad \forall x \in (0, x_0]. \quad (6.45)$$

Again, by the fact that $\tilde{w}_t \leq 1$ for all $t \in [0, T]$, it follows that $\mathbb{E}[\int_0^T \tilde{w}_t U'(t, \tilde{w}_t)\mathbf{1}_{\{\tilde{w}_t > x_0\}}dt] < +\infty$, and we also have

$$\mathbb{E}\left[\int_0^T \tilde{w}_t U'(t, \tilde{w}_t)\mathbf{1}_{\{\tilde{w}_t \leq x_0\}}dt\right] \leq -\left(\frac{\gamma}{1-\gamma}\right)\mathbb{E}\left[\int_0^T U(t, \tilde{w}_t)\mathbf{1}_{\{\tilde{w}_t \leq x_0\}}dt\right]$$

$$\leq \left(\frac{\gamma}{1-\gamma}\right)\mathbb{E}\left[\int_0^T U^-(t, \bar{x}\tilde{w}_t)dt\right] < +\infty,$$

by using the inequality (6.45), the increasing property of $U(t, x)$ with respect to x and the condition (6.14).

Case 2. If $\bar{x} > 1$, the 2nd term can be rewritten as follows:

$$\mathbb{E}\left[\int_0^T \tilde{w}_t U'(t,\tilde{w}_t)dt\right] = \mathbb{E}\left[\int_0^T \tilde{w}_t U'(t,\tilde{w}_t)\mathbf{1}_{\{\bar{x}\tilde{w}_t \leq x_0\}}dt\right]$$
$$+ \mathbb{E}\left[\int_0^T \tilde{w}_t U'(t,\tilde{w}_t)\mathbf{1}_{\{\bar{x}\tilde{w}_t > x_0\}}dt\right],$$

where x_0 is the uniform constant in Lemma 6.2 such that for all $t \in [0,T]$, the inequality (6.45) holds, and moreover, for all $t \in [0,T]$,

$$U\left(t, \frac{1}{\bar{x}}x\right) > \left(\frac{1}{\bar{x}}\right)^{-\frac{\gamma}{1-\gamma}} U(t,x), \quad \forall x \in (0, x_0]. \tag{6.46}$$

Again, the 2nd term is bounded, since $\bar{x}\tilde{w}_t \leq \bar{x}$ for all $t \in [0,T]$, and for the 1st term, we have

$$\mathbb{E}\left[\int_0^T \tilde{w}_t U'(t,\tilde{w}_t)\mathbf{1}_{\{\bar{x}\tilde{w}_t \leq x_0\}}dt\right] \leq -\left(\frac{\gamma}{1-\gamma}\right)\mathbb{E}\left[\int_0^T U(t,\tilde{w}_t)\mathbf{1}_{\{\bar{x}\tilde{w}_t \leq x_0\}}dt\right]$$
$$\leq \left(\frac{\gamma}{1-\gamma}\right)\left(\frac{1}{\bar{x}}\right)^{-\frac{\gamma}{1-\gamma}} \mathbb{E}\left[\int_0^T U^-(t,\bar{x}\tilde{w}_t)dt\right] < +\infty,$$

by using (6.45), (6.46) and the condition (6.14). Then, the 2nd term in (6.44) is finite, and hence (6.43) holds. $\square$

We here emphasize that we have to revise the classical minimax theorem based on $\mathbb{L}^1$ to derive the important conjugate duality relationship. The following minimax theorem by Kauppila (2010) (c.f. Theorem A.1 of Appendix A therein) can serve as a substitute tool on the space $\mathbb{L}_+^0$ for convexly compact sets.

Theorem 6.3 (Minimax Theorem). *Let A be a nonempty convex subset of a topological space, and B a nonempty, closed, convex and convexly compact subset of a topological vector space. Let $H: A \times B \to \mathbb{R}$ be convex on A, concave and upper-semicontinuous on B. Then, it holds that*

$$\sup_B \inf_A H = \inf_A \sup_B H.$$

We then have

Lemma 6.18. *Let Assumptions of Theorem 6.1 hold. Then, the conjugate duality results hold:*

$$\tilde{u}(x,z) = \inf_{(y,r) \in \mathcal{R}}\{\tilde{v}(y,r) + xy - zr\}, \quad \forall (x,z) \in \mathcal{H}, \tag{6.47}$$

$$\tilde{v}(y,r) = \sup_{(x,z) \in \mathcal{H}}\{\tilde{u}(x,z) - xy + zr\}, \quad \forall (y,r) \in \mathcal{R}. \tag{6.48}$$

Proof. For $n > 0$, define $\mathcal{S}_n \subset \mathbb{L}_+^0(\Omega \times [0,T], \mathcal{O}, \bar{\mathbb{P}})$ by $\mathcal{S}_n := \{\tilde{c} \in \mathbb{L}_+^0;\ 0 \leq \tilde{c} \leq n\tilde{w}\}$. Then, $\mathcal{S}_n$ is closed, convex and bounded in probability, and hence convexly compact in $\mathbb{L}_+^0$. It is easy to verify that the functional $\tilde{c} \mapsto \mathbb{E}[\int_0^T (U(t, \tilde{c}_t) - \tilde{c}_t \Gamma_t) dt]$ is upper-semicontinuous on $\mathcal{S}_n$ under convergence in measure $\bar{\mathbb{P}}$, for all $\Gamma \in \tilde{\mathcal{Y}}(y,r)$ and $(y,r) \in \mathcal{R}$. Lemma 6.12 yields that $\tilde{\mathcal{Y}}(y,r)$ is a closed convex subset of $\mathbb{L}_+^0$, we can use the above Minimax Theorem 6.3 to get the following equality, for fixed n,

$$\sup_{\tilde{c} \in \mathcal{S}_n} \inf_{\Gamma \in \tilde{\mathcal{Y}}(y,r)} \mathbb{E}\left[\int_0^T (U(t, \tilde{c}_t) - \tilde{c}_t \Gamma_t) dt\right]$$
$$= \inf_{\Gamma \in \tilde{\mathcal{Y}}(y,r)} \sup_{\tilde{c} \in \mathcal{S}_n} \mathbb{E}\left[\int_0^T (U(t, \tilde{c}_t) - \tilde{c}_t \Gamma_t) dt\right].$$

Using the bipolar relationship (6.36) and the definition, we have $\bigcup_{(x,z) \in \mathcal{H}} \tilde{\mathcal{A}}(x,z) = \bigcup_{k>0} \tilde{\mathfrak{A}}(k)$. To continue the proof, we define the auxiliary set as follows:

$$\mathfrak{A}'(k) := \left\{\tilde{c} \in \tilde{\mathfrak{A}}(k);\ \sup_{\Gamma \in \tilde{\mathcal{Y}}(y,r)} \langle \tilde{c}, \Gamma \rangle = k\right\}.$$

It is obvious to have that

$$\bigcup_{k>0} \tilde{\mathfrak{A}}(k) = \bigcup_{(x,z) \in \mathcal{H}} \tilde{\mathcal{A}}(x,z) = \bigcup_{k>0} \mathfrak{A}'(k). \tag{6.49}$$

We first prove that

$$\limsup_{n \to \infty} \inf_{\tilde{c} \in \mathcal{S}_n} \inf_{\Gamma \in \tilde{\mathcal{Y}}(y,r)} \mathbb{E}\left[\int_0^T (U(t, \tilde{c}_t) - \tilde{c}_t \Gamma_t) dt\right]$$
$$= \sup_{k>0} \sup_{\tilde{c} \in \mathfrak{A}'(k)} \inf_{\Gamma \in \tilde{\mathcal{Y}}(y,r)} \mathbb{E}\left[\int_0^T (U(t, \tilde{c}_t) - \tilde{c}_t \Gamma_t) dt\right]. \tag{6.50}$$

The direction of the inequality "$\geq$" holds by

$$\limsup_{n \to \infty} \inf_{\tilde{c} \in \mathcal{S}_n} \inf_{\Gamma \in \tilde{\mathcal{Y}}(y,r)} \mathbb{E}\left[\int_0^T (U(t, \tilde{c}_t) - \tilde{c}_t \Gamma_t) dt\right]$$
$$\geq \lim_{n \to \infty} \sup_{\tilde{c} \in \mathfrak{A}'(k) \cap \mathcal{S}_n} \inf_{\Gamma \in \tilde{\mathcal{Y}}(y,r)} \mathbb{E}\left[\int_0^T (U(t, \tilde{c}_t) - \tilde{c}_t \Gamma_t) dt\right]$$
$$= \sup_{\tilde{c} \in \mathfrak{A}'(k)} \inf_{\Gamma \in \tilde{\mathcal{Y}}(y,r)} \mathbb{E}\left[\int_0^T (U(t, \tilde{c}_t) - \tilde{c}_t \Gamma_t) dt\right], \quad \forall k > 0.$$

The other direction "$\leq$" is obvious since for any $(x,z) \in \mathcal{H}$, we have $n\tilde{w} \in \mathfrak{A}'(n\bar{p})$, and hence $\mathcal{S}_n \subset \mathfrak{A}'(n\bar{p})$.

To continue the proof, we need to prepare finiteness results as below: From definitions in Lemma 6.12 and by Lemma 6.16, it is easy to see that

$$\sup_{\tilde{c} \in \widetilde{\mathfrak{A}}(k)} \mathbb{E}\left[\int_0^T U(t, \tilde{c}_t) dt\right] = \sup_{\tilde{c} \in \mathfrak{A}(k)} \mathbb{E}\left[\int_0^T U(t, \tilde{c}_t) dt\right]$$

$$= \sup_{(x,z) \in k\mathfrak{H}(y,r)} \tilde{u}(x, z), \quad \forall k > 0. \tag{6.51}$$

Since the set $\mathcal{R}$ is open and the set $\mathfrak{H}(y,r)$ is bounded, from the concavity of $\tilde{u}$ and the fact $\tilde{u}(x,z) < \infty$ for all $(x,z) \in \mathcal{H}$, it follows that

$$\sup_{(x,z) \in k\mathfrak{H}(y,r)} \tilde{u}(x,z) < \infty, \quad \forall k > 0. \tag{6.52}$$

Now, by (6.49), (6.52) and the definition of domain $\mathcal{H}$, we further have

$$\sup_{k>0} \sup_{\tilde{c} \in \mathfrak{A}'(k)} \inf_{\Gamma \in \widetilde{\mathcal{Y}}(y,r)} \mathbb{E}\left[\int_0^T (U(t,\tilde{c}_t) - \tilde{c}_t \Gamma_t) dt\right]$$

$$= \sup_{k>0} \left\{ \sup_{\tilde{c} \in \mathfrak{A}'(k)} \mathbb{E}\left[\int_0^T U(t,\tilde{c}_t) dt\right] - k \right\}$$

$$= \sup_{k>0} \left\{ \sup_{\tilde{c} \in \widetilde{\mathfrak{A}}(k)} \mathbb{E}\left[\int_0^T U(t,\tilde{c}_t) dt\right] - k \right\}$$

$$= \sup_{k>0} \left\{ \sup_{(x,z) \in k\mathfrak{H}(y,r)} \tilde{u}(x,z) - k \right\}$$

$$= \sup_{(x,z) \in \mathcal{H}} \{\tilde{u}(x,z) - xy + zr\}.$$

On the other hand, it holds that

$$\inf_{\Gamma \in \widetilde{\mathcal{Y}}(y,r)} \sup_{\tilde{c} \in \mathcal{S}_n} \mathbb{E}\left[\int_0^T (U(t,\tilde{c}_t) - \tilde{c}_t \Gamma_t) dt\right] = \inf_{\Gamma \in \widetilde{\mathcal{Y}}(y,r)} \mathbb{E}\left[\int_0^T V^n(t, \Gamma_t, \omega) dt\right]$$

$$=: \tilde{v}^n(y,r),$$

where we define $V^n(t,y,\omega)$ according to the definition of set $\mathcal{S}_n$ as $V^n(t,y,\omega) = \sup_{0 < x \leq n\tilde{w}} (U(t,x) - xy)$. Consequently, it is enough to show that, for all $(y,r) \in \mathcal{R}$,

$$\lim_{n \to \infty} \tilde{v}^n(y,r) = \lim_{n \to \infty} \inf_{\Gamma \in \widetilde{\mathcal{Y}}(y,r)} \mathbb{E}\left[\int_0^T V^n(t, \Gamma_t, \omega) dt\right] = \tilde{v}(y,r).$$

Evidently, $\tilde{v}^n(y,r) \leq \tilde{v}(y,r)$ for all $n \geq 1$. Let $(\Gamma^n)_{n\geq 1}$ be a sequence in $\widetilde{\mathcal{Y}}(y,r)$ such that

$$\lim_{n\to\infty} \mathbb{E}\left[\int_0^T V^n(t,\Gamma_t^n)dt\right] = \lim_{n\to\infty} \tilde{v}^n(y,r).$$

There exists a sequence $h^n \in \text{conv}(\Gamma^n, \Gamma^{n+1}, \ldots)$ with $n \geq 1$, converging almost surely to a random variable Γ. Then $\Gamma \in \widetilde{\mathcal{Y}}(y,r)$ is verified because the set $\widetilde{\mathcal{Y}}(y,r)$ is closed under convergence in finite measure $\bar{\mathbb{P}}$.

We claim that the sequence of processes $(V^n(\cdot, h_\cdot^n, \omega)^-)_{n\geq 1}$ is U.I.. In fact, we can rewrite

$$(V^n(t, h_t^n, \omega))^- \quad (6.53)$$
$$= (V^n(t, h_t^n, \omega))^- \mathbf{1}_{\{h_t^n \leq U'(t,\tilde{w}_t)\}} + (V^n(t, h_t^n, \omega))^- \mathbf{1}_{\{h_t^n > U'(t,\tilde{w}_t)\}},$$

since $V^n(t,y,\omega) = V(t,y)$ for $y \geq U'(t,\tilde{w}_t) \geq U'(t, n\tilde{w}_t)$ by definition. The proof of Lemma 6.14 gives the U.I. of the sequence of processes $((V^n(\cdot, h_\cdot^n, \omega))^- \mathbf{1}_{\{h_\cdot^n > U'(\cdot, \tilde{w}_\cdot)\}})_{n\geq 1}$.

On the other hand, using the monotonicity of $(V^n)^-$ with $n > 1$, it follows that

$$(V^n(t, h_t^n, \omega))^- \mathbf{1}_{\{h_t^n \leq U'(t,\tilde{w}_t)\}} \leq (V^1(t, h_t^n, \omega))^- \mathbf{1}_{\{h_t^n \leq U'(t,\tilde{w}_t)\}}$$
$$\leq (V(t, U'(t,\tilde{w}_t)))^-.$$

By virtue of Lemma 6.17, the RHS is integrable in the product space, and hence $(V^n(\cdot, h_\cdot^n))^- \mathbf{1}_{\{h_\cdot^n \leq U'(\cdot, \tilde{w}_\cdot)\}}$, $n \geq 1$ is also U.I.. Thus, our claim holds. Moreover, we will have the following inequalities:

$$\lim_{n\to\infty} \mathbb{E}\left[\int_0^T V^n(t,\Gamma_t^n, \omega)dt\right] \geq \liminf_{n\to\infty} \mathbb{E}\left[\int_0^T V^n(t, h_t^n, \omega)dt\right]$$
$$\geq \mathbb{E}\left[\int_0^T V(t, \Gamma_t)dt\right] \geq \tilde{v}(y,r).$$

This yields that

$$\tilde{v}(y,r) = \sup_{(x,z)\in\mathcal{H}} \{\tilde{u}(x,z) - xy + zr\}. \quad (6.54)$$

The other equality (6.47) is a direct consequence of the equality (6.54) and the properties of convex conjugation, c.f. Corollary 12.2.2 and Theorem 12.2 in Rockafellar (1970). $\square$

Proof of Theorem 6.1. It is sufficient to show that the conjugate value function $\tilde{v}$ is $(-\infty, \infty]$-valued on $\mathcal{R}$. The Fenchel–Legendre transform gives that $U(t,x) \leq V(t,y) + xy$. By integration, it is easy to see that, for any $\tilde{c} \in \tilde{\mathcal{A}}(x,z)$ and $\Gamma \in \tilde{\mathcal{Y}}(y,r)$,

$$\mathbb{E}\left[\int_0^T U(t,\tilde{c}_t)dt\right] \leq \mathbb{E}\left[\int_0^T V(t,\Gamma_t)dt\right] + \mathbb{E}\left[\int_0^T \tilde{c}_t \Gamma_t dt\right].$$

Proposition 6.3 deduces that $\tilde{u}(x,z) \leq \tilde{v}(y,r) + xy - zr$. As a consequence, for all $(y,r) \in \mathcal{R}$, we have $\tilde{v}(y,r) > -\infty$ by Lemma 6.10. On the other hand, thanks to the conjugate duality (6.47) and bipolar relationship (6.36), we can follow proofs of Lemmas 6.12 and 6.18 to obtain that, for each fixed $(y,r) \in \mathcal{R}$,

$$\sup_{(x,z) \in k\mathfrak{H}(y,r)} \tilde{u}(x,z) = \inf_{s>0}\{\tilde{v}(sy,sr) + ks\}.$$

The finiteness result (6.52) for all $k > 0$ in the proof of Lemma 6.18 guarantees the existence of a constant $s(y,r) > 0$ such that $\tilde{v}(sy,sr) < +\infty$. □

6.4.2 The Proof of Second Main Result

Let us move on to the proof of Theorem 6.2. Some further lemmas and auxiliary results are needed.

Lemma 6.19. *Let Assumptions of Theorem 6.2 hold. Then, $\tilde{v}(y,r)$ is $(-\infty, \infty)$-valued on $\mathcal{R}$.*

Proof. Similar to the proof of Lemma 6.10, under the additional condition (6.8), we can show that $\tilde{v}(y,r) < +\infty$ if $\tilde{v}(sy,sr) < +\infty$ for a constant $s = s(y,r) > 0$. However, it has been shown that Theorem 6.1 admits the existence of $s = s(y,r) > 0$. □

Note that we can not mimic proofs of Lemma 6.13, 6.14 and 6.15 to obtain the existence of an optimal solution to problem (6.25). In fact, our arguments for the dual problem depend on the existence of a bounded process $\tilde{w} \in \tilde{\mathfrak{A}}(\frac{\bar{p}}{\lambda})$, which is missing in the dual space. To this end, we resort to another auxiliary optimization problem and take advantage of the bipolar results built in Lemma 6.12.

Lemma 6.20. *Define the auxiliary optimization problem to the auxiliary dual utility minimization problem (6.29) as follows:*

$$\hat{v}(k) := \inf_{\Gamma \in \tilde{\mathfrak{Y}}(k)} \mathbb{E}\left[\int_0^T V(t,\Gamma_t)dt\right], \qquad (6.55)$$

where $\widetilde{\mathfrak{Y}}(k)$ is defined in Lemma 6.12 as the bipolar set of $\widetilde{\mathcal{A}}(x,z)$ on the product space for any $(x,z) \in \mathcal{H}$. Then, for all $k > 0$, under Assumptions of Theorem 6.2, the value function $\hat{v}(k) < +\infty$, and the optimal solution $\hat{\Gamma}(k)$ exists and is unique and $\hat{\Gamma}_t(k) > 0$ for all $t \in [0,T]$. Moreover, for each $k > 0$, and any $\Gamma \in \widetilde{\mathfrak{Y}}(k)$, we have

$$\mathbb{E}\left[\int_0^T (\Gamma_t - \hat{\Gamma}_t(k))I(t,\hat{\Gamma}_t(k))dt\right] \leq 0.$$

Proof. Using the definition in Lemma 6.12, we deduce from Lemma 6.19 that, for all $k > 0$,

$$\hat{v}(k) = \inf_{\Gamma \in \widetilde{\mathfrak{Y}}(k)} \mathbb{E}\left[\int_0^T V(t,\Gamma_t)dt\right] \leq \inf_{\Gamma \in \mathfrak{Y}(k)} \mathbb{E}\left[\int_0^T V(t,\Gamma_t)dt\right]$$
$$= \inf_{(y,r) \in k\mathfrak{R}(x,z)} \tilde{v}(y,r) < +\infty.$$

Taking into account the bipolar relationship (6.38), we have $\widetilde{\mathfrak{Y}}(k)$ is convexly compact in $\mathbb{L}^0_+$, the existence and uniqueness of optimal solution $\hat{\Gamma}(k)$ will follow the similar proof of Theorem 6.1. For $k > 0$ and $\epsilon \in (0,1)$, define $\Gamma^\epsilon_t := (1-\epsilon)\hat{\Gamma}_t(k) + \epsilon\Gamma_t$ for all $t \in [0,T]$, the optimality of $\hat{\Gamma}(k)$ yields that

$$0 \leq \frac{1}{\epsilon}\mathbb{E}\left[\int_0^T \left(V(t,\Gamma^\epsilon_t) - V(t,\hat{\Gamma}_t(k))\right)dt\right]$$
$$\leq \frac{1}{\epsilon}\mathbb{E}\left[\int_0^T \left(\hat{\Gamma}_t(k) - \Gamma^\epsilon_t\right)I(t,\Gamma^\epsilon_t)dt\right]$$
$$= \mathbb{E}\left[\int_0^T \left(\hat{\Gamma}_t(k) - \Gamma_t\right)I(t,\Gamma^\epsilon_t)dt\right]. \tag{6.56}$$

We claim that the family $(\Gamma_t - \hat{\Gamma}_t(k))I(t,\Gamma^\epsilon_t))^-)_{\epsilon \in (0,1)}$ is U.I. under $\bar{\mathbb{P}}$. Note that, for all $t \in [0,T]$,

$$\left((\Gamma_t - \hat{\Gamma}_t(k))I(t,\Gamma^\epsilon_t)\right)^- \leq \hat{\Gamma}_t(k)I(t,\Gamma^\epsilon_t) \leq \hat{\Gamma}_t(k)I(t,(1-\epsilon)\hat{\Gamma}_t(k)).$$

For fixed $\epsilon_0 < 1$ and $\epsilon < \epsilon_0$, we deduce that, for all $t \in [0,T]$,

$$\left|\hat{\Gamma}_t(k)I(t,(1-\epsilon)\hat{\Gamma}_t(k))\right| \leq \left|\hat{\Gamma}_t(k)I(t,(1-\epsilon)\hat{\Gamma}_t(k))\right|\mathbf{1}_{\{\hat{\Gamma}_t(k) \leq y_1\}}$$
$$+ \left|\hat{\Gamma}_t(k)I(t,(1-\epsilon)\hat{\Gamma}_t(k))\right|\mathbf{1}_{\{\hat{\Gamma}_t(k) \geq \frac{y_2}{1-\epsilon_0}\}} \tag{6.57}$$
$$+ \left|\hat{\Gamma}_t(k)I(t,(1-\epsilon)\hat{\Gamma}_t(k))\right|\mathbf{1}_{\{y_1 < \hat{\Gamma}_t(k) < \frac{y_2}{1-\epsilon_0}\}}.$$

By applying Lemma 6.2, the reasonable asymptotic elasticity conditions $AE_0[U] < \infty$ and $AE_\infty[U] < 1$ imply that, for fixed $\mu > 0$, there exist constants $C_1 > 0$, $C_2 > 0$, $y_1 > 0$ and $y_2 > 0$ such that

$$\begin{cases} -V'(t, \mu y) < C_1 \dfrac{V(t,y)}{y}, & 0 < y \leq y_1; \\ -V'(t,y) < C_2 \dfrac{-V(t,y)}{y}, & y_2 \leq y. \end{cases} \quad (6.58)$$

Hence, the 1st term in the RHS of (6.57) is dominated by

$$\left|\hat{\Gamma}_t(k) I(t, (1-\epsilon)\hat{\Gamma}_t(k))\right| \mathbf{1}_{\{\hat{\Gamma}_t(k) \leq y_1\}} \leq \frac{C_1}{1-\epsilon_0} \left|V(t, \hat{\Gamma}_t(k))\right|.$$

The 2nd term in the RHS of (6.57) is dominated by

$$\left|\hat{\Gamma}_t(k) I(t, (1-\epsilon)\hat{\Gamma}_t(k))\right| \mathbf{1}_{\{\hat{\Gamma}_t(k) \geq \frac{y_2}{1-\epsilon_0}\}}$$

$$\leq \frac{-C_2}{1-\epsilon_0} V(t, (1-\epsilon)\hat{\Gamma}_t(k)) \mathbf{1}_{\{\hat{\Gamma}_t(k) \geq \frac{y_2}{1-\epsilon_0}\}}$$

$$\leq \frac{C_2 \left|V(t, \hat{\Gamma}_t(k))\right|}{1-\epsilon_0}.$$

These two terms are both in $\mathbb{L}^1$ by the finiteness of $\hat{v}(k)$. On the other hand, the third term $|\hat{\Gamma}_t(k) I(t, (1-\epsilon)\hat{\Gamma}_t(k))| \mathbf{1}_{\{y_1 < \hat{\Gamma}_t(k) < \frac{y_2}{1-\epsilon_0}\}}$ is dominated by $k\hat{\Gamma}_t(k) \mathbf{1}_{\{y_1 < \hat{\Gamma}_t(k) < \frac{y_2}{1-\epsilon_0}\}}$ for a constant $k > 0$, and it is obviously integrable as well. Letting $\epsilon \to 0$ and apply DCT and Fatou's lemma to obtain the stated inequality. To show that the optimal solution $\hat{\Gamma}_t(k) > 0$ for all $t \in [0, T]$, we can choose an element $\Gamma \in \widetilde{\mathfrak{Y}}(k)$ and $\Gamma_t > 0$ for all $t \in [0, T]$. The inequality (6.56) can be rewritten as:

$$0 \geq \mathbb{E}\left[\int_0^T \left(\Gamma_t - \hat{\Gamma}_t(k)\right) I(t, \Gamma_t^\epsilon) \mathbf{1}_{\{\hat{\Gamma}_t > 0\}} dt\right]$$

$$+ \mathbb{E}\left[\int_0^T \left(\Gamma_t - \hat{\Gamma}_t(k)\right) I(t, \Gamma_t^\epsilon) \mathbf{1}_{\{\hat{\Gamma}_t = 0\}} dt\right]. \quad (6.59)$$

Now, assume $\bar{\mathbb{P}}(\hat{\Gamma}_t(k) = 0) > 0$, by the uniform integrability of $(((\Gamma_t - \hat{\Gamma}_t(k)) I(t, \Gamma_t^\epsilon))^-)_{\epsilon \in (0,1)}$, the 2nd term of (6.59) goes to ∞ as $\epsilon \to 0$, since $I(t, 0) = +\infty$, and $\Gamma_t > 0$ for all $t \in [0, T]$. Then, we obtain the contradiction, and hence the conclusion holds. $\square$

Lemma 6.21. *Let Assumptions of Theorem 6.2 hold. Then, the auxiliary dual value function $\hat{v}(k)$ is continuously differentiable on $\mathbb{R}_+$, and it holds that*

$$-k\hat{v}'(k) = \mathbb{E}\left[\int_0^T \hat{\Gamma}_t(k) I(t, \hat{\Gamma}_t(k)) dt\right].$$

Proof. To show $\hat{v}(k)$ is continuously differentiable, by the convex property, it is enough to justify that its derivative exists on $\mathbb{R}_+$. Now, fix $k > 0$. We define the following function by

$$h(s) := \mathbb{E}\left[\int_0^T V\left(t, \frac{s}{k}\hat{\Gamma}_t(k)\right) dt\right], \quad \forall s > 0.$$

This function is convex. By the optimality of $\hat{\Gamma}(k)$ of problem (6.55), we have $h(s) \geq \hat{v}(s)$ for all $s > 0$, and $h(k) = \hat{v}(k)$. Again, the convexity yields that

$$\Delta^- h(k) \leq \Delta^- \hat{v}(k) \leq \Delta^+ \hat{v}(k) \leq \Delta^+ h(k),$$

where Δ^+ and Δ^- denote right- and left-derivatives, respectively. Now, it follows from MCT that

$$\Delta^+ h(k) = \lim_{\epsilon \to 0} \frac{h(k+\epsilon) - h(k)}{\epsilon}$$
$$= \lim_{\epsilon \to 0} \frac{1}{\epsilon}\mathbb{E}\left[\int_0^T \left(V(t, \frac{k+\epsilon}{k}\hat{\Gamma}_t(k)) - V(t, \hat{\Gamma}_t(k))\right) dt\right]$$
$$\leq \liminf_{\epsilon \to 0}\left(-\frac{1}{k\epsilon}\right)\mathbb{E}\left[\int_0^T \epsilon\hat{\Gamma}_t(k) I\left(t, \frac{k+\epsilon}{k}\hat{\Gamma}_t(k)\right) dt\right]$$
$$= -\frac{1}{k}\mathbb{E}\left[\int_0^T \hat{\Gamma}_t(k) I(t, \hat{\Gamma}_t(k)) dt\right].$$

Similarly, we get

$$\Delta^- h(k) \geq \limsup_{\epsilon \to 0}\mathbb{E}\left[-\int_0^T \hat{\Gamma}_t(k) I\left(t, \frac{k-\epsilon}{k}\hat{\Gamma}_t(k)\right) dt\right].$$

We can follow the same reasoning as in Lemma 6.20 to verify that the family $((\hat{\Gamma}_t(k) I(t, \frac{k-\epsilon}{k}\hat{\Gamma}_t(k))))_{\epsilon \in (0,1)}$ is U.I.. The DCT and Fatou's Lemma jointly deduce that

$$\Delta^- h(k) \geq -\frac{1}{k}\mathbb{E}\left[\int_0^T \hat{\Gamma}_t(k) I\left(t, \hat{\Gamma}_t(k)\right) dt\right].$$

This completes the proof of the lemma. $\square$

Lemma 6.22. *The auxiliary dual value function $\hat{v}(\cdot)$ has the following asymptotic property given by $-\hat{v}'(0) = +\infty$ and $-\hat{v}'(+\infty) = 0$.*

Proof. We first show $-\hat{v}'(0) = +\infty$, and toward this end, we claim that

$$\hat{v}(0+) \geq \int_0^T V(t, 0+) dt. \tag{6.60}$$

First, for any $k > 0$, it follows from definition that

$$\hat{v}(k) = \mathbb{E}\left[\int_0^T V(t, \hat{\Gamma}_t(k)) dt\right]$$

$$= \mathbb{E}\left[\int_0^T V^+(t, \hat{\Gamma}_t(k)) dt\right] - \mathbb{E}\left[\int_0^T V^-(t, \hat{\Gamma}_t(k)) dt\right].$$

Recall that $\widetilde{\mathfrak{Y}}(k) = k\widetilde{\mathfrak{Y}}(1)$, and thus $\hat{\Gamma}_t(k) = k\hat{\Gamma}_t(1)$. Now, we have from Fatou's lemma that

$$\lim_{k \to 0} \mathbb{E}\left[\int_0^T V^+(t, \hat{\Gamma}_t(k)) dt\right] \geq \mathbb{E}\left[\int_0^T V^+(t, 0+) dt\right]. \tag{6.61}$$

On the other hand, similar to the proof of Lemma 6.13, we can prove that $\mathbb{E}[\int_0^T V^-(t, \hat{\Gamma}_t(1)) dt] < +\infty$. Therefore, by the monotonicity of $V^-(t, \cdot)$ and DCT, it follows that

$$\lim_{k \to 0} \mathbb{E}\left[\int_0^T V^-(t, \hat{\Gamma}_t(k)) dt\right] = \mathbb{E}\left[\int_0^T V^-(t, 0+) dt\right],$$

which together with (6.61) imply that (6.60) holds true. Hence, if $\int_0^T V(t, 0+) dt = +\infty$, we have $\hat{v}(0+) = +\infty$, and by convexity, it follows that $\hat{v}'(0+) = -\infty$. In the case where $\int_0^T V(t, 0+) dt < +\infty$, it is not difficult to see that

$$-\hat{v}(0+) \geq \lim_{k \to 0} \frac{\hat{v}(0) - \hat{v}(k)}{k} \geq \lim_{k \to 0} \frac{\int_0^T V(t, 0+) dt - \mathbb{E}\left[\int_0^T V(t, \hat{\Gamma}_t(k)) dt\right]}{k}.$$

Hence, applying MCT to obtain that

$$-\hat{v}(0+) \geq \lim_{k \to 0} \frac{\mathbb{E}\left[\int_0^T V(t, 0+) dt\right] - \mathbb{E}\left[\int_0^T V(t, \hat{\Gamma}_t(k)) dt\right]}{k}$$

$$\geq \lim_{k \to 0} \mathbb{E}\left[\int_0^T \hat{\Gamma}_t(1) I(t, k\hat{\Gamma}_t(1)) dt\right] = +\infty.$$

For $-\hat{v}'(\infty) = 0$, since the function $-\hat{v}$ is concave and increasing, there is a finite positive limit $-\hat{v}'(\infty) := \lim_{k \to \infty} -\hat{v}'(y)$. Using the definition of Legendre-Fenchel transform, for any $y > 0$,

$$-V(t, y) \leq -U(t, x) + xy, \quad \forall x > 0.$$

Then, for any $\epsilon > 0$, we always have

$$0 \leq -\hat{v}'(\infty) = \lim_{k\to\infty} \frac{-\hat{v}(k)}{k} = \lim_{k\to\infty} \frac{\mathbb{E}\left[\int_0^T -V(t,\hat{\Gamma}_t(k))dt\right]}{k}$$

$$\leq \lim_{k\to\infty} \frac{\mathbb{E}\left[\int_0^T -U(t,\epsilon\tilde{w}_t)dt\right]}{k} + \lim_{k\to\infty} \frac{\langle \epsilon\tilde{w}, \hat{\Gamma}(k)\rangle}{k}.$$

Now, recall that for each fixed $(x,z) \in \mathcal{H}$, there exists a constant $\lambda(x,z) > 0$ such that $\tilde{w}_t \in \tilde{\mathcal{A}}(\frac{\bar{p}}{\lambda}x, \frac{\bar{p}}{\lambda}z)$, and by the definition of $\tilde{\mathfrak{Y}}(k)$, we can see the 2nd term above satisfies that

$$\lim_{k\to\infty} \frac{\langle \epsilon\tilde{w}, \hat{\Gamma}(k)\rangle}{k} \leq \lim_{k\to\infty} \frac{\epsilon\frac{\bar{p}}{\lambda}k}{k} = \epsilon\frac{\bar{p}}{\lambda}.$$

As for the 1st term, we claim that $\mathbb{E}[\int_0^T -U(t,\epsilon\tilde{w}_t)dt] < +\infty$ for each fixed ϵ small enough. Without loss of generality, it is sufficient to consider that $\epsilon < \bar{x}$, and we can apply Lemma 6.2 again. Since there exists a constant x_0 such that, for all $t \in [0,T]$,

$$U\left(t, \frac{\epsilon}{\bar{x}}x\right) > \left(\frac{\epsilon}{\bar{x}}\right)^{-\frac{\gamma}{1-\gamma}} U(t,x) \quad \forall 0 < x \leq x_0,$$

we will have

$$\mathbb{E}\left[\int_0^T -U(t,\epsilon\tilde{w}_t)dt\right] = \mathbb{E}\left[\int_0^T -U(t,\epsilon\tilde{w}_t)\mathbf{1}_{\{\bar{x}\tilde{w}_t > x_0\}}dt\right]$$

$$+ \mathbb{E}\left[\int_0^T -U(t,\epsilon\tilde{w}_t)\mathbf{1}_{\{\bar{x}\tilde{w}_t \leq x_0\}}dt\right]$$

$$\leq \mathbb{E}\left[\int_0^T -U(t,\epsilon\tilde{w}_t)\mathbf{1}_{\{\bar{x}\tilde{w}_t > x_0\}}dt\right] + \left(\frac{\epsilon}{\bar{x}}\right)^{-\frac{\gamma}{1-\gamma}} \mathbb{E}\left[\int_0^T -U(t,\bar{x}\tilde{w}_t)dt\right]$$

$$< +\infty,$$

by the fact that $\tilde{w}_t \leq 1$ for $t \in [0,T]$ and the condition (6.14). Hence $0 \leq -\hat{v}'(\infty) = \lim_{k\to\infty} \frac{-\hat{v}(k)}{k} \leq \epsilon\frac{\bar{p}}{\lambda}$, and consequently, $-\hat{v}'(\infty) = 0$ by letting ϵ go to 0. □

Lemma 6.23. *Let Assumptions of Theorem 6.2 hold. Then, for any $(x,z) \in \mathcal{H}$, suppose k satisfies $\hat{v}'(k) = -1$ with $\hat{v}(k)$ being the value function of the auxiliary dual optimization problem (6.55), then $\tilde{c}_t^*(x,z) := I(t,\hat{\Gamma}_t(k))$ is the unique (in the sense of "=" under $\bar{\mathbb{P}}$ in $\mathbb{L}_+^0$) optimal solution to problem (6.25). Moreover, we have $\tilde{c}_t^*(x,z) > 0$, $\mathbb{P}$-a.s. for all $t \in [0,T]$.*

Proof. First of all, Lemma 6.21 yields that $\langle \tilde{c}^*(x,z), \hat{\Gamma}(k) \rangle = -k\hat{v}'(k) = k$. Then, for any $\Gamma \in \widetilde{\mathfrak{Y}}(k)$, it follows from Lemma 6.20 that $\langle \tilde{c}^*(x,z), \Gamma(k) \rangle \leq \langle \tilde{c}^*(x,z), \hat{\Gamma}(k) \rangle = k$. Therefore, $\tilde{c}_t^*(x,z) \in \widetilde{\mathcal{A}}(x,z)$ by the bipolar relationship (6.38). Now, for any $\tilde{c} \in \widetilde{\mathcal{A}}(x,z)$, we have $\langle \tilde{c}, \hat{\Gamma}(k) \rangle \leq k$ and $U(t, \tilde{c}_t) \leq V(t, \hat{\Gamma}_t(k)) + \tilde{c}_t \hat{\Gamma}_t(k)$ for all $t \in [0,T]$. This implies that

$$\mathbb{E}\left[\int_0^T U(t, \tilde{c}_t) dt\right] \leq \hat{v}(k) + k = \mathbb{E}\left[\int_0^T \left(V(t, \hat{\Gamma}_t(k)) + \hat{\Gamma}_t(k) I(t, \hat{\Gamma}_t(k))\right) dt\right]$$

$$= \mathbb{E}\left[\int_0^T U(t, I(\hat{\Gamma}_t(k))) dt\right] = \mathbb{E}\left[\int_0^T U(t, \tilde{c}_t^*) dt\right],$$

which infers the optimality of $\tilde{c}^*$. The uniqueness of the optimal solution follows from the strict concavity of the utility function U. Under Assumptions of Theorem 6.2, for any pair $(x,z) \in \mathcal{H}$, since $\widetilde{\mathfrak{Y}}(k)$ is convexly compact and $\hat{\Gamma}_t(k)$ is bounded in probability, we have the optimal solution $\tilde{c}_t^*(x,z) > 0$, $\mathbb{P}$-a.s. for all $t \in [0,T]$ since $\hat{\Gamma}_t(k)$ is bounded in probability if and only if $\hat{\Gamma}_t(k)$ is finite $\bar{\mathbb{P}}$-a.s. and by definition, we have $I(t,x) > 0$ for $x < +\infty$. □

Let $(x,z) \in \text{cl}\mathcal{H}$, the proof of Lemma 6.7 shows that, there exists $\tilde{c} \in \widetilde{\mathcal{A}}(x,z)$ such that $\bar{\mathbb{P}}[\tilde{c} > 0] > 0$. Similar to the proof of Lemma 12 in Hugonnier and Kramkov (2004), we have

Lemma 6.24. *Assume that conditions of Proposition 6.3 hold, and let $(y^n, r^n) \in \mathcal{R}$ and $\Gamma^n \in \widetilde{\mathcal{Y}}(y^n, r^n)$ converge to $(y,r) \in \mathbb{R}^2$ and $\Gamma \in \mathbb{L}_+^0$, as $n \to \infty$, respectively. If Γ is a strictly positive random variable, then $(y,r) \in \mathcal{R}$ and $\Gamma \in \widetilde{\mathcal{Y}}(y,r)$.*

The following lemma is the last result we need to prepare to proceed to the proof of Theorem 6.2.

Lemma 6.25. *Under Assumption 6.3, we have $\bar{\mathbb{P}}[\tilde{c}^*(x_1, z_1) \neq \tilde{c}^*(x_2, z_2)] > 0$ for two distinct points $(x_i, z_i) \in \mathcal{H}$ with $i = 1, 2$.*

Proof. Assume that there exist two distinct pairs (x_1, z_1) and (x_2, z_2) in $\mathcal{H}$ and $\bar{\mathbb{P}}[\tilde{c}^*(x_1,z_1) \neq \tilde{c}^*(x_2,z_2)] = 0$. The definition of set $\widetilde{\mathcal{A}}(x_1,z_1)$ implies that $\langle \tilde{c}^*(x_2,z_2), \Gamma \rangle \leq x_1 - z_1 \langle \tilde{w}, \Gamma \rangle$ for all $\Gamma \in \widetilde{\mathcal{M}}$. However, we also know that $\tilde{c}^*(x_2, z_2) \in \widetilde{\mathcal{A}}(x_2, z_2)$, which deduces that $x_2 - z_2 \langle \tilde{w}, \Gamma \rangle \leq x_1 - z_1 \langle \tilde{w}, \Gamma \rangle$ for all $\Gamma \in \widetilde{\mathcal{M}}$. On the other hand, by symmetry and replacing $\tilde{c}^*(x_2, z_2)$ by $\tilde{c}^*(x_1, z_1)$, we conclude that $x_1 - z_1 \langle \tilde{w}, \Gamma \rangle \leq x_2 - z_2 \langle \tilde{w}, \Gamma \rangle$ for all $\Gamma \in \widetilde{\mathcal{M}}$. Therefore, we must have $x_1 - z_1 \langle \tilde{w}, \Gamma \rangle = x_2 - z_2 \langle \tilde{w}, \Gamma \rangle$ for all $\Gamma \in \widetilde{\mathcal{M}}$. This is a contradiction to Assumption 6.3, since we can obtain a constant $K = \frac{x_1 - x_2}{z_1 - z_2}$ and $\mathbb{E}^{\mathbb{Q}}[E] = \langle w, Y \rangle = \langle \tilde{w}, \Gamma \rangle = \frac{x_1 - x_2}{z_1 - z_2}$ for all $\mathbb{Q} \in \mathcal{M}$. □

Proof of Theorem 6.2. We first prove that the dual value function $\tilde{v}(y,z)$ is continuously differentiable on $\mathcal{R}$. Theorems 4.1.1 and 4.1.2 in Hiriart-Urruty and Lemaréchal (2001) give the equivalence between the above statement and the fact that the value function $\tilde{u}(x,z)$ is strictly concave on $\mathcal{H}$. Note that U is a strictly concave function. To show that the value function is strictly concave is equivalent to show that, for any two distinct points $(x_i, z_i) \in \mathcal{H}$ with $i = 1, 2$, the optimal consumption policies are different, i.e., $\bar{\mathbb{P}}[\tilde{c}^*(x_1, z_1) \neq \tilde{c}^*(x_2, z_2)] > 0$, which is the consequence of Lemma 6.25. To continue the remaining part, it amounts to show that the assertion (ii) holds. Recall that $\hat{\Gamma}(k)$ is the optimal solution of the auxiliary dual problem (6.55)) such that, for all $t \in [0, T]$,

$$\hat{\Gamma}_t(k) = U'(t, \tilde{c}_t^*(x, z)), \quad k = \langle \tilde{c}^*(x, z), \hat{\Gamma}(k) \rangle.$$

Since $\widetilde{\mathfrak{Y}}(k)$ is closed under convergence in measure $\bar{\mathbb{P}}$, there exists a sequence $(y^n, r^n) \in k\mathfrak{R}(x, z)$ such that $\Gamma^n \in \widetilde{\mathcal{Y}}(y^n, r^n)$ and Γ^n converges to $\hat{\Gamma}(k)$ as $n \to \infty$, $\bar{\mathbb{P}}$-a.s. by passing to a subsequence if necessary. Since the set $k\mathfrak{R}(x, z)$ is bounded, there exists a further subsequence (y^n, r^n) converging to $(y, r) \in \mathbb{R}^2$. By passing to this further subsequence, as we have shown $\bar{\mathbb{P}}[\hat{\Gamma}(k) > 0] = 1$, it follows that $(y, r) \in k\mathfrak{R}(x, z)$ such that $\hat{\Gamma}(k) \in \widetilde{\mathcal{Y}}(y, r)$ due to Lemma 6.24. Moreover, for this pair $(y, r) \in \mathcal{R}$, by Fatou's lemma and Proposition 6.3, we have

$$xy - zr = k = \langle \tilde{c}^*(x, z), \hat{\Gamma}(k) \rangle. \tag{6.62}$$

The corresponding optimizer $\Gamma_t^*(y, r)$ of (6.29) then verifies that, for all $t \in [0, T]$,

$$\Gamma_t^*(y, r) = \hat{\Gamma}_t(k) = U'(t, \tilde{c}^*(x, z)). \tag{6.63}$$

To see this, on one hand, we have $\hat{\Gamma}(k) \in \widetilde{\mathcal{Y}}(y, r)$, and hence

$$\mathbb{E}\left[\int_0^T V(t, \Gamma_t^*(y, r))dt\right] = \inf_{\Gamma \in \widetilde{\mathcal{Y}}(y,r)} \mathbb{E}\left[\int_0^T V(t, \Gamma_t(y, r))dt\right]$$

$$\leq \mathbb{E}\left[\int_0^T V(t, \hat{\Gamma}_t(k))dt\right].$$

Additionally, we have

$$\mathbb{E}\left[\int_0^T V(t, \hat{\Gamma}_t(y, r))dt\right] = \inf_{\Gamma \in \widetilde{\mathfrak{Y}}(k)} \mathbb{E}\left[\int_0^T V(t, \Gamma_t(y, r))dt\right]$$

$$\leq \inf_{\Gamma \in \widetilde{\mathcal{Y}}(y,r)} \mathbb{E}\left[\int_0^T V(t, \Gamma_t(y, r))dt\right] = \mathbb{E}\left[\int_0^T V(t, \Gamma_t^*(y, r))dt\right].$$

Using the equality $U(t, \tilde{c}_t^*(x,z)) = V(t, \hat{\Gamma}_t(k)) + \tilde{c}_t^*(x,z)\hat{\Gamma}_t(k)$ for all $t \in [0,T]$, we can conclude $(y,r) \in \partial \tilde{u}(x,z)$ by Theorem 23.5 in Rockafellar (1970), since

$$\tilde{u}(x,z) = \tilde{v}(y,z) + xy - zr. \tag{6.64}$$

In particular, it implies that

$$\partial \tilde{u}(x,z) \cap \mathcal{R} \neq \varnothing. \tag{6.65}$$

Similar to the proof of Theorem 2 in Hugonnier and Kramkov (2004), it is not difficult to show that $\partial \tilde{u}(x,z) \subset \mathcal{R}$. For any $(y,r) \in \partial \tilde{u}(x,z)$, there exists a sequence $(y^n, r^n) \in \partial \tilde{u}(x,z) \cap \mathcal{R}$ converging to (y,r) as $n \to \infty$ by (6.65) and the fact that $\partial \tilde{u}(x,z)$ is closed and convex. Since $U'(\cdot, \tilde{c}_t^*(x,z))$ is strictly positive and $U'(\cdot, \tilde{c}_t^*(x,z)) \in \widetilde{\mathcal{Y}}(y,r)$, Lemma 6.24 infers that $(y,r) \in \mathcal{R}$. Conversely, for any $(y,r) \in \partial \tilde{u}(x,z)$, we have

$$\mathbb{E}\left[\int_0^T |V(t, \Gamma_t^*(y,r)) + \tilde{c}_t^*(x,z)\Gamma_t^*(y,r) - U(t, \tilde{c}_t^*(x,z))|dt\right]$$

$$= \mathbb{E}\left[\left(\int_0^T V(t, \Gamma_t^*(y,r)) + \tilde{c}_t^*(x,z)\Gamma_t^*(y,r) - U(t, \tilde{c}_t^*(x,z))dt\right)\right]$$

$$\leq \tilde{v}(y,r) + xy - zr - \tilde{u}(x,z) = 0.$$

This concludes (6.62) and (6.63). □

Chapter 7

Optimal Tracking Problem in Portfolio Optimization*

Optimal portfolio allocation with benchmark performance has been studied in some literature Browne (2000), Gaivoronski et al. (2005), Yao et al. (2006), Strub and Baumann (2018) and many others. The target benchmark is usually a prescribed capital process or a specific portfolio in the financial market, and the goal is to choose the portfolio in some risky assets to track the return or the value of the benchmark process. In practice, both professional and individual investors may measure their portfolio performance using different benchmarks, such as S&P500 index, Goldman Sachs commodity index, special liability, inflation and exchange rates.

Tracking and monotone follower problem. The existing research mainly focuses on mathematical problems that minimize the difference between the controlled portfolio and the benchmark, which are formulated as either a linear quadratic control problem using the mean-variance analysis or a utility maximization problem at the terminal time. This chapter aims to enrich the study of optimal tracking by proposing a different tracking procedure and analyzing the associated control problem. We are particularly interested in the fund management when the fund manager can dynamically inject capital to keep the total fund capital above a non-decreasing benchmark process at each intermediate time. The control problem involves the regular portfolio control and the singular capital injection control together with American type floor constraints. The optimality is attained when the cost of the accumulated capital injection is minimized.

The alternative well-known optimal tracking problem in the literature is the monotone follower problem, see for instance Karatzas and Shreve (1984) and Bayraktar and Egami (2008), in which one needs to choose a monotone process as a singular control to closely track a given diffusion

---

*This chapter is based on the work Bo et al. (2021).

process such as a Brownian motion with drift. This paper investigates the opposite direction as we look for a regular control such that the controlled diffusion process can closely follow a given monotone process. Our mathematical problem is also motivated by some stochastic control problems with minimum guaranteed floor constraints, which are conventionally defined as utility maximization problems such that the controlled wealth processes dominate an exogenous target process at the terminal time or at each intermediate time. See some related studies among El Karoui et al. (2005), El Karoui and Meziou (2006), Bouchard et al. (2010), Di Giacinto et al. (2011), Sekine (2012), Di Giacinto et al. (2014) and Chow et al. (2020), in which European type or American type floor constraints have been examined in various market models. In the aforementioned research, some typical techniques to handle the floor constraints are to introduce the option based portfolio or the insured portfolio allocation such that the floor constraints can be guaranteed. We instead reformulate the optimal tracking problem with dynamic floor constraints to a constraint-free stochastic control problem under a running maximum cost criterion, see Lemma 7.1.

Stochastic control with running maximum cost. Stochastic control with a running maximum cost or a running maximum process is itself an interesting topic and attracted a lot of attention in the past decades, see Barron and Ishii (1989), Barron (1993), Barles et al. (1994), Bokanowski et al. (2015) and Kröner et al. (2018), in which the viscosity solution approach plays the key role. In contrast, we can take advantage of the specific payoff function and state processes to conclude the existence of a classical solution to the HJB equation. See also some recent control problems on optimal consumption in Guasoni et al. (2020) and Deng et al. (2020), in which the utility function depends on the running maximum of the control and the value function can be obtained explicitly. However, as opposed to Guasoni et al. (2020) and Deng et al. (2020), the running maximum of a controlled diffusion process appears in our finite horizon control problem and complicates the analysis. We choose to work with an auxiliary state process with reflection similar to Weerasinghe and Zhu (2016) to reformulate the control problem again, which corresponds to a nonlinear HJB equation with a Neumann boundary condition. By using the heuristic dual transform, we can apply the probabilistic representation and some stochastic flow arguments to first establish the existence and uniqueness of the classical solution to the dual PDE. In fact, our primal value function is not strictly concave and the inverse transform needs to be carefully carried out in a restricted domain. To this end, we derive an explicit threshold for the initial wealth, beyond which the ratcheting benchmark process is dynamically superhedgeable by the portfolio process and no capital needs to be injected to catch up with the benchmark. This threshold facilitates the full characterization of the primal value function on the whole domain using some delicate continuity and convergence analysis based on the probabilistic representation results. The piecewise feedback optimal portfolio across different regions can also be derived and rigorously verified. Even if the primary focus of the chapter is to track a non-decreasing benchmark process, our approach and theoretical results can be applied in some market index tracking problems when the index process is not monotone. In Section 7.7, we present some examples when the index process follows a geometric Brownian motion (GBM), and show that the market index tracking problem can be transformed into an equivalent problem with a non-decreasing benchmark. In the model with infinite time horizon, the value function and the optimal portfolio can be obtained fully explicitly.

The rest of the chapter is organized as follows. Section 7.1 introduces the market and a non-decreasing benchmark process. Section 7.2 formulates the optimal tracking portfolio problem. Section 7.3 transforms the prime control problem into an auxiliary one with a controlled state process with reflection. This leads to a nonlinear HJB equation with a Neumann boundary condition. Section 7.4 linearizes the HJB equation using the dual transform. A probabilistic solution to the linear dual equation is provided in Section 7.5. The feedback optimal portfolio and the proof of the verification theorem are given in Section 7.6. Section 7.7 presents some illustrative examples when the index process follows a GBM.

7.1 The Market Model

In this section, we describe the market model used in this chapter. Let $(\Omega, \mathcal{F}, \mathbb{F}, \mathbb{P})$ be a filtered probability space with the reference filtration $\mathbb{F} = (\mathcal{F}_t)_{t \in [0,T]}$ satisfying the usual conditions, where $T > 0$ is a finite terminal date. This filtered probability space supports a d-dimensional Brownian motion $W = (W^1, \ldots, W^d)^\top = (W_t^1, \ldots, W_t^d)^\top_{t \in [0,T]}$. Consider a financial market consists of d risky assets whose price dynamics is given by, for $t \in [0, T]$,

$$\frac{dS_t^i}{S_t^i} = \mu_i dt + \sum_{j=1}^d \sigma_{ij} dW_t^j, \quad i = 1, \ldots, d, \tag{7.1}$$

where, the constant drift $\mu_i \in \mathbb{R}$ and constant volatility $\sigma_{ij} \in \mathbb{R}$ for $i, j = 1, \ldots, d$. We assume that the interest rate level $r = 0$ which amounts to the change of numéraire. To this end, all processes including the wealth process and the benchmark process are defined after the change of numéraire.

For $t \in [0,T]$, denote by θ_t^i the amount of wealth (as an $\mathbb{F}$-adapted process) that the fund manager allocates in asset $S^i = (S_t^i)_{t\in[0,T]}$ at time t. The self-financing wealth process under the portfolio $\theta = (\theta_t^1,\ldots,\theta_t^d)^\top_{t\in[0,T]}$ is given by, for all $t \in [0,T]$,

$$V_t^\theta = v + \int_0^t \theta_s^\top \mu ds + \int_0^t \theta_s^\top \sigma dW_s, \qquad (7.2)$$

where the initial wealth $V_0^\theta = v \geq 0$, the return vector $\mu = (\mu_1,\ldots,\mu_d)^\top$ and the volatility matrix $\sigma = (\sigma_{ij})_{i,j=1}^d$ which is assumed to be invertible (its inverse is denoted by σ^{-1}).

We consider the portfolio allocation by a fund manager that is to optimally track an exogenous non-decreasing capital benchmark process $A = (A_t)_{t\in[0,T]}$ taking the absolutely continuous form that

$$A_t := a + \int_0^t f(s, Z_s) ds. \qquad (7.3)$$

Here, $a \geq 0$ stands for the initial benchmark that the fund manager needs to track at time $t = 0$. The function $f(\cdot,\cdot)$, representing the benchmark growth rate, is assumed to satisfy the following condition:

Assumption 7.1. the function $f : [0,T] \times \mathbb{R} \to \mathbb{R}_+$ is continuous and for $t \in [0,T]$, $f(t,\cdot) \in C^2(\mathbb{R})$ with bounded first and second order derivatives.

The stochastic factor process $Z = (Z_t)_{t\in[0,T]}$ in (7.3) satisfies the SDE:

$$dZ_t = \mu_Z(Z_t)dt + \sigma_Z(Z_t)dW_t^\gamma \qquad (7.4)$$

with the initial value $Z_0 = z \in \mathbb{R}$ and $W^\gamma = (W_t^\gamma)_{t\in[0,T]}$ is a linear combination of the d-dimensional Brownian motion $(W^1,\ldots,W^d)$ with weights $\gamma = (\gamma_1,\ldots,\gamma_d)^\top \in [-1,1]^d$, which itself is a Brownian motion (for the case where the dimension $d = 1$, the weight would reduce to $\gamma \in \{-1,1\}$).

We impose the following conditions on coefficients $\mu_Z(\cdot)$ and $\sigma_Z(\cdot)$ that

Assumption 7.2. the coefficients $\mu_Z, \sigma_Z : \mathbb{R} \to \mathbb{R}$ is in $C^2(\mathbb{R})$ with bounded first and second order derivatives.

If Z is an OU process or a geometric Brownian motion (GBM), the assumption 7.2 clearly holds. The reasons for us to consider the non-decreasing benchmark process $A = (A_t)_{t\in[0,T]}$ in (7.3) are twofold. First, the process A may refer to some non-decreasing growth rate process after the change of numéraire, which generalizes the deterministic growth rate benchmark considered in Yao et al. (2006) for their optimal tracking problem. This non-decreasing benchmark process can also describe some consumer price index or higher education price index, which is observed to be non-decreasing over the past decades. That is, our portfolio management problem might be suitable to model some pension fund management or education savings fund management when the aim is to track some non-decreasing price index affected by the stochastic factor $Z = (Z_t)_{t \in [0,T]}$. Second, one key step in our approach is the dual transform (c.f. Section 7.4), which relies crucially on the concavity of the primal value function and the convexity of the solution to the dual PDE problem.

7.2 Formulation of Optimal Tracking Problem

This section formulates a type of optimal tracking problems given the non-decreasing benchmark process A defined by (7.3). More precisely, the optimal tracking problem considered in this chapter is formulated that combines the portfolio control with another capital injection singular control together with dynamic floor constraints.

We assume that the fund manager can inject capital to the fund account from time to time whenever it is necessary such that the total capital dynamically dominates the benchmark floor process A. That is, the fund manager optimally tracks the process A by choosing the regular control θ as the dynamic portfolio in risky assets and the singular control $C = (C_t)_{t \in [0,T]}$ as the cumulative capital injection such that $C_t + V_t^{\theta} \geq A_t$ at each intermediate time $t \in [0, T]$. This can be displayed in Figure 7.1. The goal of the

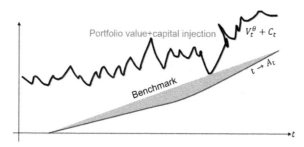

Fig. 7.1 The portfolio optimization with capital injection singular control and dynamic floor constraints.

optimal tracking problem is to minimize the expected cost of the discounted cumulative capital injection under American-type floor constraints that

$$\begin{cases} u(a,\text{v},z) := \inf_{C,\theta} \mathbb{E}\left[C_0 + \int_0^T e^{-\rho t} dC_t\right], \\ \text{s.t. } A_t \leq C_t + V_t^\theta \text{ at each } t \in [0,T], \end{cases} \quad (7.5)$$

where the constant $\rho \geq 0$ is the discount rate and $C_0 = (a - \text{v})^+$ is the initial injected capital to match with the initial benchmark.

Recall that we consider the model after the change of numéraire. The non-negative discount rate $\rho \geq 0$ is equivalent to the assumption that the discount rate in the original market dominates the interest rate before the change of numéraire. For the symmetric case, one can simply consider problem (7.5) with $\rho = 0$. On the other hand, for a large initial wealth $\text{v} \gg a$ and some special choices of $f(t,z)$ and $Z = (Z_t)_{t\in[0,T]}$, it is possible that the benchmark process $A = (A_t)_{t\in[0,T]}$ is dynamically superhedgeable by a portfolio in risky assets at each time $t \in [0,T]$. That is, there exists a portfolio θ^* such that $V_t^{\theta^*} \geq A_t$, for any $t \in [0,T]$. Then $C_t^* \equiv 0$ for any $t \in [0,T]$ is an admissible capital injection control and $(0,\theta^*)$ is an optimal control for problem (7.5) and the value function $u(a,\text{v},z) \equiv 0$. We will characterize the region for v explicitly in Remark 7.4 such that there is no need to inject capital for problem (7.5).

To handle the problem (7.5) with dynamic floor constraints, our first step is to reformulate it based on the observation that, for a fixed control θ, the optimal C is always the smallest adapted right-continuous and non-decreasing process that dominates $A - V^\theta$ (c.f. Figure 7.2). Let $\mathcal{U}$ be the

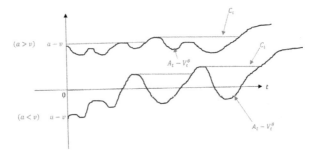

Fig. 7.2 A sample path of the optimal cumulative capital process $t \to C_t$ given paths of the wealth process $t \to V_t^\theta$ and benchmark process $t \to A_t$ under a portfolio $\theta = (\theta_t)_{t\in[0,T]}$.

set of regular $\mathbb{F}$-adapted control processes $\theta = (\theta_t)_{t \in [0,T]}$ such that (7.2) is well-defined. The following lemma gives an equivalent formulation of (7.5).

Lemma 7.1. *For each fixed regular control θ, the optimal singular control $C^{\theta,*} = (C_t^{\theta,*})_{t \in [0,T]}$ satisfies that*

$$C_t^{\theta,*} = 0 \vee \sup_{s \in [0,t]} (A_s - V_s^\theta), \quad \forall t \in [0,T]. \tag{7.6}$$

The problem (7.5) with the American-type floor constraints $A_t \leq C_t + V_t^\theta$ for all $t \in [0,T]$, admits the equivalent formulation as a unconstrained control problem under a running maximum cost that

$$u(a, v, z) = (a - v)^+ + \inf_{\theta \in \mathcal{U}} \mathbb{E}\left[\int_0^T e^{-\rho t} d\left(0 \vee \sup_{s \in [0,t]} (A_s - V_s^\theta)\right)\right]. \tag{7.7}$$

Proof. It is easy to see that, for all $(a, v, z) \in [0, \infty)^2 \times \mathbb{R}$,

$$u(a, v, z) = \inf_\theta \inf_C \mathbb{E}\left[C_0 + \int_0^T e^{-\rho t} dC_t\right].$$

The integration by parts gives that $C_0 + \int_0^T e^{-\rho t} dC_t = e^{-\rho T} C_T + \rho \int_0^T e^{-\rho t} C_t dt$. For each fixed $\theta = (\theta_t)_{t \in [0,T]}$, we need to choose the optimal singular control $C = (C_t)_{t \in [0,T]}$ to minimize

$$\inf_C F(C), \quad \text{where } F(C) := \mathbb{E}\left[e^{-\rho T} C_T + \rho \int_0^T e^{-\rho t} C_t dt\right],$$

subjecting to $C_t \geq A_t - V_t^\theta$ at each $t \in [0,T]$. Note that the cost functional $F(C)$ is strictly increasing in C. That is, if $C^1 \leq C^2$ and $C^1 \neq C^2$, then we have $F(C^1) < F(C^2)$. Therefore, the optimal choice of the control C is the minimal non-negative and non-decreasing process C_t such that $C_t \geq A_t - V_t^\theta$ for $t \in [0,T]$. We claim that the minimal process is the non-decreasing envelope $C_t^* := 0 \vee \sup_{s \leq t}(A_s - V_s^\theta)$. Note that C_t^* is non-negative and satisfies the dynamic floor constraint. Let $\widetilde{C}$ be another non-negative and non-decreasing process satisfying $\widetilde{C}_t \geq A_t - V_t^\theta$, $t \in [0,T]$. Suppose that $\widetilde{C} \leq C^*$ and $\widetilde{C} \neq C^*$. That is, there exists a $t \in [0,T]$ and a set O with $\mathbb{P}(O) > 0$ such that $\widetilde{C}_t(\omega) < C_t(\omega)$ for $\omega \in O$. By definition, we have

$$A_t(\omega) - V_t^\theta(\omega) \leq \widetilde{C}_t(\omega) < C_t(\omega) = \sup_{s \leq t}(A_s(\omega) - V_s^\theta(\omega)).$$

For each fixed $\omega \in O$, let $t^* < t$ be the time such that $A_{t^*}(\omega) - V_{t^*}^\theta(\omega) = \sup_{s \leq t}(A_s(\omega) - V_s^\theta(\omega))$. It follows that $\widetilde{C}_t(\omega) < A_{t^*}(\omega) - V_{t^*}^\theta(\omega) \leq \widetilde{C}_{t^*}(\omega)$.

We obtain a contradiction that the process $\widetilde{C}$ is non-decreasing. Therefore, the original problem can be written as

$$u(a, \mathrm{v}, z) = C_0^* + \inf_\theta \mathbb{E}\left[\int_0^T e^{-\rho t} dC_t^*\right]$$

$$= (a - \mathrm{v})^+ + \inf_\theta \mathbb{E}\left[\int_0^T e^{-\rho t} d\left(0 \vee \sup_{s \leq t}(A_s - V_s^\theta)\right)\right].$$

This completes the proof of the lemma. □

To handle the running maximum term in the objective function, one can choose the monotone running maximum process as a controlled state process as in Barles et al. (1994) and Kröner et al. (2018) to derive the HJB equation with a free boundary condition. One can also choose the distance between the underlying process and its running maximum as a reflected state process as in Weerasinghe and Zhu (2016) and derive the HJB equation with a Neumann boundary condition. Herein, we follow the second method that allows us to prove the existence of a classical solution using the probabilistic representation. Note that our problem mathematically differs from the one in Weerasinghe and Zhu (2016) because we consider the control in both the drift and volatility of the state process together with a stochastic factor process affecting the benchmark capital.

7.3 Auxiliary Stochastic Control Problem

This section introduces an auxiliary controlled state process to replace the process $V^\theta = (V_t^\theta)_{t \in [0,T]}$ given in (7.2) and formulates an associated stochastic control problem. Toward this end, we define the following difference process by

$$D_t := A_t - V_t^\theta + \mathrm{v} - a, \quad \forall t \in (0, T], \quad D_0 = 0.$$

Let $x \geq 0$. Consider its running maximum process $L = (L_t)_{t \in [0,T]}$ given by

$$L_t := x \vee \sup_{s \in [0, t]} D_s \geq 0, \quad \forall t \in [0, T]. \tag{7.8}$$

Obviously, the initial value $L_0 = x \geq 0$. Therefore, we can easily see that $(a-\mathrm{v})^+ - u(a, \mathrm{v}, z)$ with $u(a, \mathrm{v}, z)$ given in (7.7) is equivalent to the auxiliary stochastic control problem:

$$\sup_{\theta \in \mathcal{U}} \mathbb{E}\left[-\int_0^T e^{-\rho s} dL_s\right], \tag{7.9}$$

when we set the initial level $L_0 = x = (v-a)^+$. Furthermore, we can start with the introduction of our new controlled state process $X = (X_t)_{t\in[0,T]}$ for problem (7.9), which is defined as the reflected process $X_t := L_t - D_t$ for $t \in [0,T]$ that satisfies the SDE, for all $t \in [0,T]$,

$$X_t = -\int_0^t f(s, Z_s)ds + \int_0^t \theta_s^\top \mu ds + \int_0^t \theta_s^\top \sigma dW_s + L_t \qquad (7.10)$$

with the initial value $X_0 = L_0 = x \geq 0$. In particular, the running maximum process L_t increases if and only if $X_t = 0$, i.e., $L_t = D_t$. In light of the "Skorokhod problem", it satisfies the following representation, for all $t \in [0,T]$,

$$L_t = x \vee \int_0^t \mathbf{1}_{\{X_s=0\}} dL_s.$$

We will change the notation from L_t to L_t^X from this point onwards to emphasize its dependence on the new state process X given in (7.10). Moreover, the stochastic factor process $Z = (Z_t)_{t\in[0,T]}$ defined in (7.4) is chosen as another state process.

For ease of presentation, denote by the domain $\Theta_T := [0,T] \times \mathbb{R} \times [0,\infty)$. Let $\mathcal{U}_t$ be the set of admissible controls taking the feedback form as $\theta_s = \theta(s, Z_s, X_s)$ for $s \in [t,T]$, where $\theta : \Theta_T \to \mathbb{R}^n$ is a measurable function such that the following reflected SDE has a weak solution that

$$X_s = -\int_t^s f(r, Z_r) dr + \int_t^s \theta(r, Z_r, X_r)^\top \mu dr$$
$$+ \int_t^s \theta(r, Z_r, X_r)^\top \sigma dW_r + L_s^X, \quad \forall s \in [t,T] \qquad (7.11)$$

with $X_t = x \geq 0$. Here, $L_s^X = x \vee \int_t^s \mathbf{1}_{\{X_r=0\}} dL_r^X$ is a continuous, non-negative and non-decreasing process, which increases only when the state process $(X_s)_{s\in[t,T]}$ hits the level 0. For $(t, z, x) \in \Theta_T$, the dynamic version of the auxiliary problem (7.9) is given by

$$v(t,z,x) := \sup_{\theta \in \mathcal{U}_t} J(\theta; t, z, x) := \sup_{\theta \in \mathcal{U}_t} \mathbb{E}_{t,z,x}\left[-\int_t^T e^{-\rho(s-t)} dL_s^X\right]. \qquad (7.12)$$

Here $\mathbb{E}_{t,z,x}[\cdot] := \mathbb{E}[\cdot|Z_t = z, X_t = x]$.

It is important to note the equivalence that $v(0, z, (v-a)^+) = (a-v)^+ - u(a,v,z)$, i.e., we have

$$u(a,v,z) = \begin{cases} a - v - v(0,z,0), & \text{if } a \geq v, \\ -v(0,z,v-a), & \text{if } a < v, \end{cases}$$

where $u(a, v, z)$ is the value function of the original optimal tracking problem defined by (7.5), and a (resp. v) represents the initial benchmark level (resp. the initial wealth). We now mainly focus on the auxiliary control problem (7.12) and seek to obtain its optimal portfolio in a feedback form.

The following result gives some preliminary properties of the value function v defined in (7.12). The proof is standard by the following solution representation of "the Skorokhod problem" and it is hence omitted.

Lemma 7.2. *Let* $(t, z, x) \in \Theta_T$. *Then, the value function* $v(t, z, x)$ *defined by (7.12) is non-decreasing in* $x \geq 0$. *Furthermore, for all* $(t, z) \in [0, T] \times \mathbb{R}$,
$$|v(t, z, x_1) - v(t, z, x_2)| \leq |x_1 - x_2|, \quad \forall x_1, x_2 \geq 0.$$

We first have the following remark on Lemma 7.2:

Remark 7.1. For $(t, z) \in [0, T] \times \mathbb{R}$, if $x \to v(t, z, x)$ is $C^1([0, \infty))$, Lemma 7.2 implies that $0 \leq v_x(t, z, x) \leq 1$ for all $(t, z, x) \in \Theta_T$. Hereafter, we use $v_x := \partial_x v$, $v_t := \partial_t v$, $v_{xx} = \partial_{xx}^2 v$, $v_{zx} := \partial_{zx}^2 v$ and $v_{zz} = \partial_{zz}^2$ to denote the (first, second order or mixed) partial derivatives of the value function v w.r.t. its arguments, if exist.

By using the dynamic program, the value function v defined in (7.12) formally satisfies the following HJB equation, for $(t, z, x) \in [0, T] \times \mathbb{R} \times \mathbb{R}_+$,

$$\begin{cases} v_t + \sup_{\theta \in \mathbb{R}^n} \left[v_x \theta^\top \mu + \dfrac{v_{xx}}{2} \theta^\top \sigma \sigma^\top \theta + v_{xz} \sigma_Z(z) \theta^\top \sigma \gamma \right] \\ \qquad + v_z \mu_Z(z) + v_{zz} \dfrac{\sigma_Z^2(z)}{2} - f(t, z) v_x = \rho v; \\ v(T, z, x) = 0, \quad \forall (z, x) \in \mathbb{R} \times [0, \infty); \\ v_x(t, z, 0) = 1, \quad \forall (t, z) \in [0, T] \times \mathbb{R}. \end{cases} \quad (7.13)$$

The Neumann boundary condition $v_x(t, z, 0) = 1$ in (7.13) stems from the martingale optimality condition because the process L_s^X increases whenever the process X_s visits the value 0 for $s \in [t, T]$. Suppose $v_{xx} < 0$ on $[0, T) \times \mathbb{R} \times \mathbb{R}_+$, the feedback optimal control determined by (7.13) is obtained by, for $(t, z, x) \in \Theta_T$,

$$\theta^*(t, z, x) = -(\sigma \sigma^\top)^{-1} \dfrac{v_x(t, z, x) \mu + v_{xz}(t, z, x) \sigma_Z(z) \sigma \gamma}{v_{xx}(t, z, x)}. \quad (7.14)$$

Plugging (7.14) into the HJB equation (7.13), we have that, for $(t, z, x) \in [0, T) \times \mathbb{R} \times \mathbb{R}_+$,

$$v_t - \rho v - \alpha \dfrac{v_x^2}{v_{xx}} + \dfrac{\sigma_Z^2(z)}{2} \left(v_{zz} - \dfrac{v_{xz}^2}{v_{xx}} \right) - \phi(z) \dfrac{v_x v_{xz}}{v_{xx}}$$

$$+ \mu_Z(z)v_z - f(t,z)v_x = 0, \qquad (7.15)$$

where the coefficients are given by

$$\alpha := \frac{1}{2}\mu^\top(\sigma\sigma^\top)^{-1}\mu, \quad \phi(z) := \sigma_Z(z)\mu^\top(\sigma\sigma^\top)^{-1}\sigma\gamma, \quad \forall z \in \mathbb{R}. \qquad (7.16)$$

Note that the HJB equation (7.13) is fully nonlinear. To study the existence of a classical solution to Eq. (7.13), we will first apply the heuristic dual transform to linearize the original HJB equation (7.13) and then establish the existence and uniqueness of a classical solution to the dual PDE using the probabilistic representation and stochastic flow analysis in the next section.

7.4 The Linearization of HJB Equation

This section linearizes the nolinear HJB equation (7.15) by applying Fenchel-Legendre transform.

To introduce Fenchel-Legendre (FL) transform, we first assume that the value function $v \in C^{1,2,2}([0,T) \times \mathbb{R} \times [0,\infty)) \cap C(\Theta_T)$ and $v_{xx} < 0$ on $[0,T) \times \mathbb{R} \times \mathbb{R}_+$, which will be discussed and verified in detail later (Section 7.6). More precisely, for $(t,z,y) \in [0,T] \times \mathbb{R} \times \mathbb{R}_+$, the FL transform is only applied with respect to x that

$$\hat{v}(t,z,y) := \sup_{x>0}\{v(t,z,x) - xy\} \text{ and } x^*(t,z,y) := v_x(t,z,\cdot)^{-1}(y), \qquad (7.17)$$

where $y \mapsto v_x(t,z,\cdot)^{-1}(y)$ denotes the inverse function of $x \mapsto v_x(t,z,x)$, and $x^* = x^*(t,z,y)$ in (7.17) satisfies the equation:

$$v_x(t,z,x^*) = y, \quad \forall (t,z) \in [0,T] \times \mathbb{R}. \qquad (7.18)$$

On the other hand, in light of Lemma 7.2 and Remark 7.1, the variable y in fact only takes values in $(0,1)$. It follows from (7.17) that, for $(t,z,y) \in [0,T] \times \mathbb{R} \times (0,1)$,

$$\hat{v}(t,z,y) = v(t,z,x^*) - x^*y. \qquad (7.19)$$

Taking the derivative with respect to y on both sides of (7.19) and (7.18), we deduce

$$\hat{v}_y(t,z,y) = v_x(t,z,x^*)x_y^* - x_y^*y - x^* = yx_y^* - x_y^*y - x^* = -x^*, \qquad (7.20)$$

and also $v_{xx}(t,z,x^*)x_y^* = 1$ that yields $x_y^* = \frac{1}{v_{xx}(t,z,x^*)}$. Because of (7.20), we obtain

$$\hat{v}_{yy}(t,z,y) = -x_y^* = -\frac{1}{v_{xx}(t,z,x^*)}, \quad x_z^* = -\frac{v_{xz}(t,z,x^*)}{v_{xx}(t,z,x^*)}. \qquad (7.21)$$

It follows by (7.18) and (7.19) that
$$\hat{v}_t(t,z,y) = v_t(t,z,x^*), \quad \hat{v}_z(t,z,y) = v_z(t,z,x^*),$$
$$\hat{v}_{zz}(t,z,y) = v_{zz}(t,z,x^*) - \frac{v_{xz}(t,z,x^*)^2}{v_{xx}(t,z,x^*)}. \quad (7.22)$$

By applying the 2nd equality in (7.21) and (7.22), we further have
$$\hat{v}_{yz}(t,z,y) = v_{xz}(t,z,x^*)x_y^* = \frac{v_{xz}(t,z,x^*)}{v_{xx}(t,z,x^*)}. \quad (7.23)$$

Using (7.15) and (7.19), it holds that
$$v_t(t,z,x^*) - \rho v(t,z,x^*) - \alpha \frac{v_x(t,z,x^*)^2}{v_{xx}(t,z,x^*)} - \frac{\sigma_Z^2(z)}{2} \frac{v_{xz}(t,z,x^*)^2}{v_{xx}(t,z,x^*)}$$
$$- \phi(z)\frac{v_x(t,z,x^*)v_{xz}(t,z,x^*)}{v_{xx}(t,z,x^*)} + \mu_Z(z)v_z(t,z,x^*)$$
$$+ \frac{\sigma_Z^2(z)}{2}v_{zz}(t,z,x^*) - f(t,z)v_x(t,z,x^*) = 0. \quad (7.24)$$

Plugging (7.18), (7.21), (7.22) and (7.23) into (7.24), we derive that, for $(t,z,y) \in [0,T] \times \mathbb{R} \times (0,1)$,
$$\hat{v}_t(t,z,y) - \rho \hat{v}(t,z,y) + \rho y \hat{v}_y(t,z,y) + \alpha y^2 \hat{v}_{yy}(t,z,y) + \mu_Z(z)\hat{v}_z(t,z,y)$$
$$+ \frac{\sigma_Z^2(z)}{2}\hat{v}_{zz}(t,z,y) - \phi(z)y\hat{v}_{yz}(t,z,y) - f(t,z)y = 0. \quad (7.25)$$

In addition, the terminal condition $v(T,z,x) = 0$ of the HJB equation (7.13) then yields that, for all $(z,y) \in \mathbb{R} \times (0,1)$,
$$\hat{v}(T,z,y) = \sup_{x>0}\{v(T,z,x) - xy\} = \sup_{x>0}\{-xy\} = 0. \quad (7.26)$$

Note that $x_y^* = \frac{1}{v_{xx}(t,z,x^*)} < 0$, and for each $(t,z) \in [0,T] \times \mathbb{R}$, the map $y \mapsto x^*(t,z,y) := v_x(t,z,\cdot)^{-1}(y)$ is one to one. By using the Neumann boundary condition of the HJB equation (7.13), we deduce from (7.18) that $v_x(t,z,0) = 1$ and $x^*(t,z,1) = 0$. Therefore, in view of (7.20), we have, for all $(t,z) \in [0,T] \times \mathbb{R}$,
$$\hat{v}_y(t,z,1) = -x^*(t,z,1) = 0. \quad (7.27)$$

In summary, the HJB equation (7.13) can be transformed into the linear dual PDE of $\hat{v}$ that, for $(t,z,y) \in [0,T] \times \mathbb{R} \times (0,1)$,
$$\begin{cases} \hat{v}_t + \alpha y^2 \hat{v}_{yy} + \rho y \hat{v}_y - \rho \hat{v} - \phi(z)y\hat{v}_{yz} + \mu_Z(z)\hat{v}_z \\ \qquad + \frac{\sigma_Z^2(z)}{2}\hat{v}_{zz} - f(t,z)y = 0, \\ \hat{v}(T,z,y) = 0, \quad \forall (z,y) \in \mathbb{R} \times [0,1], \\ \hat{v}_y(t,z,1) = 0, \quad \forall (t,z) \in [0,T] \times \mathbb{R}. \end{cases} \quad (7.28)$$

We will establish the probabilistic solution of linear dual HJB equation (7.28) in the forthcoming section.

7.5 Probabilistic Solution of Linear Dual HJB Equation

In this section, we solve the linear dual HJB equation (7.28) in the spirit of Feymann-Kac's formula.

To solve Eq. (7.28), we define the following function by, for all $(t, z, u) \in \Theta_T$,

$$h(t, z, u) := -\mathbb{E}\left[\int_t^T e^{-\rho s} f(s, M_s^{t,z}) e^{-R_s^{t,u}} ds\right]. \tag{7.29}$$

Here, the process $M^{t,z} = (M_s^{t,z})_{s \in [t,T]}$ with $(t, z) \in [0, T] \times \mathbb{R}$ satisfies the following SDE, for $s \in [t, T]$,

$$M_s^{t,z} = z + \int_t^s \mu_Z(M_r^{t,z}) dr + \varrho \int_t^s \sigma_Z(M_r^{t,z}) dB_r^1$$
$$+ \sqrt{1 - \varrho^2} \int_t^s \sigma_Z(M_r^{t,z}) dB_r^2. \tag{7.30}$$

The processes $B^1 = (B_t^1)_{t \in [0,T]}$ and $B^2 = (B_t^2)_{t \in [0,T]}$ are two standard Brownian motions with a specific correlation coefficient

$$\varrho := \frac{(\sigma^{-1}\mu)^\top}{|\sigma^{-1}\mu|}\gamma. \tag{7.31}$$

The process $R^{t,u} = (R_s^{t,u})_{s \in [t,T]}$ with $(t, u) \in [0, T] \times [0, \infty)$ is a reflected Brownian motion with drift defined by, for all $s \in [t, T]$,

$$R_s^{t,u} := u + \sqrt{2\alpha}\int_t^s dB_r^1 + \int_t^s (\alpha - \rho) dr + \int_t^s dL_r^{t,R} \geq 0, \tag{7.32}$$

where $[t, T] \ni s \mapsto L_s^{t,R}$ is a continuous and non-decreasing process that increases only on $\{s \in [t, T]; R_s^{t,u} = 0\}$ with $L_t^{t,R} = 0$. In lieu of the solution representation of the "Skorokhod problem", it follows that, for all $(s, u) \in [t, T] \times [0, \infty)$,

$$L_s^{t,R} = 0 \vee \left\{-u + \max_{r \in [t,s]}\left[-\sqrt{2\alpha}(B_r^1 - B_t^1) - (\alpha - \rho)(r - t)\right]\right\}. \tag{7.33}$$

By Assumptions 7.1 and 7.2, for all $(t, z, u) \in \Theta_T$,

$$|h(t, z, u)| = \mathbb{E}\left[\int_t^T e^{-\rho s} f(s, M_s^{t,z}) e^{-R_s^{t,u}} ds\right]$$
$$\leq C\mathbb{E}\left[\int_t^T e^{-\rho s}(1 + |M_s^{t,z}|) ds\right]$$

$$\leq C(T-t) + C(T-t)\mathbb{E}\left[\sup_{s\in[t,T]} |M_s^{t,z}|\right]$$
$$\leq C(T-t)(1+|z|), \tag{7.34}$$

for some constant $C = C_f > 0$. This yields that, the function h given in (7.29) is well-defined.

In the sequel, we explore the regularity of the function h defined in (7.29) in the following result.

Proposition 7.1. *Under Assumptions 7.1 and 7.2, the function $h \in C^{1,2,2}(\Theta_T)$. We also have, for all $(t,z,u) \in \Theta_T$,*

$$h_u(t,z,u)$$
$$= \mathbb{E}\left[\int_t^T e^{-\rho s} f(s, M_s^{t,z}) e^{-R_s^{t,u}} \mathbf{1}_{\{\max_{r\in[t,s]}[-\sqrt{2\alpha}(B_r^1 - B_t^1) - (\alpha-\rho)(r-t)] \leq u\}} ds\right]$$
$$= \mathbb{E}\left[\int_t^{\tau_u^t \wedge T} e^{-\rho s} f(s, M_s^{t,z}) e^{-R_s^{t,u}} ds\right] \tag{7.35}$$

with $\tau_u^t := \inf\{s \geq t;\ \sqrt{2\alpha}(B_s^1 - B_t^1) - (\alpha-\rho)(s-t) = u\}$ and $\inf \emptyset := +\infty$ by convention.

Proof. We first derive the representation of the partial derivative h_u of the function h w.r.t. the variable u. Let $(t,z) \in [0,T] \times \mathbb{R}$ be fixed. For any $u_2 > u_1 \geq 0$, it follows from (7.29) that

$$\frac{h(t,z,u_2) - h(t,z,u_1)}{u_2 - u_1} = -\int_t^T \mathbb{E}\left[e^{-\rho s} f(s, M_s^{t,z}) \frac{e^{-R_s^{t,u_2}} - e^{-R_s^{t,u_1}}}{u_2 - u_1}\right] ds.$$

A direct calculation yields that, for $s \in [t,T]$,

$$\lim_{u_2 \downarrow u_1} \frac{e^{-R_s^{t,u_2} - \rho s} - e^{-R_s^{t,u_1} - \rho s}}{u_2 - u_1}$$
$$= \begin{cases} -e^{-R_s^{t,u_1} - \rho s}, & \max_{r\in[t,s]}\left[-\sqrt{2\alpha}(B_r^1 - B_t^1) - (\alpha-\rho)(r-t)\right] \leq u_1, \\ 0, & \max_{r\in[t,s]}\left[-\sqrt{2\alpha}(B_r^1 - B_t^1) - (\alpha-\rho)(r-t)\right] > u_1. \end{cases}$$

Since $\sup_{(s,u_1,u_2)\in[t,T]\times[0,\infty)^2} \left|\frac{e^{-R_s^{t,u_2} - \rho s} - e^{-R_s^{t,u_1} - \rho s}}{u_2 - u_1}\right| \leq 1$, the DCT yields that

$$\lim_{u_2 \downarrow u_1} \frac{h(t,z,u_2) - h(t,z,u_1)}{u_2 - u_1}$$

$$= \mathbb{E}\left[\int_t^T e^{-\rho s} f(s, M_s^{t,z}) e^{-R_s^{t,u_1}} \mathbf{1}_{\{\max_{r \in [t,s]}[-\sqrt{2\alpha}(B_r^1 - B_t^1) - (\alpha - \rho)(r-t)] \leq u_1\}} ds\right]$$

$$= \mathbb{E}\left[\int_t^{\tau_{u_1}^t \wedge T} e^{-\rho s} f(s, M_s^{t,z}) e^{-R_s^{t,u_1}} ds\right], \tag{7.36}$$

where $\tau_{u_1}^t := \inf\{s \geq t; -\sqrt{2\alpha}(B_s^1 - B_t^1) - (\alpha-\rho)(s-t) = u_1\}$. On the other hand, for the case $u_1 > u_2 \geq 0$, similar to the computation of (7.36), we have $\lim_{u_2 \uparrow u_1} \frac{h(t,z,u_2) - h(t,z,u_1)}{u_2 - u_1} = \lim_{u_2 \downarrow u_1} \frac{h(t,z,u_2) - h(t,z,u_1)}{u_2 - u_1}$. Hence, (7.35) holds.

We next derive the representation of h_{uu}. Let $(t,z) \in [0,T] \times \mathbb{R}$. For any $u_0, u_n \geq 0$ and $u_n \to u_0$ as $n \to \infty$, we have, for $n \geq 1$,

$$\Delta_n = \frac{h_u(t,z,u_n) - h_u(t,z,u_0)}{u_n - u_0} = \mathbb{E}\left[\frac{1}{u_n - u_0} \int_{\tau_0}^{\tau_n} e^{-\rho s} f(s, M_s^{t,z}) e^{-R_s^{t,u_0}} ds\right]$$

$$+ \mathbb{E}\left[\frac{1}{u_n - u_0} \int_t^{\tau_0} e^{-\rho s} f(s, M_s^{t,z}) \left(e^{-R_s^{t,u_n}} - e^{-R_s^{t,u_0}}\right) ds\right]$$

$$+ \mathbb{E}\left[\frac{1}{u_n - u_0} \int_{\tau_0}^{\tau_n} e^{-\rho s} f(s, M_s^{t,z}) \left(e^{-R_s^{t,u_n}} - e^{-R_s^{t,u_0}}\right) ds\right]$$

$$:= \Delta_n^{(1)} + \Delta_n^{(2)} + \Delta_n^{(3)} \tag{7.37}$$

with $\tau_0 := \tau_{u_0}^t \wedge T$ and $\tau_n := \tau_{u_n}^t \wedge T$. To deal with $\Delta_n^{(1)}$, we first focus on the case when $u_n \downarrow u_0$ as $n \to \infty$.

We next introduce the following Brownian motion with drift $\widetilde{B}^t = (\widetilde{B}_s^t)_{s \in [t,T]}$ given by

$$\widetilde{B}_s^t := -\sqrt{2\alpha}(B_s^1 - B_t^1) - (\alpha - \rho)(s-t), \quad \forall s \in [t, T].$$

It follows from (7.32) and (7.33) that

$$R_s^{t,u_0} = u_0 - \widetilde{B}_s^t + \left(-u_0 + \max_{r \in [t,s]} \widetilde{B}_r^t\right)^+, \quad \forall s \in [t, T].$$

This yields that

$$\Delta_n^{(1)} = \mathbb{E}\left[\frac{1}{u_n - u_0} \int_{\tau_0}^{\tau_n} e^{-\rho s} f(s, M_s^{t,z}) e^{-R_s^{t,u_0}} ds\right]$$

$$= \mathbb{E}\left[\frac{1}{u_n - u_0} \int_t^T e^{-\rho s} f(s, M_s^{t,z}) e^{-R_s^{t,u_0}} \mathbf{1}_{\{\tau_0 \leq s \leq \tau_n\}} ds\right]$$

$$= \mathbb{E}\left[\frac{1}{u_n - u_0} \int_t^T e^{-\rho s} f(s, M_s^{t,z}) e^{-R_s^{t,u_0}} \mathbf{1}_{\{u_0 \leq \max_{r \in [t,s]} \widetilde{B}_s^t \leq u_n\}} ds\right]$$

$$= \frac{1}{u_n - u_0} \int_t^T \int_{-\infty}^{\infty} \int_{u_0}^{u_n} \int_{-\infty}^{y} e^{-\rho s} f(s,m) e^{-u_0 + x - (-u_0 + y)^+}$$
$$\times \psi_{t,z}(s,x,y,m) dx dy dm ds. \quad (7.38)$$

Here, $\psi_{t,z}(s,x,y,m)$ is the joint density function of the three-dimensional random variable $(M_s^{t,z}, \widetilde{B}_s^1, \max_{r \in [t,s]} \widetilde{B}_r^1)$. Thus, we have from (7.38) that

$$\lim_{n \to \infty} \Delta_n^{(1)} \quad (7.39)$$
$$= \int_t^T \int_{-\infty}^{\infty} \int_{-\infty}^{u_0} e^{-\rho s} f(s,m) e^{-u_0 + x} \psi(t,s,z,x,u_0,m) dx dm ds.$$

For the case where $u_0 > 0$, $u_n \uparrow u_0$ as $n \to \infty$, we can follow the similar argument to get that

$$\lim_{n \to +\infty} \Delta_n^{(1)} = \int_t^T \int_{-\infty}^{\infty} \int_{-\infty}^{u_0} e^{-\rho s} f(s,m) e^{-u_0 + x} \psi_{t,z}(s,x,u_0,m) dx dm ds.$$

Similar to the derivation of (7.35), we also have

$$\lim_{n \to +\infty} \Delta_n^{(2)} = \lim_{n \to +\infty} \mathbb{E}\left[\frac{1}{u_n - u_0} \int_t^{\tau_0} e^{-\rho s} f(s, M_s^{t,z}) \left(e^{-R_s^{t,u_n}} - e^{-R_s^{t,u_0}}\right) ds\right]$$
$$= \mathbb{E}\left[\int_t^{\tau_0} e^{-\rho s} f(s, M_s^{t,z}) e^{-R_s^{t,u_0}} ds\right].$$

Finally, we can show by Assumption 7.1 and DCT that

$$|\Delta_n^{(3)}| \leq \mathbb{E}\left[\int_{\tau_0}^{\tau_n} e^{-\rho s} f(s, M_s^{t,z}) \frac{\left|e^{-R_s^{t,u_n}} - e^{-R_s^{t,u_0}}\right|}{u_n - u_0} ds\right]$$
$$\leq \mathbb{E}\left[\int_{\tau_0}^{\tau_n} e^{-\rho s} f(s, M_s^{t,z}) ds\right] \to 0, \text{ as } n \to \infty.$$

Hence, $\lim_{n \to +\infty} |\Delta_n^{(3)}| = 0$. Putting all the pieces together, we can derive from the decomposition (7.37) that

$$h_{uu}(t,z,u_0) = \int_t^T \int_{-\infty}^{\infty} \int_{-\infty}^{u_0} e^{-\rho s} f(s,m) e^{-u_0 + x} \psi_{t,z}(s,x,u_0,m) dx dm ds$$
$$+ \mathbb{E}\left[\int_t^{\tau_0} e^{-\rho s} f(s, M_s^{t,z}) e^{-R_s^{t,u_0}} ds\right]. \quad (7.40)$$

where $\tau_0 := \tau_{u_0}^t \wedge T$.

We next derive the representations of h_y, h_{yy} and h_{yu}. Fix $(t,u) \in [0,T] \times [0,\infty)$. In light of Assumption 7.2, Theorem 3.3.2 in Kunita (2019) yields that, for $s \in [t,T]$, the family $(M_s^{t,z})_{z \in \mathbb{R}}$ admits a modification which is continuously differentiable w.r.t. z. Moreover, $\partial_z M_s^{t,z}$ is continuous in z and satisfies the following SDE for $s \in [t, T]$ that

$$\partial_z M_s^{t,z} = 1 + \int_t^s \mu_Z'(M_r^{t,z})\partial_z M_r^{t,z} dr + \varrho \int_t^s \sigma_Z'(M_r^{t,z})\partial_z M_r^{t,z} dB_r^1$$
$$+ \sqrt{1-\varrho^2} \int_t^s \sigma_Z'(M_r^{t,z})\partial_z M_r^{t,z} dB_r^2. \tag{7.41}$$

For any $p \geq 2$, the following moment estimate holds:

$$\sup_{z \in \mathbb{R}} \mathbb{E}\left[\max_{s \in [t,T]} |\partial_z M_s^{t,z}|^p\right] < +\infty. \tag{7.42}$$

Thanks to (7.29), for distinct $z, \hat{z} \in \mathbb{R}$ and some constant $C > 0$, we have

$$\frac{h(t,z,u) - h(t,\hat{z},u)}{z - \hat{z}}$$
$$= -\mathbb{E}\left[\int_t^T e^{-\rho s - R_s^{t,u}} \frac{f(s, M_s^{t,z}) - f(s, M_s^{t,\hat{z}})}{z - \hat{z}} ds\right]. \tag{7.43}$$

By using Assumption 7.1, the following results hold with $s \in [t, T]$, $\mathbb{P}$-a.s.

$$\begin{cases} \dfrac{f(s, M_s^{t,z}) - f(s, M_s^{t,\hat{z}})}{z - \hat{z}} \xrightarrow{\hat{z} \to z} f'(s, M_s^{t,z})\partial_z M_s^{t,z}, \\ \left|\dfrac{f(s, M_s^{t,z}) - f(s, M_s^{t,\hat{z}})}{z - \hat{z}}\right| \leq C \left|\dfrac{M_s^{t,z} - M_s^{t,\hat{z}}}{z - \hat{z}}\right|. \end{cases}$$

We have from (7.42) that, for any $p \geq 2$, $\sup_{\hat{z} \neq z} \mathbb{E}[|\frac{M_s^{t,z} - M_s^{t,\hat{z}}}{z - \hat{z}}|^p] < +\infty$. This implies that $(\frac{M_s^{t,z} - M_s^{t,\hat{z}}}{z - \hat{z}})_{\hat{z} \neq z}$ is U.I.. Thus, in view of (7.43), we arrive at

$$h_z(t,z,u) = -\lim_{\hat{z} \to z} \mathbb{E}\left[\int_t^T e^{-\rho s - R_s^{t,u}} \frac{f(s, M_s^{t,z}) - f(s, M_s^{t,\hat{z}})}{z - \hat{z}} ds\right]$$
$$= -\mathbb{E}\left[\int_t^T e^{-\rho s - R_s^{t,u}} f'(s, M_s^{t,z})\partial_z M_s^{t,z} ds\right], \tag{7.44}$$

where $f'(t,z)$ denotes the partial derivative of f w.r.t. z. Similarly, we can obtain

$$h_{zu}(t,z,u) = \mathbb{E}\left[\int_t^T e^{-\rho s - R_s^{t,u}} f'(s, M_s^{t,z})\partial_z M_s^{t,z}\right.$$
$$\left. \times \mathbf{1}_{\{\max_{r \in [t,s]}[-\sqrt{2\alpha}(B_r^1 - B_t^1) - (\alpha-\rho)(r-t)] \leq u\}} ds\right]. \tag{7.45}$$

We next derive the expression of h_{yy}. To this end, we need the dynamics of $\partial_{zz}^2 M_s^{t,z}$ for $s \in [t,T]$. Following a similar argument of Theorem 3.4.2 in Kunita (2019), we can deduce

$$\partial_{zz}^2 M_s^{t,z} = \int_t^s \left\{ \mu_Z''(M_r^{t,z})|\partial_z M_r^{t,z}|^2 + \mu_Z'(M_r^{t,z})\partial_{zz}^2 M_r^{t,z} \right\} dr$$

$$+ \varrho \int_t^s \left\{ \sigma_Z''(M_r^{t,z})|\partial_z M_r^{t,z}|^2 + \sigma_Z'(M_r^{t,z})\partial_{zz}^2 M_r^{t,z} \right\} dB_r^1$$

$$+ \sqrt{1-\varrho^2} \int_t^s \left\{ \sigma_Z''(M_r^{t,z})|\partial_z M_r^{t,z}|^2 + \sigma_Z'(M_r^{t,z})\partial_{zz}^2 M_r^{t,z} \right\} dB_r^2.$$

Furthermore, for any $p \geq 1$, we have

$$\sup_{z \in \mathbb{R}} \mathbb{E}\left[\max_{s \in [t,T]} |\partial_{zz}^2 M_s^{t,z}|^{2p} \right] < +\infty.$$

The chain rule with Assumption 7.1 yields that, $\mathbb{P}$-a.s.

$$\frac{f'(s, M_s^{t,z})\partial_z M_s^{t,z} - f'(s, M_s^{t,\hat{z}})\partial_z M_s^{t,\hat{z}}}{z - \hat{z}} \xrightarrow{\hat{z} \to z} f''(s, M_s^{t,z})|\partial_z M_s^{t,z}|^2$$

$$+ f'(s, M_s^{t,z})\partial_{zz}^2 M_s^{t,z},$$

and there exists a constant $C > 0$ such that, for all $\hat{z} \neq z$,

$$\left| \frac{f'(s, M_s^{t,z})\partial_z M_s^{t,z} - f'(s, M_s^{t,\hat{z}})\partial_z M_s^{t,\hat{z}}}{z - \hat{z}} \right| \leq C \left| \frac{M_s^{t,z} - M_s^{t,\hat{z}}}{z - \hat{z}} \partial_z M_s^{t,z} \right|$$

$$+ C \left| \frac{\partial_z M_s^{t,z} - \partial_z M_s^{t,\hat{z}}}{z - \hat{z}} \right|. \tag{7.46}$$

Define $I(s; z, \hat{z}) := \mathbb{E}\left[\max_{r \in [t,s]} |\partial_z M_r^{t,z} - \partial_z M_r^{t,\hat{z}}|^{2p} \right]$ for $(s, z, \hat{z}) \in [t,T] \times \mathbb{R}^2$. It follows from (7.41) and Assumption 7.2 that, for some $C > 0$,

$I(s; z, \hat{z})$

$$\leq C \int_t^s \left\{ I(r; z, \hat{z}) + \sup_{u \in \mathbb{R}} \mathbb{E}\left[\max_{s \in [t,T]} |\partial_z M_s^{t,u}|^{2p} \right] \mathbb{E}\left[|M_r^{t,z} - M_r^{t,\hat{z}}|^{2p} \right] \right\} dr$$

$$\leq C \int_t^s \left\{ I(r; z, \hat{z}) + |z - \hat{z}|^{2p} \right\} dr.$$

The Gronwall's lemma yields that

$$\sup_{\hat{z} \neq z} \mathbb{E}\left[\sup_{r \in [t,s]} \left| \frac{\partial_z M_r^{t,z} - \partial_z M_r^{t,\hat{z}}}{z - \hat{z}} \right|^{2p} \right] < +\infty, \quad \forall s \in [t,T].$$

Thus, the LHS of (7.46) with $\hat{z} \neq z$ and $s \in [t,T]$ is U.I.. Then

$$h_{zz}(t, z, u) \tag{7.47}$$

$$= -\lim_{\hat{z} \to z} \mathbb{E}\left[\int_t^T e^{-\rho s - R_s^{t,u}} \frac{f'(s, M_s^{t,z}) \partial_z M_s^{t,z} - f'(s, M_s^{t,\hat{z}}) \partial_z M_s^{t,\hat{z}}}{z - \hat{z}} ds\right]$$

$$= -\mathbb{E}\left[\int_t^T e^{-\rho s - R_s^{t,u}} \left(f''(s, M_s^{t,z}) |\partial_z M_s^{t,z}|^2 + f'(s, M_s^{t,z}) \partial_{zz}^2 M_s^{t,z}\right) ds\right],$$

where $f''(t, z)$ denotes the second-order partial derivative of f w.r.t. z.

We then move on to the expression of h_t. Let us consider the solutions $M^{t,y} = (M_s^{t,z})_{s \in [t,T]}$ and $M^{\hat{t},z} = (M_s^{\hat{t},z})_{s \in [\hat{t},T]}$ of SDE (7.30) with parameters $(t, z) \in [0, T] \times \mathbb{R}$ and $(\hat{t}, z) \in [0, T] \times \mathbb{R}$ respectively. Moreover, for $r \geq 0$, we introduce $\mathcal{F}_r^t := \mathcal{F}_{t+r}$, $B_r^{1,t} := B_{t+r}^1 - B_t^1$, and $B_r^{2,t} := B_{t+r}^2 - B_t^2$ and define $\mathcal{F}_r^{\hat{t}}, B_r^{i,\hat{t}}, i = 1, 2$ for $r \geq 0$ in a similar way. It is not difficult to check that

$$(M_{t+r}^{t,z}, B_r^{1,t}, B_r^{2,t})_{r \geq 0} \stackrel{d}{=} (M_{\hat{t}+r}^{\hat{t},z}, B_r^{1,\hat{t}}, B_r^{2,\hat{t}})_{r \geq 0}. \tag{7.48}$$

For any $\delta \in [0, T - t]$, it holds that

$$h(t + \delta, z, u) = -\mathbb{E}\left[\int_{t+\delta}^T e^{-\rho s} f(s, M_s^{t+\delta, z}) e^{-R_s^{t+\delta, u}} ds\right]$$

$$= -\mathbb{E}\left[e^{-\rho \delta} \int_t^{T-\delta} e^{-\rho s} f(s, M_s^{t,z}) e^{-R_s^{t,u}} ds\right].$$

The DCT yields that

$$\lim_{\delta \downarrow 0} \frac{1}{\delta}(h(t + \delta, z, u) - h(t, z, u)) = \lim_{\delta \downarrow 0} \mathbb{E}\left[\frac{e^{-\rho \delta}}{\delta} \int_{T-\delta}^T e^{-\rho s} f(s, M_s^{t,z}) e^{-R_s^{t,u}} ds\right.$$

$$\left. + \frac{1 - e^{-\rho \delta}}{\delta} \int_t^T e^{-\rho s} f(s, M_s^{t,z}) e^{-R_s^{t,u}} ds\right]$$

$$= \mathbb{E}\left[e^{-\rho T} f(T, M_T^{t,z}) e^{-R_T^{t,u}} + \rho \int_t^T e^{-\rho s} f(s, M_s^{t,z}) e^{-R_s^{t,u}} ds\right].$$

Similarly, for $t \in (0, T]$, we have

$$\lim_{\delta \downarrow 0} \frac{1}{\delta}(h(t, z, u) - h(t - \delta, z, u))$$

$$= \mathbb{E}\left[e^{-\rho T} f(T, M_T^{t,z}) e^{-R_T^{t,u}} + \rho \int_t^T e^{-\rho s} f(s, M_s^{t,z}) e^{-R_s^{t,u}} ds\right].$$

Thus, we conclude that, for $(t, z, u) \in \Theta_T$,

$$h_t(t, z, u) \tag{7.49}$$

$$= \mathbb{E}\left[e^{-\rho T} f(T, M_T^{t,z}) e^{-R_T^{t,u}} + \rho \int_t^T e^{-\rho s} f(s, M_s^{t,z}) e^{-R_s^{t,u}} ds\right].$$

Lastly, we verify the continuity of h_t, h_{uu}, h_{yu} and h_{yy} in (t,y,u) using expressions (7.49), (7.40), (7.45) and (7.47). In fact, by Theorem 3.4.3 of Kunita (2019), we have $M_s^{t,z}$, $\partial_z M_s^{t,z}$ and $\partial_{zz}^2 M_s^{t,z}$ for $s \in [t,T]$ admit the respective modifications which are continuous in (t,z,s), $\mathbb{P}$-a.s.. Moreover, by (7.32) and (7.33), $R_s^{t,u}$ is also continuous in (t,u,s), $\mathbb{P}$-a.s.. We can conclude by the DCT that h_t, h_{zu} and h_{zz} are continuous in (t,z,u). For the continuity of h_{uu}, in view of (7.48), for any $\epsilon \in \mathbb{R}$ satisfying $\tau_0^\epsilon := \tau_{u_0}^t \wedge (T-\epsilon) \in [t,T]$,

$$h_{uu}(t+\epsilon, z, u_0) = \mathbb{E}\left[e^{-\rho \tau_0^\epsilon} f(\tau_0^\epsilon, M_{\tau_0^\epsilon}^{t,z}) \Gamma(\tau_0^\epsilon)\right]$$
$$+ \mathbb{E}\left[\int_t^{\tau_0^\epsilon} e^{-\rho s} f(s, M_s^{t,z}) e^{-R_s^{t,u_0}} ds\right]. \quad (7.50)$$

Here, for $t \in [0,T]$,

$$\Gamma(t) := \int_0^{T-t} \frac{1}{\sqrt{4\alpha \pi s}} e^{-\frac{\tilde{\mu}^2}{4\alpha} s} ds + (T-t) \int_{T-t}^{+\infty} \frac{1}{\sqrt{4\alpha \pi s^3}} e^{-\frac{\tilde{\mu}^2}{4\alpha} s} ds. \quad (7.51)$$

Note that τ_0^ϵ is continuous in (u_0, ϵ), $\mathbb{P}$-a.s.. The continuity of h_{uu} in (t,z,u) follows from (7.50) and DCT, which completes the entire proof. $\square$

Using Proposition 7.1, we have the following key result.

Theorem 7.1. *Let Assumptions 7.1 and 7.2 hold. Then, the mapping h defined in (7.29) solves the following Neumann problem, for $(t,z,u) \in [0,T) \times \mathbb{R} \times \mathbb{R}_+$,*

$$\begin{cases} h_t + \alpha h_{uu} + (\alpha - \rho) h_u + \phi(z) h_{uz} \\ \qquad + \mu_Z(z) h_z + \frac{\sigma_Z^2(z)}{2} h_{zz} = f(t,z) e^{-u-\rho t}, \\ h(T,z,u) = 0, \quad \forall (z,u) \in \mathbb{R} \times [0,\infty), \\ h_u(t,z,0) = 0, \quad \forall (t,z) \in [0,T] \times \mathbb{R}. \end{cases} \quad (7.52)$$

On the other hand, if some function h defined on Θ_T with a polynomial growth is a classical solution to Neumann problem (7.52), then h has the probabilistic representation (7.29).

Proof. For $(t,u) \in [0,T] \times [0,\infty)$, define $B_s^{t,u} := u + \sqrt{2\alpha}(B_s^1 - B_t^2) + (\alpha - \rho)(s-t)$ for $s \in [t,T]$, and $M^{t,z} = (M_s^{t,z})_{s \in [t,T]}$ for $(t,z) \in [0,T] \times \mathbb{R}$ is the unique (strong) solution of SDE (7.30) under Assumption 7.2. Consider Remark 4.17 of Chapter 5 in Karatzas and Shreve (1991). Assumption 7.2 guarantees that the time-homogeneous martingale problem on $(M^{t,z}, B^{t,u}) = (M_s^{t,z}, B_s^{t,u})_{s \in [t,T]}$ is well posed (the definition of well-posedness of a time-homogeneous martingale problem can be found in Definition 4.15 of Chapter 5 in Karatzas and Shreve (1991), p.320). By applying Theorem 5.4.20 in Karatzas and Shreve (1991), $(M^{t,z}, B^{t,u})$ is a strong Markov process with $(t, z, u) \in [0, T] \times \mathbb{R} \times \mathbb{R}_+$. For $\varepsilon \in (0, u)$, let us define

$$\tau_\varepsilon^t := \inf\left\{s \geq t;\ \left|B_s^{t,u} - u\right| \geq \varepsilon \text{ or } \left|M_s^{t,z} - z\right| \geq \varepsilon\right\} \wedge T. \tag{7.53}$$

Since the paths of $(M^{t,z}, B^{t,u})$ are continuous, we have $\tau_\varepsilon^t > t$, $\mathbb{P}$-a.s.. For any $\hat{t} \in [t, T]$, it holds that

$$B_{\hat{t} \wedge \tau_\varepsilon^t}^{t,u} = R_{\hat{t} \wedge \tau_\varepsilon^t}^{t,u}. \tag{7.54}$$

In fact, if $\tau_\varepsilon^t \leq \hat{t}$, the following two cases may happen:

(i) $\left|B_{\tau_\varepsilon^t}^{t,u} - u\right| < \varepsilon$: this yields that $0 < -\varepsilon + u < B_{\tau_\varepsilon^t}^{t,u} < \varepsilon + u$, and hence (7.54) holds.

(ii) $\left|B_{\tau_\varepsilon^t}^{t,u} - u\right| \geq \varepsilon$: this implies that $B_{\tau_\varepsilon^t}^{t,u} = u + \varepsilon > 0$ or $B_{\tau_\varepsilon^t}^{t,u} = u - \varepsilon > 0$, and again (7.54) holds.

If $\hat{t} < \tau_\varepsilon^t$, then $\left|B_{\hat{t}}^{t,u} - u\right| < \varepsilon$ and $\left|M_{\hat{t}}^{t,z} - z\right| < \varepsilon$. This implies that $0 < -\varepsilon + u < B_{\hat{t}}^{t,u} < \varepsilon + u$, and hence (7.54) holds. Using the strong Markov property with (7.54), we get

$$-\mathbb{E}\left[\int_{\hat{t} \wedge \tau_\varepsilon^t}^T f(s, M_s^{t,z}) e^{-R_s^{t,u} - \rho s} ds \bigg| \mathcal{F}_{\hat{t} \wedge \tau_\varepsilon^t}\right] = h\left(\hat{t} \wedge \tau_\varepsilon^t, M_{\hat{t} \wedge \tau_\varepsilon^t}^{t,z}, B_{\hat{t} \wedge \tau_\varepsilon^t}^{t,u}\right),$$

where $h(t, z, u) = -\mathbb{E}[\int_t^T e^{-\rho s} f(s, M_s^{t,z}) e^{-B_s^{t,u} - L_s^{t,R}} ds]$. So that, for $(t, y, u) \in \Theta_T$, it holds that

$$h(t, z, u) \tag{7.55}$$
$$= \mathbb{E}\left[h\left(\hat{t} \wedge \tau_\varepsilon^t, M_{\hat{t} \wedge \tau_\varepsilon^t}^{t,z}, B_{\hat{t} \wedge \tau_\varepsilon^t}^{t,u}\right) - \int_t^{\hat{t} \wedge \tau_\varepsilon^t} e^{-\rho s} f(s, M_s^{t,z}) e^{-B_s^{t,u} - L_s^{t,R}} ds\right].$$

From Proposition 7.1 and Itô's rule, it follows that

$$\frac{1}{\hat{t} - t} \mathbb{E}\left[\int_t^{\hat{t} \wedge \tau_\varepsilon^t} e^{-\rho s} f(s, M_s^{t,z}) e^{-B_s^{t,u} - L_s^{t,R}} ds\right]$$
$$= \frac{1}{\hat{t} - t} \mathbb{E}\left[h\left(\hat{t} \wedge \tau_\varepsilon^t, M_{\hat{t} \wedge \tau_\varepsilon^t}^{t,z}, B_{\hat{t} \wedge \tau_\varepsilon^t}^{t,u}\right) - h(t, z, u)\right]$$
$$= \frac{1}{\hat{t} - t} \mathbb{E}\left[\int_t^{\hat{t} \wedge \tau_\varepsilon^t} (h_t + \mathcal{L}h)(s, M_s^{t,z}, B_s^{t,u}) ds\right]$$

$$+ \frac{1}{\hat{t}-t}\mathbb{E}\left[\int_t^{\hat{t}\wedge\tau_\varepsilon^t} h_u\left(s, M_s^{t,z}, B_s^{t,u}\right) dL_s^{t,R}\right], \tag{7.56}$$

where the operator $\mathcal{L}$ acted on $C^2(\mathbb{R}\times[0,\infty))$ is defined by, for $g\in C^2(\mathbb{R}\times[0,\infty))$,

$$\mathcal{L}g := \alpha g_{uu} + (\alpha-\rho)g_u + \phi(y)g_{uz} + \mu_Z(z)g_z + \frac{\sigma_Z^2(z)}{2}g_{zz}. \tag{7.57}$$

Together with (7.53), Assumption 7.1 yields that, the process $e^{-\rho s}f(s, M_s^{t,z})e^{-B_s^{t,u}-L_s^{t,R}}$ for $s\in[t,\hat{t}\wedge\tau_\varepsilon^t]$ is bounded. The BCT yields that

$$\lim_{\hat{t}\downarrow t}\frac{1}{\hat{t}-t}\mathbb{E}\left[\int_t^{\hat{t}\wedge\tau_\varepsilon^t} e^{-\rho s}f(s, M_s^{t,z})e^{-B_s^{t,u}-L_s^{t,R}} ds\right] = f(t,z)e^{-u-\rho t}.$$

Similarly, we have

$$\lim_{\hat{t}\downarrow t}\frac{1}{\hat{t}-t}\mathbb{E}\left[\int_t^{\hat{t}\wedge\tau_\varepsilon^t} (h_t+\mathcal{L}h)\left(s, M_s^{t,z}, B_s^{t,u}\right) ds\right] = (h_t+\mathcal{L}h)(t,z,u).$$

Note that $R_s^{t,u}>0$ on $s\in[t,\hat{t}\wedge\tau_\varepsilon^t]$ for all $(t,u)\in[0,T]\times[0,\infty)$. We then have

$$\frac{1}{\hat{t}-t}\mathbb{E}\left[\int_t^{\hat{t}\wedge\tau_\varepsilon^t} h_u\left(s, M_s^{t,z}, B_s^{t,u}\right) dL_s^{t,R}\right] = 0.$$

By applying (7.56), we obtain $(h_t+\mathcal{L}h)(t,z,u) = f(t,z)e^{-u-\rho t}$ on $(t,z,u)\in[0,T)\times\mathbb{R}\times\mathbb{R}_+$.

We next verify that the function h in (7.29) satisfies the boundary conditions of the Neumann problem (7.52). By using the representation form (7.29), it is easy to see that $h(T,z,u)=0$ for all $(z,u)\in\mathbb{R}\times[0,\infty)$. It remains to show the validity of homogeneous Neumann boundary condition. In fact, for any positive sequence $(u_n)_{n\geq 1}$ satisfying $u_n\downarrow 0$ as $n\to\infty$,

$$\bigcup_{n\geq 1} A_s^{t,u_n} := \bigcup_{n\geq 1}\left\{\max_{r\in[t,s]}\left[-\sqrt{2\alpha}(B_r^1-B_t^1)-(\alpha-\rho)(r-t)\right] > u_n\right\}\in\mathcal{F}_s,$$

we have

$$\mathbb{P}\left(\bigcup_{n\geq 1} A_s^{t,u_n}\right) = 1. \tag{7.58}$$

In light of (7.35) in Proposition 7.1 and Assumption 7.2, it follows from DCT that, for $(t,z)\in[0,T]\times\mathbb{R}$,

$$h_u(t,z,0) = \lim_{n\to\infty}\int_t^T \mathbb{E}\left[f(s, M_s^{t,z})e^{-R_s^{t,u}-\rho s}\mathbf{1}_{\{(A_s^{t,u_n})^c\}}\right] ds = 0. \tag{7.59}$$

That is, the Neumann boundary condition in (7.52) holds.

We next assume that the Neumann problem (7.52) has a classical solution h with a polynomial growth. For $n \in \mathbb{N}$ and $t \in [0,T]$, we define $\tau_n^t := \inf\{s \geq t; |M_s^{t,z}| \geq n \text{ or } |R_s^{t,u}| \geq n\} \wedge T$. Itô's rule gives that, for $(t,z,u) \in \Theta_T$,

$$\mathbb{E}\left[h\left(\tau_n^t, M_{\tau_n^t}^{t,z}, R_{\tau_n^t}^{t,u}\right)\right] = h(t,z,u) + \mathbb{E}\left[\int_t^{\tau_n^t}(h_t + \mathcal{L}h)\left(r, M_r^{t,z}, R_r^{t,u}\right)dr\right]$$

$$+ \mathbb{E}\left[\int_t^{\tau_n^t} h_u(r, M_r^{t,z}, R_r^{t,u})\mathbf{1}_{\{R_r^{t,u}=0\}} dL_r^{t,R}\right]$$

$$= h(t,z,u) + \mathbb{E}\left[\int_t^{\tau_n^t} f(r, M_r^{t,z})e^{-R_r^{t,u}-\rho r} dr\right]. \quad (7.60)$$

Moreover, the polynomial growth of h implies the existence of a constant $C = C_T > 0$ such that, for some $p \geq 1$,

$$\left|h(\tau_n^t, M_{\tau_n^t}^{t,z}, R_{\tau_n^t}^{t,u})\right| \leq C\left\{1 + \max_{r \in [t,T]}|M_r^{t,z}|^p + \max_{r \in [t,T]}|R_r^{t,u}|^p\right\}.$$

Note that $\lim_{n \to \infty} h(\tau_n^t, M_{\tau_n^t}^{t,z}, R_{\tau_n^t}^{t,u}) = h(T, M_T^{t,z}, R_T^{t,u}) = 0$ in view of (7.52). Letting $n \to \infty$ on both sides of (7.60), we obtain from DCT and MCT that the representation (7.29) holds. $\square$

Remark 7.2. Under Assumptions 7.1 and 7.2, it follows from (7.35) and (7.40) that, for all $(t,z,u) \in \Theta_T$,

$$h_u(t,z,u) + h_{uu}(t,z,u) \quad (7.61)$$

$$= \mathbb{E}\left[e^{-\rho\tau_u^t} f(\tau_u^t, M_{\tau_u^t}^{t,z})\Gamma(\tau_u^t) + 2\int_t^{\tau_u^t} e^{-\rho s} f(s, M_s^{t,z}) e^{-R_s^{t,u}} ds\right].$$

Here, the stopping time τ_u^t is given in Proposition 7.1 and the function $\Gamma(t)$ for $t \in [0,T]$ is given by (7.51). Note that $f > 0$ in Assumption 7.1 guarantees that $h_{uu} + h_u > 0$ for $(t,z,u) \in [0,T) \times \mathbb{R} \times [0,\infty)$. This implies that $\hat{v}(t,z,y)$ in (6.55) is strictly convex in $y \in (0,1]$.

The well-posedness of problem (7.28) is now given in the following result.

Corollary 7.1. *Under Assumptions of Theorem 7.1, the Neumann problem (7.28) has a unique classical solution $\hat{v}$ such that, for all $(t,z,y) \in [0,T] \times \mathbb{R} \times (0,1]$,*

$$|\hat{v}(t,z,y)| \leq C(1 + |z|^p + |\ln y|^p), \quad \exists p > 1, \quad (7.62)$$

and the following function defined on $(t, z, u) \in \Theta_T$,
$$h(t, z, u) := e^{-\rho t} \hat{v}(t, z, e^{-u}) \tag{7.63}$$
has the probabilistic representation (7.29). Moreover, for each $(t, z) \in [0, T] \times \mathbb{R}$, the solution $(0, 1] \ni y \mapsto \hat{v}(t, z, y)$ is strictly convex.

Proof. We first show the existence of a classical solution to the Neumann problem (7.28). By applying Theorem 7.1, the mapping h defined by (7.29) solves problem (7.52). It readily follows that, for $(t, z, y) \in [0, T] \times \mathbb{R} \times (0, 1]$, $\hat{v}(t, z, y) := e^{\rho t} h(t, z, -\ln y)$ solves the Neumann problem (7.28). The existence of a classical solution to problem (7.28) then follows by Proposition 7.1. For the uniqueness, let $\hat{v}^{(i)}$ for $i = 1, 2$ be two classical solutions to the Neumann problem (7.28) such that $h^{(i)}(t, z, u) := e^{-\rho t} \hat{v}^{(i)}(t, z, e^{-u})$ for $(t, z, u) \in \Theta_T$ satisfies the polynomial growth for $i = 1, 2$. Theorem 7.1 yields that both $h^{(1)}$ and $h^{(2)}$ admit the probabilistic representation (7.29), and hence $h^1 = h^2$ on Θ_T. Therefore, it holds that $\hat{v}^{(1)}(t, z, y) = e^{\rho t} h^{(1)}(t, z, -\ln y) = e^{\rho t} h^{(2)}(t, z, -\ln y) = \hat{v}^{(2)}(t, z, y)$ for $(t, z, y) \in [0, T] \times \mathbb{R} \times (0, 1]$. Furthermore, the strict convexity of $(0, 1] \ni y \to \hat{v}(t, z, y)$ for fixed $(t, z) \in [0, T) \times \mathbb{R}$ follows from the fact that $\hat{v}_{yy} = \frac{e^{\rho t}}{y^2}[h_{uu} + h_u] > 0$ in Remark 7.2 as it is assumed that $f > 0$. Thus, we complete the proof of the corollary. $\square$

7.6 Verification Results

The previous Corollary 7.1 provides the well-posedness of the dual equation (7.28) in classical sense. In other words, Eq. (7.28) has a unique classical solution $\hat{v}(t, z, y)$ for $(t, z, y) \in [0, T] \times \mathbb{R} \times (0, 1]$, which is strictly convex in $y \in (0, 1]$.

We next recover the classical solution $v(t, z, x)$ of the primal HJB equation (7.15) via $\hat{v}(t, z, y)$ using the inverse transform and prove the verification theorem for problem (7.12).

Theorem 7.2 (Verification Theorem). *Under Assumption 7.1 and Assumption 7.2, we have*

(i) *The primal HJB equation (7.15) has a solution $v \in C^{1,2,2}([0, T) \times \mathbb{R} \times [0, \infty)) \cap C(\Theta_T)$. Moreover, for $(t, z, x) \in \Theta_T$, the solution v of HJB equation (7.15) can be written as:*

$$v(t, z, x) = \begin{cases} \inf_{y \in (0, 1]} \{\hat{v}(t, z, y) + xy\}, & \text{if } (t, z, x) \in \mathcal{C}_T \text{ or } x = 0, \\ 0, & \text{if } (t, z, x) \in \mathcal{C}_T^c \cap \Theta_T. \end{cases} \tag{7.64}$$

Here, the region $\mathcal{C}_T$ in (7.64) is given by
$$\mathcal{C}_T := \{(t,z,x) \in [0,T] \times \mathbb{R} \times \mathbb{R}_+;\ x \in (0, \xi(t,z))\}, \quad (7.65)$$
where the function $\xi(t,z)$ with $(t,z) \in [0,T] \times \mathbb{R}$ is defined by
$$\xi(t,z) := \mathbb{E}\left[\int_t^T e^{-\rho(s-t)} f(s, M_s^{t,z}) e^{\sqrt{2\alpha}(B_s^1 - B_t^1) + (\alpha - \rho)(s-t)} ds\right], \quad (7.66)$$
and the process $(M_s^{t,z})_{s \in [t,T]}$ with $(t,z) \in [0,T] \times \mathbb{R}$ is the strong solution to SDE (7.30). Since $f > 0$ in (7.3), we get $\xi(\cdot,\cdot) > 0$ so that $\mathcal{C}_T \neq \emptyset$. Here, for $(t,z,y) \in [0,T] \times \mathbb{R} \times (0,1]$, the function $\hat{v}(t,z,y) = e^{\rho t}h(t,z,-\ln y)$ solves the dual PDE (7.28) with Neumann boundary condition and $\hat{v}(t,z,y)$ is strictly convex in $y \in (0,1]$.

(ii) For $(t,z,x) \in \Theta_T$, define the feedback control function as follows:
$$\theta^*(t,z,x) \quad (7.67)$$
$$:= \begin{cases} -(\sigma\sigma^\top)^{-1} \dfrac{v_x(t,z,x)\mu + v_{xz}(t,z,x)\sigma_Z(z)\sigma\gamma}{v_{xx}(t,z,x)}, \\ \qquad\qquad\qquad\qquad\qquad \text{if } (t,z,x) \in \mathcal{C}_T \text{ or } x = 0, \\ -\mu(\sigma\sigma^\top)^{-1} \lim_{y\downarrow 0} y\hat{v}_{yy}(t,z,y) \quad \text{if } (t,z,x) \in \mathcal{C}_T^c \cap \Theta_T. \\ + (\sigma\sigma^\top)^{-1}\sigma_Z(z)\sigma\gamma \lim_{y\downarrow 0} \hat{v}_{yz}(t,z,y), \end{cases}$$
Given the processes $(Z,X) = (Z_t, X_t)_{t\in[0,T]}$ in (7.11), we define $\theta_t^* := \theta^*(t, Z_t, X_t)$ for $t \in [0,T]$. Then $\theta^* = (\theta_t^*)_{t \in [0,T]} \in \mathcal{U}_t$ is an optimal strategy. Moreover, for all $\theta \in \mathcal{U}_t$, it holds that $J(\theta; t, z, x) \leq e^{-\rho t} v(t,z,x)$ with $(t,z,x) \in [0,T] \times \mathbb{R} \times [0,\infty)$.

Before proving Theorem 7.2, we first have the following remarks:

Remark 7.3. We explain here the role of the critical function $\xi(t,z)$ defined by (7.66) in Theorem 7.2. In fact, for $(t,z) \in [0,T) \times \mathbb{R}$ and $x \geq \xi(t,z)$, it follows from Theorem 7.2-(i) that the value function $v(t,z,x) = 0$. Then, by Theorem 7.2-(ii), we have that, for the strategy $\theta^* \in \mathcal{U}_t$ given by (7.67),
$$\mathbb{E}_{t,z,x}\left[-\int_t^T e^{-\rho s} dL_s^{X^*}\right] = 0,$$
where the process $(L_s^{X^*})_{s\in[t,T]}$ is the reflected term of the process $(X_s^*)_{s\in[t,T]}$ in (7.11) with θ replaced by θ^*. It follows from integration by parts that $e^{-\rho T} L_T^{X^*} + \rho \int_t^T e^{-\rho s} L_s^{X^*} ds = x$, $\mathbb{P}$-a.s., and hence $L_T^{X^*} = L_t^{X^*} = x$, $\mathbb{P}$-a.s. since $\xi(t,z) > 0$ for $(t,z) \in [0,T) \times \mathbb{R}$.

Therefore, with the strategy $\theta^* \in \mathcal{U}_t$, the non-negative process X^* is given by

$$X_s^* = x - \int_t^s f(r, Z_r)dr + \int_t^s (\theta_r^*)^\top \mu dr + \int_t^s (\theta_r^*)^\top \sigma dW_r.$$

On the other hand, for $0 \leq x < \xi(t, z)$, we have $v_x(t, z, x) > 0$, and hence $w(t, z, x) < 0$. This implies that, for this initial value x at time t, the reflected term $L_s^{X^*}$ is strictly increasing in $s \in [t, T)$ with a positive probability.

Remark 7.4. Note that $u(a, \mathrm{v}, z) = -v(0, z, \mathrm{v} - a)$ with $\mathrm{v} > a$, where $u(a, \mathrm{v}, z)$ is the value function for problem (7.5). According to Remark 7.3, if the initial wealth v is sufficiently large such that $\mathrm{v} - a > \xi(0, z)$ for the given $f(\cdot, \cdot)$ and $\mu_Z(\cdot)$, $\sigma_Z(\cdot)$, we can conclude that $u(a, \mathrm{v}, z) = 0$ and the optimal singular control $C_t^* \equiv 0$ for $t \in [0, T]$. Therefore, A_t is dynamically superhedgeable that $A_t \leq V_t^{\theta^*}$ for $t \in [0, T]$.

Proof of Theorem 7.2. We first prove the claim (i). It follows from Assumption 7.1 that $\xi(t, z) > 0$ for $(t, z) \in [0, T) \times \mathbb{R}$, according to its definition. Thanks to the probabilistic representation of derivatives of h in the proof of Proposition 7.1, we have

$$\begin{cases}
\hat{v}_y(t, z, 1) = -e^{\rho t} h_u(t, z, 0) = 0, \\
\hat{v}_y(T, z, y) = y^{-1} e^{\rho t} h_u(T, z, -\ln y) = 0, \quad z \in (0, \infty), \\
\hat{v}_{yy}(t, z, e^{-u}) = e^{\rho t + 2u}(h_u(t, z, u) + h_{uu}(t, z, u)) \geq 0, \quad u \in [0, \infty), \\
\quad (\text{``} > \text{''holds for } t \in [0, T)), \\
\lim_{y \to 0} \hat{v}_y(t, z, y) = \lim_{u \to +\infty} \hat{v}_y(t, z, e^{-u}) = -\lim_{u \to +\infty} e^{\rho t + u} h_u(t, z, u) = -\xi(t, z), \\
\lim_{y \to 0} \hat{v}_{yy}(t, z, y) = \lim_{u \to +\infty} \hat{v}_{yy}(t, z, e^{-u}) \\
\qquad\qquad\qquad\quad = \lim_{u \to +\infty} e^{\rho t + 2u}(h_u(t, z, u) + h_{uu}(t, z, u)) = +\infty, \\
\lim_{y \to 0} \hat{v}_{zz}(t, z, y) = \lim_{u \to +\infty} e^{\rho t} h_{zz}(t, z, u) = 0, \\
\lim_{y \to 0} |\hat{v}_{yz}(t, z, y)| = \lim_{u \to +\infty} e^{\rho t + u} |h_{zu}(t, z, u)| < +\infty.
\end{cases}$$

In light of (7.65) and (7.66), the region $\mathcal{C}_T$ has a boundary that is at least C^1.

We next consider the original HJB equation (7.13), however, restricted to the domain $(t, y, z) \in \mathcal{C}_T$ that

$$\begin{cases} v_t + \sup_{\theta \in \mathbb{R}^n} \left[v_x \theta^\top \mu + \frac{v_{xx}}{2} \theta^\top \sigma \sigma^\top \theta + v_{xz} \sigma_Z(z) \theta^\top \sigma \gamma \right] \\ \quad + v_z \mu_Z(z) + v_{zz} \frac{\sigma_Z^2(z)}{2} - f(t,z) v_x = \rho v, \quad (t, z, x) \in \mathcal{C}_T; \\ v_x(t, z, 0) = 1, \quad \forall (t, z) \in [0, T) \times \mathbb{R}. \end{cases} \quad (7.68)$$

For $(t, z, x) \in \mathcal{C}_T$, let us define $y^* = y^*(t, z, x) \in (0, 1]$ that satisfies

$$\hat{v}_y(t, z, y^*) = -x. \quad (7.69)$$

Thanks to (7.69), we have, for all $(t, z, x) \in \mathcal{C}_T$,

$$v(t, z, x) = \inf_{y \in (0,1]} \{ \hat{v}(t, z, y) + xy \}$$
$$= \hat{v}(t, z, y^*(t, z, x)) + xy^*(t, z, x). \quad (7.70)$$

Note that $(0, 1] \ni y \to \hat{v}_y(t, z, y)$ is strictly increasing for fixed $(t, z) \in [0, T] \times \mathbb{R}$, as well as $\hat{v}_y(t, z, 1) = 0$ and $\lim_{y \to 0} \hat{v}_y(t, z, y) = -\xi(t, z)$, we have that $x \to y^*(t, z, x)$ is decreasing, $\lim_{x \to 0} y^*(t, z, x) = 1$ as well as $\lim_{x \to \xi(t,z)} y^*(t, z, x) = 0$. It follows from the implicit function theorem that y^* is C^1 on $\mathcal{C}_T$. Hence, v in (7.70) is well defined, and it is $C^{1,2,2}$ on $\mathcal{C}_T$. On the other hand, a direct calculation yields that, for $(t, z, x) \in \mathcal{C}_T$,

$$y^*(t, z, x) = v_x(t, z, x), \quad \hat{v}_t(t, z, y^*(t, z, x)) = v_t(t, z, x),$$
$$\hat{v}_z(t, z, y^*(t, z, x)) = v_z(t, z, x),$$
$$\hat{v}_{yy}(t, z, y^*(t, z, x)) = -\frac{1}{v_{xx}(t, z, x)}, \quad \hat{v}_{zy}(t, z, y^*(t, z, x)) = \frac{v_{xz}(t, z, x)}{v_{xx}(t, z, x)},$$
$$\hat{v}_{zz}(t, z, y^*(t, z, x)) = \left(v_{zz} - \frac{v_{xz}^2}{v_{xx}} \right)(t, z, x). \quad (7.71)$$

Recall that $v_{xx}(t, z, x) < 0$ for $(t, z, x) \in \mathcal{C}_T$. Plugging (7.71) into (7.28), we deduce that v defined in (7.70) solves the dual PDE (7.68).

We next study the behavior of v on $\mathcal{C}_T^c \cap ([0, T) \times \mathbb{R} \times \mathbb{R}_+)$. To this end, for $(t, z) \in [0, T) \times \mathbb{R}$, let $(t_n, z_n, x_n) \in \mathcal{C}_T$ for $n \geq 1$ be a sequence such that $(t_n, z_n, x_n) \to (t, z, \xi(t, z))$. We claim that

$$\lim_{n \to +\infty} y^*(t_n, z_n, x_n) = 0. \quad (7.72)$$

We prove (7.72) by contradiction. Assume that, up to a subsequence, there exists a constant $\delta > 0$ such that $\lim_{n \to +\infty} y^*(t_n, z_n, x_n) = \delta$. By using (7.69), it yields that

$$\hat{v}_y(t, z, \delta) = \lim_{n \to +\infty} \hat{v}_y(t_n, z_n, y^*(t_n, z_n, x_n)) = -\lim_{n \to +\infty} x_n = -\xi(t, z),$$

which contradicts the definition (7.66) of $\xi(t,z)$ and the fact that $\lim_{y \to 0} \hat{v}_y(t,z,y) = -\xi(t,z)$. Furthermore, it follows from (7.72) that

$$\lim_{n \to +\infty} v(t_n, z_n, x_n) = \lim_{n \to +\infty} \{\hat{v}(t_n, z_n, y^*(t_n, z_n, x_n)) + x_n y^*(t_n, z_n, x_n)\}$$
$$= 0. \tag{7.73}$$

Using (7.72) again, we also have

$$\lim_{n \to +\infty} v_t(t_n, z_n, x_n) = \lim_{n \to +\infty} \hat{v}_t(t_n, z_n, y^*(t_n, z_n, x_n)) = 0,$$

$$\lim_{n \to +\infty} v_z(t_n, z_n, x_n) = \lim_{n \to +\infty} \hat{v}_z(t_n, z_n, y^*(t_n, z_n, x_n)) = 0,$$

$$\lim_{n \to +\infty} v_x(t_n, z_n, x_n) = \lim_{n \to +\infty} y^*(t_n, z_n, x_n) = 0, \tag{7.74}$$

$$\lim_{n \to +\infty} v_{xx}(t_n, z_n, x_n) = -\lim_{n \to +\infty} \hat{v}_{yy}(t_n, z_n, y^*(t_n, z_n, x_n))^{-1} = 0,$$

$$\lim_{n \to +\infty} v_{xz}(t_n, z_n, x_n) = -\lim_{n \to +\infty} \frac{\hat{v}_{yz}}{\hat{v}_{yy}}(t_n, z_n, y^*(t_n, z_n, x_n)) = 0,$$

$$\lim_{n \to +\infty} v_{zz}(t_n, z_n, x_n) = \lim_{n \to +\infty} \left(\hat{v}_{zz} - \frac{\hat{v}_{yz}^2}{\hat{v}_{yy}}\right)(t_n, z_n, y^*(t_n, z_n, x_n)) = 0.$$

Let us define $v(t,z,x) = 0$ for $(t,z,x) \in \mathcal{C}_T^c \cap ([0,T) \times \mathbb{R} \times \mathbb{R}_+)$. Using (7.73) and (7.74), we have that v given by (7.70) and its partial derivatives up to order two are continuous on $\partial \mathcal{C}_T \cap ([0,T) \times \mathbb{R} \times \mathbb{R}_+)$. Hence, v is $C^{1,2,2}$ on $[0,T) \times \mathbb{R} \times \mathbb{R}_+$. Moreover, using (7.71) and (7.28) on $[0,T] \times \mathbb{R} \times \mathbb{R}_+$, we have that v given by (7.64) solves the following HJB equation:

$$\begin{cases} v_t + \sup_{\theta \in \mathbb{R}^n} \left[v_x \theta^\top \mu + \frac{v_{xx}}{2}\theta^\top \sigma \sigma^\top \theta + v_{xz}\sigma_Z(z)\theta^\top \sigma \gamma\right] \\ \quad + v_z \mu_Z(z) + v_{zz}\frac{\sigma_Z^2(z)}{2} - f(t,z)v_x = \rho v, \\ v_x(t,z,0) = 1, \quad \forall (t,z) \in [0,T) \times \mathbb{R}. \end{cases} \tag{7.75}$$

On the other hand, note that $v_x \geq 0$ on $(t,z,x) \in [0,T) \times \mathbb{R} \times \mathbb{R}_+$, and $v(t,z,x) \to 0$ as $x \to +\infty$. It follows from (7.12) that $v(t,z,0) \leq 0$. As a consequence, $v(l,z,x) \leq 0$ for $(t,z,x) \in [0,T) \times \mathbb{R} \times [0,\infty)$ and it follows from (7.34) that there exists a constant $C > 0$ independent of T that

$$|v(t,z,x)| = -v(t,z,x) = \sup_{y \in (0,1]} \{-\hat{v}(t,z,y) - xy\} \leq \sup_{y \in (0,1]} \{-\hat{v}(t,z,y)\}$$
$$= \sup_{y \in (0,1]} \{-e^{\rho t} h(t,z,-\ln y)\} \leq e^{\rho t} C(T-t)(1+|z|), \tag{7.76}$$

for $(t,z,x) \in [0,T) \times \mathbb{R} \times [0,\infty)$, where $h(t,z,u)$ is given by (7.29).

We next prove the continuity of v on the boundary of $[0,T)\times\mathbb{R}\times[0,+\infty)$. Note that $v(t,z,0) = \hat{v}(t,z,1)$ and let us consider $(t,z) \in [0,T)\times\mathbb{R}$ and $(t_n, z_n, x_n) \in [0,T)\times\mathbb{R}\times\mathbb{R}_+$ satisfying $(t_n, z_n, x_n) \to (t, z, 0)$ as $n \to \infty$. By mimicking the proof showing (7.72), one can obtain

$$\lim_{n\to\infty} y^*(t_n, z_n, x_n) = 1. \quad (7.77)$$

An application of L'Hospital's rule gives that

$$\lim_{x\downarrow 0} \frac{1}{x}(v(t,z,x) - v(t,z,0))$$

$$= \lim_{x\downarrow 0} \frac{1}{x}(\hat{v}(t,z,y^*(t,z,x)) + xy^*(t,z,x) - \hat{v}(t,z,1))$$

$$= \lim_{x\downarrow 0} y^*(t,z,x) - \lim_{x\downarrow 0} \frac{\hat{v}_y(t,z,y^*(t,z,x)) - \hat{v}(t,z,1)}{y^*(t,z,x) - 1} \times \lim_{x\downarrow 0} \frac{y^*(t,z,x) - 1}{x}$$

$$= 1 - \hat{v}_y(t,z,1) \times \left(\lim_{x\downarrow 0} y_x^*(t,z,x)\right) = 1.$$

Moreover, as $\lim_{n\to\infty} v_x(t_n, z_n, x_n) = \lim_{n\to\infty} y^*(t_n, z_n, x_n) = 1$, it holds that

$$\lim_{n\to\infty} v_x(t_n, z_n, x_n) = v_x(t, z, 0). \quad (7.78)$$

Similarly, we also have

$$\lim_{x\downarrow 0} \frac{1}{x}(v_x(t,z,x) - v_x(t,z,0)) = \lim_{x\downarrow 0} \frac{1}{x}(y^*(t,z,x) - 1) = \lim_{x\downarrow 0} y_x^*(t,z,x)$$

$$= -\hat{v}_{yy}(t,z,1)^{-1},$$

and $\lim_{n\to\infty} v_{xx}(t_n, z_n, x_n) = -\lim_{n\to+\infty} \hat{v}_{yy}(t_n, z_n, y^*(t_n, z_n, x_n))^{-1} = -\hat{v}_{yy}(t,z,1)^{-1}$. As a consequence

$$\lim_{n\to+\infty} v_{xx}(t_n, z_n, x_n) = v_{xx}(t, z, 0). \quad (7.79)$$

In a similar fashion, the limits (7.78) and (7.79) also hold for v_z, v_{xz}, and v_{zz}. Hence, we conclude that $v \in C^{1,2,2}([0,T)\times\mathbb{R}\times[0,\infty))$.

On the other hand, for $(z,x) \in \mathbb{R}\times[0,+\infty)$, we define $v(T,z,x) = 0$, and consider $(t_n, z_n, x_n) \in [0,T)\times\mathbb{R}\times[0,+\infty)$ satisfying $(t_n, z_n, x_n) \to (T, z, x)$ as $n \to +\infty$. In view of (7.76), we have $\lim_{n\to+\infty} v(t_n, z_n, x_n) = 0$, which yields that $v \in C(\Theta_T)$. By combining Eq. (7.75), we deduce that $v \in C^{1,2,2}([0,T)\times\mathbb{R}\times[0,+\infty)) \cap C(\Theta_T)$ and v satisfies that

$$\begin{cases} v_t + \sup_{\theta\in\mathbb{R}^n}\left[v_x\theta^\top\mu + \frac{v_{xx}}{2}\theta^\top\sigma\sigma^\top\theta + v_{xz}\sigma_Z(z)\theta^\top\sigma\gamma\right] \\ \qquad + v_z\mu_Z(z) + v_{zz}\frac{\sigma_Z^2(z)}{2} - f(t,z)v_x = \rho v, \\ v_x(t,z,0) = 1, \quad \forall (t,z)\in[0,T)\times\mathbb{R}, \\ v(T,z,x) = 0, \quad \forall (z,x)\in\mathbb{R}\times[0,+\infty), \end{cases} \quad (7.80)$$

and the estimate (7.76) holds for $(t,z,x)\in[0,T]\times\mathbb{R}\times[0,+\infty)$.

We next prove the item (ii). We first show the continuity of $\theta^*(t,z,x)$ on $(t,z,x) \in \Theta_T$, which verifies the admissibility of $\theta_t^* = \theta^*(t, Z_t, X_t)$ for $t \in [0,T]$ (i.e., $\theta^* \in \mathcal{U}_t$). Let us define $y^*(t,z,0) = 1$. Thanks to (7.77), y^* is continuous at $(t,z,0)$. For $(t,z,x) \in \Theta_T$, we rewrite (7.67) as:

$$\theta^*(t,z,x) = -\mu(\sigma\sigma^\top)^{-1} y^*(t,z,x)\hat{v}_{yy}(t,z,y^*(t,z,x))$$
$$+ (\sigma\sigma^\top)^{-1}\sigma_Z(z)\sigma\gamma\hat{v}_{yz}(t,z,y^*(t,z,x)). \qquad (7.81)$$

It is easy to see that $\theta^*(t,z,x)$ is continuous for $(t,z,x) \in \mathcal{C}_T \cup \{(t,z,0); (t,z) \in [0,T) \times \mathbb{R}\}$. Thus, it remains to prove that

$$\lim_{n \to +\infty} \theta^*(t_n, z_n, x_n) = \theta^*(t,z,x), \qquad (7.82)$$

where $x = \xi(t,z)$, and $(t_n, z_n, x_n) \in \mathcal{C}_T$, $\lim_{n \to +\infty}(t_n, z_n, x_n) = (t,z,x)$. By virtue of (7.81), we have

$$\theta^*(t_n, z_n, x_n) = -\mu(\sigma\sigma^\top)^{-1} y^*(t_n, z_n, x_n)\hat{v}_{yy}(t,z,y^*(t_n,z_n,x_n))$$
$$+ (\sigma\sigma^\top)^{-1}\sigma_Z(z)\sigma\gamma\hat{v}_{yz}(t,z,y^*(t_n,z_n,x_n)). \qquad (7.83)$$

Note that $x^*(t_n, z_n, x_n) \to 0$ as $n \to +\infty$. By tending n to $+\infty$ on both sides of (7.83), we deduce that

$$\lim_{n \to \infty} \theta^*(t_n, z_n, x_n) = -\mu(\sigma\sigma^\top)^{-1} \lim_{y \downarrow 0} x\hat{v}_{yy}(t,z,y)$$
$$+ (\sigma\sigma^\top)^{-1}\sigma_Z(z)\sigma\gamma \lim_{y \downarrow 0} \hat{v}_{yz}(t,z,y) = \theta^*(t,z,x).$$

Following a similar argument, we can establish the convergence (7.82) for any $(t,z,x) \in \partial \mathcal{C}_T \cap ([0,T] \times \mathbb{R} \times [0,+\infty))$ and $(t_n, z_n, x_n) \in \mathcal{C}_T$. Hence, $\theta^*(t,y,z)$ is continuous for $(t,z,x) \in [0,T] \times \mathbb{R} \times [0,+\infty)$. Furthermore, one can see from (6.55), (7.67) and (7.81) that there exists constant $C > 0$ such that

$$|\theta^*(t,z,x)| \leq C(1+|z|), \quad \forall (t,z,x) \in \Theta_T. \qquad (7.84)$$

With the continuity of θ^* on Θ_T and the estimate (7.84), we can apply Theorems 2.2, 2.4 and Remark 2.1 in Chapter 4 of Ikeda and Watanabe (1989) to conclude that the following SDE admits a weak solution that, for $t \in [0,T]$,

$$\begin{cases} \tilde{X}_s = -\int_t^s f(r, Z_r)dr + \int_t^s \theta^*(r, Z_r, \Phi^t(\tilde{X})_r)^\top \mu dr \\ \qquad + \int_t^s \theta^*(r, Z_r, \Phi^t(\tilde{X})_r)^\top \sigma dW_r, \\ dZ_s = \mu_Z(Z_s)ds + \sigma_Z(Z_s)dW_s^\gamma, \quad s \in [t,T]. \end{cases} \qquad (7.85)$$

Here, the mapping $\Phi^t : C([t,T];\mathbb{R}) \to C([t,T];\mathbb{R})$ satisfies that, for all $\varphi \in C([t,T];\mathbb{R})$,

(i) $\Phi^t(\varphi)_s = \varphi_s + \eta_s$ for $s \in [t, T]$, and $\Phi(\varphi)_t = \varphi_t$.
(ii) $\Phi^t(\varphi)_s \geq 0$ for $s \in [t, T]$.
(iii) $s \to \eta_s$ is continuous, non-negative and non-decreasing, and $\eta_s = x \vee \int_t^s \mathbf{1}_{\{\Phi^t(\varphi)_r = 0\}} d\eta_r$ for $s \in [t, T]$.

Define $X^* := \Phi^t(\tilde{X})$ and $L^* := \Phi^t(\tilde{X}) - \tilde{X}$. Then (X^*, L^*, W) solves the SDE that

$$X_s^* = -\int_t^s f(r, Z_r) dr + \int_t^s \theta^*(r, Z_r, X_r^*)^\top \mu dr$$
$$+ \int_t^s \theta^*(r, Z_r, X_r^*)^\top \sigma dW_r + L_s^*, \quad \forall s \in [t, T],$$

where L^* satisfies (iii). This shows that $\theta^* \in \mathcal{U}_t$ is admissible.

Let us fix $(t, z, x) \in [0, T) \times \mathbb{R} \times [0, \infty)$, and $\theta \in \mathcal{U}_t$. For any $n > T^{-1}$, we define

$$\tau_n^t := \left(T - \frac{1}{n}\right) \wedge \inf\{s \geq t : |Z_s| + |X_s| > n\}. \tag{7.86}$$

It holds that $\tau_n^t \uparrow T$ as $n \to \infty$, $\mathbb{P}$-a.s.. Itô's formula yields that

$$\mathbb{E}_{t,z,x}\left[-\int_t^{\tau_n^t} e^{-\rho s} dL_s^X + e^{-\rho \tau_n^t} v(\tau_n^t, Z_{\tau_n^t}, X_{\tau_n^t}) - e^{-\rho t} v(t, Z_t, X_t)\right]$$
$$= \mathbb{E}_{t,z,x}\left[\int_t^{\tau_n^t} e^{-\rho s} (v_t + \mathcal{L}^{\theta_s} v)(s, Z_s, X_s) ds\right]$$
$$+ \mathbb{E}_{t,z,x}\left[\int_t^{\tau_n^t} e^{-\rho s} (v_z(s, Z_s, X_s) - 1) dL_s^X\right], \tag{7.87}$$

where, for $\theta \in \mathbb{R}^n$, the operator $\mathcal{L}_t^\theta$ acted on $C^2(\mathbb{R} \times [0, \infty))$ is defined by

$$\mathcal{L}_t^\theta \varphi(z, x) := \varphi_x(z, x) \theta^\top \mu + \frac{\varphi_{xx}(z, x)}{2} \theta^\top \sigma \sigma^\top \theta + \varphi_{xz}(z, x) \sigma_Z(z) \theta^\top \sigma \gamma$$
$$+ \varphi_z(z, x) \mu_Z(z) + \varphi_{zz}(z, x) \frac{\sigma_Z^2(z)}{2} - f(t, z) \varphi_x(z, x) - \rho \varphi(z, x),$$

for all $\varphi \in C^2(\mathbb{R} \times [0, \infty))$. Then, SDE (7.11) and the boundary condition in (7.80) yield that

$$\mathbb{E}_{t,z,x}\left[\int_t^{\tau_n^t} e^{-\rho s} (v_x(s, Z_s, X_s) - 1) dL_s^X\right]$$
$$= \mathbb{E}_{t,z,x}\left[\int_t^{\tau_n^t} e^{-\rho s} (v_x(s, Z_s, X_s) - 1) \mathbf{1}_{\{X_s = 0\}} dL_s^X\right] = 0.$$

On the other hand, the HJB equation (7.80) satisfied by v also gives that, for all $t \in [0,T]$, $\mathbb{P}$-a.s.

$$\left(v_t + \mathcal{L}_t^\theta v\right)(t, Z_t, X_t) \leq 0, \tag{7.88}$$

where the equality holds in (7.88) if $\theta = \theta^*$. Hence, we deduce from (7.88) that

$$\mathbb{E}_{t,z,x}\left[-\int_t^{\tau_n^t} e^{-\rho s} dL_s^X\right] \leq e^{-\rho t} v(t, z, x)$$

$$- \mathbb{E}_{t,z,x}\left[e^{-\rho \tau_n^t} v(\tau_n^t, Z_{\tau_n^t}, X_{\tau_n^t})\right]. \tag{7.89}$$

By applying the estimate (7.76), we have $|v(\tau_n^t, Z_{\tau_n^t}, X_{\tau_n^t})| \leq e^{\rho \tau_n^t} C(T - \tau_n^t)\{1 + \sup_{s\in[t,T]} |Z_s|^p\}$, $\mathbb{P}$-a.s. Sending $n \to +\infty$ and noting that $\tau_n^t \uparrow T$, $\mathbb{P}$-a.s., we can deduce from DCT that

$$\lim_{n\to+\infty} \mathbb{E}_{t,z,x}\left[e^{-\rho \tau_n} v(\tau_n^t, Z_{\tau_n^t}, X_{\tau_n^t})\right] = 0. \tag{7.90}$$

Therefore, as $n \to +\infty$ in (7.89), we have that, for all $\theta \in \mathcal{U}_t$,

$$J(\theta; t, z, x) = \mathbb{E}_{t,z,x}\left[-\int_t^T e^{-\rho s} dL_s^X\right] \leq e^{-\rho t} v(t, z, x), \tag{7.91}$$

where the equality in (7.91) holds for $\theta = \theta^*$. This finally verifies that $\theta^* \in \mathcal{U}_t$ is an optimal strategy. $\square$

7.7 The GBM Benchmark Process

The benchmark process considered in (7.5) is restricted to be non-decreasing. This section illustrates the application to some market index tracking problems when the index process follows a GBM. In both cases of finite horizon and infinite horizon, we show that the market index tracking problem can actually be transformed into an equivalent optimal tracking problem with a non-decreasing benchmark process. The semi-closed-form results can be derived.

As in Guasoni et al. (2011), we consider the market index such as S&P500 or Nasdaq 100 with the price process $I = (I_t)_{t\in[0,T]}$ satisfying

$$\frac{dI_t}{I_t} = \mu_I dt + \sigma_I dW_t^\gamma, \tag{7.92}$$

where $I_0 = z > 0$, the constant return rate $\mu_I \in \mathbb{R}$ and the volatility $\sigma_I > 0$. Note that the index process in (7.92) is not monotone. Hence, it does not fit directly into our framework in Section 7.1.

7.7.1 Finite Horizon Case

For the given market index process I, consider the optimal tracking problem similar to (7.5):

$$\begin{cases} u(a, \mathrm{v}, z) := \inf_{C,\theta} \mathbb{E}\left[C_0 + \int_0^T e^{-\rho t} dC_t\right] \\ \text{s.t. } I_t \leq V_t^\theta + C_t \text{ at each } t \in [0, T]. \end{cases} \quad (7.93)$$

To exclude the trivial case, we assume that I can not be dynamically replicated by some portfolio θ, which amounts to the condition that

$$\lambda := \mu_I - \sigma_I \gamma^\top \sigma^{-1} \mu \neq 0. \quad (7.94)$$

It is not difficult to see that Lemma 7.1 still holds when A_t is replaced by the market index process I_t. We can therefore follow the argument in Section 7.3 and introduce the following reflected state process:

$$\begin{aligned} X_t &= -(I_t - I_0) + \int_0^t \theta_s^\top \mu\, ds + \int_0^t \theta_s^\top \sigma\, dW_s + L_t^X \\ &= -\int_0^t f(I_s)ds + \int_0^t \bar{\theta}_s^\top \mu\, ds + \int_0^t \bar{\theta}_s^\top \sigma\, dW_s + L_t^X, \end{aligned} \quad (7.95)$$

where the running maximum process $L^X = (L_t^X)_{t \in [0,T]}$ with $L_0^X = x \geq 0$ is defined as in (7.8) with A being replaced by I. In (7.95), we used the notations $f(z) := \lambda z$ for $z > 0$, and

$$\bar{\theta}_t^\top := \theta_t^\top - \sigma_I \gamma^\top \sigma^{-1} I_t, \quad \forall t \in [0, T]. \quad (7.96)$$

It follows that $(a - \mathrm{v})^+ - u(a, \mathrm{v}, z)$ with $u(a, \mathrm{v}, z)$ given in (7.93) is equivalent to the following auxiliary control problem

$$\sup_{\bar{\theta} \in \bar{\mathcal{U}}_0} \mathbb{E}\left[-\int_0^T e^{-\rho s} dL_s^X\right], \quad (7.97)$$

where the initial level $L_0^X = x = (\mathrm{v} - a)^+$.

For $t \in [0,T]$, denote by $\bar{\mathcal{U}}_t$ the set of admissible controls taking the feedback form $\bar{\theta}_s = \bar{\theta}(s, I_s, X_s)$ for $s \in [t, T]$, where $\bar{\theta} : \Theta_T \to \mathbb{R}^n$ is a measurable function such that the following reflected SDE has a weak solution:

$$\begin{aligned} X_s &= -\int_t^s f(I_r)dr + \int_t^s \bar{\theta}(r, I_r, X_r)^\top \mu\, dr \\ &\quad + \int_0^t \bar{\theta}(r, I_r, X_r)^\top \sigma\, dW_r + L_s^X, \quad \forall s \in [t, T] \end{aligned} \quad (7.98)$$

with $X_t = x \geq 0$. Here, $L_s^X = x \vee \int_t^s \mathbf{1}_{\{X_r=0\}} dL_r^X$ is a continuous, non-negative and non-decreasing process, which increases only when the state process X_s hits the level 0 for $s \in [t,T]$. For $(t,z,x) \in \Theta_T$, the dynamic version of the auxiliary problem (7.97) is given by

$$v(t,z,x) := \sup_{\bar{\theta} \in \bar{\mathcal{U}}_t} \mathbb{E}_{t,z,x}\left[-\int_t^T e^{-\rho(s-t)} dL_s^X\right], \qquad (7.99)$$

where $\mathbb{E}_{t,z,x}[\cdot] := \mathbb{E}[\cdot | I_t = z, X_t = x]$. Again, we shall consider $v(t,z,x) := e^{\rho t} \mathrm{w}(t,z,x)$ as the solution to the primal HJB equation.

If the coefficient λ in (7.94) satisfies $\lambda < 0$, the optimal strategy is actually trivial that $\bar{\theta}^* \equiv 0$, and the associated value function becomes $v(t,z,x) = 0$ for all $(t,z,x) \in [0,T] \times \mathbb{R}_+^2$. In fact, let $X^* = (X_s^*)_{s \in [t,T]}$ be the state process (7.98) with $\bar{\theta}^* = (\bar{\theta}_s^*)_{s \in [t,T]} \in \bar{\mathcal{U}}_t$. Then, for $\lambda < 0$, it holds that $X_s^* > 0$ for all $s \in [t,T]$. This yields that $L_s^{X^*} = L_t^{X^*} = x > 0$ for all $s \in [t,T]$, a.s., and hence $e^{-\rho t} v(t,z,x) \leq 0 = \mathbb{E}_{t,z,x}[-\int_t^T e^{-\rho s} dL_s^{X^*}]$.

If, on the other hand, the coefficient $\lambda > 0$, i.e., $f(z) > 0$, the auxiliary stochastic control problem (7.99) with the state process (7.95) falls into the framework in Sections 7.3–7.6 under the assumption that $f > 0$. Hence, all results still hold and Theorem 7.2 gives the characterization of the value function and the optimal portfolio for the index tracking problem (7.99).

To further explore more explicit results, let us focus on the special case $\sigma_I = 0$, i.e., the benchmark process $I_t = z e^{\mu t}$ for $t \in [0,T]$ is the so-called "growth rate benchmark" that has been studied in Yao et al. (2006). With this deterministic benchmark $I = (I_t)_{t \in [0,T]}$, the dual HJB equation (7.25) can be reduced to

$$\hat{v}_t(t,z,y) - \rho \hat{v}(t,z,y) + \rho y \hat{v}_y(t,z,y) + \alpha y^2 \hat{v}_{yy}(t,z,y) \\ + \mu_I z \hat{v}_z(t,z,y) = \lambda z y. \qquad (7.100)$$

In view of (6.55), it holds that $\hat{v}(t,z,x) = e^{\rho t} h(t,z,-\ln x)$ for $(t,z,x) \in [0,T] \times \mathbb{R}_+^2$. It follows from (7.29) that, for $(t,z,u) \in \Theta_T$,

$$h(t,z,u) = -\lambda z e^{-\mu_I t} \mathbb{E}\left[\int_t^T e^{(\mu_I - \rho)s - R_s^{t,u}} ds\right]$$

$$= -\lambda z e^{-\mu_I t} \int_t^T e^{(\mu_I - \rho)s} \mathbb{E}\left[e^{-R_s^{t,u}}\right] ds. \qquad (7.101)$$

Here, we recall that $(R_s^{t,u})_{s \in [t,T]}$ with $(t,u) \in [0,T] \times [0,\infty)$ is a reflected Brownian motion with drift defined by (7.32).

We now compute the term $\mathbb{E}[e^{-R_s^{t,u}}]$ in (7.101). Using Harrison (1985) on page 49, for $(t,m) \in \mathbb{R}_+ \times \mathbb{R}$, we have

$$\mathbb{P}\left(R_t^{0,u} \leq m\right) = \Phi\left(\frac{-u+m-(\alpha-\rho)t}{\sqrt{2\alpha t}}\right) - e^{\frac{(\alpha-\rho)m}{\alpha}}$$
$$\times \Phi\left(\frac{-u-m-(\alpha-\rho)t}{\sqrt{2\alpha t}}\right), \quad (7.102)$$

where $\Phi(m) = \int_{-\infty}^{m} \frac{1}{\sqrt{2\pi}} e^{-\frac{u^2}{2}} du$, $m \in \mathbb{R}$, denotes the standard normal cumulative distribution function. It follows from the Markov property that, for $(s,m) \in (t,T] \times \mathbb{R}_+$,

$$\frac{\mathbb{P}(R_s^{t,u} \in dm)}{dm} = \frac{1}{\sqrt{2\alpha(s-t)}} \Phi'\left(\frac{-u+m-(\alpha-\rho)(s-t)}{\sqrt{2\alpha(s-t)}}\right) - \frac{\alpha-\rho}{\alpha} e^{\frac{(\alpha-\rho)m}{\alpha}}$$
$$\times \Phi\left(\frac{-u-m-(\alpha-\rho)(s-t)}{\sqrt{2\alpha(s-t)}}\right)$$
$$+ \frac{1}{\sqrt{2\alpha(s-t)}} e^{\frac{(\alpha-\rho)m}{\alpha}} \Phi'\left(\frac{-u-m-(\alpha-\rho)(s-t)}{\sqrt{2\alpha(s-t)}}\right)$$
$$=: \psi(m,u,s-t). \quad (7.103)$$

The expectation can therefore be explicitly written as:

$$\mathbb{E}\left[e^{-R_s^{t,u}}\right] = \int_0^\infty e^{-m} \mathbb{P}(R_s^{t,u} \in dm) = \int_0^\infty e^{-m} \psi(m,u,s-t) dm.$$

By virtue of (7.101), the dual HJB equation (7.100) admits the solution in an integral form that

$$\hat{v}(t,z,x) = e^{\rho t} h(t,z,-\ln x) \quad (7.104)$$
$$= -\lambda z e^{(\rho-\mu_I)t} \int_t^T \int_0^\infty e^{(\mu_I-\rho)s-m} \psi(m,-\ln x,s-t) dm ds.$$

The critical point defined by (7.66) in Theorem 7.2 can also be explicitly computed that

$$\xi(t,z) = \lambda z \mathbb{E}\left[\int_t^T e^{(\alpha+\mu_I-2\rho)(s-t)} e^{\sqrt{2\alpha} B_{s-t}^1} ds\right] = \lambda z \int_t^T e^{(2\alpha+\mu_I-2\rho)(s-t)} ds$$

$$= \begin{cases} \frac{\lambda z}{2\alpha+\mu_I-2\rho} \left[e^{(2\alpha+\mu_I-2\rho)(T-t)} - 1\right], & 2\alpha+\mu_I-2\rho \neq 0, \\ \lambda z(T-t), & 2\alpha+\mu_I-2\rho = 0. \end{cases}$$

7.7.2 Infinite Horizon Case

This section considers the same optimal tracking problem (7.93) with the infinite time horizon (i.e., $T \to \infty$), in which the value function and optimal portfolio can be obtained explicitly.

For the infinite horizon control problem, we note that $v(z,x)$ is the solution to the stationary HJB equation (7.15) that

$$-\rho v - \alpha \frac{v_x^2}{v_{xx}} + \frac{\sigma_I^2}{2} z^2 \left(v_{zz} - \frac{v_{xz}^2}{v_{xx}} \right) - \sigma_I \mu^\top (\sigma \sigma^\top)^{-1} \sigma \gamma z \frac{v_x v_{xz}}{v_{xx}}$$
$$+ \mu_I z v_z - \lambda z v_x = 0 \qquad (7.105)$$

with the Neumann boundary condition $v_x(z,0) = 1$. The dual PDE for (7.105) becomes that, for $(z,y) \in \mathbb{R}_+ \times (0,1)$,

$$-\rho \hat{v}(z,y) + \rho y \hat{v}_y(z,y) + \alpha y^2 \hat{v}_{yy}(z,y) + \mu_I z \hat{v}_z(z,y) + \frac{\sigma_I^2}{2} z^2 \hat{v}_{zz}(z,y)$$
$$- \sigma_I \mu^\top (\sigma \sigma^\top)^{-1} \sigma \gamma y z \hat{v}_{yz}(z,y) - \lambda z y = 0. \qquad (7.106)$$

In light of (7.27), we have that $\hat{v}_y(z,1) = 0$. It is easy to verify that the dual equation (7.106) has the following general solution given by, for all $(z,y) \in \mathbb{R}_+ \times (0,1]$,

$$\hat{v}(z,y) = \lambda z h(y) = \lambda z \left(\frac{1}{\lambda} y + C_1 y^{\gamma_1} + C_2 y^{\gamma_2} \right),$$

where $C_1, C_2 \in \mathbb{R}$ are unknown constants that can be determined later and γ_i, $i = 1,2$, satisfy

$$(\mu_I - \rho) + (\rho - \sigma_I \gamma^\top \sigma^{-1} \mu) \gamma_i + \alpha \gamma_i (\gamma_i - 1) = 0, \quad i = 1, 2.$$

Let us assume that the discount factor $\rho > \mu_I$. It holds that $-\infty < \gamma_1 < 0 < \gamma_2 < 1$. Note that it is required that $h(0) \leq 0$ and h is convex. Using $h(1) = \frac{1}{\lambda} + C_1 + C_2$, we must have that $C_1 = 0$. It then follows from $\hat{v}_y(z,1) = 0$ that $C_2 = -\frac{1}{\gamma_2 \lambda}$. Hence, we conclude that

$$\hat{v}(z,y) = yz - \frac{1}{\gamma_2} y^{\gamma_2} z, \quad \forall (z,y) \in \mathbb{R}_+ \times (0,1]. \qquad (7.107)$$

As $T \to \infty$, it is easy to verify that the counterpart of $\xi(t,z)$ in (7.66) is given by

$$\xi(z) := \mathbb{E} \left[\int_0^\infty e^{-\rho s} f(M_s^{0,z}) e^{\sqrt{2\alpha} B_s^1 + (\rho - \alpha)s} ds \right] = z \int_0^{+\infty} e^{\lambda s} ds = +\infty.$$

Using the same argument to derive (7.64) with the closed-form solution (7.107), the solution of the primal HJB equation (7.105) can be obtained by, for all $(z, x) \in \mathbb{R}_+^2$,

$$v(z, x) = \inf_{y \in (0,1]} \{\hat{v}(z, y) + xy\} = \hat{v}(z, y^*(z, x)) + xy^*(z, x),$$

where $y^*(z, x) = (1 + \frac{x}{z})^{\frac{1}{\gamma_2 - 1}} < 1$, and hence $y^*(z, x)^{\gamma_2 - 1} = 1 + \frac{x}{z}$. A straightforward calculation yields that, for $(z, x) \in \mathbb{R}_+^2$,

$$v(z, x) = z\left(1 + \frac{x}{z}\right)^{\frac{1}{\gamma_2 - 1}}\left(1 - \frac{1}{\gamma_2} - \frac{1}{\gamma_2}\frac{x}{z}\right) + x\left(1 + \frac{x}{z}\right)^{\frac{1}{\gamma_2 - 1}}$$

$$= z\frac{\gamma_2 - 1}{\gamma_2}\left(1 + \frac{x}{z}\right)^{\frac{\gamma_2}{\gamma_2 - 1}} < 0. \tag{7.108}$$

It follows that $v_x(z, x) = (1 + \frac{x}{z})^{\frac{1}{\gamma_2 - 1}}$ and $v_x(z, 0) = 1$ for all $z \in \mathbb{R}_+$, i.e., the Neumann boundary condition holds. By the similar argument to derive (7.67), the feedback optimal portfolio is obtained by

$$\bar{\theta}^*(z, x) = -(\sigma\sigma^\top)^{-1}\frac{v_x(z, x)\mu + zv_{xz}(z, x)\sigma_I\sigma\gamma}{v_{xx}(z, x)} \tag{7.109}$$

$$= -(\gamma_2 - 1)(x + z)(\sigma\sigma^\top)^{-1}\mu + (\gamma_2 - 1)\sigma_I\left(\frac{z^3}{x} + z^2\right)\sigma\gamma.$$

We can show the optimality of $\bar{\theta}^*$ by the verification argument similar to Theorem 7.2. In particular, since $\rho > \mu_I$, the counterpart of (7.90) for the infinite horizon case can be checked by

$$\lim_{s \to +\infty} \mathbb{E}\left[e^{-\rho s}|v(I_s, X_s)|\right] = \frac{1 - \gamma_2}{\gamma_2}\lim_{s \to +\infty} \mathbb{E}\left[e^{-\rho s}I_s\left(1 + \frac{X_s}{I_s}\right)^{\frac{\gamma_2}{\gamma_2 - 1}}\right]$$

$$\leq \frac{1 - \gamma_2}{\gamma_2}\lim_{s \to +\infty} \mathbb{E}\left[e^{-\rho s}I_s\right] = \frac{1 - \gamma_2}{\gamma_2}z\lim_{s \to +\infty}e^{-(\rho - \mu_I)s} = 0.$$

Lastly, similar to the finite horizon case, we can also consider the simple case when $\sigma_I = 0$, in which the deterministic benchmark process $I_t = ze^{\mu_I t}$ describes a growth rate as studied in Yao et al. (2006). From (7.108) and (7.109), it follows that, the value function and the optimal feedback strategy can be further simplified to

$$\begin{cases} v(z, x) = z\frac{\gamma_0 - 1}{\gamma_0}\left(1 + \frac{x}{z}\right)^{\frac{\gamma_0}{\gamma_0 - 1}}, \\ \bar{\theta}^*(z, x) = -(\gamma_0 - 1)(x + z)(\sigma\sigma^\top)^{-1}\mu, \end{cases} \tag{7.110}$$

where the constant $\gamma_0 > 0$ is given by
$$\gamma_0 = \frac{\alpha - \rho + \sqrt{(\rho - \alpha)^2 + 4\alpha(\rho - \mu_I)}}{2\alpha},$$
with α being defined in (7.16).

Section 1.1.3 in Chapter 1 discussed the (infinite-horizon) optimal tracking portfolio problem when the dimension of Brownian motion may be different from the one of risky assets, which is usually referred as the incomplete market model. However, we here stress that the finite-horizon optimal tracking problem in the incomplete market model is still an open problem.

Chapter 8

Extended Merton's Problem with Relaxed Benchmark Tracking[*]

The continuous time optimal portfolio-consumption problem has been extensively studied in different models since the seminal work Merton (1969, 1971). This chapter aims to study this problem from a new perspective by simultaneously considering the wealth tracking with respect to an exogenous benchmark process. Similar to a large body of literature on optimal tracking portfolio, see, for instance, Browne (1999a,b, 2000), Teplá (2001), Gaivoronski et al. (2005), Yao et al. (2006), Strub and Baumann (2018), the goal of tracking is to ensure the agent's wealth level being close to a targeted benchmark such as the market index, the inflation rate or the consumption index. However, unlike the conventional formulation of optimal tracking portfolio in the aforementioned studies, we adopt the relaxed tracking formulation proposed in Chapter 7 using capital injection such that the benchmark process is regarded as a minimum floor constraint of the total capital.

The rest of the chapter is organized as follows. Section 8.1 introduces the market model and formulates the extended Merton's problem with relaxed benchmark tracking. In Section 8.2, an auxiliary state process with reflection are proposed and the associated HJB equation with two Neumann boundary conditions for the equivalent control problem is derived. In Section 8.3, we propose a tailor-made decomposition-homogenization technique (see also Bo et al. (2024)), by this decomposition-homogenization method, we address the solvability of the dual PDE based on stochastic flow analysis. The verification theorem on the optimal feedback control is presented in Section 8.4 together with the technical proofs on the strength of stochastic flow analysis and estimations of the optimal control. It is also verified therein that the expected total capital injection is bounded.

---

[*]This chapter is based on the work Bo et al. (2024).

8.1 The Market Model

Let $(\Omega, \mathcal{F}, \mathbb{F}, \mathbb{P})$ be a filtered probability space with the filtration $\mathbb{F} = (\mathcal{F}_t)_{t \geq 0}$ satisfying the usual conditions, which supports a d-dimensional Brownian motion $(W^1, \ldots, W^d) = (W^1_t, \ldots, W^d_t)_{t \geq 0}$. Similarly to (7.1) in Chapter 7, we consider a market model consisting of d risky assets, whose price dynamics are described by

$$\frac{dS^i_t}{S^i_t} = \mu_i dt + \sum_{j=1}^{d} \sigma_{ij} dW^j_t, \quad i = 1, \ldots, d \tag{8.1}$$

with the return rate $\mu_i \in \mathbb{R}$, $i = 1, \ldots, d$, and the volatility $\sigma_{ij} \in \mathbb{R}$, $i, j = 1, \ldots, d$. Let us denote $\mu := (\mu_1, \ldots, \mu_d)^\top$ with $\top$ representing the transpose operator, and $\sigma := (\sigma_{ij})_{d \times d}$. It is assumed that σ is invertible. We also assume that the riskless interest rate $r = 0$ that amounts to the change of numéraire and μ is not a zero vector. From this point onwards, all values are defined after the change of numéraire. At time $t \geq 0$, let θ^i_t be the amount of wealth that a fund manager allocates in asset $S^i = (S^i_t)_{t \geq 0}$ and c_t be the consumption rate. The self-financing wealth process of the agent satisfies the controlled SDE:

$$V^{\theta, c}_t = v + \int_0^t \theta_s^\top \mu ds + \int_0^t \theta_s^\top \sigma dW_s - \int_0^t c_s ds, \quad \forall t \geq 0, \tag{8.2}$$

where $v \geq 0$ represents the initial wealth level of the agent.

8.1.1 The Benchmark Process

To incorporate the wealth tracking into our optimal consumption problem, let us consider a general type of benchmark processes $M = (M_t)_{t \geq 0}$, which is described by

$$M_t = m_t + Z_t, \quad \forall t \geq 0, \tag{8.3}$$

where $Z_t = z + \int_0^t \mu_Z Z_s ds + \int_0^t \sigma_Z Z_s dW^\eta_s$ is a GBM and $m_t := \max\{m, \sup_{s \leq t} B_s\}$ is the running maximum process of the drifted Brownian motion $B_t = b + \mu_B t + \sigma_B W^\gamma_t$. Here, the model parameters $z, m \geq 0$, $b \in \mathbb{R}$, $\mu_Z, \mu_B \in \mathbb{R}$ and $\sigma_Z, \sigma_B \geq 0$. For the correlative vector $\gamma = (\gamma_1, \ldots, \gamma_d)^\top \in [-1, 1]^d$, the process $W^\gamma = (W^\gamma_t)_{t \geq 0}$ is a linear combination of the d-dimensional Brownian motion $(W^1, \ldots, W^d)$ with weights γ, which itself is a Brownian motion. Similarly, the process $W^\eta = (W^\eta_t)_{t \geq 0}$ is a linear combination of the d-dimensional Brownian motion $(W^1, \ldots, W^d)$ with weights $\eta = (\eta_1, \ldots, \eta_d) \in [-1, 1]^d$. The benchmark process in the general form of (8.3) can effectively capture the long-term increasing trend of many typical benchmark processes, such as S&P 500, NASDAQ and Dow Jones, or the movements of CPI index and higher education costs in the long run. Figure 8.1-(a) illustrates the increasing trend of simulated sample paths of (8.3), which is consistent to Figure 8.1-(b) that displays the long term growing trend of the observed data of S&P500, NASDAQ and Dow Jones from April 1, 2010 to November 01, 2020. Similarly, Figure 8.1-(c) plots the Consumer Price Index for Urban Wage Earners and Clerical Workers (CPI-W) from 1985 to 2021, and Figure 8.1-(d) plots the total cost of U.S. undergraduate students over time from 1970 to 2021, which both exhibit the same increasing trend in the long run.

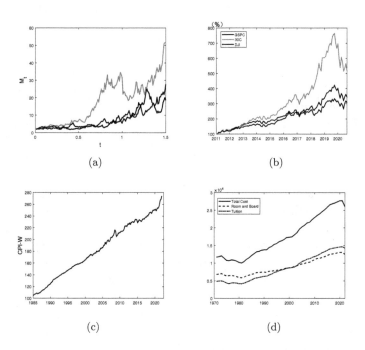

Fig. 8.1 (a): Simulated sample paths of the benchmark process $t \to M_t$ via Monte Carlo with dimension $d = 1$. The model parameters are set to be $z = 0.8$, m = 0, $b = 1$, $\mu_Z = 2$, $\sigma_Z = 1$, $\mu_B = 2$, $\sigma_B = 0$, $\gamma = \eta = 1$. (b): The price movements of market indices S&P500 (GSPC), NASDAQ (IXIC) and Dow Jones (DJI) based on observed data (April 1, 2011 to November 01, 2020) from Yahoo Finance. (c): Consumer Price Index for the US's Urban Wage Earners and Clerical Workers (CPI-W) from 1985 to 2021, available from https://www.ssa.gov/oact/STATS/cpiw_graph.html. (d) Total cost of college in the U.S. for undergraduate students from 1970 to 2021, available from https://www.bestcolleges.com/research/college-costs-over-time/.

8.1.2 Formulation of Extended Merton's Problem

We consider the relaxed benchmark tracking using the capital injection. At any time $t \geq 0$, it is assumed that the fund manager can strategically inject capital A_t such that the total wealth $V_t + A_t$ stays above the benchmark process M_t. In the objective function, in addition to the expected utility on consumption, the fund manager also needs to take into account the cost of total capital injection. Mathematically speaking, the fund manager now aims to maximize the following objective function under dynamic floor constraint that, for all $(v, m, z, b) \in \overline{\mathbb{R}}_+ \times \overline{\mathbb{R}}_+ \times \overline{\mathbb{R}}_+ \times \mathbb{R}$,

$$\begin{cases} w(v, m, z, b) \\ := \sup_{(\theta, c, A) \in \mathbb{U}} \mathbb{E}\left[\int_0^\infty e^{-\rho t} U(c_t) dt - \beta\left(A_0 + \int_0^\infty e^{-\rho t} dA_t\right)\right], \\ \text{s.t. } M_t \leq A_t + V_t^{\theta, c} \text{ at each } t \geq 0, \end{cases} \quad (8.4)$$

where $\rho > 0$ is the discount rate and $\beta > 0$ is the utility per injected capital, which can also be interpreted as the weight of relative importance between the consumption performance and the cost of capital injection. Here, $(\theta, c, A) \in \mathbb{U}$ denotes an admissible control where $(\theta, c) = (\theta_t, c_t)_{t \geq 0}$ is an $\mathbb{F}$-adapted process taking values on $\mathbb{R}^d \times \mathbb{R}_+$, and $A = (A_t)_{t \geq 0}$ is a right-continuous, non-decreasing and $\mathbb{F}$-adapted process. Herein, the utility function is considered as the power utility $U(x) = \frac{1}{p}x^p$, $x \in \mathbb{R}_+$, with the risk aversion parameter $p \in (-\infty, 0) \cup (0, 1)$. In fact, one can see that problem (8.4) shall reduce to the classical Merton's problem as in Merton (1969, 1971) when the benchmark process $M_t \equiv 0$ and $\beta \to \infty$. As a consequence, we call problem (8.4) the *extended Merton's problem*.

Stochastic control problems with minimum guaranteed floor constraints have been studied in different contexts, see among El Karoui et al. (2005), El Karoui and Meziou (2006), Di Giacinto et al. (2011, 2014), Sekine (2012), and Chow et al. (2020) and references therein. In previous studies, the minimum guaranteed level is usually chosen as constant or deterministic level and some typical techniques to handle the floor constraints are to introduce the option based portfolio or the insured portfolio allocation such that the floor constraints can be guaranteed. However, if we consider the Merton's problem under the strict floor constraint on wealth that $V_t^{\theta, c} \geq M_t$ a.s. for all $t \geq 0$, the set of admissible controls might be empty due to the more complicated benchmark process $M = (M_t)_{t \geq 0}$ in (8.3). In this regard, we introduce the singular control of capital injection $A = (A_t)_{t \geq 0}$ in our relaxed tracking formulation such that the admissible set can be enlarged and the optimal control problem can become solvable. By minimizing the cost of capital injection, the controlled wealth process $V_t^{\theta,c}$ stays very close to the benchmark process M_t as desired.

8.1.3 Unconstrained Control Problem

To address the dynamic floor constraint in (8.4), our first step is to reformulate it into an unconstrained control problem. By applying Lemma 7.1, for each fixed regular control (θ,c), the optimal singular control $A_t^{(\theta,c),*}$ satisfies the form that

$$A_t^{(\theta,c),*} = 0 \vee \sup_{s \leq t}(M_s - V_s^{\theta,c}), \quad \forall t \geq 0. \tag{8.5}$$

Thus, the original problem (8.4) with the constraint $M_t \leq A_t + V_t^{\theta,c}$ for all $t \geq 0$ admits an equivalent formulation as an unconstrained utility maximization problem with a running maximum cost that

$$w(v, m, z, b) = -\beta(m \vee b + z - v)^+ \tag{8.6}$$
$$+ \sup_{(\theta,c)\in\mathbb{U}^r} \mathbb{E}\left[\int_0^\infty e^{-\rho t}U(c_t)dt - \beta\int_0^\infty e^{-\rho t}d\left(0 \vee \sup_{s \leq t}(M_s - V_s^{\theta,c})\right)\right].$$

Here, $\mathbb{U}^r$ denotes the set of regular $\mathbb{F}$-adapted admissible strategies $(\theta,c) = (\theta_t, c_t)_{t \geq 0}$ taking values on $\mathbb{R}^d \times \mathbb{R}_+$ such that, for any $T > 0$, the SDE (8.2) admits a weak solution on $[0,T]$.

It is worth noting that some existing studies can be found in stochastic control problems with a running maximum cost, see Barron and Ishii (1989), Barles et al. (1994), Bokanowski et al. (2015), Weerasinghe and Zhu (2016) and Kröner et al. (2018), where the viscosity solution approach usually plays the key role. In our optimal control problem (8.6), two fundamental questions need to be addressed: (i) Can we characterize the optimal portfolio and consumption control pair (θ^*, c^*) in the feedback form if it exists? (ii) Whether the relaxed tracking formulation is well-defined in the sense that the expected total capital injection $\mathbb{E}[\int_0^\infty e^{-\rho t}d(0 \vee \sup_{s \leq t}(M_s - V_s^{\theta^*,c^*}))]$ is finite? We will verify that our problem formulation does not require the injection of infinitely large capital to meet the tracking goal. The present paper contributes positive answers to both questions.

In solving the stochastic control problem (8.6) with a running maximum cost, we introduce two auxiliary state processes with reflections and study an auxiliary stochastic control problem, which gives rise to the HJB equation with two Neumann boundary conditions. By applying the dual transform and stochastic flow analysis, we can conjecture and carefully verify that the classical solution of the dual PDE satisfies a separation form of three terms, all of which admit probabilistic representations involving some dual reflected diffusion processes and/or the local time at the reflection boundary. We stress that the main challenge is to prove the smoothness of the conditional expectation of the integration of an exponential-like functional of the reflected drifted-Brownian motion (RDBM) with respect to the local time of another correlated RDBM. We propose a new method of decomposition-homogenization to the dual PDE, which allows us to show the smoothness of the conditional expectation of the integration of exponential-like functional of the RDBM with respect to the local time of an independent RDBM.

By using the classical solution to the dual PDE with Neumann boundary conditions and establishing some technical estimations of candidate optimal controls, we can address the previous question (i) and rigorously characterize the optimal control pair (θ^*, c^*) in a feedback form in the verification theorem. Based on our estimations of the optimal control processes, we can further answer the previous question; (ii) and verify that the expected total capital injection $\mathbb{E}[\int_0^\infty e^{-\rho t} d(0 \vee \sup_{s \leq t}(M_s - V_s^{\theta^*, c^*}))]$ is indeed bounded, and hence our problem (8.4) in a relaxed tracking formulation using the additional singular control is well defined. Moreover, it is also shown that $\mathbb{E}[\int_0^\infty e^{-\rho t} d(0 \vee \sup_{s \leq t}(M_s - V_s^{\theta^*, c^*}))]$ is bounded below by a positive constant, indicating that the capital injection is necessary for the well-posedness for the control problem. We also note that $A_t^* = \sup_{s \leq t}(V_s^{\theta^*, c^*} - M_s)^-$ records the largest shortfall when the wealth process $V_s^{\theta^*, c^*}$ falls below the benchmark process m_s up to time t. As a manner of risk management, the finite expectation $\mathbb{E}[\int_0^\infty e^{-\rho t} d(0 \vee \sup_{s \leq t}(M_s - V_s^{\theta^*, c^*}))]$ can quantitatively reflect the expected largest shortfall of the wealth management with respect to the benchmark in a long run.

8.2 Equivalent Stochastic Control Problem

In this section, we formulate and study a more tractable auxiliary stochastic control problem, which is mathematically equivalent to the unconstrained optimal control problem (8.6). To this end, we first introduce a new auxiliary state process to replace the wealth process $V^{\theta,c} = (V_t^{\theta,c})_{t \geq 0}$ given in (8.2). Let us first define

$$D_t := M_t - V_t^{\theta,c} + \mathrm{v} - \mathrm{m} \vee b - z, \quad \forall t \geq 0, \tag{8.7}$$

where $M = (M_t)_{t\geq 0}$ is defined by (8.3), and it is clear that $D_0 = 0$. Moreover, for any $x \geq 0$, we define the running maximum process of the process $D = (D_t)_{t\geq 0}$ that

$$L_t := x \vee \sup_{s\leq t} D_s - x \geq 0, \quad \forall t \geq 0 \tag{8.8}$$

with the initial value $L_0 = 0$. The auxiliary state process $X = (X_t)_{t\geq 0}$ is then defined as the reflected process $X_t := L_t - D_t$ for $t \geq 0$ that satisfies the SDE that for all $t > 0$,

$$X_t = x + \int_0^t \theta_s^\top \mu ds + \int_0^t \theta_s^\top \sigma dW_s - \int_0^t c_s ds$$
$$- \int_0^t \mu_Z Z_s ds - \int_0^t \sigma_Z Z_s dW_s^\eta - \int_0^t dm_s + L_t \tag{8.9}$$

with the initial value $X_0 = x \geq 0$. In particular, X_t hits 0 if the running maximum process L_t increases. We will change the notation from L_t to L_t^X from this point onwards to emphasize its dependence on the new state process X given in (8.9).

On the other hand, for the running maximum process $m = (m_t)_{t\geq 0}$ in (8.3), we also introduce a second auxiliary state process $I_t := m_t - B_t$ for all $t \geq 0$. As a result, I_t hits 0 if m_t increases, and we have

$$I_t = I_0 + \int_0^t dm_s - \int_0^t \mu_B ds - \int_0^t \sigma_B dW_s^\gamma, \quad \forall t \geq 0, \tag{8.10}$$

where the initial state value $I_0 = m \vee b - b \geq 0$. The problem (8.6) can be solved by studying the auxiliary problem that, for all $(x, y, z) \in \overline{\mathbb{R}}_+^3$,

$$\begin{cases} u(x, h, z) := \sup_{(\theta,c)\in\mathbb{U}^r} J(x, h, z; \theta, c) \\ \quad = \sup_{(\theta,c)\in\mathbb{U}^r} \mathbb{E}_{x,h,z} \left[\int_0^\infty e^{-\rho t} U(c_t) dt - \beta \int_0^\infty e^{-\rho t} dL_t^X \right], \\ \text{s.t. } (X, I, Z) \text{ satisfies (8.9), (8.10) and (8.3),} \end{cases} \tag{8.11}$$

where $\mathbb{E}_{x,h,z}[\cdot] := \mathbb{E}[\cdot | X_0 = x, I_0 = h, Z_0 = z]$. Note the equivalence that

$$w(v, m, z, b) = \begin{cases} u(v - m \vee b - z, m \vee b - b, z), & \text{if } v \geq m \vee b + z, \\ u(0, m \vee b - b, z) - \beta(m \vee b + z - v), & \text{if } v < m \vee b + z, \end{cases} \tag{8.12}$$

where $w(v, m, z, b)$ is given by (8.6).

We have the following property of the value function u in (8.11), whose proof is omit here.

Lemma 8.1. *The function* $x \to u(x,h,z)$ *and* $h \to u(x,h,z)$ *are non-decreasing. Moreover, it holds that,* $|u(x_1,h_1,z_1) - u(x_2,h_2,z_2)| \leq \beta|x_1 - x_2|$ *for all* $(x_1,x_2,h_1,h_2,z_1,z_2) \in \overline{\mathbb{R}}_+^6$, *where* $\beta > 0$ *is the utility parameter related to the capital injection appeared in* (8.4).

By dynamic program, we can derive the associated HJB equation that, for $(x,h,z) \in \mathbb{R}_+^3$,

$$\begin{cases} \sup_{\theta \in \mathbb{R}^d} \left[\theta^\top \mu u_x + \frac{1}{2}\theta^\top \sigma\sigma^\top \theta u_{xx} + \sigma_Z \theta^\top \sigma \eta z(u_{xx} - u_{xz}) - \sigma_B \theta^\top \sigma\gamma u_{xh} \right] \\ + \sup_{c \geq 0} \left(\frac{c^p}{p} - cu_x \right) + \frac{1}{2}\sigma_B^2 u_{hh} - \mu_B u_h + \frac{1}{2}\sigma_Z^2 z^2 (u_{zz} + u_{xx} - 2u_{xz}) \\ + \mu_Z z(u_z - u_x) + \sigma_Z \sigma_B z\eta^\top \gamma(u_{xh} - u_{hz}) = \rho u, \\ u_x(0,h,z) = \beta, \quad \forall (h,z) \in \overline{\mathbb{R}}_+^2, \\ u_h(x,0,z) = u_x(x,0,z), \quad \forall (x,z) \in \overline{\mathbb{R}}_+^2. \end{cases}$$
(8.13)

Here, the first Neumann boundary condition in (8.13) stems from the fact that $X_t = 0$ when L_t increases; while the second Neumann boundary condition in (8.13) comes from the fact that $I_t = 0$ when m_t increases.

Assuming that HJB equation (8.13) admits a unique classical solution u satisfying $u_{xx} < 0$ and $u_x \geq 0$, which will be verified in later sections, the first-order condition yields the candidate optimal feedback control that

$$\theta^* = -(\sigma\sigma^\top)^{-1} \frac{\mu u_x - \sigma_B \sigma\gamma u_{xh} + \sigma_Z \sigma\eta z(u_{xx} - u_{xz})}{u_{xx}}, \quad c^* = (u_x)^{\frac{1}{p-1}}.$$

Plugging the above results into (8.13), we apply Lemma 8.1 and the Legendre-Fenchel transform of the solution u only with respect to x that $\hat{u}(y,h,z) := \sup_{x \geq 0}\{u(x,h,z) - yx\}$ for all $(y,h,z) \in (0,\beta] \times \overline{\mathbb{R}}_+^2$. Equivalently, $u(x,h,z) = \inf_{y \in (0,\beta]}\{\hat{u}(y,h,z) - xy\}$ for all $(x,h,z) \in \overline{\mathbb{R}}_+^3$. The dual transform can linearize the HJB equation (8.13), and we obtain the following dual PDE, for $\hat{u}(y,h,z)$ that, for all $(y,h,z) \in (0,\beta] \times \mathbb{R}_+^2$,

$$\frac{\alpha^2}{2}y^2 \hat{u}_{yy} + \rho y \hat{u}_y + \frac{\sigma_B^2}{2}\hat{u}_{hh} - \mu_B \hat{u}_h + \frac{1}{2}\sigma_Z^2 z^2 \hat{u}_{zz}$$
$$+ \mu_Z z \hat{u}_z + \kappa_1 y \hat{u}_{yh} - \kappa_2 zy \hat{u}_{yz} + \sigma_Z \sigma_B \gamma^\top \eta z \hat{u}_{zh} \qquad (8.14)$$
$$+ (\kappa_2 - \mu_Z)zy + \left(\frac{1-p}{p}\right) y^{-\frac{p}{1-p}} = \rho \hat{u},$$

where the coefficients

$$\alpha := (\mu^\top (\sigma\sigma^\top)^{-1} \mu)^{\frac{1}{2}} > 0, \quad \kappa_1 := \sigma_B \mu^\top (\sigma\sigma^\top)^{-1} \sigma\gamma,$$

$$\kappa_2 := \sigma_Z \mu^\top (\sigma\sigma^\top)^{-1} \sigma\eta. \tag{8.15}$$

Correspondingly, the first Neumann boundary condition in (8.13) is transformed to the Neumann boundary condition that

$$\hat{u}_y(\beta, h, z) = 0, \quad \forall (h,z) \in \overline{\mathbb{R}}_+^2. \tag{8.16}$$

The second Neumann boundary condition in (8.13) is transformed to the Neumann boundary condition that

$$\hat{u}_z(y, 0, z) = y, \quad \forall (y,z) \in (0,\beta] \times \overline{\mathbb{R}}_+. \tag{8.17}$$

8.3 Solvability of the Dual PDE

This section examines the existence of solution to PDE (8.14) with two Neumann boundary conditions (8.16) and (8.17) in the classical sense using the probabilistic approach. Before stating the main result of this section, let us first introduce the following function that, for all $(r,h,z) \in \mathbb{R}_+^3$,

$$v(r,h,z) := \frac{1-p}{p}\beta^{-\frac{p}{1-p}}\mathbb{E}\left[\int_0^\infty e^{-\rho s + \frac{p}{1-p}R_s^r}ds\right] \tag{8.18}$$
$$+ \beta(\kappa_2 - \mu_Z)\mathbb{E}\left[\int_0^\infty e^{-\rho s - R_s^r}N_s^z ds\right] - \beta\mathbb{E}\left[\int_0^\infty e^{-\rho s - R_s^r}dK_s^h\right].$$

Here, the processes $R^r = (R_t^r)_{t\geq 0}$ and $H^h = (H_t^h)_{t\geq 0}$ with $(r,h) \in \overline{\mathbb{R}}_+^2$ are two reflected processes satisfying that, for all $t \geq 0$,

$$R_t^r = r + \left(\frac{\alpha^2}{2} - \rho\right)t + \alpha B_t^1 + L_t^r \geq 0, \tag{8.19}$$

$$H_t^h = h - \mu_B t - \varrho_1 \sigma_B B_t^1 - \sqrt{1-\varrho_1^2}\sigma_B B_t^0 + K_t^h \geq 0, \tag{8.20}$$

and the process $N^z = (N_t^z)_{t\geq 0}$ is a geometric Brownian motion given by

$$dN_t^z = \mu_Z N_t^z dt + \varrho_2 \sigma_Z N_t^z dB_t^1 + \sqrt{1-\varrho_2^2}\sigma_Z N_t^z B_t^2, \quad N_0^z = z,$$

where $B^0 = (B_t^0)_{t\geq 0}$, $B^1 = (B_t^1)_{t\geq 0}$ and $B^2 = (B_t^2)_{t\geq 0}$ are three independent scalar Brownian motions; while $L^r = (L_t^r)_{t\geq 0}$ (resp. $K^h = (K_t^h)_{t\geq 0}$) is a continuous and non-decreasing process that increases only on the set $\{t \geq 0;\ R_t^r = 0\}$ with $L_0^r = 0$ (resp. $\{t \geq 0;\ H_t^h = 0\}$ with $K_0^h = 0$) such that $R_t^r \geq 0$ (resp. $H_t^h \geq 0$) a.s. for $t \geq 0$, the correlative coefficients are respectively given by

$$\varrho_1 := \frac{\mu^\top(\sigma\sigma^\top)^{-1}\sigma\gamma}{\alpha}, \quad \varrho_2 := \frac{\mu^\top(\sigma\sigma^\top)^{-1}\sigma\eta}{\alpha}. \tag{8.21}$$

Note that $(R^r_t)_{t\geq 0}$ in (8.19) and $(H^h_t)_{t\geq 0}$ in (8.20) are RDBMs, and the processes $L^r = (L^r_t)_{t\geq 0}$ and $K^h = (K^h_t)_{t\geq 0}$ are uniquely determined by the above properties (Harrison (1985)). Using the solution representation of "the Skorokhod problem", it follows that, for all $t \geq 0$,

$$\begin{cases} L^r_t = 0 \vee \left\{ -r + \max_{s\in[0,t]} \left[-\alpha B^1_s - \left(\frac{\alpha^2}{2} - \rho\right) s \right] \right\}, \\ K^h_t = 0 \vee \left\{ -h + \max_{s\in[0,t]} \left(\mu_B s + \sigma_B B^3_s \right) \right\}, \end{cases} \quad (8.22)$$

where the process $B^3 = (B^3_t)_{t\geq 0} = (\varrho_1 B^1_t + \sqrt{1 - \varrho_1^2} B^0_t)_{t\geq 0}$ is a scalar Brownian motion. The main result of this section is stated as follows:

Theorem 8.1. *Assume $\rho > \alpha^2 p/(2-2p)$ and $\mu_Z > \kappa_2$. Consider the function $v(r, h, z)$ for $(r, h, z) \in \overline{\mathbb{R}}^3_+$ defined by the probabilistic representation (8.18). For $(y, h, z) \in (0, \beta] \times \overline{\mathbb{R}}^2_+$, let us define*

$$\hat{u}(y, h, z) := v\left(-\ln\frac{y}{\beta}, h, z\right). \quad (8.23)$$

Then, for each $(h, z) \in \overline{\mathbb{R}}^2_+$, we have

(i) *the function $(0, \beta] \ni y \mapsto \hat{u}(y, h, z)$ is strictly convex.*
(ii) *the function $\hat{u}(y, h, z)$ is a classical solution of PDE (8.14) with Neumann boundary conditions (8.16) and (8.17).*

On the other hand, if the Neumann problem (8.14)–(8.17) has a classical solution $\hat{u}(y, h, z)$ satisfying $|\hat{u}(y, h, z)| \leq C(1 + |y|^{-q} + z^q)$ for some $q > 1$ and some constant constant $C > 0$, then $v(r, h, z) := \hat{u}(\beta e^{-r}, h, z)$ admits the probabilistic representation (8.18).

Theorem 8.1 provides a probabilistic presentation of the classical solution to the PDE (8.14) with Neumann boundary conditions (8.16) and (8.17). Our method in the proof of Theorem 8.1 is completely from a probabilistic perspective. More precisely, we start with the proof of the smoothness of the function v by applying properties of reflected processes (R^r, H^h, N^z), the homogenization technique of the Neumann problem and the stochastic flow analysis. Then, we show that v solves a linear PDE by verifying two related Neumann boundary conditions at $r = 0$ and $h = 0$, respectively.

The next result deals with the first two terms of the function v given in (8.18), whose proof is similar to that of Bo et al. (2021, Theorem 4.2) after minor modifications. For the completeness, we provide a sketch of the proof in Appendix.

Lemma 8.2. *Assume $\rho > \alpha^2|p|/(2-2p) + \mu_Z$ and $\mu_Z > \kappa_2$. For any $(r,z) \in \overline{\mathbb{R}}_+^2$, denote by $l(r,z)$ the sum of the first term and the second term of the function v given in (8.18) that*

$$l(r,z) := \frac{1-p}{p}\beta^{-\frac{p}{1-p}}\mathbb{E}\left[\int_0^\infty e^{-\rho s + \frac{p}{1-p}R_s^r}ds\right]$$
$$+ \beta(\kappa_2 - \mu_Z)\mathbb{E}\left[\int_0^\infty e^{-\rho s - R_s^r}N_s^z ds\right]. \qquad (8.24)$$

Then, the function $l(r,z)$ is a classical solution to the following Neumann problem with Neumann boundary condition at $r = 0$:

$$\begin{cases} \dfrac{\alpha^2}{2}l_{rr} + \left(\dfrac{\alpha^2}{2} - \rho\right)l_r + \dfrac{1}{2}\sigma_Z^2 z^2 l_{zz} + \mu_Z z l_z + \kappa_2 z l_{rz} \\ \quad + (\kappa_2 - \mu_Z)\beta z e^{-r} + \dfrac{1-p}{p}\beta^{-\frac{p}{1-p}}e^{\frac{p}{1-p}r} = \rho l, \text{ on } \mathbb{R}_+^2, \\ l_r(0,z) = 0, \quad \forall z \in \overline{\mathbb{R}}_+. \end{cases} \qquad (8.25)$$

On the other hand, if the Neumann problem (8.25) has a classical solution $l(r,z)$ for $r \in \mathbb{R}_+$ satisfying $|l(r,z)| \leq C(1 + e^{qr} + z^q)$ for some $q > 1$ and a constant $C > 0$ depending on $(\mu, \sigma, \mu_Z, \sigma_Z, p)$, then this solution $l(r,z)$ admits the probabilistic representation (8.24). Moreover, $l(r,z)$ has the following explicit form, for $(r,z) \in \overline{\mathbb{R}}_+^2$,

$$l(r,z) = C_1\beta^{-\frac{p}{1-p}}e^{\frac{p}{1-p}r} + C_2\beta e^{-r} + z\left(\beta e^{-r} - \frac{\beta}{\ell}e^{-\ell r}\right), \qquad (8.26)$$

where $C_1, C_2 > 0$ are two positive constants defined by

$$C_1 := \frac{2(1-p)^3}{p(2\rho(1-p) - \alpha^2 p)}, \quad C_2 := \frac{2(1-p)^2}{2\rho(1-p) - \alpha^2 p}\beta^{-\frac{1}{1-p}}, \qquad (8.27)$$

and the constant ℓ is the positive root of the quadratic equation given by

$$\frac{1}{2}\alpha^2\ell^2 + \left(\rho - \kappa_2 - \frac{1}{2}\alpha^2\right)\ell + \mu_Z - \rho = 0, \qquad (8.28)$$

which is given by

$$\ell = \frac{-(\rho - \kappa_2 - \frac{1}{2}\alpha^2) + \sqrt{(\rho - \kappa_2 - \frac{1}{2}\alpha^2)^2 + 2\alpha^2(\rho - \mu_Z)}}{\alpha^2} > 0. \qquad (8.29)$$

The challenging step in our problem is to handle the last term of the function v given in (8.18), which differs substantially from the first two terms of v as it now involves both the reflected process R^r and the local time term K^h of the reflected process H^h. In particular, we highlight that the reflected process R^r is not independent of the local time process K^h. As a preparation step to handle the smoothness of the second term of the function v given in (8.18), let us first discuss the case when the reflected process R^r is independent of the local time K^h.

Lemma 8.3. *Let us consider the function that, for all $(r,h) \in \overline{\mathbb{R}}_+^2$,*

$$\varphi(r,h) := -\beta \mathbb{E}\left[\int_0^\infty e^{-\rho s - R_s^r} dG_s^h\right], \qquad (8.30)$$

where the reflected process $R^r = (R_t^r)_{t \geq 0}$ with $r \in \overline{\mathbb{R}}_+$ is given by (8.19), and the process $(P^h, G^h) = (P_t^h, G_t^h)_{t \geq 0}$ satisfies the reflected SDE:

$$P_t^h = h - \int_0^t \mu_B ds - \int_0^t \sigma_B dB_s^0 + \int_0^t dG_s^h \geq 0. \qquad (8.31)$$

Here, $G^h = (G_t^h)_{t \geq 0}$ is a continuous and non-decreasing process that increases only on the time set $\{t \in \mathbb{R}_+;\ P_t^h = 0\}$ with $G_0^h = 0$ and such that $P_t^h \geq 0$ a.s. for $t \geq 0$. Then, the processes $G^h = (G_t^h)_{t \geq 0}$ and $R^r = (R_t^r)_{t \geq 0}$ are independent. Moreover, the function $\varphi(r,h)$ is a classical solution to the following PDE with Neumann boundary conditions at $r=0$ and $h=0$:

$$\begin{cases} \dfrac{\alpha^2}{2}\varphi_{rr} + \left(\dfrac{\alpha^2}{2} - \rho\right)\varphi_r + \dfrac{\sigma_B^2}{2}\varphi_{hh} - \mu_B \varphi_h = \rho\varphi, & \text{on } \mathbb{R}_+^2, \\ \varphi_r(0,h) = 0, \quad \forall h \in \overline{\mathbb{R}}_+, \\ \varphi_h(r,0) = \beta e^{-r}, \quad \forall r \in \overline{\mathbb{R}}_+. \end{cases} \qquad (8.32)$$

On the other hand, if the Neumann problem (8.32) has a classical solution $\varphi(r,h)$ satisfying $|\varphi(r,h)| \leq C$ for some constant $C > 0$ depending on $(\mu, \sigma, \mu_B, \sigma_B)$, then this solution $\varphi(r,h)$ satisfies the probabilistic representation (8.30).

The following result guarantees the smoothness of the function $\varphi(r,h)$ for $(r,h) \in \overline{\mathbb{R}}_+^2$ defined by (8.30).

Lemma 8.4. *Consider the function $\varphi(r,h)$ for $(r,h) \in \overline{\mathbb{R}}_+^2$ defined by (8.30). Assume $\sigma_B \neq 0$, then it holds that $\varphi \in C^{2,2}(\overline{\mathbb{R}}_+^2)$. Moreover, for all $(r,h) \in \overline{\mathbb{R}}_+^2$, we have*

$$\varphi_r(r,h) = \beta \mathbb{E}\left[\int_0^{\tau_r} e^{-\rho s - R_s^r} dG_s^h\right], \quad \varphi_h(r,h) = \beta \mathbb{E}\left[e^{-\rho \eta_h - R_{\eta_h}^r}\right], \qquad (8.33)$$

$$\varphi_{rr}(r,h) = \beta \int_0^\infty \int_{-\infty}^r e^{-\rho s - r + x} \phi_1(s,x,r) dx d\mathbb{E}[G_s^h] - \beta \mathbb{E}\left[\int_0^{\tau_r} e^{-\rho s - R_s^r} dG_s^h\right], \tag{8.34}$$

$$\varphi_{rh}(r,h) = -\beta \mathbb{E}\left[e^{-\rho \eta_h - R_{\eta_h}^r} \mathbf{1}_{\eta_h < \tau_r}\right]$$
$$= -\beta \int_0^\infty \int_0^r \int_{-\infty}^y e^{-\rho s - r + x} \phi_1(s,x,y) \phi_2(s,h) dx dy ds. \tag{8.35}$$

Here $\tau_r := \inf\{s \geq 0; \ -\alpha B_s^1 - (\alpha^2/2 - \rho)s = r\}$, $\eta_h := \inf\{s \geq 0; \ \sigma_B B_s^0 + \mu_B s = h\}$ (with $\inf \emptyset = +\infty$ by convention), and functions $\phi_1(s,x,y), \phi_2(s,h)$ are respectively given by, for all $(s,x,y,h) \in \mathbb{R}_+^4$,

$$\begin{aligned}\phi_1(s,x,y) &= \frac{2(2y-x)}{\hat{\sigma}^2 \sqrt{2\hat{\sigma}^2 \pi s^3}} \exp\left(\frac{\hat{\mu}}{\hat{\sigma}}x - \frac{1}{2}\hat{\mu}^2 s - \frac{(2y-x)^2}{2\hat{\sigma}^2 s}\right), \\ \phi_2(s,h) &= \frac{h}{\sqrt{2\sigma_B^2 \pi s^3}} \exp\left(-\frac{(h-\mu_B s)^2}{2\sigma_B^2 s}\right),\end{aligned} \tag{8.36}$$

where parameters $\hat{\mu} := \alpha/2 - \rho/\alpha$ and $\hat{\sigma} := \alpha$.

Proof. The independence between the reflected process $R^r = (R_t^r)_{t \geq 0}$ and the process $G^h = (G_t^h)_{t \geq 0}$ plays an important role in the proof below. Before calculating the partial derivatives of $\varphi(r,h)$, we first claim that, for all $(r,h) \in \mathbb{R}_+^2$,

$$\varphi(r,h) = -\beta \int_0^\infty \mathbb{E}\left[e^{-\rho s - R_s^r}\right] d\mathbb{E}[G_s^h]. \tag{8.37}$$

In fact, fix $(T,r,h) \in \mathbb{R}_+^3$, and let $n \geq 1$, $s_i = iT/n$ with $i = 0, 1, \ldots, n$. Then, it holds that

$$\mathbb{E}\left[\int_0^T e^{-\rho s - R_s^r} dG_s^h\right] = \mathbb{E}\left[\lim_{n \to \infty} \sum_{i=1}^n e^{-\rho s_i - R_{s_i}^r} \left(G_{s_i}^h - G_{s_{i-1}}^h\right)\right].$$

Note that $e^{-\rho s - R_s^r} \leq 1$ a.s., for all $s \in [0,T]$. Then $\sum_{i=1}^n e^{-\rho s_i - R_{s_i}^r}(G_{s_i}^h - G_{s_{i-1}}^h) \leq \sum_{i=1}^n (G_{s_i}^h - G_{s_{i-1}}^h) = G_T^h$. Thus, from the dominated convergence theorem (DCT) and the independence between B^0 and B^1, it follows that

$$\begin{aligned}\mathbb{E}\left[\int_0^T e^{-\rho s - R_s^r} dG_s^h\right] &= \lim_{n \to \infty} \mathbb{E}\left[\sum_{i=1}^n e^{-\rho s_i - R_{s_i}^r} \left(G_{s_i}^h - G_{s_{i-1}}^h\right)\right] \\ &= \lim_{n \to \infty} \sum_{i=1}^n \mathbb{E}\left[e^{-\rho s_i - R_{s_i}^r}\right] \left\{\mathbb{E}[G_{s_i}^h] - \mathbb{E}[G_{s_{i-1}}^h]\right\} \\ &= \int_0^T \mathbb{E}\left[e^{-\rho s - R_s^r}\right] d\mathbb{E}[G_s^h]. \tag{8.38}\end{aligned}$$

Letting $T \to \infty$ on both side of (8.38) and applying MCT, it follows that
$$\mathbb{E}\left[\int_0^\infty e^{-\rho s - R_s^r} dG_s^h\right] = \int_0^\infty \mathbb{E}\left[e^{-\rho s - R_s^r}\right] d\mathbb{E}[G_s^h].$$
This verifies the claim (8.37).

Let $h \in \mathbb{R}_+$ be fixed. First of all, we consider the case with arbitrary $r_2 > r_1 \geq 0$. It follows from (8.30) that
$$\frac{\varphi(r_2, h) - \varphi(r_1, h)}{r_2 - r_1} = -\beta \int_0^\infty \mathbb{E}\left[e^{-\rho s} \frac{e^{-R_s^{r_2}} - e^{-R_s^{r_1}}}{r_2 - r_1} dG_s^h\right].$$
A direct calculation yields that, for all $s \geq 0$,
$$\lim_{r_2 \downarrow r_1} \frac{e^{-R_s^{r_2} - \rho s} - e^{-R_s^{r_1} - \rho s}}{r_2 - r_1}$$
$$= \begin{cases} -e^{-R_s^{r_1} - \rho s}, & \max_{q \in [0,s]}\{-\alpha B_q^1 - (\frac{1}{2}\alpha^2 - \rho)q\} \leq r_1, \\ 0, & \max_{q \in [0,s]}\{-\alpha B_q^1 - (\frac{1}{2}\alpha^2 - \rho)q\} > r_1. \end{cases}$$
Note that $\sup_{(r_1, r_2) \in \mathbb{R}_+^2} |(e^{-R_s^{r_2} - \rho s} - e^{-R_s^{r_1} - \rho s})/(r_2 - r_1)| \leq e^{-\rho s}$. Then, the DCT yields that
$$\lim_{r_2 \downarrow r_1} \frac{\varphi(r_2, h) - \varphi(r_1, h)}{r_2 - r_1}$$
$$= \beta \mathbb{E}\left[\int_0^\infty e^{-\rho s - R_s^{r_1}} \mathbf{1}_{\{\max_{q \in [0,s]}\{-\alpha B_q^1 - (\frac{1}{2}\alpha^2 - \rho)q\} \leq r_1\}} dG_s^h\right]$$
$$= \beta \mathbb{E}\left[\int_0^{T_{r_1}} e^{-\rho s - R_s^{r_1}} dG_s^h\right].$$
For the case $r_1 > r_2 \geq 0$, similar to the computations for (8.39), we can show that
$$\lim_{r_2 \uparrow r_1} \frac{\varphi(r_2, h) - \varphi(r_1, h)}{r_2 - r_1} = \lim_{r_2 \downarrow r_1} \frac{\varphi(r_2, h) - \varphi(r_1, h)}{r_2 - r_1}.$$
Thus, the representation (8.33) holds.

For any real numbers $r_0, r_n \geq 0$ with $r_n \to r_0$ as $n \to \infty$, we have from (8.33) that, for all $n \geq 1$,
$$\Delta_n := \frac{\varphi_r(r_n, h) - \varphi_r(r_0, h)}{r_n - r_0}$$
$$= \beta \mathbb{E}\left[\frac{1}{r_n - r_0} \int_{T_{r_0}}^{T_{r_n}} e^{-\rho s - R_s^{r_0}} dG_s^h\right]$$
$$+ \beta \mathbb{E}\left[\frac{1}{r_n - r_0} \int_0^{T_{r_0}} e^{-\rho s} \left(e^{-R_s^{r_n}} - e^{-R_s^{r_0}}\right) dG_s^h\right] \quad (8.39)$$
$$+ \beta \mathbb{E}\left[\frac{1}{r_n - r_0} \int_{T_{r_0}}^{T_{r_n}} e^{-\rho s} \left(e^{-R_s^{r_n}} - e^{-R_s^{r_0}}\right) dG_s^h\right]$$
$$:= \Delta_n^{(1)} + \Delta_n^{(2)} + \Delta_n^{(3)}.$$

In order to handle the term $\Delta_n^{(1)}$, we first focus on the case with $r_n \downarrow r_0$ as $n \to \infty$. Let us define the drifted-Brownian motion $\tilde{W}_s := -\alpha B_s^1 - (\frac{\alpha^2}{2} - \rho)s$ for all $s \in \mathbb{R}_+$. In view of (8.19) and (8.22), it holds that

$$R_s^{r_0} = r_0 - \tilde{W}_s + \left(\sup_{q \in [0,s]} \tilde{W}_q - r_0\right)^+, \quad \forall s \geq 0.$$

Note that $\phi_1(s,x,y)$ defined by (8.36) is the joint probability density of two-dimensional random variable $(\tilde{W}_s, \max_{q \in [0,s]} \tilde{W}_q)$ for any $s \geq 0$. Then, we have

$$\mathbb{E}\left[\frac{1}{r_n - r_0} \int_{T_{r_0}}^{T_{r_n}} e^{-\rho s - R_s^{r_0}} dG_s^h\right]$$

$$= \mathbb{E}\left[\frac{1}{r_n - r_0} \int_0^\infty e^{-\rho s - R_s^{r_0}} \mathbf{1}_{\{T_{r_0} < s \leq T_{r_n}\}} dG_s^h\right]$$

$$= \int_0^\infty \int_{r_0}^{r_n} \int_{-\infty}^y \frac{e^{-\rho s - r_0 + x - (y - r_0)^+}}{r_n - r_0} \phi_1(s,x,y) dx dy d\mathbb{E}[G_s^h]. \quad (8.40)$$

For $(s,y) \in \mathbb{R}_+^2$, set $g(s,y) := \int_{-\infty}^y e^{-\rho s - r_0 + x - (y - r_0)^+} \phi_1(s,x,y) dx$. Then, by the continuity of $y \to g(s,y)$, we have

$$\lim_{n \to \infty} \frac{1}{r_n - r_0} \int_{r_0}^{r_n} g(s,y) dy = g(s, r_0). \quad (8.41)$$

It follows from (8.40), (8.41) and DCT that

$$\lim_{n \to \infty} \mathbb{E}\left[\frac{1}{r_n - r_0} \int_{T_{r_0}}^{T_{r_n}} e^{-\rho s - R_s^{r_0}} dG_s^h\right]$$

$$= \int_0^\infty \int_{-\infty}^{r_0} e^{-\rho s - r_0 + x} \phi_1(s,x,r_0) dx d\mathbb{E}[G_s^h].$$

For the case where $r_0 > 0$ and $r_n \geq 0$ with $r_n \uparrow r_0$ as $n \to \infty$, using a similar argument as above, we can derive that

$$\lim_{n \to \infty} \Delta_n^{(1)} = \beta \int_0^\infty \int_{-\infty}^{r_0} e^{-\rho s - r_0 + x} \phi_1(s,x,r_0) dx d\mathbb{E}[G_s^h].$$

In a similar fashion as in the derivation of (8.39), we also have

$$\lim_{n \to \infty} \Delta_n^{(2)} = \beta \lim_{n \to \infty} \mathbb{E}\left[\frac{1}{r_n - r_0} \int_0^{T_{r_0}} e^{-\rho s} \left(e^{-R_s^{r_n}} - e^{-R_s^{r_0}}\right) dG_s^h\right]$$

$$= -\beta \mathbb{E}\left[\int_0^{T_{r_0}} e^{-\rho s - R_s^{r_0}} dG_s^h\right].$$

At last, we can also obtain

$$\left|\Delta_n^{(3)}\right| = \beta \mathbb{E}\left[\frac{1}{r_n - r_0}\left|\int_{T_{r_0}}^{T_{r_n}} e^{-\rho s}\left(e^{-R_s^{r_n}} - e^{-R_s^{r_0}}\right) dG_s^h\right|\right]$$

$$\leq \beta \mathbb{E}\left[\frac{G_{T_{r_n}} - G_{T_{r_0}}}{r_n - r_0} \sup_{s \geq 0}\left|e^{-\rho s}\left(e^{-R_s^{r_n}} - e^{-R_s^{r_0}}\right)\right|\right].$$

Note that $\sup_{s \geq 0}|(e^{-R_s^{r_n}} - e^{-R_s^{r_0}})/(r_n - r_0)| \leq 1$, $\mathbb{P}$-a.s., and the fact that $G_{T_{r_n}} \to G_{T_{r_0}}$, a.s., as $n \to \infty$, the DCT yields that $\lim_{n \to \infty} |\Delta_n^{(3)}| = 0$. Putting all the pieces together, we arrive at

$$\varphi_{rr}(r_0, h) = \beta \int_0^\infty \int_{-\infty}^{r_0} e^{-\rho s - r_0 + x} \phi_1(s, x, r_0) dx d\mathbb{E}[G_s^h]$$

$$- \beta \mathbb{E}\left[\int_0^{T_{r_0}} e^{-\rho s - R_s^{r_0}} dG_s^h\right]. \tag{8.42}$$

We next derive the representation of the partial derivative $\varphi_h(r, h)$. For any $h_2 > h_1 \geq 0$, it follows from (8.30) that

$$\frac{\varphi(r, h_2) - \varphi(r, h_1)}{h_2 - h_1} = -\beta \int_0^\infty \mathbb{E}\left[e^{-\rho s - R_s^r}\right] d\left(\frac{G_s^{h_2} - G_s^{h_1}}{h_2 - h_1}\right).$$

In lieu of (8.22), it holds that, for $i = 1, 2$, $G_s^{h_i} = h_i \vee \{\max_{l \in [0,s]}(\mu_B l + \sigma_B B_l^0)\} - h_i$ for $s \geq 0$. For $h > 0$, we introduce $\eta_h := \inf\{s \geq 0; \sigma_B W_s + \mu_B s = h\}$ with $\inf \emptyset = +\infty$ by convention. A direct calculation yields that, for all $s \geq 0$,

$$\frac{G_s^{h_2} - G_s^{h_1}}{h_2 - h_1} = \begin{cases} 0, & s < \eta_{h_1}, \\ -\dfrac{G_s^{h_1}}{h_2 - h_1}, & \eta_{h_1} \leq s < \eta_{h_2}, \\ -1, & s \geq \eta_{h_2}. \end{cases}$$

Then, it holds that

$$\mathbb{E}\left[\int_0^\infty e^{-\rho s - R_s^r} d\left(\frac{G_s^{h_2} - G_s^{h_1}}{h_2 - h_1}\right)\right] = -\mathbb{E}\left[\int_{\eta_{h_1}}^{\eta_{h_2}} e^{-\rho s - R_s^r} d\left(\frac{G_s^{h_1}}{h_2 - h_1}\right)\right]$$

$$= -\mathbb{E}\left[\int_{\eta_{h_1}}^{\eta_{h_2}}\left(e^{-\rho s - R_s^r} - e^{-\rho \eta_{h_1} - R_{\eta_{h_1}}^r}\right) d\left(\frac{G_s^{h_1}}{h_2 - h_1}\right)\right] - \mathbb{E}\left[e^{-\rho \eta_{h_1} - R_{\eta_{h_1}}^r}\right].$$

Note that, as $h_2 \downarrow h_1$, we have

$$\mathbb{E}\left[\int_{\eta_{h_1}}^{\eta_{h_2}}\left(e^{-\rho s - R_s^r} - e^{-\rho \eta_{h_1} - R_{\eta_{h_1}}^r}\right) d\left(\frac{G_s^{h_1}}{h_2 - h_1}\right)\right]$$

$$\leq \mathbb{E}\left[\sup_{s\in[\eta_{h_1},\eta_{h_2}]}\left|e^{-\rho s - R_s^r} - e^{-\rho\eta_{h_1}-R_{\eta_{h_1}}^r}\right|\right] \to 0.$$

This yields that

$$\lim_{h_2\downarrow h_1}\mathbb{E}\left[\int_0^\infty e^{-\rho s - R_s^r} d\left(\frac{G_s^{h_2}-G_s^{h_1}}{h_2-h_1}\right)\right] = -\mathbb{E}\left[e^{-\rho\eta_{h_1}-R_{\eta_{h_1}}^r}\right]. \tag{8.43}$$

Similarly, using the above argument, we also obtain

$$\lim_{h_2\uparrow h_1}\frac{\varphi(r,h_2)-\varphi(r,h_1)}{h_2-h_1} = \lim_{h_2\downarrow h_1}\frac{\varphi(r,h_2)-\varphi(r,h_1)}{h_2-h_1}.$$

Thus, we can claim that, for all $(r,h)\in\overline{\mathbb{R}}_+^2$,

$$\varphi_h(r,h) = \beta\mathbb{E}\left[e^{-\rho\eta_h-R_{\eta_h}^r}\right]. \tag{8.44}$$

Furthermore, in view of Proposition 2.5 in Abraham (2000), it follows that

$$\mathbb{E}[G_s^h] = \int_0^s p(0,h;l,0)dl, \quad \forall s\geq 0, \tag{8.45}$$

where $p(0,h_0;s,h) = \mathbb{P}(P_s^{h_0}\in dh)/dh$ is the condition density function of the reflected drifted Brownian motion $P_s^{h_0}$ at time $s\geq 0$ (Veestraeten (2004)). Hence, we deduce that, for all $(r,h)\in\overline{\mathbb{R}}_+^2$,

$$\varphi(r,h) = -\beta\int_0^\infty \mathbb{E}\left[e^{-\rho s-R_s^r}\right]p(0,h;s,0)ds.$$

In view that for every fixed $s\geq 0$, the function $h\to p(0,h;s,0)$ belongs to $C^2(\overline{\mathbb{R}}_+)$ (Veestraeten (2004)), we have, for all $(r,h)\in\overline{\mathbb{R}}_+^2$,

$$\varphi_{hh}(r,h) = -\beta\int_0^\infty \mathbb{E}\left[e^{-\rho s-R_s^r}\right]\frac{\partial^2 p(0,h;s,0)}{\partial h^2}ds. \tag{8.46}$$

Following (8.44) and a similar argument as in the proof of (8.39), we obtain

$$\varphi_{rh}(r,h) = -\beta\mathbb{E}\left[e^{-\rho\eta_h-R_{\eta_h}^r}\mathbf{1}_{\eta_h<\tau_r}\right]$$
$$= -\beta\mathbb{E}\left[\exp\left(-\rho\eta_h - r + \tilde{W}_{\eta_h} - \left(\sup_{q\in[0,\eta_h]}\tilde{W}_q - r\right)^+\right)\mathbf{1}_{\sup_{q\in[0,\eta_h]}\tilde{W}_q<r}\right]$$
$$= -\beta\int_0^\infty\int_0^r\int_{-\infty}^y e^{-\rho s-r+x}\phi_1(s,x,y)\phi_2(s,h)dxdyds. \tag{8.47}$$

The desired conclusion on the smoothness of $\varphi(r,h)$ in the lemma follows by combining expressions in (8.39), (8.42), (8.44), (8.46) and (8.47). $\square$

We next give the proof of Lemma 8.3.

Proof of Lemma 8.3. In lieu of Lemma 8.4, we first prove that the function $\varphi(r, h)$ defined by (8.30) satisfies the Neumann problem (8.32). In fact, for all $(r, h) \in \overline{\mathbb{R}}_+^2$ and $t \geq 0$, we introduce

$$\hat{B}_t^r := r + \left(\frac{\alpha^2}{2} - \rho\right)t + \alpha B_t^1, \quad \tilde{B}_t^h := h - \mu_B t - \sigma_B B_t^0,$$

where, we recall that $B^1 = (B_t^1)_{t \geq 0}$ and $B^0 = (B_t^0)_{t \geq 0}$ are two independent scalar Brownian motions, which are given in (8.19) and (8.20). For any $\epsilon \in (0, r \wedge h)$, let us define

$$\tau_\epsilon := \inf\{t \geq 0; \ |\hat{B}_t^r - r| \geq \epsilon \text{ or } |\tilde{B}_t^h - h| \geq \epsilon\}.$$

We can easily check that, for any $\hat{t} \geq 0$, $\hat{R}_{\hat{t} \wedge \tau_\epsilon}^r = \hat{B}_{\hat{t} \wedge \tau_\epsilon}^r$ and $P_{\hat{t} \wedge \tau_\epsilon}^h = \tilde{B}_{\hat{t} \wedge \tau_\epsilon}^h$. Then, by (8.20), (8.22) and the strong Markov property, we get

$$-\beta \mathbb{E}\left[\int_{\hat{t} \wedge \tau_\epsilon}^\infty e^{-\rho s - R_s^r} dG_s^h \bigg| \mathcal{F}_{\hat{t} \wedge \tau_\epsilon}\right] = e^{-\rho(\hat{t} \wedge \tau_\epsilon)} \varphi\left(R_{\hat{t} \wedge \tau_\epsilon}^r, P_{\hat{t} \wedge \tau_\epsilon}^h\right),$$

where $\varphi(r, h) = -\beta \mathbb{E}[\int_0^\infty e^{-\rho s - R_s^r} dG_s^h]$ using (8.53). Consider using Lemma 8.4, we can verify that the function φ satisfies $\varphi_r(0, h) = 0$, and $\varphi_h(r, 0) = \beta e^{-r}$ for all $(r, h) \in \mathbb{R}_+^2$. Therefore, for $(r, h) \in \overline{\mathbb{R}}_+^2$, it holds that

$$\varphi(r, h) = \mathbb{E}\left[e^{-\rho(\hat{t} \wedge \tau_\epsilon)} \varphi\left(R_{\hat{t} \wedge \tau_\epsilon}^r, P_{\hat{t} \wedge \tau_\epsilon}^h\right) - \beta \int_0^{\hat{t} \wedge \tau_\epsilon} e^{-\rho s - R_s^r} dG_s^h\right].$$

By using Lemma 8.4 and Itô's formula, we have

$$\frac{\beta}{\hat{t}} \mathbb{E}\left[\int_0^{\hat{t} \wedge \tau_\epsilon} e^{-\rho s - R_s^r} dG_s^h\right] = \frac{1}{\hat{t}} \mathbb{E}\left[e^{-\rho(\hat{t} \wedge \tau_\epsilon)} \varphi\left(R_{\hat{t} \wedge \tau_\epsilon}^r, P_{\hat{t} \wedge \tau_\epsilon}^h\right) - \varphi(r, h)\right]$$

$$= \frac{1}{\hat{t}} \mathbb{E}\left[\int_0^{\hat{t} \wedge \tau_\epsilon} e^{-\rho s} (\mathcal{L}\varphi - \rho\varphi)(R_s^r, P_s^h) ds\right] + \frac{1}{\hat{t}} \mathbb{E}\left[\int_0^{\hat{t} \wedge \tau_\epsilon} e^{-\rho s} \varphi_r(R_s^r, P_s^h) dL_s^r\right]$$

$$+ \frac{1}{\hat{t}} \mathbb{E}\left[\int_0^{\hat{t} \wedge \tau_\epsilon} e^{-\rho s} \varphi_h(R_s^r, P_s^h) dG_s^h\right], \tag{8.48}$$

where the operator $\mathcal{L}$ acted on $C^2(\overline{\mathbb{R}}_+^2)$ is defined by

$$\mathcal{L}g := \frac{\alpha^2}{2} g_{rr} + \left(\frac{\alpha^2}{2} - \rho\right) g_r + \frac{1}{2} \sigma_B^2 g_{hh} - \mu_B g_h, \quad \forall g \in C^2(\overline{\mathbb{R}}_+^2).$$

The DCT yields that

$$\lim_{\hat{t} \downarrow 0} \frac{1}{\hat{t}} \mathbb{E}\left[\int_0^{\hat{t} \wedge \tau_\epsilon} e^{-\rho s} (\mathcal{L}\varphi - \rho\varphi)(R_s^r, P_s^h) ds\right] = (\mathcal{L}\varphi - \rho\varphi)(r, h).$$

Note that, for all $r \in \mathbb{R}_+$, $\varphi_h(r,0) = \beta e^{-r}$ and $R_s^r > 0$ on $s \in [0, \hat{t} \wedge \tau_\epsilon]$. We then have

$$\frac{1}{t}\mathbb{E}\left[\int_0^{\hat{t}\wedge\tau_\epsilon} \varphi_r(R_s^r, P_s^h) dL_s^r\right] = 0, \quad \varphi_h(R_s^r, P_s^h)\mathbf{1}_{\{P_s^h = 0\}} = \beta e^{-R_s^r}.$$

By using (8.48), we obtain $(\mathcal{L}\varphi - \rho\varphi)(r,h) = 0$ on $(r,h) \in \overline{\mathbb{R}}_+^2$.

Next, we assume that the PDE (8.32) admits a classical solution φ satisfying $|\varphi(r,h)| \leq C$ for some constant $C > 0$ depending on $(\mu, \sigma, \mu_B, \sigma_B)$ only. Using the Neumann problem (8.32), the Itô's formula gives that, for all $(T, r, h) \in \overline{\mathbb{R}}_+^3$,

$$\mathbb{E}\left[e^{-\rho T}\varphi(R_T^r, P_T^h)\right] = \varphi(r,h) + \mathbb{E}\left[\int_0^T e^{-\rho s}(\mathcal{L}\varphi - \rho\varphi)(R_s^r, P_s^h) ds\right]$$

$$+ \mathbb{E}\left[\int_0^T e^{-\rho s}\varphi_r(R_s^r, P_s^h)\mathbf{1}_{\{R_s^r = 0\}} dL_s^r\right]$$

$$+ \mathbb{E}\left[\int_0^T e^{-\rho s}\varphi_h(R_s^r, P_s^h)\mathbf{1}_{\{P_s^h = 0\}} dG_s^h\right]$$

$$= \varphi(r,h) + \beta\mathbb{E}\left[\int_0^T e^{-R_s^r - \rho s} dG_s^h\right]. \tag{8.49}$$

Moreover, by applying DCT and the boundedness of φ on $\overline{\mathbb{R}}_+^2$, we obtain $\lim_{T\to\infty} \mathbb{E}[e^{-\rho T}\varphi(R_T^r, P_T^h)] = 0$. Letting $T \to \infty$ on both sides of (8.49) and using DCT and MCT, we obtain the representation (8.53) of the solution $\varphi(r,h)$, which completes the proof. □

Next, we propose a homogenization method of Neumann boundary conditions to study the smoothness of the last term in the probabilistic representation (8.18) together with the application of the result obtained in Lemma 8.3.

Proposition 8.1. *Define the function $\xi(r,h)$ by, for $(r,h) \in \overline{\mathbb{R}}_+^2$,*

$$\xi(r,h) := -\kappa_1 \mathbb{E}\left[\int_0^\infty e^{-\rho s}\varphi_{rh}(R_s^r, H_s^h) ds\right], \tag{8.50}$$

where we recall that the reflected processes $R^r = (R_t^r)_{t\geq 0}$ and $H^h = (H_t^h)_{t\geq 0}$ with $(r,h) \in \overline{\mathbb{R}}_+^2$ are given in (8.19) and (8.20), respectively. Moreover, let us define that

$$\psi(r,h) := \varphi(r,h) + \xi(r,h), \quad \forall (r,h) \in \overline{\mathbb{R}}_+, \tag{8.51}$$

where the function $\varphi(r,h) \in C^2(\overline{\mathbb{R}}_+^2)$ is given by (8.30) in Lemma 8.3. Then, the function $\psi(r,h)$ is a classical solution to the following Neumann problem with Neumann boundary conditions at $r=0$ and $h=0$:

$$\begin{cases} \dfrac{\alpha^2}{2}\psi_{rr} + \left(\dfrac{\alpha^2}{2}-\rho\right)\psi_r + \dfrac{\sigma_B^2}{2}\psi_{hh} - \mu_B \psi_h \\ \qquad -\kappa_1 \psi_{rh} = \rho\psi, \text{ on } \mathbb{R}_+^2, \\ \psi_r(0,h) = 0, \quad \forall h \in \overline{\mathbb{R}}_+, \\ \psi_h(r,0) = \beta e^{-r}, \quad \forall r \in \overline{\mathbb{R}}_+. \end{cases} \qquad (8.52)$$

On the other hand, if the Neumann problem (8.52) has a classical solution $\psi(r,h)$ satisfying $|\psi(r,h)| \leq C$ for some constant $C>0$ depending on $(\mu, \sigma, \mu_B, \sigma_B, \gamma)$, then this solution $\psi(r,h)$ satisfies the following probabilistic representation:

$$\psi(r,h) = -\beta \mathbb{E}\left[\int_0^\infty e^{-\rho s - R_s^r} dK_s^h\right]. \qquad (8.53)$$

Proof. Note that the integral in the expectation is the Lebesgue integral in (8.50). Then, using the stochastic flow argument (c.f. the argument used in Bo et al. (2021, Theorem 4.2)), it is not difficult to verify that $\xi(r,h) \in C^2(\overline{\mathbb{R}}_+^2)$ is a classical solution to the following Neumann problem with homogeneous Neumann boundary conditions:

$$\begin{cases} \dfrac{\alpha^2}{2}\xi_{rr} + \left(\dfrac{\alpha^2}{2}-\rho\right)\xi_r + \dfrac{\sigma_B^2}{2}\xi_{hh} - \mu_B \xi_h - \kappa_1 \xi_{rh} \\ \qquad -\rho\xi = \kappa_1 \varphi_{rh}, \text{ on } \mathbb{R}_+^2, \\ \xi_r(0,h) = 0, \quad \forall h \in \overline{\mathbb{R}}_+, \\ \xi_h(r,0) = 0, \quad \forall r \in \overline{\mathbb{R}}_+. \end{cases} \qquad (8.54)$$

From Eq. (8.32) in Lemma 8.3, it follows that

$$\begin{cases} \dfrac{\alpha^2}{2}\varphi_{rr} + \left(\dfrac{\alpha^2}{2}-\rho\right)\varphi_r + \dfrac{\sigma_B^2}{2}\varphi_{hh} - \mu_B \varphi_h = \rho\varphi, \text{ on } \mathbb{R}_+^2, \\ \varphi_r(0,h) = 0, \quad \forall h \in \overline{\mathbb{R}}_+, \\ \varphi_h(r,0) = \beta e^{-r}, \quad \forall r \in \overline{\mathbb{R}}_+. \end{cases}$$

In terms of the two Neumann problems above, we can conclude that $\psi(r,h) = \varphi(r,h) + \xi(r,h)$ satisfies the Neumann problem (8.52). This shows the first part of the proposition.

We next prove the second part of the proposition. To do it, we assume that the Neumann problem (8.32) admits a classical solution ψ satisfying $|\psi(r,h)| \leq C$ for some constant $C > 0$ depending on $(\mu, \sigma, \mu_B, \sigma_B, \gamma)$ only. Then, the Itô's formula gives that, for all $(T, r, h) \in \overline{\mathbb{R}}_+^3$,

$$\mathbb{E}\left[e^{-\rho T}\psi(R_T^r, H_T^h)\right] = \psi(r,h) + \mathbb{E}\left[\int_0^T e^{-\rho s}(\mathcal{L}\psi - \rho\psi)(R_s^r, H_s^h)ds\right]$$

$$+ \mathbb{E}\left[\int_0^T e^{-\rho s}\psi_r(R_s^r, H_s^h)\mathbf{1}_{\{R_s^r=0\}}dL_s^r\right]$$

$$+ \mathbb{E}\left[\int_0^T e^{-\rho s}\psi_h(R_s^r, H_s^h)\mathbf{1}_{\{H_s^h=0\}}dK_s^h\right], \qquad (8.55)$$

where the operator $\mathcal{L}$ acted on $C^2(\overline{\mathbb{R}}_+^2)$ is given by

$$\mathcal{L}g := \frac{\alpha^2}{2}g_{rr} + \left(\frac{\alpha^2}{2} - \rho\right)g_r + \frac{\sigma_B^2}{2}g_{hh} - \mu_B g_h - \kappa_1 g_{rh}, \quad \forall g \in C^2(\overline{\mathbb{R}}_+^2).$$

It follows from the first equation in (8.52) that $\mathbb{E}[\int_0^T e^{-\rho s}(\mathcal{L}\psi - \rho\psi)(R_s^r, H_s^h)ds] = 0$. Using the Neumann boundary conditions in (8.52), we obtain

$$\mathbb{E}\left[\int_0^T e^{-\rho s}\psi_r(R_s^r, H_s^h)\mathbf{1}_{\{R_s^r=0\}}dL_s^r\right] = \mathbb{E}\left[\int_0^T e^{-\rho s}\psi_r(0, H_s^h)dL_s^r\right] = 0,$$

and

$$\mathbb{E}\left[\int_0^T e^{-\rho s}\psi_h(R_s^r, H_s^h)\mathbf{1}_{\{H_s^h=0\}}dK_s^h\right] = \mathbb{E}\left[\int_0^T e^{-\rho s}\psi_h(R_s^r, 0)dK_s^h\right]$$

$$= \beta\mathbb{E}\left[\int_0^T e^{-\rho s - R_s^r}dK_s^h\right].$$

This yields from (8.55) that

$$\mathbb{E}\left[e^{-\rho T}\psi(R_T^r, H_T^h)\right] = \psi(r,h) + \beta\mathbb{E}\left[\int_0^T e^{-R_s^r - \rho s}dK_s^h\right]. \qquad (8.56)$$

Moreover, we have the boundedness of ψ on $\overline{\mathbb{R}}_+^2$. In fact, the function φ is bounded on $\mathbb{R}_+^2$ via (8.30); while φ_{rh} is also bounded by applying (8.35) in Lemma 8.4. This gives from DCT that $\lim_{T\to\infty}\mathbb{E}[e^{-\rho T}\psi(R_T^r, H_T^h)] = 0$. Letting $T \to \infty$ on both sides of (8.55), using DCT and MCT, we obtain the representation (8.53) of the solution $\psi(r,h)$, which completes the proof. □

Remark 8.1. Proposition 8.1 has proved that, the function ψ given by (8.53) is in $C^2(\overline{\mathbb{R}}_+^2)$. Moreover, it also holds that

$$\psi_r(r,h) = \beta \mathbb{E}\left[\int_0^{\tau_r} e^{-\rho s - R_s^r} dK_s^h\right], \quad \psi_h(r,h) = \beta \mathbb{E}\left[e^{-\rho \zeta_h - R_{\zeta_h}^r}\right], \quad (8.57)$$

$$\psi_{rr}(r,h) = \beta \lim_{\Delta r \to 0} \mathbb{E}\left[\frac{1}{\Delta r}\int_{\tau_r}^{\tau_r + \Delta r} e^{-\rho s - R_s^r} dK_s^h\right]$$
$$- \beta \mathbb{E}\left[\int_0^{\tau_r} e^{-\rho s - R_s^r} dK_s^h\right], \quad (8.58)$$

$$\psi_{rh}(r,h) = -\beta \mathbb{E}\left[e^{-\rho \zeta_h - R_{\zeta_h}^r} \mathbf{1}_{\zeta_h < \tau_r}\right], \quad (8.59)$$

where τ_r is defined in Lemma 8.4, and $\zeta_h := \inf\{s \geq 0;\ \sigma_B B_s^3 + \mu_B s = h\}$ with convention $\inf \emptyset = +\infty$. From these representations, it follows that $\psi_r(r,h) + \psi_{rr}(r,h) > 0$ for all $(r,h) \in \overline{\mathbb{R}}_+^2$.

We can finally present the proof of the main result in this section, i.e., Theorem 8.1.

Proof of Theorem 8.1. By applying Lemma 8.2 and Proposition 8.1, the function $v(r,h,z)$ defined by (8.18) is a classical solution to the following Neumann problem:

$$\begin{cases} \dfrac{\alpha^2}{2} v_{rr} + \left(\dfrac{\alpha^2}{2} - \rho\right) v_r + \dfrac{1}{2}\sigma_B^2 v_{hh} - \mu_B v_h + \dfrac{1}{2}\sigma_Z^2 z^2 v_{zz} \\ \quad + \mu_Z z v_z - \kappa_1 v_{rh} + \kappa_2 z v_{rz} + \sigma_Z \sigma_B \eta^\top \gamma z v_{zh} \\ \quad + (\kappa_2 - \mu_Z)\beta z e^{-r} + \left(\dfrac{1-p}{p}\right)\beta^{-\frac{p}{1-p}} e^{\frac{p}{1-p}r} = \rho v, \text{ on } \mathbb{R}_+^3, \\ v_r(0,h,z) = 0, \quad \forall (h,z) \in \overline{\mathbb{R}}_+^2, \\ v_h(r,0,z) = \beta e^{-r}, \quad \forall (r,z) \in \overline{\mathbb{R}}_+^2. \end{cases} \quad (8.60)$$

Then, we can verify that $\hat{u}(y,h,z) = v(-\ln(y/\beta), h, z)$ for $(y,h,z) \in (0,\beta] \times \overline{\mathbb{R}}_+^2$ is a classical solution of the Neumann problem (8.14) with Neumann boundary conditions (8.16) and (8.17). Moreover, the strict convexity of $(0,\beta] \ni y \to \hat{u}(y,h,z)$ for fixed $(h,z) \in \overline{\mathbb{R}}_+^2$ follows from the fact that $\hat{u}_{yy} = (v_{rr} + v_r)/y^2 > 0$ by applying Lemma 8.2, Lemma 8.4 and Remark 8.1. On the other hand, in a similar fashion of Proposition 8.1's proof, we can verify that if the Neumann problem (8.14)-(8.17) has a classical solution $\hat{u}(y,h,z)$ satisfying $|\hat{u}(y,h,z)| \leq C(1 + |y|^{-q} + z^q)$ for some $q > 1$ and some constant constant $C > 0$, then $v(r,h,z) := \hat{u}(\beta e^{-r}, h, z)$ has the probabilistic representation (8.18). □

8.4 Verification Theorem

Theorem 8.1 establishes existence and uniqueness of the classical solution $\hat{u}$ on $(0, \beta] \times \overline{\mathbb{R}}_+^2$ to the dual PDE (8.14) with Neumann boundary conditions (8.16) and (8.17). Moreover, this solution $\hat{u}(y, h, z)$ is strictly convex in $y \in (0, \beta]$. The following verification theorem will recover the classical solution u of the primal HJB equation (8.13) via the inverse transform of $\hat{u}$, and provide the optimal (admissible) portfolio-consumption control in the feedback form to the primal stochastic control problem (8.11).

Theorem 8.2. *Let $\rho_0 > 0$ be the constant depending on model parameters $(\mu, \sigma, \mu_B, \sigma_B, \mu_Z, \sigma_Z, \gamma, p, \beta)$ explicitly specified later in (8.101). For the discount rate $\rho > \rho_0$, it holds that*

(i) *Consider the function v on $\overline{\mathbb{R}}_+^3$ defined by (8.18). Let $\hat{u}(y, h, z) = v(-\ln\frac{y}{\beta}, h, z)$ for $(y, h, z) \in (0, \beta] \times \overline{\mathbb{R}}_+^2$. Introduce that*

$$u(x, h, z) = \inf_{y \in (0,\beta]} \{\hat{u}(y, h, z) + yx\}, \quad \forall (x, h, z) \in \overline{\mathbb{R}}_+^3. \tag{8.61}$$

Then, the function $u(x, h, z)$ is a classical solution to the following HJB equation with Neumann boundary conditions:

$$\begin{cases} \sup_{\theta \in \mathbb{R}^d} \left[\theta^\top \mu u_x + \frac{1}{2} \theta^\top \sigma \sigma^\top \theta u_{xx} + \sigma_Z \theta^\top \sigma \eta z (u_{xx} - u_{xz}) - \sigma_B \theta^\top \sigma \gamma u_{xh} \right] \\ + \sup_{c \geq 0} \left(\frac{c^p}{p} - cu_x \right) + \frac{1}{2} \sigma_B^2 u_{hh} - \mu_B u_h + \frac{1}{2} \sigma_Z^2 z^2 (u_{zz} + u_{xx} - 2u_{xz}) \\ + \mu_Z z (u_z - u_x) + \sigma_Z \sigma_B z \eta^\top \gamma (u_{xh} - u_{hz}) = \rho u, \\ u_x(0, h, z) = \beta, \quad \forall (h, z) \in \overline{\mathbb{R}}_+^2, \\ u_h(x, 0, z) = u_x(x, 0, z), \quad \forall (x, z) \in \overline{\mathbb{R}}_+^2. \end{cases} \tag{8.62}$$

(ii) *Define the optimal feedback control function that, for $(x, h, z) \in \overline{\mathbb{R}}_+^3$,*

$$\begin{cases} \theta^*(x, h, z) := -(\sigma \sigma^\top)^{-1} \dfrac{\mu u_x - \sigma_B \sigma \gamma u_{xh} + \sigma_Z \sigma \eta z (u_{xx} - u_{xz})}{u_{xx}}, \\ c^*(x, h, z) := u_x^{\frac{1}{p-1}}. \end{cases} \tag{8.63}$$

For $(x, h, z) \in \overline{\mathbb{R}}_+^3$, consider the controlled reflected process $(X^*, I, Z) = (X_t^*, I_t, Z_t)_{t \geq 0}$ given by, for $t \geq 0$,

$$\begin{cases} X_t^* = x + \int_0^t \theta^*(X_s^*, I_s, Z_s)^\top \mu ds + \int_0^t \theta^*(X_s^*, I_s, Z_s)^\top \sigma dW_s \\ \quad - \int_0^t c^*(X_s^*, I_s, Z_s) ds - \int_0^t \mu_Z Z_s ds \\ \quad - \int_0^t \sigma_Z Z_s dW_s^\eta - \int_0^t dm_s + L_t^{X^*}, \\ I_t = h - \int_0^t \mu_B ds - \int_0^t \sigma_B dW_s^\gamma + \int_0^t dm_s, \\ Z_t = z + \int_0^t \mu_Z Z_s ds + \int_0^t \sigma_Z Z_s dW_s^\eta. \end{cases} \quad (8.64)$$

Above, the running maximum process $m = (m_t)_{t \geq 0}$ is given in (8.3) and $L_0^{X^*} = 0$. Define $\theta_t^* = \theta^*(X_t^*, I_t, Z_t)$ and $c_t^* = c^*(X_t^*, I_t, Z_t)$ for all $t \geq 0$. Then, $(\theta^*, c^*) = (\theta_t^*, c_t^*)_{t \geq 0} \in \mathbb{U}^r$ is an optimal investment-consumption strategy. Moreover, for any admissible strategy $(\theta, c) \in \mathbb{U}^r$, we have

$$\mathbb{E}\left[\int_0^\infty e^{-\rho t} \frac{(c_t)^p}{p} dt - \beta \int_0^\infty e^{-\rho t} dL_t^X\right] \leq u(x, h, z), \quad \forall (x, h, z) \in \overline{\mathbb{R}}_+^3,$$

where the equality holds when $(\theta, c) = (\theta^*, c^*)$.

Proof. We first show item (i). For $(x, h, z) \in \overline{\mathbb{R}}_+^3$, define $y^* = y^*(x, h, z) \in (0, \beta]$ satisfying $\hat{u}_y(y^*, h, z) = -x$. Then, we have

$$u(x, h, z) = \inf_{y \in (0, \beta]} \{\hat{u}(y, h, z) + yx\}$$
$$= \hat{u}(y^*(x, h, z), h, z) + xy^*(x, h, z). \quad (8.65)$$

By applying Lemmas 8.2 and 8.4, it follows that

$$\hat{u}_y(y, h, z) = -\frac{1}{y} v_r\left(-\ln\frac{y}{\beta}, h, z\right)$$
$$= -\frac{1}{y}\left[l_r\left(-\ln\frac{y}{\beta}, z\right) + \psi_r\left(-\ln\frac{y}{\beta}, h\right)\right] \leq 0. \quad (8.66)$$

Then, $(0, \beta] \ni y \to \hat{u}(y, h, z)$ is decreasing for fixed $(z, h) \in \mathbb{R}_+^2$. Moreover, note that $\hat{u}_y(\beta, h, z) = 0$, and hence

$$\lim_{y \to 0} \hat{u}_y(y, h, z) = \lim_{r \to +\infty} \hat{u}_y(\beta e^{-r}, h, z) = -\lim_{r \to +\infty} e^r v_r(r, h, z) = -\infty.$$

Thus, y^* and u defined by (8.61) is well-defined on $\overline{\mathbb{R}}_+^2$. Moreover, it follows from Theorem 8.1 that $y \mapsto \hat{u}(y, h, z)$ is strictly convex, which implies that

$x \mapsto u(x, h, z)$ is strictly concave. Thus, a direct calculation yields that u solves the primal HJB equation (8.13).

We next prove item (ii). It follows from Theorem 8.1 that $\theta^*(x, h, z)$ and $c^*(x, h, z)$ given by (8.63) are continuous on $\mathbb{R}^3_+$. We then claim that, there exists a pair of positive constants (C_o, C_q) such that, for all $(x, h, z) \in \mathbb{R}^3_+$,

$$|\theta^*(x, h, z)| \leq C_o(1 + x + z), \quad |c^*(x, h, z)| \leq C_q(1 + x). \tag{8.67}$$

Let $\gamma_1 := |(\sigma\sigma^\top)^{-1}\mu|$, $\gamma_2 := |(\sigma\sigma^\top)^{-1}\sigma_B\sigma\gamma|$ and $\gamma_3 := |(\sigma\sigma^\top)^{-1}\sigma_Z\sigma\eta|$. In view of the duality transform, we arrive at

$$|\theta^*(x, z)| \leq \gamma_1 \left|\frac{u_x}{u_{xx}}(x, h, z)\right| + \gamma_2 \left|\frac{u_{xz}}{u_{xx}}(x, h, z)\right| + \gamma_3 \left|\frac{zu_{xz}}{u_{xx}}(x, h, z)\right| + \gamma_3 z$$

$$= \gamma_1 y^*(x, h, z)\hat{u}_{yy}(y^*(x, h, z), h, z) + \gamma_2 |\hat{u}_{yh}(y^*(x, h, z), h, z)|$$
$$+ \gamma_3 |z\hat{u}_{yz}(y^*(x, h, z), h, z)| + \gamma_3 z$$

$$= \gamma_1 y^*(x, h, z) v_{rr}\left(-\ln\frac{y^*(x, h, z)}{\beta}, h, z\right) + \gamma_2 \left|v_{rh}\left(-\ln\frac{y^*(x, h, z)}{\beta}, h, z\right)\right|$$
$$+ \gamma_3 \left|zv_{rz}\left(-\ln\frac{y^*(x, h, z)}{\beta}, h, z\right)\right| + \gamma_3 z$$

$$= \gamma_1 x + \frac{\gamma_1}{y^*(x, h, z)}(l_{rr} + \psi_{rr})\left(-\ln\frac{y^*(x, h, z)}{\beta}, h, z\right)$$
$$+ \frac{\gamma_2}{y^*(x, h, z)} \left|\psi_{rh}\left(-\ln\frac{y^*(x, h, z)}{\beta}, h\right)\right|$$
$$+ \frac{\gamma_3}{y^*(x, h, z)} \left|zl_{rz}\left(-\ln\frac{y^*(x, h, z)}{\beta}, z\right)\right| + \gamma_3 z, \tag{8.68}$$

where the last equality holds since

$$x = -\hat{u}_y(y^*(x, h, z), h, z) = \frac{1}{y^*(x, h, z)} v_r\left(-\ln\frac{y^*(x, h, z)}{\beta}, h, z\right)$$
$$= \frac{1}{y^*(x, h, z)}(l_r + \psi_r)\left(-\ln\frac{y^*(x, h, z)}{\beta}, h, z\right).$$

It follows from (8.57) and (8.58) that, for all $(r, h) \in \mathbb{R}^2_+$,

$$\psi_r(r, h) + \psi_{rr}(r, h) = \beta \lim_{\Delta r \to 0} \mathbb{E}\left[\frac{1}{\Delta r}\int_{T_r}^{T_r + \Delta r} e^{-\rho s - R_s^r} dK_s^h\right]. \tag{8.69}$$

Using the representation (8.22), we deduce that, for all $h \geq 0$ and $t \geq s \geq 0$,

$$K_t^h - K_s^h$$
$$= 0 \vee \left\{-h + \max_{l \in [0,t]}(\mu_B l + \sigma_B B_l^2)\right\} - 0 \vee \left\{-h + \max_{l \in [0,s]}(\mu_B l + \sigma_B B_l^2)\right\}$$

$$= h \vee \left\{ \max_{l \in [0,t]} (\mu_B l + \sigma_B B_l^2) \right\} - h \vee \left\{ \max_{l \in [0,s]} (\mu_B l + \sigma_B B_l^2) \right\}.$$

If $\max_{l \in [0,t]}(\mu_B l + \sigma_B B_l^2) \leq h$, it holds that $K_t^h - K_s^h = 0 \leq K_t^0 - K_s^0$; while, if $\max_{l \in [0,t]}(\mu_B l + \sigma_B B_l^0) > h$, we also have

$$K_t^h - K_s^h \leq \max_{l \in [0,t]} (\mu_B l + \sigma_B B_l^2) - h \vee \left\{ \max_{l \in [0,s]} (\mu_B l + \sigma_B B_l^2) \right\}$$

$$\leq \max_{l \in [0,t]} (\mu_B l + \sigma_B B_l^2) - \max_{l \in [0,s]} (\mu_B l + \sigma_B B_l^2) = K_t^0 - K_s^0.$$

Hence, we can deduce that $K_t^0 - K_t^h \geq K_s^0 - K_s^h$, i.e., the process $\{K_t^0 - K_t^h\}_{t \geq 0}$ is non-decreasing. This implies that, for all $h \geq 0$,

$$\psi_r(r,h) + \psi_{rr}(r,h) = \beta \lim_{\Delta r \to 0} \mathbb{E}\left[\frac{1}{\Delta r} \int_{T_r}^{T_r + \Delta r} e^{-\rho s - R_s^r} dK_s^h \right]$$

$$= \beta \lim_{\Delta r \to 0} \mathbb{E}\left[\frac{1}{\Delta r} \int_{T_r}^{T_r + \Delta r} e^{-\rho s - R_s^r} dK_s^0 \right]$$

$$- \beta \lim_{\Delta r \to 0} \mathbb{E}\left[\frac{1}{\Delta r} \int_{T_r}^{T_r + \Delta r} e^{-\rho s - R_s^r} d(K_s^0 - K_s^h) \right] \quad (8.70)$$

$$\leq \beta \lim_{\Delta r \to 0} \mathbb{E}\left[\frac{1}{\Delta r} \int_{T_r}^{T_r + \Delta r} e^{-\rho s - R_s^r} dK_s^0 \right]$$

$$= \psi_r(r,0) + \psi_{rr}(r,0).$$

Note that, it follows from Proposition 8.1 that

$$\psi_{rr}\left(-\ln \frac{y^*(x,h,z)}{\beta}, h\right)$$

$$\leq (\psi_r + \psi_{rr})\left(-\ln \frac{y^*(x,h,z)}{\beta}, 0\right) - \psi_r\left(-\ln \frac{y^*(x,z)}{\beta}, h\right) \quad (8.71)$$

$$\leq (\psi_r + \psi_{rr})\left(-\ln \frac{y^*(x,h,z)}{\beta}, 0\right)$$

$$= (\varphi_r + \varphi_{rr} + \xi_r + \xi_{rr})\left(-\ln \frac{y^*(x,h,z)}{\beta}, 0\right).$$

In view of Lemma 8.4, we obtain that

$$\varphi_r(r,0) + \varphi_{rr}(r,0) = \beta \int_0^\infty \int_{-\infty}^r e^{-\rho s - r + x} \phi_1(s,x,r) dx d\mathbb{E}\left[G_s^0\right].$$

For $r \in [0,1]$, by the continuity of $r \to \int_0^\infty \int_{-\infty}^r e^{-\rho s + x} \phi_1(s,x,r) dx d\mathbb{E}[G_s^0]$, we can obtain

$$\int_0^\infty \int_{-\infty}^r e^{-\rho s + x} \phi_1(s,x,r) dx d\mathbb{E}[G_s^0]$$

$$\leq \max_{r \in [0,1]} \int_0^\infty \int_{-\infty}^r e^{-\rho s + x} \phi_1(s, x, r) dx d\mathbb{E}[G_s^0] < +\infty. \tag{8.72}$$

For the other case $r > 1$, we obtain from (8.36) that

$$\int_0^1 \int_{-\infty}^r e^{-\rho s + x} \phi_1(s, x, r) dx d\mathbb{E}[G_s^0]$$

$$= \int_0^1 \int_{-\infty}^r e^{-\rho s + x} \frac{2(2r - x)}{\sqrt{2\hat{\sigma}^2 \pi s^3}} \exp\left(\frac{\hat{\mu}}{\hat{\sigma}} x - \frac{1}{2}\hat{\mu}^2 s - \frac{(2r-x)^2}{2\hat{\sigma}^2 s}\right) dx d\mathbb{E}[G_s^0]$$

$$\stackrel{y=r-x}{=} \int_0^1 \int_0^\infty e^{-\rho s + r - y} \frac{2(r+y)}{\sqrt{2\hat{\sigma}^2 \pi s^3}} \exp\left(\frac{\hat{\mu}}{\hat{\sigma}}(r-y) - \frac{1}{2}\hat{\mu}^2 s - \frac{(r+y)^2}{2\hat{\sigma}^2 s}\right) dy d\mathbb{E}[G_s^0]$$

$$\leq \int_0^1 e^{-\rho s} \frac{2 \exp\left(-\frac{1}{2\hat{\sigma}^2 s}\right)}{\sqrt{2\hat{\sigma}^2 \pi s^3}}$$

$$\times \int_0^\infty (r+y) \exp\left(\left(\frac{\hat{\mu}}{\hat{\sigma}} + 1\right) r + \frac{y^2}{2\hat{\sigma}^2 s} - \frac{(r+y)^2 - 1}{2\hat{\sigma}^2 s}\right) dy d\mathbb{E}[G_s^0]$$

$$\leq 9\hat{\sigma}^{\frac{11}{2}} \int_0^1 e^{-\rho s} d\mathbb{E}[G_s^0] \int_0^\infty (r+y) \exp\left(\left(\frac{\hat{\mu}}{\hat{\sigma}} + 1\right) r - \frac{y-1}{\hat{\sigma}^2}\right) dy \tag{8.73}$$

$$\leq 9\hat{\sigma}^{\frac{11}{2}} \mathbb{E}\left[\int_0^1 dG_s^0\right] \int_0^\infty (1+y) \exp\left(-\frac{y-1}{\hat{\sigma}^2}\right) dy \leq 18\hat{\sigma}^5 e^{\hat{\sigma}^{-2}} (|\mu_B| + 3\sigma_B).$$

Moreover, using the fact $\frac{1}{2}\hat{\mu}^2 s + \frac{(2r-x)^2}{2\hat{\sigma}^2 s} \geq \frac{\hat{\mu}}{\hat{\sigma}}(r-x)$, it holds that

$$\int_1^\infty \int_{-\infty}^r e^{-\rho s + x} \frac{2(2r-x)}{\sqrt{2\hat{\sigma}^2 \pi s^3}} \exp\left(\frac{\hat{\mu}}{\hat{\sigma}} x - \frac{1}{2}\hat{\mu}^2 s - \frac{(2r-x)^2}{2\hat{\sigma}^2 s}\right) dx d\mathbb{E}[G_s^0]$$

$$\leq \int_1^\infty \int_{-\infty}^r e^{-\rho s + x} \frac{2(2r-x)}{\sqrt{2\hat{\sigma}^2 \pi}} \exp\left(\frac{2\hat{\mu}}{\hat{\sigma}} r\right) dx d\mathbb{E}[G_s^0]$$

$$\stackrel{y=r-x}{=} \int_1^\infty \int_0^\infty e^{-\rho s + r - y} \frac{2(r+y)}{\sqrt{2\hat{\sigma}^2 \pi}} \exp\left(\frac{2\hat{\mu}}{\hat{\sigma}} r\right) dy d\mathbb{E}[G_s^0]$$

$$\leq \frac{4}{\sqrt{2\hat{\sigma}^2 \pi}} \int_0^\infty e^{-\rho s} d\mathbb{E}[G_s^0] \leq \frac{4(|\mu_B| + 3\sigma_B)}{\sqrt{2\hat{\sigma}^2 \pi}}. \tag{8.74}$$

Thus, it follows from (8.69)–(8.74) that

$$\frac{1}{y^*(x,z)}(\varphi_{rr} + \varphi_r)\left(-\ln\frac{y^*(x,z)}{\beta}, 0\right) \leq M_1(1+x), \tag{8.75}$$

where the positive constant M_1 is defined by

$$M_1 := 18\hat{\sigma}^5 e^{\hat{\sigma}^{-2}}(|\mu_B| + 3\sigma_B) + \frac{4(|\mu_B| + 3\sigma_B)}{\sqrt{2\hat{\sigma}^2 \pi}}$$

$$+ \max_{r \in [0,1]} \int_0^\infty \int_{-\infty}^r e^{-s+x} \phi_1(s,x,r) dx d\mathbb{E}[G_s^0]. \tag{8.76}$$

In what follows, let us define that, for $(r,h) \in \overline{\mathbb{R}}_+^2$,

$$f(r,h) := \varphi_{rh}(r,h) = -\beta \mathbb{E}\left[e^{-\rho \eta_h - R^r_{\eta_h}} \mathbf{1}_{\eta_h < \tau_r}\right]$$
$$= -\beta \int_0^\infty \int_0^r \int_{-\infty}^y e^{-\rho s - r + x} \phi_1(s,x,y) \phi_2(s,h) dx dy ds.$$

It follows from Propositions 8.1 and 4.1 in Bo et al. (2021) that

$$\xi_r(r,0) = -\kappa \mathbb{E}\left[\int_0^{\tau_r} e^{-\rho s} f_r(R^r_s, H^0_s) ds\right], \tag{8.77}$$

$$\xi_{rr}(r,0) = -\kappa \Gamma \mathbb{E}\left[e^{-\rho \tau_r} f_r(0, H^0_{\tau_r})\right]$$
$$-\kappa \mathbb{E}\left[\int_0^{\tau_r} e^{-\rho s} f_{rr}(R^r_s, H^0_s) ds\right] \tag{8.78}$$

with $\Gamma := \int_0^\infty e^{-\hat{\mu}^2 s/(2\hat{\sigma}^2)}/\sqrt{2\hat{\sigma}^2 \pi s} ds$ being a positive constant. It follows from a direct calculation that, for all $(r,h) \in \overline{\mathbb{R}}_+^2$,

$$f_r(r,h) = \beta \int_0^\infty \int_0^r \int_{-\infty}^y e^{-\rho s - r + x} \phi_1(s,x,y) \phi_2(s,h) dx dy ds$$
$$- \beta \int_0^\infty \int_{-\infty}^r e^{-\rho s - r + x} \phi_1(s,x,r) \phi_2(s,h) dx ds,$$

$$f_{rr}(r,h) = -f_r(r,h) + \beta \int_0^\infty \int_{-\infty}^r e^{-\rho s - r + x} \phi_1(s,x,r) \phi_2(s,h) dx ds$$
$$- \beta \int_0^\infty \int_{-\infty}^r e^{-\rho s - r + x} \frac{\partial \phi_1}{\partial y}(s,x,r) \phi_2(s,h) dx ds$$
$$- \beta \int_0^\infty e^{-\rho s} \phi_1(s,r,r) \phi_2(s,h) ds.$$

Note that, we have from (8.36) that

$$|f_r(r,h)| \leq \beta e^{-r} \int_0^\infty \int_0^\infty \int_{-\infty}^y e^{-s+x} \phi_1(s,x,y) \phi_2(s,h) dx dy ds$$
$$+ \beta e^{-r} \int_0^\infty \int_{-\infty}^\infty e^{-s+x} \phi_1(s,x,r) \phi_2(s,h) dx ds,$$

$$|f_r(r,h) + f_{rr}(r,h)| \leq \beta e^{-r} \int_0^\infty \int_{-\infty}^r e^{-s+x} \phi_1(s,x,r) \phi_2(s,h) dx ds$$
$$+ \beta e^{-r} \int_0^\infty \int_{-\infty}^\infty e^{-s+x} \frac{\partial \phi_1}{\partial y}(s,x,r) \phi_2(s,h) dx ds$$
$$+ \beta e^{-r} \int_0^\infty e^{-s+r} \phi_1(s,r,r) \phi_2(s,h) ds.$$

In a similar fashion of (8.72)–(8.76), we deuce that $|f_r(r,h)| \leq M_2\beta e^{-r}$ and $|f_r(r,h) + f_{rr}(r,h)| \leq M_3\beta e^{-r}$, where the finite positive constants are given by

$$M_2 := \sup_{h\in\mathbb{R}_+} \int_0^\infty \int_0^\infty \int_{-\infty}^y e^{-s+x} \phi_1(s,x,y)\phi_2(s,h) dx dy ds$$
$$+ \sup_{(r,h)\in\mathbb{R}_+^2} \int_0^\infty \int_{-\infty}^r e^{-s+x} \phi_1(s,x,r)\phi_2(s,h) dx ds < +\infty, \quad (8.79)$$

$$M_3 := \sup_{(r,h)\in\mathbb{R}_+} \int_0^\infty \int_{-\infty}^r e^{-s+x} \phi_1(s,x,r)\phi_2(s,h) dx ds$$
$$+ \sup_{(r,h)\in\mathbb{R}_+^2} \int_0^\infty \int_{-\infty}^\infty e^{-s+x} \frac{\partial \phi_1}{\partial y}(s,x,r)\phi_2(s,h) dx ds \quad (8.80)$$
$$+ \sup_{(r,h)\in\mathbb{R}_+^2} \int_0^\infty e^{-s+r} \phi_1(s,r,r)\phi_2(s,h) ds < +\infty.$$

Therefore, it holds that

$$(\xi_r + \xi_{rr})(r,0)$$
$$\leq |\kappa|\mathbb{E}\left[\int_0^{T_r} e^{-\rho s} |f_r + f_{rr}|(R_s^r, H_s^0) ds\right] + \beta|\kappa|\Gamma M_2 \mathbb{E}\left[e^{-\rho \tau_r}\right]$$
$$\leq |\kappa|\beta M_3 \mathbb{E}\left[\int_0^{T_r} e^{-\rho s - R_s^r} ds\right] + \beta|\kappa|\Gamma M_2 \mathbb{E}\left[e^{-\rho \tau_r}\right]$$
$$= \beta e^{-r} M_3 |\kappa| \int_0^\infty \int_0^r \int_{-\infty}^y e^{-\rho s - x} \phi_1(s,x,y) dx dy ds + \beta|\kappa|\Gamma M_2 \mathbb{E}\left[e^{-\rho \tau_r}\right]$$
$$= \beta e^{-r} M_3 M_2 |\kappa| + \beta|\kappa|\Gamma M_2 e^{-r} \leq M_4 \beta e^{-r}, \quad (8.81)$$

where the positive constant M_4 is defined by

$$M_4 := M_2(M_3|\kappa| + |\kappa|\Gamma). \quad (8.82)$$

Thus, we deduce from (8.77)–(8.82) that

$$\frac{1}{y^*(x,h,z)}(\psi_{rr} + \psi_r)\left(-\ln\frac{y^*(x,h,z)}{\beta}, 0\right) \leq (M_4 + M_1)(1+x). \quad (8.83)$$

On the other hand, using Lemma 8.2, we have

$$l_{rr}(r,z) = C_1 \beta^{-\frac{p}{1-p}} \left(\frac{p}{1-p}\right)^2 e^{\frac{p}{1-p}r} + C_2 \beta e^{-r} + z(\beta e^{-r} - \beta \ell e^{-\ell r})$$
$$= \frac{p}{1-p} l_r(r) + \frac{\beta e^{-r}}{1-p} C_2 + \frac{1}{1-p} z\beta e^{-r} - z\beta e^{-\ell r}\left(\ell + \frac{p}{1-p}\right).$$

Then, by using the condition $\rho > (\alpha^2 p + 1)/(2 - 2p)$, we deduce that

$$\frac{1}{y^*(x,h,z)} l_{rr}\left(-\ln\frac{y^*(x,h,z)}{\beta}, z\right)$$
$$\leq \frac{1}{y^*(x,h,z)}\left[\frac{1}{1-p}(l_r + \varphi_r)\left(-\ln\frac{y^*(x,h,z)}{\beta}, h, z\right)\right.$$
$$\left. + \frac{y^*(x,h,z)}{1-p}C_2 + \frac{1}{1-p}zy^*(x,h,z)\right]$$
$$= \frac{1}{1-p}x + \frac{1}{1-p}C_2 + \frac{1}{1-p}z \leq \frac{1}{1-p}(x+z) + 2(1-p)\beta^{-\frac{1}{1-p}}. \quad (8.84)$$

By using Lemma 8.2 again, we have $|zl_{rz}(r,z)| = z\beta(e^{-\ell r} - e^{-r}) \leq l_r(r,z)$ for all $(r,z) \in \mathbb{R}_+^2$. Thus, it holds that

$$\frac{1}{y^*(x,h,z)}\left|zl_{rz}\left(-\ln\frac{y^*(x,h,z)}{\beta}, z\right)\right|$$
$$\leq \frac{1}{y^*(x,h,z)}(l_r + \varphi_r)\left(-\ln\frac{y^*(x,h,z)}{\beta}, h, z\right) = x. \quad (8.85)$$

Moreover, note that

$$|\psi_{rh}(r,z)| = \beta\mathbb{E}\left[e^{-\rho\zeta_h - R^r_{\zeta_h}}\mathbf{1}_{\zeta_h<\tau_r}\right]$$
$$\leq \beta\mathbb{E}\left[\exp\left(-\rho\zeta_h - r - \left(\frac{\alpha^2}{2} - \rho\right)\zeta_h - \alpha B^1_{\zeta_h} - L^r_{\zeta_h}\right)\right] \quad (8.86)$$
$$\leq \beta e^{-r}\mathbb{E}\left[\exp\left(-\frac{\alpha^2}{2}\zeta_h - \alpha B^1_{\zeta_h}\right)\right] = \beta e^{-r}.$$

In lieu of (8.68), (8.71), (8.75), (8.84) (8.85) and (8.86), we deduce that $|\theta^*(x,h,z)| \leq C_o(1 + x + z)$, where the positive constant C_o is defined by

$$C_o := 1 + \left(\frac{1}{1-p} + 2(1-p)\beta^{-\frac{1}{1-p}} + M_1 + M_4\right)\gamma_1 + \gamma_2. \quad (8.87)$$

Here, the constant C_2 is given by (8.27) and constants M_1 and M_4 are defined as (8.76) and (8.82).

Next, we show the linear growth of $c^*(x,h,z)$ on $(x,h,z) \in \overline{\mathbb{R}}_+^3$. Note that $y^*(x,h,z) = u_x(x,h,z)$ for all $(x,h,z) \in \overline{\mathbb{R}}_+^3$, we arrive at

$$|c^*(x,h,z)| = u_x(x,h,z)^{\frac{1}{p-1}} = \left(\frac{1}{y^*(x,h,z)}\right)^{\frac{1}{1-p}}. \quad (8.88)$$

Using the relationship $-x = v_y(y^*(x,h,z), h, z)$, one has

$$x = -v_y(y^*(x,z), h, z) = \frac{1}{y^*(x,h,z)}(l_r + \psi_r)\left(-\ln\frac{y^*(x,h,z)}{\beta}, h, z\right)$$

$$\geq \frac{1}{y^*(x,h,z)} l_r \left(-\ln \frac{y^*(x,h,z)}{\beta}, z \right). \tag{8.89}$$

As a result, Lemma 8.2 yields that, for all $r \geq 1$,

$$l_r(r,z) = \beta^{-\frac{p}{1-p}} \frac{pC_1}{1-p} e^{\frac{pr}{1-p}} - C_2 \beta e^{-r} + z\beta(e^{-\ell r} - e^{-r})$$

$$\geq (\beta e^{-r})^{-\frac{p}{1-p}} \left(\frac{pC_1}{1-p} - C_2 \beta^{\frac{1}{1-p}} e^{-\frac{r}{1-p}} \right)$$

$$= (\beta e^{-r})^{-\frac{p}{1-p}} \frac{2(1-p)}{2\rho(1-p) - \alpha^2 p} \left(1 - e^{-\frac{1}{1-p} r} \right)$$

$$\geq (\beta e^{-r})^{-\frac{p}{1-p}} \frac{2(1-p)}{2\rho(1-p) - \alpha^2 p} \left(1 - e^{-\frac{1}{1-p}} \right).$$

This yields that, for the case where $y^*(x,h,z) \leq \beta e^{-1}$,

$$l_r \left(-\ln \frac{y^*(x,h,z)}{\beta}, z \right)$$

$$\geq \frac{2(1-p)}{2\rho(1-p) - \alpha^2 p} \left(1 - e^{-\frac{1}{1-p}} \right) \left(\frac{1}{y^*(x,h,z)} \right)^{\frac{p}{1-p}}. \tag{8.90}$$

For the case with $y^* > \beta e^{-1}$, we have $0 < y^*(x,h,z)^{-\frac{1}{1-p}} \leq \beta^{-\frac{1}{1-p}} e^{\frac{1}{1-p}}$ for all $(x,h,z) \in \overline{\mathbb{R}}_+^3$. Therefore $(x,h,z) \to y^*(x,h,z)^{-\frac{1}{1-p}}$ is bounded. Thus, it follows from (8.89) and (8.90) that, for all $(x,h,z) \in \overline{\mathbb{R}}_+^3$,

$$0 < \left(\frac{1}{y^*(x,h,z)} \right)^{\frac{1}{1-p}} \leq C_q(1+x), \tag{8.91}$$

where the positive constant C_q is specified as

$$C_q := \beta^{-\frac{1}{1-p}} e^{\frac{1}{1-p}} + \frac{(2\rho(1-p) - \alpha^2 p)}{2(1-p)} \left(1 - e^{-\frac{1}{1-p}} \right)^{-1}. \tag{8.92}$$

We can then conclude from (8.88) and (8.91) that $|c^*(x,z)| \leq C_q(1+x)$ for $(x,h,z) \in \overline{\mathbb{R}}_+^3$. Hence, it follows from (8.67) that, for any $T \geq 0$, the SDE (8.64) satisfied by X^* admits a weak solution on $[0,T]$ (Laukajtys and Słomiński (2013)), which gives that $(\theta^*, c^*) \in \mathbb{U}^r$.

On the other hand, fix $(T,x,h,z) \in \overline{\mathbb{R}}_+^4$ and $(\theta, c) = (\theta_t, c_t)_{t \geq 0} \in \mathbb{U}^r$. By applying Itô's formula to $e^{-\rho T} u(X_T, I_T, Z_T)$, we arrive at

$$e^{-\rho T} u(X_T, I_T, Z_T) + \int_0^T e^{-\rho s} \frac{(c_s)^p}{p} ds = u(x,h,z)$$

$$+ \int_0^T e^{-\rho s} u_x(X_s, I_s, Z_s) \theta_s^\top \sigma dW_s + \int_0^T e^{-\rho s} \sigma_B u_z(X_s, I_s, Z_s) dW_s^\gamma \tag{8.93}$$

$$+ \int_0^T e^{-\rho s} \sigma_Z Z_s u_z(X_s, I_s, Z_s) dW_s^\eta + \int_0^T e^{-\rho s}(u_z - u_x)(X_s, I_s, Z_s) dm_s$$

$$+ \int_0^T e^{-\rho s} u_x(X_s, I_s, Z_s) dL_s^X + \int_0^T e^{-\rho s}(\mathcal{L}^{\theta_s, c_s} u - \rho u)(X_s, I_s, Z_s) ds,$$

where the operator $\mathcal{L}^{\theta,c}$ with $(\theta, c) \in \mathbb{R}^d \times \mathbb{R}_+$ is defined on $C^2(\overline{\mathbb{R}}_+^2)$ that

$$\mathcal{L}^{\theta,c} g := \theta^\top \mu g_x + \frac{1}{2} \theta^\top \sigma \sigma^\top \theta g_{xx} + \sigma_Z \theta^\top \sigma \eta z (g_{xx} - g_{xz}) - \sigma_B \theta^\top \sigma \gamma g_{xh} + \frac{c^p}{p}$$

$$- c g_x + \frac{1}{2} \sigma_B^2 g_{hh} - \mu_B g_h + \frac{1}{2} \sigma_Z^2 z^2 (g_{zz} + g_{xx} - 2 g_{xz})$$

$$+ \mu_Z z (g_z - g_x) + \sigma_Z \sigma_B z \eta^\top \gamma (g_{xh} - g_{hz}), \quad \forall g \in C^2(\overline{\mathbb{R}}_+^2).$$

Taking the expectation on both sides of the equality (8.93), we deduce from the Neumann boundary condition $u_x(0, h, z) = \beta$ and $u_h(x, 0, z) = u_x(x, 0, z)$ that

$$\mathbb{E}\left[\int_0^T e^{-\rho s} \frac{(c_s)^p}{p} ds - \beta \int_0^T e^{-\rho s} dL_s^X\right]$$

$$= u(x, h, z) - \mathbb{E}\left[e^{-\rho T} u(X_T, I_T, Z_T)\right]$$

$$+ \mathbb{E}\left[\int_0^T e^{-\rho s}(\mathcal{L}^{\theta_s, c_s} u - \rho u)(X_s, I_s, Z_s) ds\right]$$

$$\leq u(x, h, z) - \mathbb{E}\left[e^{-\rho T} u(X_T, I_T, Z_T)\right], \quad (8.94)$$

where the last inequality in (8.94) holds true due to $(\mathcal{L}^{\theta,c} u - \rho u)(x, h, z) \leq 0$ for all $(x, h, z) \in \overline{\mathbb{R}}_+^3$ and $(\theta, c) \in \mathbb{R}^d \times \mathbb{R}_+$.

We next verify the validity of the so-called transversality conditions:

$$\limsup_{T \to \infty} \mathbb{E}\left[e^{-\rho T} u(X_T, I_T, Z_T)\right] \geq 0, \quad (8.95)$$

$$\lim_{T \to \infty} \mathbb{E}\left[e^{-\rho T} u(X_T^*, I_T, Z_T)\right] = 0. \quad (8.96)$$

In view of Lemma 8.2 and Proposition 8.1, it follows that $x \to u(x, h, z)$, $h \to u(x, h, z)$ and $z \to u(x, h, z)$ are non-decreasing. Thus, we obtain

$$\limsup_{T \to \infty} \mathbb{E}\left[e^{-\rho T} u(X_T, I_T, Z_T)\right] \geq \limsup_{T \to \infty} \mathbb{E}\left[e^{-\rho T} u(0, 0, 0)\right] = 0. \quad (8.97)$$

Using Lemma 8.2 and Proposition 8.1 again, it holds that $|u_x(x, h, z)| \leq \beta$, $|u_h(x, h, z)| \leq \beta$ and $|u_z(x, h, z)| \leq \beta/\ell$ for all $(x, h, z) \in \overline{\mathbb{R}}_+^3$. Thus, we can see that, for all $(x, z, h) \in \overline{\mathbb{R}}_+^3$,

$$|u(x, h, z)| \leq |u(x, h, z) - u(x, h, 0)| + |u(x, h, 0) - u(x, 0, 0)|$$
$$+ |u(x, 0, 0) - u(0, 0, 0)| + |u(0, 0, 0)| \quad (8.98)$$
$$\leq \beta(x + h) + \frac{\beta}{\ell} z + |u(0, 0, 0)|.$$

By applying Itô's formula to $|I_t|^2$ and $|X_t|^2$, it follows from (8.67) and the Gronwall's lemma that, for all $t \geq 0$,

$$\mathbb{E}[|I_t|^2] \leq h^2 + (\sigma_B^2 + \mu_B^2)te^t, \tag{8.99}$$

$$\mathbb{E}\left[|X_t^*|^2\right] \leq x^2 + 2Kt\left[1 + Ke^{Kt} + z^2 e^{(2|\mu_Z| + 3\sigma_Z^2)t}\left(1 + \frac{Ke^{Kt}}{2|\mu_Z| + 3\sigma_Z^2}\right)\right]$$
$$\times (1 + x^2),$$

where the positive constant K is specified as

$$K := 4C_o|\mu| + 2C_o^2|\sigma\sigma^\top| + |\mu_Z| + \sigma_Z^2 + 4C_o|\sigma_Z\sigma\eta|. \tag{8.100}$$

Let us define the constant

$$\rho_0 := \frac{\alpha^2|p| + 1}{2(1-p)} + K + 2|\mu_Z| + 3\sigma_Z^2 + 1 \tag{8.101}$$

with K being given in (8.100). Then, using estimates (8.98) and (8.99), it follows that, for the discount rate $\rho > \rho_0$,

$$\lim_{T \to \infty} \mathbb{E}\left[e^{-\rho T}|u(X_T^*, I_T, Z_T)|\right] \leq \beta \lim_{T \to \infty} \mathbb{E}\left[e^{-\rho T}(X_T^* + I_T + Z_T)\right] = 0.$$

Finally, letting $T \to \infty$ in (8.94), we obtain from (8.98) and DCT that, for any $(\theta, c) \in \mathbb{U}^r$ and $(x, h, z) \in \overline{\mathbb{R}}_+^3$,

$$\mathbb{E}\left[\int_0^\infty e^{-\rho s}\frac{(c_s)^p}{p}ds - \beta \int_0^\infty e^{-\rho s}dL_s^X\right] \leq u(x, h, z),$$

where the equality holds when $(\theta, c) = (\theta^*, c^*)$. Thus, the proof of the verification theorem is complete. □

Remark 8.2. Note that, the state processes of the primal control problem (8.4) and the auxiliary control problem (8.11) satisfy the following relationship, for $t \geq 0$,

$$X_t = x + V_t^{\theta,c} - (m_t + Z_t - m_0 - z)$$
$$+ \sup_{s \in [0,t]} \left(-x - V_s^{\theta,c} + (m_s + Z_s - m_0 - z)\right)^+, \tag{8.102}$$

$$I_t = h + (m_t - m_0) - B_t. \tag{8.103}$$

As a consequence, we can obtain the auxiliary state process $(X, I, Z) = (X_t, I_t, Z_t)_{t \geq 0}$ by using the process $(V^{\theta,c}, B, m, Z) = (V_t^{\theta,c}, B_t, m_t, Z_t)_{t \geq 0}$. However, from (8.102) and (8.103), we can also see that different primal state processes $(V^{\theta,c}, B, m, Z)$ may correspond to the same auxiliary state process (X, I, Z). Theorem 8.2 gives the optimal feedback control (θ^*, c^*) in terms of (X, I, Z) but not by $(V^{\theta,c}, B, m, Z)$. This is an important reason why we introduce the auxiliary state process (X, I, Z) and study the auxiliary optimal control problem instead, which allows us to characterize the optimal control (θ^*, c^*) in the feedback form.

The following lemma shows that the expectation of the total optimal capital injection is always positive and finite.

Lemma 8.5. *Consider the optimal investment-consumption strategy $(\theta^*, c^*) = (\theta_t^*, c_t^*)_{t \geq 0}$ provided in Theorem 8.2. Then, we have*

(i) *The expectation of the discounted capital injection under the optimal strategy (θ^*, c^*) is finite. Namely, for $\rho > \rho_0$ with $\rho_0 > 0$ being given in Theorem 8.2,*

$$\mathbb{E}\left[\int_0^\infty e^{-\rho t} dA_t^*\right] < +\infty. \qquad (8.104)$$

(ii) *The expectation of the discounted capital injection under the optimal strategy (θ^*, c^*) is positive. Namely, for $\rho > \rho_0$ with $\rho_0 > 0$ being given in Theorem 8.2, it holds that*

$$\mathbb{E}\left[\int_0^\infty e^{-\rho t} dA_t^*\right] \geq z \frac{1-\kappa}{\kappa}\left(1+\frac{x}{z}\right)^{\frac{\kappa}{\kappa-1}} > 0. \qquad (8.105)$$

Here, the optimal capital injection under the optimal strategy (θ^*, c^*) is given by

$$A_t^* = 0 \vee \sup_{s \leq t}(M_s - V_s^{\theta^*, c^*}), \quad \forall t \geq 0. \qquad (8.106)$$

Proof. We first show item (i). For $(v, m, z, b) \in \overline{\mathbb{R}}_+^3 \times \mathbb{R}$, we have from (8.4) that

$$\beta \mathbb{E}\left[\int_0^\infty e^{-\rho t} dA_t^*\right] = \mathbb{E}\left[\int_0^\infty e^{-\rho t} \frac{(c_t^*)^p}{p} dt\right] - w(v, m, z, b). \qquad (8.107)$$

Thus, to prove (8.104), it suffices to show that

$$\mathbb{E}\left[\int_0^\infty e^{-\rho t} \frac{(c_t^*)^p}{p} dt\right] < +\infty. \qquad (8.108)$$

By using (8.107), $\mathbb{E}[\int_0^\infty e^{-\rho t}(c_t^*)^p/p\, dt]$ can not be $-\infty$ because $w(v, m, z, b)$ is finite and $\mathbb{E}[\int_0^\infty e^{-\rho t} dA_t^*]$ is nonnegative. The estimate (8.108) obviously holds for the case $p < 0$ as $\mathbb{E}[\int_0^\infty e^{-\rho t} \frac{(c_t^*)^p}{p} dt]$ is negative in this case. Therefore, we only focus on the case with $p \in (0, 1)$. For $p \in (0, 1)$, it follows from (8.67) and (8.99) that

$$\mathbb{E}\left[\int_0^\infty e^{-\rho t}\frac{(c_t^*)^p}{p}dt\right] \leq \frac{1}{p}(C_q)^p \mathbb{E}\left[\int_0^\infty e^{-\rho t}(1+|X_t^*|)^p dt\right]$$

$$\leq \frac{(C_q)^p}{p}\int_0^\infty e^{-\rho t}(1+\mathbb{E}[|X_t^*|])dt \leq \frac{(C_q)^p}{p}\int_0^\infty e^{-\rho t}[1+(1+\mathbb{E}[|X_t^*|^2])]dt$$

$$\leq K(1+x^2) < +\infty,$$

where $x = (\mathrm{v} - m \vee b - z)^+$ via (8.12), $C_q > 0$ is the constant given by (8.92), and $K > 0$ is a constant depending on model parameters $(\mu, \sigma, \mu_Z, \sigma_Z, \mu_B, \sigma_B, p, \beta, \rho)$ only.

Next, we verify item (ii). For any admissible portfolio $\theta = (\theta_t)_{t \geq 0}$, we introduce, for all $t \geq 0$,

$$\tilde{V}_t^\theta = \mathrm{v} + \int_0^t \theta_s^\top \mu ds + \int_0^t \theta_s^\top \sigma dW_s, \quad \tilde{A}_t^\theta = 0 \vee \sup_{s \leq t}(Z_s - \tilde{V}_s^\theta). \quad (8.109)$$

Note that, $c_t^* > 0$ for all $t \geq 0$. Then, it follows from (8.2), (8.106) and (8.109) that $\tilde{V}_t^{\theta^*} \geq V_t^{\theta^*, c^*}$ for all $t \geq 0$, and hence

$$\mathbb{E}\left[\int_0^\infty e^{-\rho t} dA_t^*\right] > \mathbb{E}\left[\int_0^\infty e^{-\rho t} d\tilde{A}_t^{\theta^*}\right] \geq \inf_\theta \mathbb{E}\left[\int_0^\infty e^{-\rho t} d\tilde{A}_t^\theta\right] =: \tilde{w}(\mathrm{v}, z). \quad (8.110)$$

It is not difficult to verify that, for all $(\mathrm{v}, z) \in \overline{\mathbb{R}}_+ \times \mathbb{R}_+$,

$$\tilde{w}(\mathrm{v}, z) = z \frac{1-\ell}{\ell}\left(1 + \frac{(\mathrm{v}-z)^+}{z}\right)^{\frac{\ell}{\ell-1}} > 0, \quad (8.111)$$

where the constant $\ell \in (0,1)$ is given by (8.29). Thus, we deduce from (8.110) and (8.111) that

$$\mathbb{E}\left[\int_0^\infty e^{-\rho t} dA_t^*\right] \geq z \frac{1-\ell}{\ell}\left(1 + \frac{(\mathrm{v}-z)^+}{z}\right)^{\frac{\kappa}{\kappa-1}} > 0, \quad (8.112)$$

which completes the proof of the lemma. □

Chapter 9

Reinforcement Learning for Optimal Tracking Portfolio in Incomplete Markets*

This chapter studies the reinforcement learning (RL) for infinite horizon optimal tracking portfolio problem proposed in Chapter 7 but in incomplete markets with unknown parameters. In particular, the benchmark process is modeled by a geometric Brownian motion with zero drift driven by some unhedgeable risk. The relaxed tracking formulation is adopted where the fund account compensated by the injected capital needs to outperform the benchmark process at any time, and the goal is to minimize the cost of the discounted total capital injection. When model parameters are known, we formulate the equivalent auxiliary control problem with reflected state dynamics, for which the classical solution of the HJB equation with Neumann boundary condition is obtained explicitly. When model parameters are unknown, we introduce the exploratory formulation for the auxiliary control problem with entropy regularization and develop the continuous-time q-learning algorithm in models of reflected diffusion processes. In some illustrative numerical example, we show the satisfactory performance of the q-learning algorithm.

For continuous-time stochastic control problems, Wang et al. (2020), Jia and Zhou (2022a,b, 2023) develop the theoretical foundation for reinforcement learning in the continuous time exploration formulation with continuous state space and action space. Wang et al. (2020) firstly study the exploratory learning formulation by incorporating the entropy regularization to encourage the policy exploration, and the optimal policy has been shown as a Gaussian type for LQ control problems. Jia and Zhou (2022a) examine the policy evaluation problem and proposed the martingale loss to facilitate the algorithm design. The policy gradient problem is then studied in Jia and Zhou (2022b) in which the martingale approach in Jia and

---

*This chapter is based on the work Bo et al. (2025).

Zhou (2022a) can be adopted for the policy gradient and some Actor-Critic algorithms can be devised to learn the value function and the stochastic policies alternatively. Later, Jia and Zhou (2023) establish a continuous time q-learning theory by considering the first-order approximation of the conventional discrete-time Q-function. We also refer to some recent subsequent studies in different contexts such as Wang et al. (2023) in applying the continuous-time Actor-Critic algorithm for solving the optimal execution problem in the Almgren-Chriss model; the application of policy improvement algorithm in Bai et al. (2023) for solving the optimal dividend problem in diffusion models; the generalization of continuous-time q-learning in Wei and Yu (2023) for mean-field control problems in McKean-Vlasov diffusion models. The occupation measure considered in Zhao et al. (2023) facilitates the derivation of performance-difference and local-approximation formulas for continuous-time policy gradient (PG) and trust region policy optimization (TRPO) algorithms.

This chapter aims to further extend the studies on the relaxed tracking portfolio using capital injection in Chapter 7 and Chapter 8 in incomplete market models in which there exists some unhedgeable risk driving the external benchmark process (see its definition in (9.4)). As a direct consequence, the previous methodology based on the dual transform and the probabilistic representations for the dual PDE therein fails because the dual HJB equation can no longer be linearized. We consider the benchmark process governed by a geometric Brownian motion with zero drift in the present paper. With knowledge of market parameters, we formulate an equivalent auxiliary stochastic control problem similar to Chapter 7, which results in a stochastic control problem with the underlying state process exhibiting reflections at the boundary zero. Leveraging the benchmark dynamics as a geometric Brownian motion (GBM) facilitates the dimension reduction of the control problem. We then derive the explicit classical solution to the associated HJB equation with a Neumann boundary condition. Furthermore, we consider the realistic situation when all market model parameters are unknown, for which we are interested in developing the continuous-time reinforcement learning approach. We are interested in generalizing the continuous-time q-learning from the standard diffusion models in Jia and Zhou (2023) to solve our auxiliary stochastic control problem with the reflection boundary at 0. In particular, for the stochastic control problems with reflections, it is shown in the present paper that the value function and the associated q-function can also be characterized by the martingale condition of some stochastic processes involving the reflection term or the local time at the barrier zero; see Proposition 9.2 and Theorem 9.2. In addition, some extra transversality conditions (see (9.38) and (9.32)) are also needed in our infinite horizon setting. Building upon our martingale characterization in Theorem 9.2, we can devise the offline q-learning algorithm for the targeted stochastic control problems with reflections using the stochastic approximation resulting from the martingale orthogonality condition. As a new theoretical contribution to the literature, in the context of the infinite horizon stochastic control problems with reflections, we establish the convergence of the stochastic approximation algorithms for time discretization and horizon truncation when $\Delta t \to 0$ and $T \to \infty$; see Theorem 9.3. To illustrate the efficiency of our q-learning algorithm, we provide a simulation example of our optimal tracking portfolio problem with a particular choice of the temperature parameter $\gamma = \rho/d$. Here, ρ stands for the subjective discount rate and d is the dimension of stocks, both are known to the agent. In this case, we present an example beyond the linear-quadratic (LQ) control framework such that the exploratory HJB equation with a Neumann boundary condition can be solved fully explicitly, allowing us to obtain the exact parameterization of the optimal value function and the optimal q-function. For some initial inputs and choices of learning rates, we illustrate the very satisfactory performance of the iteration convergence of learned parameters towards their true values.

The remainder of this chapter is organized as follows. Section 9.1 introduces the incomplete market model and the benchmark process as well as the portfolio optimization problem with the relaxed tracking. Section 9.2 delves into the classical control problem when the model is known. By adopting an auxiliary control formulation and conducting dimension reduction, the HJB equation is solved explicitly. In Section 9.3, when market parameters are unknown, the continuous-time reinforcement learning approach is proposed for the exploratory formulation of the control problem with entropy regularization. In particular, the q-learning algorithm is developed together with some convergence results on time discretization and horizon truncation. Section 9.4 provides some numerical examples by implementing the q-learning algorithm. In addition, we present the comparison between the q-learning algorithm and the maximum likelihood estimation (MLE) for some real market data to demonstrate the outperformance of our algorithm. Lastly, all proofs of main results in previous sections are collected in Section 9.5.

9.1 The Market Model

In this section, we describe the incomplete market model considered in the chapter. To this purpose, let $(\Omega, \mathcal{F}, \mathbb{F}^W, \mathbb{P}^W)$ be a filtered probability space with the filtration $\mathbb{F}^W = (\mathcal{F}_t^W)_{t \geq 0}$ satisfying the usual conditions, which supports a $d+1$-dimensional Brownian motion $(W^0, W^1, \ldots, W^d) = (W_t^0, W_t^1, \ldots, W_t^d)_{t \geq 0}$. As in (7.1) in Chapter 7, we consider a financial market consisting of d risky assets whose price dynamics are driven by the d-dimensional Brownian motion $(W^1, \ldots, W^d)$ satisfying

$$\frac{dS_t^i}{S_t^i} = \mu_i dt + \sum_{j=1}^d \sigma_{ij} dW_t^j, \quad i = 1, \ldots, d, \quad t \geq 0 \tag{9.1}$$

with the mean return rate $\mu_i \in \mathbb{R}$ and the volatility $\sigma_{ij} \in \mathbb{R}$. Hereafter, denote by $\mu = (\mu_1, \ldots, \mu_d)^\top$ the vector of return rate and by $\sigma = (\sigma_{ij})_{d \times d}$ the volatility matrix. Assume that the riskless interest rate $r = 0$ that amounts to the change of numéraire and the return rate $\mu \neq \mathbf{0}$. From this point onwards, all processes including the wealth process and the benchmark process are defined after the change of numéraire.

9.1.1 The Value of Portfolio

For $t \geq 0$, let θ_t^i be the amount of wealth (the resulting process $\theta^i = (\theta_t^i)_{t \geq 0}$ is assumed to be $\mathbb{F}^W$-adapted) that the fund manager allocates in asset $S^i = (S_t^i)_{t \geq 0}$ at time t. The self-financing wealth process under the portfolio control $\theta = (\theta_t^1, \ldots, \theta_t^d)_{t \geq 0}^\top$ is given by, for $t \geq 0$,

$$V_t^\theta = \mathrm{v} + \int_0^t \theta_s^\top \mu ds + \int_0^t \theta_s^\top \sigma dW_s, \quad V_0^\theta = \mathrm{v} \in \overline{\mathbb{R}}_+. \tag{9.2}$$

We consider the portfolio decision making by the fund manager whose goal is to optimally track a stochastic benchmark process $Z = (Z_t)_{t \geq 0}$. In the present chapter, the benchmark process satisfying the geometric Brownian motion (GBM) with zero drift as studied in Björk and Murgoci (2014) that may refer to futures price process of a commodity, a stock index, or an interest rate index. As in Björk and Murgoci (2014), the benchmark process $Z = (Z_t)_{t \geq 0}$ is assumed to satisfy the following exponential martingale form given by

$$\frac{dZ_t}{Z_t} = \sigma_Z dW_t^\kappa, \quad \forall t > 0, \tag{9.3}$$

with the initial value $Z_0 = z > 0$ and the volatility $\sigma_Z > 0$.

For the nonzero correlative coefficient $\kappa \in [-1,1]$, the process $W^\kappa = (W_t^\kappa)_{t\geq 0}$ appeared in (9.3) is a standard Brownian motion, which is defined by
$$W_t^\kappa := \kappa W_t^0 + \sqrt{1-\kappa^2} W_t^\eta, \quad t \geq 0. \tag{9.4}$$
Here, $W^\eta = (W_t^\eta)_{t\geq 0}$ is a linear combination of d-dimensional Brownian motion $W = (W^1, \ldots, W^d)$ with weights $\eta = (\eta_1, \ldots, \eta_d)^\top \in [-1,1]^d$. We also recall that $W^0 = (W_t^0)_{t\geq 0}$ is a Brownian motion independent of the d-dimensional Brownian motion W, which stands for the unhedgeable risk. As a consequence, the market model is *incomplete* which is different from the complete market model considered in Chapter 7.

9.1.2 Formulation of Relaxed Tracking Problem

For the benchmark process $Z = (Z_t)_{t\geq 0}$ described as (9.3) and (9.4), an optimal tracking portfolio problem is considered that combines the portfolio control with another singular control as capital injection together with dynamic floor constraints. To be precise, we assume that the fund manager can strategically inject capital $A = (A_t)_{t\geq 0}$ into the fund account from time to time whenever it is necessary such that the total capital dynamically dominates the benchmark process Z, i.e., $A_t + V_t^\theta \geq Z_t$ for all $t \geq 0$.

The goal of the optimal tracking problem is to minimize the expected cost of the discounted total capital injection under dynamic floor constraints that
$$\begin{cases} \mathrm{w}(\mathrm{v},z) := \inf_{A,\theta} \mathbb{E}\left[A_0 + \int_0^\infty e^{-\rho t} dA_t\right] \\ \text{s.t. } Z_t \leq A_t + V_t^\theta \text{ at each } t \geq 0, \end{cases} \tag{9.5}$$
where the constant $\rho > 0$ is the discount rate and $A_0 = (z - \mathrm{v})^+$ is the initial injected capital to match with the initial benchmark.

To cope with the problem (9.5) with dynamic floor constraints, we observe that, for a fixed control θ, the optimal A is the smallest adapted right-continuous and non-decreasing process that dominates $Z - V^\theta$. Let $\mathbb{U}$ be the set of regular $\mathbb{F}$-adapted control processes $\theta = (\theta_t)_{t\geq 0}$ such that (9.2) is well-defined. Similar to Chapter 7 or Bo et al. (2021), for each fixed regular control θ, the optimal singular control A_t^* satisfies that $A_t^* = 0 \vee \sup_{s\in[0,t]}(Z_s - V_s^\theta)$, $\forall t \geq 0$. As a result, the problem (9.5) with floor constraints admits the equivalent formulation as a unconstrained control problem but with a running maximum cost that
$$\mathrm{w}(\mathrm{v},z) = (z-\mathrm{v})^+ + \inf_{\theta \in \mathbb{U}} \mathbb{E}\left[\int_0^\infty e^{-\rho t} d\left(0 \vee \sup_{s\in[0,t]}(Z_s - V_s^\theta)\right)\right]. \tag{9.6}$$

In next section, we will focus on the solvability of problem (9.6) in the sense of strong control by introducing an auxiliary control problem.

9.2 The Equivalent Control Problem

To formulate the auxiliary control problem, we impose a new controlled state process to replace the process $V^\theta = (V_t^\theta)_{t\geq 0}$ given in (9.2). To do it, we first define the following difference process by $D_t := Z_t - V_t^\theta + \mathrm{v} - z$, $\forall t \geq 0$. It is obvious that $D_0 = 0$. Moreover, for any $x \geq 0$, let us consider the running maximum process of $D = (D_t)_{t\geq 0}$ defined by

$$L_t := x \vee \sup_{s\in[0,t]} D_s \geq 0, \quad \forall t > 0, \tag{9.7}$$

with the initial value $L_0 = x \geq 0$. One can easily see that $(z-\mathrm{v})^+ - \mathrm{w}(z,\mathrm{v})$ with the value function $\mathrm{w}(\mathrm{v},z)$ given in (9.6) is equivalent to the auxiliary control problem:

$$\sup_{\theta\in\mathbb{U}} \mathbb{E}\left[-\int_0^\infty e^{-\rho t} dL_t\right], \tag{9.8}$$

when we set the initial level $L_0 = x = (\mathrm{v}-z)^+$.

We start with the introduction of a new controlled state process $X = (X_t)_{t\geq 0}$ for problem (9.8), which is defined as the reflected process $X_t := L_t - D_t$ for $t \geq 0$ that satisfies the SDE:

$$X_t = -\int_0^t \sigma_Z Z_s dW_s^\kappa + \int_0^t \theta_s^\top \mu ds + \int_0^t \theta_s^\top \sigma dW_s + L_t, \quad \forall t \geq 0 \tag{9.9}$$

with the initial value $X_0 = x \geq 0$. In particular, the running maximum process L_t increases if and only if $X_t = 0$, i.e., $L_t = D_t$. We will change the notation from L_t to L_t^X from this point onwards to emphasize its dependence on the new state process X given in (9.9). The benchmark process $Z = (Z_t)_{t\geq 0}$ defined in (9.3) is chosen as another state process.

Let $\mathbb{U}^r$ be the set of admissible portfolios (controls) such that the reflected SDE (9.9) has a unique strong solution. Then, we propose the following stochastic control problem, for $(x,z) \in \overline{\mathbb{R}}_+ \times \mathbb{R}_+$,

$$\hat{\mathrm{w}}(x,z) := \sup_{\theta\in\mathbb{U}^r} J(\theta;x,z) := \sup_{\theta\in\mathbb{U}^r} \mathbb{E}_{x,z}\left[-\int_0^\infty e^{-\rho t} dL_t^X\right], \tag{9.10}$$

where $\mathbb{E}_{x,z}[\cdot] := \mathbb{E}[\cdot|X_0 = x, Z_0 = z]$. Using the definition (9.10) of $\hat{\mathrm{w}}(x,z)$, it is not difficult to check that (i) the value function $x \to \hat{\mathrm{w}}(x,z)$ is non-decreasing; (ii) it holds that $\sup_{x_1,x_2\geq 0} \left|\frac{\hat{\mathrm{w}}(x_1,z)-\hat{\mathrm{w}}(x_2,z)}{x_1-x_2}\right| \leq 1$, that is, the value function $x \to \hat{w}(x,z)$ is Lipschitz continuous with Lipschitz coefficient being 1. These properties will be applied in the remaining sections.

We next apply the change of measure to simplify the stochastic control problem in (9.10) by reducing the dimension. To this end, let us introduce the normalized processes of triplet (X,θ,L^X) by the benchmark process $Z = (Z_t)_{t\geq 0}$ defined by

$$Y_t := \frac{X_t}{Z_t}, \quad \tilde{\theta}_t := \frac{\theta_t}{Z_t}, \quad L_t^Y := \int_0^t \frac{dL_s^X}{Z_s}, \quad \forall t \geq 0. \tag{9.11}$$

Note that the process Z is strictly positive in light of (9.3). Then, $t \to L_t^Y$ is a non-decreasing process and satisfies that, a.s.,

$$\int_0^t \mathbf{1}_{Y_s=0} dL_s^Y = \int_0^t \mathbf{1}_{X_s=0} \frac{dL_s^X}{Z_s} = \int_0^t \frac{dL_s^X}{Z_s} = L_t^Y, \quad \forall t \geq 0.$$

This yields that $L^Y = (L_t^Y)_{t\geq 0}$ is the local time process for the process $Y = (Y_t)_{t\geq 0}$ at the reflecting boundary 0. Then, the Itô's rule yields that

$$dY_t = \sigma_Z^2(Y_t+1)dt - \sigma_Z(Y_t+1)dW_t^\kappa + \tilde{\theta}_t^\top \mu dt + \tilde{\theta}_t^\top \sigma dW_t + dL_t^Y. \tag{9.12}$$

In the sequel, we introduce the following change of measure specified by

$$\left.\frac{d\mathbb{Q}^W}{d\mathbb{P}^W}\right|_{\mathcal{F}_t} = Z_t, \quad \forall t \geq 0. \tag{9.13}$$

We then consider the following stochastic control problem formulated as:

$$u(y) := \sup_{\theta \in \tilde{\mathbb{U}}^r} \mathbb{E}^{\mathbb{Q}^W}\left[-\int_0^\infty e^{-\rho s} dL_s^Y\right], \quad \forall y \in \mathbb{R}_+, \tag{9.14}$$

where the normalized admissible set $\tilde{\mathbb{U}}^r := \{(\frac{\theta_t}{Z_t})_{t\geq 0};\ \theta \in \mathbb{U}^r\}$, and $\mathbb{E}^{\mathbb{Q}}[\cdot]$ is the expectation under the probability measure $\mathbb{Q}^W \sim \mathbb{P}^W$. It follows from (9.12) that the underlying state process Y has the following dynamics under $\mathbb{Q}^W$ given by

$$dY_t = -\sigma_Z(Y_t+1)d\tilde{W}_t^\kappa + \tilde{\theta}_t^\top \tilde{\mu} dt + \tilde{\theta}_t^\top \sigma d\tilde{W}_t + dL_t^Y, \tag{9.15}$$

where $\tilde{W}_t^\kappa := W_t^\kappa - \sigma_Z t$ and $\tilde{W}_t := (W_t^1 - \sqrt{1-\kappa^2}\sigma_Z\eta_1 t,\ldots,W^d - \sqrt{1-\kappa^2}\sigma_Z\eta_d t)$, for $t \geq 0$ are $\mathbb{Q}^W$-Brownian motions and $\tilde{\mu} = \mu + \sqrt{1-\kappa^2}\sigma_Z\sigma\eta$. For the case $\kappa = 1$, the process $\tilde{W}^\kappa = (\tilde{W}_t^\kappa)_{t\geq 0}$ is also independent of $W = (W^1,\ldots,W^d)^\top$.

The following lemma gives the relationship between the value function $u(y)$ for $y \geq 0$ defined by (9.14) and the value function $\hat{w}(x,z)$ for $(x,z) \in \mathbb{R}_+ \times \mathbb{R}_+$ defined by (9.10), whose proof is straightforward, and hence omitted.

Lemma 9.1. It holds that $\hat{w}(x,z) = zu\left(\frac{x}{z}\right)$ for all $(x,z) \in \overline{\mathbb{R}}_+ \times \mathbb{R}_+$.

Lemma 9.1 implies that we may solve the simplified stochastic control problem (9.14) with one-dimensional state process with reflection instead of problem (9.10). Then, the corresponding value function satisfies the following HJB equation given by, for $y > 0$,

$$\sup_{\tilde{\theta} \in \mathbb{R}^n} \left\{ u'(y)\tilde{\theta}^\top \tilde{\mu} + u''(y)\tilde{\theta}^\top \sigma\sigma^\top \tilde{\theta} - \sqrt{1-\kappa^2}\sigma_Z(y+1)\tilde{\theta}^\top \sigma\eta u''(y) \right\}$$
$$+ \frac{1}{2}\sigma_Z^2(y+1)^2 u''(y) = \rho u(y) \qquad (9.16)$$

with Neumann boundary condition $u'(0) = 1$. If the value function u is strictly concave, the optimal feedback portfolio is given by

$$\tilde{\theta}^*(y) = -(\sigma\sigma^\top)^{-1} \frac{u'(y)\tilde{\mu} - \sqrt{1-\kappa^2}\sigma_Z(y+1)\sigma\eta u''(y)}{u''(y)}, \quad \forall y \geq 0. \quad (9.17)$$

Plugging (9.17) into (9.16), we arrive at

$$-\alpha \frac{(u'(y))^2}{u''(y)} + \sqrt{1-\kappa^2}\zeta(y+1)u'(y)$$
$$+ \frac{1}{2}\sigma_Z^2 \kappa^2 (y+1)^2 u''(y) = \rho u(y), \qquad (9.18)$$

where the coefficients are defined by $\alpha := \frac{1}{2}\tilde{\mu}^\top(\sigma\sigma^\top)^{-1}\tilde{\mu}$ and $\zeta := \sigma_Z\eta^\top\sigma^{-1}\tilde{\mu}$. Consider the candidate solution of Eq. (9.18) given by

$$u(y) = \frac{\lambda-1}{\lambda}(1+y)^{\frac{\lambda}{\lambda-1}}, \quad \forall y \geq 0. \qquad (9.19)$$

Here, $\lambda \in (0,1)$ is the unique solution to the following equation:

$$\ell(\lambda) := \alpha\lambda(\lambda-1)^2 + \rho(\lambda-1)^2 - \sqrt{1-\kappa^2}\zeta\lambda(\lambda-1)$$
$$- \frac{1}{2}\kappa^2\sigma_Z^2\lambda = 0. \qquad (9.20)$$

Consequently, Eq. (9.20) has a unique root $\lambda \in (0,1)$. Based on the above results, we have the following verification result for the strong control problem, whose proof is standard and hence omitted.

Proposition 9.1. *The function u given by (9.19) is a classical solution of the HJB equation (9.16). Define the following optimal feedback control function by, for all $y \in \overline{\mathbb{R}}_+$,*

$$\tilde{\theta}^*(y) = (1-\lambda)(y+1)(\sigma\sigma^\top)^{-1}\mu + \sqrt{1-\kappa^2}\sigma_Z(y+1)(\sigma\sigma^\top)^{-1}\sigma\eta. \quad (9.21)$$

Consider the controlled state process $Y^* = (Y_t^*)_{t\geq 0}$ that obeys the following SDE, for all $t \geq 0$,

$$Y_t^* = -\int_0^t \sigma_Z(Y_s+1)d\tilde{W}_s^\kappa + \int_0^t \tilde{\theta}^*(Y_s^*)^\top \mu ds$$
$$+ \int_0^t \tilde{\theta}^*(Y_s^*)^\top \sigma dW_s + L_t^{Y^*} \qquad (9.22)$$

with $L_0^{Y^*} = y$. Let $\tilde{\theta}_t^* = \tilde{\theta}^*(Y_t^*)$ for all $t \geq 0$. Then, $\tilde{\theta}^* = (\tilde{\theta}_t^*)_{t\geq 0} \in \tilde{\mathbb{U}}^r$ is an optimal portfolio strategy.

9.3 The Continuous Time Reinforcement Learning (RL)

In this section, the model is assumed to be unknown to the fund manager, i.e., all model parameters $\mu = (\mu_1, \ldots, \mu_d)^\top$, $\sigma = (\sigma_{ij})_{d\times d}$, σ_Z, κ and $\eta = (\eta_1, \ldots, \eta_d)^\top \in [-1,1]^d$ are unknown. Thereby, the classical control approach in the previous section is not applicable. Our goal is to develop a continuous-time reinforcement learning algorithm to find the optimal tracking portfolio in problem (9.14).

Reinforcement learning allows the decision maker to learn the optimal action in the unknown environment through the interactions with the environment as the repeated trial-and-error procedure. Specifically, the agent exercises a sequel of actions $\tilde{\theta} = (\tilde{\theta}_t)_{t\geq 0}$ and observe responses from $(Y^{\tilde{\theta}}, L^Y) = (Y_t^{\tilde{\theta}}, L_t^Y)_{t\geq 0}$ along with a stream of discounted running rewards $(-e^{-\rho t}dL_t^Y)_{t\geq 0}$, and continuously update and improve his actions based on these observations.

To describe the exploration step in reinforcement learning, we randomize the action $\tilde{\theta}$ and consider its distribution. Assume that the probability space is rich enough to support the uniformly distributed random variable on $[0,1]$ that is independent of $(\tilde{W}^\kappa, W^1, \ldots, W^d)$ and can be used to generate other random variables with specified density functions. Let $K = (K_t)_{t\in[0,T]}$ be a process of mutually independent copies of a uniform random variable on $[0,1]$ which is also independent of the BMs $(W^0, W^1, \ldots, W^d)$, the construction of which requires a suitable extension of probability space (c.f. Sun (2006)). We then further expand the filtered probability space to $(\Omega, \mathcal{F}, \mathbb{F}, \mathbb{Q})$ where $\mathbb{F} = (\mathcal{F}_t^W \vee \sigma(K_s, 0 \leq s \leq t))_{t\geq 0}$ and the probability measure $\mathbb{Q}$, now defined on $\mathbb{F}$, is an extension from $\mathbb{Q}^W$ (i.e. the two probability measures coincide when restricted to $\mathbb{F}^W$). In particular, let $\boldsymbol{\pi} : y \in \overline{\mathbb{R}}_+ \to \boldsymbol{\pi}(\cdot \mid y) \in \mathcal{P}(\mathcal{A})$ be a given (feedback) policy with $\mathcal{A} := \mathbb{R}^d$, and $\mathcal{P}(\mathcal{A})$ is a suitable collection of probability density functions. At each time $t \geq 0$, an action $\tilde{\theta}_t^\pi$ is generated from the joint density $\boldsymbol{\pi}(\cdot \mid Y_t)$. Fix a policy $\boldsymbol{\pi}$ and an initial state $y \geq 0$, we can consider the following reflected SDE given by

$$dY_t^\pi = -\sigma_Z(Y_t^\pi + 1)d\tilde{W}_t^\kappa + (\tilde{\theta}_t^\pi)^\top \tilde{\mu} dt \\ + (\tilde{\theta}_t^\pi)^\top \sigma d\tilde{W}_t + dL_t^{Y^\pi}, \quad t > 0 \tag{9.23}$$

with $Y_0^\pi = y$. Here, $L^{Y^\pi} = (L_t^{Y^\pi})_{t\geq 0}$ denotes the local time of the state process $Y^\pi = (Y_t^\pi)_{t\geq 0}$ at the level 0.

To encourage exploration, we adopt the Shannon's entropy regularizer suggested in Wang et al. (2020) that leads to the following value function associated to a given policy π that

$$J(y;\pi) \qquad (9.24)$$
$$= \mathbb{E}^{\mathbb{Q}}\left[-\gamma \int_0^\infty e^{-\rho t}\ln\pi\left(\tilde{\theta}_t^\pi \mid Y_t^\pi\right)dt - \int_0^\infty e^{-\rho t}dL_t^{Y^\pi}\bigg|Y_0^\pi = y\right],$$

where $\mathbb{E}^{\mathbb{Q}}$ is the expectation with respect to both the Brownian motion and the action randomization. In the above, $\gamma > 0$ is a given parameter indicating the level of exploration, also known as the temperature parameter.

Similar to Wang et al. (2020), we can show that the average of the sample trajectories $(Y_t^\pi)_{t\geq 0}$ converges to $(\tilde{Y}_t^\pi)_{t\geq 0}$, which satisfies the following reflected SDE:

$$d\tilde{Y}_t^\pi = \tilde{b}(\tilde{Y}_t^\pi, \pi(\cdot \mid \tilde{Y}_t^\pi))dt + \tilde{\sigma}(\tilde{Y}_t^\pi, \pi(\cdot \mid \tilde{Y}_t^\pi))dB_t + dL_t^\pi, \qquad (9.25)$$

where $B = (B_t)_{t\geq 0}$ is a standard BM independent of $(\tilde{W}^\kappa, W)$, the coefficient functions $(\tilde{b}, \tilde{\sigma})$ are respectively defined by, for $(y, \pi) \in \overline{\mathbb{R}}_+ \times \mathcal{P}(\mathcal{A})$,

$$\begin{cases} \tilde{b}(y,\pi) := \int_{\mathcal{A}} \tilde{\theta}^\top \tilde{\mu}\pi(\tilde{\theta}\mid y)d\tilde{\theta}, \\ \tilde{\sigma}(y,\pi) := \sqrt{\int_{\mathcal{A}}\left(\frac{1}{2}\tilde{\theta}^\top\sigma\sigma^\top\tilde{\theta} - \sqrt{1-\kappa^2}\sigma_Z(y+1)\tilde{\theta}^\top\sigma\eta + \frac{\sigma_Z^2}{2}(y+1)^2\right)\pi(\tilde{\theta}\mid y)d\tilde{\theta}}, \end{cases}$$

where $L^\pi = (L_t^\pi)_{t\geq 0}$ denotes the local time process of the state process $\tilde{Y}^\pi = (\tilde{Y}_t^\pi)_{t\geq 0}$ at the level 0, namely, $L^\pi = (L_t^\pi)_{t\geq 0}$ is a continuous nondecreasing process satisfying $\int_0^t \mathbf{1}_{\tilde{Y}_s^\pi=0}dL_s^\pi = L_t^\pi$ for all $t \geq 0$. It follows from the property of Markovian projection in Brunick and Shreve (2013) that Y_t^π and $\tilde{Y}_t^\pi$ have the same distribution for any $t \geq 0$. As a consequence, the value function (9.24) is equivalent to the following relaxed stochastic control form (see also Kushner and Dupuis (1991) and Kushner (1998)) given by

$$J(y;\pi) = \mathbb{E}^{\mathbb{Q}^W}\bigg[-\gamma\int_0^\infty e^{-\rho t}\int_{\mathcal{A}}\ln\pi\left(\tilde{\theta}\mid\tilde{Y}_t^\pi\right)\pi\left(\tilde{\theta}\mid\tilde{Y}_t^\pi\right)d\tilde{\theta}dt$$
$$-\int_0^\infty e^{-\rho t}dL_t^\pi\bigg|\tilde{Y}_0^\pi = y\bigg]. \qquad (9.26)$$

Reinforcement Learning for Optimal Tracking Portfolio in Incomplete Markets 281

The task of reinforcement learning is to find the optimal policy to attain the maximum of the value function that
$$v(y) = \max_{\pi \in \Pi} J(y; \pi), \quad y \geq 0, \tag{9.27}$$
where the set Π stands for the set of admissible (random) policies. The following gives the precise definition of admissible policies.

Definition 9.1. A policy π is called admissible, that is, $\pi \in \Pi$, if

(i) π takes the feedback form as $\pi_t = \pi(\cdot | Y_t)$ for $t \geq 0$, where $\pi(\cdot|\cdot) : \mathcal{A} \times \overline{\mathbb{R}_+} \to \mathbb{R}$ is a measurable function and $\pi(\cdot|y) \in \mathcal{P}(\mathcal{A})$ for all $y \geq 0$;
(ii) the SDE (9.25) admits a unique strong solution for any initial $y \geq 0$.

Assume that the value function (9.26) under a given admissible policy $\pi \in \Pi$ is smooth enough, we can see that it satisfies the following Neumann problem, for $\pi \in \mathcal{P}(\mathcal{A})$,

$$\begin{cases} \int_{\mathcal{A}} \left[H(\tilde{\theta}, y, J_y(y; \pi), J_{yy}(y; \pi)) - \gamma \ln \pi(\tilde{\theta} \mid y) \right] \pi(\tilde{\theta} \mid y) d\tilde{\theta} \\ \quad + \frac{\sigma_Z^2}{2}(y+1)^2 J_{yy}(y; \pi) = \rho J(y; \pi), \\ J_y(0; \pi) = 1. \end{cases} \tag{9.28}$$

On the other hand, the value function defined by (9.27) satisfies the following exploratory HJB equation with a Neumann boundary condition that, for $y \geq 0$,

$$\begin{cases} \sup_{\pi \in \mathcal{P}(\mathcal{A})} \int_{\mathcal{A}} \left[H(\tilde{\theta}, y, v_y(y), v_{yy}(y)) - \gamma \ln \pi(\tilde{\theta}) \right] \pi(\tilde{\theta}) d\tilde{\theta} \\ \quad + \frac{\sigma_Z^2}{2}(y+1)^2 v_{yy}(y) = \rho v(y), \\ v_y(0) = 1, \end{cases} \tag{9.29}$$

where the Hamilton operator $H : \mathcal{A} \times \overline{\mathbb{R}_+} \times \mathbb{R}^2 \to \mathbb{R}$ is defined by

$$H(\tilde{\theta}, y, P, Q) := \tilde{\theta}^\top \tilde{\mu} P + \frac{1}{2} \tilde{\theta}^\top \sigma \sigma^\top \tilde{\theta} Q - \sqrt{1 - \kappa^2} \sigma_Z (y+1) \tilde{\theta}^\top \sigma \eta Q. \tag{9.30}$$

Moreover, by assuming $v_{yy} < 0$, the candidate optimal policy is a Gaussian measure after normalization that

$$\pi^*(\tilde{\theta} \mid y) = \frac{\exp\left\{ \frac{1}{\gamma} H(\tilde{\theta}, y, v_y(y), v_{yy}(y)) \right\}}{\int_{\mathcal{A}} \exp\left\{ \frac{1}{\gamma} H(\tilde{\theta}, y, v_y(y), v_{yy}(y)) \right\} d\tilde{\theta}}. \tag{9.31}$$

Next, we first establish the following so-called policy improvement theorem.

Theorem 9.1 (Policy Improvement Theorem). *For given $\pi \in \Pi$, assume that the value function $J(\cdot; \pi) \in C^2(\mathbb{R}_+)$ satisfies Eq. (9.28) with the Neumann boundary condition and $J_{yy}(\cdot; \pi) < 0$ for any $y \geq 0$. We consider the following mapping $\mathcal{I}$ on Π given by*

$$\mathcal{I}(\pi) := \frac{\exp\left\{\frac{1}{\gamma} H(\cdot, y, J_y(y;\pi), J_{yy}(y;\pi))\right\}}{\int_{\mathcal{A}} \exp\left\{\frac{1}{\gamma} H(\tilde{\theta}, y, J_y(y;\pi), J_{yy}(y;\pi))\right\} d\tilde{\theta}}$$

$$= \mathcal{N}\left(\tilde{\theta} \,\bigg|\, -(\sigma\sigma^\top)^{-1} \frac{\mu J_y(y;\pi) - \sqrt{1-\kappa^2}\sigma_Z(y+1)\sigma\eta J_{yy}(y;\pi)}{J_{yy}(y;\pi)}, \right.$$

$$\left. -(\sigma\sigma^\top)^{-1} \frac{\gamma}{J_{yy}(y;\pi)}\right),$$

where denote by $\mathcal{N}(\tilde{\theta}|a, b)$ the Gaussian density function with mean vector $a \in \mathbb{R}^d$ and covariance matrix $b \in \mathbb{R}^{d \times d}$. Then, we have

(i) *Denote by $\pi' = \mathcal{I}(\pi)$. If $\pi' \in \Pi$ satisfies*

$$\lim_{T\to\infty} \mathbb{E}^{\mathbb{Q}^W}\left[e^{-\rho T} J(\tilde{Y}_T^{\pi'}; \pi')\right] = 0, \quad (9.32)$$

it holds that $J(y; \pi') \geq J(y; \pi)$ for all $y \geq 0$.

(ii) *If the map $\mathcal{I}$ has a fixed point $\pi^* \in \Pi$ satisfying*

$$\begin{cases} \lim_{T\to\infty} \mathbb{E}^{\mathbb{Q}^W}\left[e^{-\rho T} J(\tilde{Y}_T^{\pi^*}; \pi^*)\right] = 0, \\ \limsup_{T\to\infty} \mathbb{E}^{\mathbb{Q}^W}\left[e^{-\rho T} J(\tilde{Y}_T^{\pi}; \pi^*)\right] \geq 0, \ \forall \pi \in \Pi, \end{cases} \quad (9.33)$$

then π^* is the optimal policy that $v(y) = \max_{\pi \in \Pi} J(y; \pi) = J(y; \pi^*)$.

(iii) *In particular, if we choose the temperature parameter $\gamma = \frac{\rho}{d}$, the map $\mathcal{I}$ has an explicit fixed point $\pi^* \in \Pi$ given by*

$$\pi^*(\cdot|y) \quad (9.34)$$

$$= \mathcal{N}\left(y \,\bigg|\, (\sigma\sigma^\top)^{-1}(\tilde{\mu} + \sigma_z\sqrt{1-\kappa^2}\sigma\eta)(1+y), (\sigma\sigma^\top)^{-1}\gamma(1+y)^2\right).$$

Theorem 9.1-(i) provides a theoretical result for the policy improvement iteration; while Theorem 9.1-(ii) shows that the optimal policy is characterized by a fixed point of the iteration operator. In fact, Theorem 9.1-(i) holds true for general Hamilton operator H under some mild assumptions. However, it is usually difficult to verify the existence of fixed points and the transversality conditions (9.33) for the general case. To address this issue, we choose a specific temperature parameter $\gamma = \rho/d$ in Theorem 9.1-(iii), which significantly simplifies the finding of a fixed point of the mapping $\mathcal{I}$ by using the explicit expression (9.34). The proof of Theorem 9.1-(iii) also shows that if we start with a special Gaussian policy as $\pi_0(\cdot|y) \sim \mathcal{N}(\cdot\, |c_1(1+y), c_2(1+y)^2)$, then just after two steps of iterations, the value function can no longer be improved, i.e., the fixed point is attained. We will mainly focus on this special choice of $\gamma = \rho/d$ and verify the transversality conditions (9.33) in Section 9.4, which implies that the fixed point (9.34) is the optimal policy.

Note that the policy improvement iteration in Theorem 9.1 depends on the knowledge of the model parameters, which are not known in the reinforcement learning procedure. Thus, in order to design an implementable algorithms, we turn to generalize the q-leaning theory initially proposed in Jia and Zhou (2023) for our purpose.

9.3.1 q-Function and Martingale Characterization

The aim of this section is to derive the q-function of our optimal tracking problem in continuous time and provide martingale characterizations of the q-function. We first give the definition of the q-function as the counterpart of the Q-function in the continuous time framework (c.f. Jia and Zhou (2023)).

Definition 9.2 (q-function). The q-function of problem (9.23)–(9.24) associated with a given policy $\pi \in \Pi$ is defined as, for all $(y, \tilde{\theta}) \in \mathbb{R}_+ \times \mathcal{A}$,

$$q(y, \tilde{\theta}; \pi) := H\left(\tilde{\theta}, y, J_y(y; \pi), J_{yy}(y; \pi)\right) + \frac{\tilde{\sigma}_Z^2(y+1)^2}{2} J_{yy}(y; \pi)$$
$$- \rho J(y; \pi). \tag{9.35}$$

We also note that the difference between the q-function and the Hamiltonian is independent from θ, and this allows one to use the q-function for a policy improvement theorem. Thus, the policy improvement mapping $\mathcal{I}$ in Theorem 9.1 can be expressed in terms of the q-function by $\mathcal{I}(\pi) = \frac{\exp\{\frac{1}{\gamma} q(y, \cdot; \pi)\}}{\int_{\mathcal{A}} \exp\{\frac{1}{\gamma} q(y, \tilde{\theta}; \pi)\} d\tilde{\theta}}$. This implies that the policy improvement iteration in Theorem 9.1 can be conducted by learning the q-function.

The following proposition gives the martingale condition to characterize the q-function for a given policy π when the value function is given.

Proposition 9.2. *Let a policy $\pi \in \Pi$, its value function $J \in C^2(\overline{\mathbb{R}}_+)$ satisfying Eq. (9.28) and a continuous function $\hat{q} : \overline{\mathbb{R}}_+ \times \mathcal{A} \to \mathbb{R}$ be given. Then, $\hat{q}(y, \tilde{\theta}) = q(y, \tilde{\theta}; \pi)$ for all $(y, \tilde{\theta}) \in \overline{\mathbb{R}}_+ \times \mathcal{A}$ if and only if for all $y \in \overline{\mathbb{R}}_+$, the following process*

$$e^{-\rho t} J\left(Y_t^\pi; \pi\right) - \int_0^t e^{-\rho s} \hat{q}\left(Y_s^\pi, \tilde{\theta}_s^\pi\right) ds - \int_0^t e^{-\rho s} dL_s^\pi, \quad t \geq 0 \quad (9.36)$$

is an $(\mathbb{F}, \mathbb{Q})$-martingale, where $Y^\pi = (Y_t^\pi)_{t\geq 0}$ is the solution to Eq. (9.23) with $Y_0^\pi = y$.

In the following, we strengthen Proposition 9.2 and characterize the q-function and the value function associated with a given policy π simultaneously. This result is the crucial theoretical tool for designing the q-learning algorithm.

Theorem 9.2. *Let a policy $\pi \in \Pi$, a function $\hat{J} \in C^2(\overline{\mathbb{R}}_+)$ and a continuous function $\hat{q} : \overline{\mathbb{R}}_+ \times \mathcal{A} \to \mathbb{R}$ be given such that*

$$\lim_{T \to \infty} \mathbb{E}^{\mathbb{Q}}\left[e^{-\rho T} \hat{J}(Y_T^\pi)\right] = 0, \quad (9.37)$$

$$\int_{\mathcal{A}} [\hat{q}(y, \tilde{\theta}) - \gamma \ln \pi(\tilde{\theta}|y)] \pi(\tilde{\theta}|y) d\tilde{\theta} = 0, \quad \forall y \geq 0. \quad (9.38)$$

Then, $\hat{J}$ and $\hat{q}$ are respectively the value function satisfying Eq. (9.28) and the q-function associated with π if and only if for all $y \in \overline{\mathbb{R}}_+$, the following process

$$e^{-\rho t} \hat{J}(Y_t^\pi) - \int_0^t e^{-\rho s} \hat{q}\left(Y_s^\pi, \tilde{\theta}_s^\pi\right) ds - \int_0^t e^{-\rho s} dL_s^\pi, \quad t \geq 0$$

is an $(\mathbb{F}, \mathbb{Q})$-martingale, where $Y^\pi = (Y_t^\pi)_{t\geq 0}$ is the solution to Eq. (9.23) with $Y_0^\pi = y$. If it holds further that $\pi(\tilde{\theta}|y) = \frac{\exp\{\frac{1}{\gamma}\hat{q}(y,\tilde{\theta})\}}{\int_{\mathcal{A}} \exp\{\frac{1}{\gamma}\hat{q}(y,\tilde{\theta})\} d\tilde{\theta}}$ satisfying

$$\limsup_{T \to \infty} \mathbb{E}^{\mathbb{Q}}\left[e^{-\rho T} \hat{J}(Y_T^{\pi'})\right] \geq 0, \quad \forall \pi' \in \Pi, \quad (9.39)$$

then π is the optimal policy and $\hat{J}$ is the value function.

9.3.2 Continuous-time q-Learning Algorithm

In this subsection, we design q-learning algorithms to simultaneously learn and update the parameterized value function and the policy based on the martingale condition in Theorem 9.2.

Given a policy $\pi \in \Pi$, we parameterize the value function by a family of functions $J^\xi(\cdot)$, where $\xi \in \Theta \subset \mathbb{R}^{L_\xi}$ and L_ξ is the dimension of the parameter, and parameterize the q-function by a family of functions $q^\psi(\cdot, \cdot)$, where $\psi \in \Psi \subset \mathbb{R}^{L_\psi}$ and L_ψ is the dimension of the parameter. Moreover, we shall make the following assumptions for the parameterized family $\{J^\xi\}$ and $\{q^\psi\}$.

($\mathbf{A}_\xi$) The family of parameterized value functions $J^\xi(\cdot)$ is C^1 in ξ and satisfies $\lim_{t \to \infty} \mathbb{E}^{\mathbb{Q}}[e^{-\rho t} J^\xi(Y_t^\pi)] = 0$ and the Neumann condition $J_y^\xi(0) = 1$. Moreover, there exist continuous functions $G_J(\cdot)$ and $\tilde{J}(\cdot)$ such that $|J^\xi(y)| + |J_y^\xi(y)| + |J_{yy}^\xi(y)| \leq G_J(\xi)\tilde{J}(y)$ for all $(\xi, y) \in \Theta \times \mathbb{R}_+$.

($\mathbf{A}_\psi$) The family of parameterized q-functions $q^\psi(\cdot, \cdot)$ is C^1 in ψ and satisfies that

$$\int_{\mathcal{A}} [q^\psi(y, \tilde{\theta}) - \gamma \ln \pi(\tilde{\theta} \mid y)] \pi(\tilde{\theta} \mid y) d\tilde{\theta} = 0, \quad \forall y \in \mathbb{R}_+. \quad (9.40)$$

Furthermore, there exist continuous functions $G_q(\cdot)$ and $\tilde{q}(\cdot, \cdot)$ such that

$$|q^\psi(y, \tilde{\theta})| \leq G_q(\psi) \tilde{q}(y, \tilde{\theta}), \quad \forall (\psi, y, \tilde{\theta}) \in \Psi \times \mathbb{R}_+ \times \mathcal{A}. \quad (9.41)$$

Then, the learning task is to find the "optimal" (in some sense) parameters ξ and ψ. The key step in the algorithm design is to enforce the martingale condition stipulated in Theorem 9.2. More precisely, let $M = (M_t)_{t \geq 0}$ be the martingale given in Theorem 9.2, i.e.,

$$M_t = e^{-\rho t} J(Y_t^\pi) - \int_0^t e^{-\rho s} q\left(Y_s^\pi, \tilde{\theta}_s^\pi\right) ds - \int_0^t e^{-\rho s} dL_s^\pi, \quad (9.42)$$

and $M^{\xi,\psi} = (M_t^{\xi,\psi})_{t \geq 0}$ be its parameterized process defined by

$$M_t^{\xi,\psi} = e^{-\rho t} J^\xi(Y_t^\pi) - \int_0^t e^{-\rho s} q^\psi\left(Y_s^\pi, \tilde{\theta}_s^\pi\right) ds - \int_0^t e^{-\rho s} dL_s^\pi, \quad (9.43)$$

where $Y^\pi = (Y_t^\pi)_{t \geq 0}$ is the solution to (9.23) with $Y_0^\pi = y$.

It follows from the martingale orthogonality condition that, for any test adapted continuous process $\varsigma = (\varsigma_t)_{t \geq 0}$ with $\mathbb{E}^{\mathbb{Q}}[\int_0^\infty |\varsigma_t|^2 d\langle M \rangle_t] < \infty$,

$$\mathbb{E}^{\mathbb{Q}}\left[\int_0^\infty \varsigma_t dM_t\right] = 0. \quad (9.44)$$

In fact, the following result shows that this is a necessary and sufficient condition for the parameterized process $M^{\xi,\psi}$ to be a martingale. Its proof is omitted because it is similar to the one of Proposition 4 in Jia and Zhou (2022b).

Proposition 9.3. *The parameterized process $M^{\xi,\psi} = (M_t^{\xi,\psi})_{t\geq 0}$ given by (9.43) is a martingale if and only if*

$$\mathbb{E}^{\mathbb{Q}}\left[\int_0^\infty \varsigma_t dM_t^{\xi,\psi}\right] = 0, \qquad (9.45)$$

for any ς with $\mathbb{E}^{\mathbb{Q}}[\int_0^\infty |\varsigma_t|^2 d\langle M\rangle_t] < \infty$.

Proposition 9.3 tells that, to find the "optimal" parameters ξ and ψ, it is enough to explore the solution (ξ^*, ψ^*) of the martingale orthogonality equation (9.45). This can be implemented by using stochastic approximation to update parameters as, for $\chi \in \{\xi, \psi\}$,

$$\chi \leftarrow \chi + \alpha_\chi \int_0^\infty \varsigma_t dM_t^{\xi,\psi}, \qquad (9.46)$$

where $\alpha_\chi > 0$ is the learning rate. However, the martingale orthogonality equation (9.45) and the associated update rule (9.46) may not be directly applicable due to its requirement of having full trajectories of the infinite horizon. To overcome this difficulty, we first truncate the martingale orthogonality equation at a sufficiently long time T:

$$\mathbb{E}^{\mathbb{Q}}\left[\int_0^T \varsigma_t dM_t^{\xi,\psi}\right] = 0. \qquad (9.47)$$

Note that the truncated martingale orthogonality equation (9.47) involves integration in continuous time, which still cannot be directly applicable. Thus, we turn to consider truncated discrete-time martingale orthogonality equation. Let $K \in \mathbb{N}$ and $\Delta t = T/K$, consider the partition $0 = t_0 < t_1 < t_2 < \cdots < t_K = T$ with $t_k - t_{k-1} = \Delta t$ for $k = 1, \ldots, K$. In light of (9.47), we present the truncated discretized martingale orthogonality equation as

$$\mathbb{E}^{\mathbb{Q}}\left[\sum_{k=0}^{K-1} \varsigma_{t_k}\left(M_{t_{k+1}}^{\xi,\psi} - M_{t_k}^{\xi,\psi}\right)\right] = 0. \qquad (9.48)$$

The next theorem states the convergence of the stochastic approximation algorithms when $\Delta t \to 0$ and $T \to \infty$.

Theorem 9.3. *For a continuous test function $\varsigma = (\varsigma_t)_{t\geq 0}$ satisfying $\mathbb{E}^{\mathbb{Q}}[\int_0^\infty |\varsigma_t|^2 d\langle M\rangle_t] < \infty$, consider the martingale orthogonality condition equations (9.45), (9.47) and (9.48). Then, we have*

(i) *any convergent subsequence of the solution* $(\xi_{\mathrm{ML}}^*(T), \psi_{\mathrm{ML}}^*(T))$ *that solves the truncated martingale orthogonality equation* (9.47) *converges to the solution of the martingale orthogonality equation* (9.45), *that is*

$$\lim_{T \to \infty} (\xi_{\mathrm{ML}}^*(T), \psi_{\mathrm{ML}}^*(T)) = (\xi_{\mathrm{ML}}^*, \psi_{\mathrm{ML}}^*)$$

solves Eq. (9.45).

(ii) *any convergent subsequence of the solution* $(\xi_{\mathrm{ML}}^*(\Delta t; T), \psi_{\mathrm{ML}}^*(\Delta t; T))$ *that solves the truncated discretized martingale orthogonality equation* (9.48) *converges to the solution of the truncated martingale orthogonality equation* (9.47), *that is*

$$\lim_{\Delta t \to 0} (\xi_{\mathrm{ML}}^*(\Delta t; T), \psi_{\mathrm{ML}}^*(\Delta t; T)) = (\xi_{\mathrm{ML}}^*(T), \psi_{\mathrm{ML}}^*(T))$$

solves Eq. (9.47).

A direct result is that if the solution $(\xi_{\mathrm{ML}}^*(\Delta t; T), \psi_{\mathrm{ML}}^*(\Delta t; T))$ of the truncated discretized martingale orthogonality equation (9.48) converges, then it holds that

$$\lim_{T \to \infty} \lim_{\Delta t \to 0} (\xi_{\mathrm{ML}}^*(\Delta t; T), \psi_{\mathrm{ML}}^*(\Delta t; T)) = (\xi_{\mathrm{ML}}^*, \psi_{\mathrm{ML}}^*)$$

solves the martingale orthogonality equation (9.45). Therefore, it provides a theoretical foundation for implementing the truncation and discretization in the learning algorithm.

Next, we use the truncated discretized martingale orthogonality equation (9.48), Theorems 9.1, 9.2, 9.3 to design the q-learning algorithm. To learn the optimal value function and q-function, we choose the proper parameterized function approximators J^ξ and q^ψ for $\xi \in \Theta \subset \mathbb{R}^{L_\xi}, \psi \in \Psi \subset \mathbb{R}^{L_\psi}$, which satisfy the assumptions $(\mathbf{A}_\xi)$ and $(\mathbf{A}_\psi)$. Theorem 9.1 and the definition of q-function the parameterized form of the optimal policy as

$$\pi^\psi(\tilde{\theta} \mid y) = \frac{\exp\left\{\frac{1}{\gamma} q^\psi(y, \tilde{\theta})\right\}}{\int_{\mathcal{A}} \exp\left\{\frac{1}{\gamma} q^\psi(y, \tilde{\theta})\right\} d\tilde{\theta}}, \quad \forall (y, \tilde{\theta}) \in \overline{\mathbb{R}}_+ \times \mathcal{A}. \tag{9.49}$$

Moreover, it can be easily see that with the policy π^ψ given in (9.49), the consistency condition in the assumption $(\mathbf{A}_\psi)$ trivially holds that

$$\int_{\mathcal{A}} \left[q^\psi(y, \tilde{\theta}) - \gamma \ln \pi^\psi(\tilde{\theta} \mid y)\right] \pi^\psi(\tilde{\theta} \mid y) d\tilde{\theta} = 0. \tag{9.50}$$

Based on the truncated discretized martingale orthogonality equation (9.48), we can apply the stochastic approximation method Robbins and Monro (1951) to update the parameters in the following manner, which corresponds to the TD(0) algorithm (c.f. Sutton (1988) and Jia and Zhou (2022b)):

$$\xi \leftarrow \xi + \alpha_\xi \sum_{k=0}^{K-1} \iota_{t_k} G_k, \quad \psi \leftarrow \psi + \alpha_\psi \sum_{k=0}^{K-1} \varsigma_{t_k} G_k, \quad (9.51)$$

where $\alpha_\xi, \alpha_\psi > 0$ are the learning rates, ι, ς are test functions, and for $k = 0, 1, \ldots, K-1$, the quantity G_k is given by

$$G_k := e^{-\rho t_{k+1}} J^\xi \left(Y_{t_{k+1}}^{\pi^\psi}\right) - e^{-\rho t_k} J^\xi \left(Y_{t_k}^{\pi^\psi}\right) - e^{-\rho t_k} q^\psi \left(Y_{t_k}^{\pi^\psi}, \theta_{t_k}^{\pi^\psi}\right) \Delta t$$
$$- e^{-\rho t_k} \left(L_{t_{k+1}}^{\pi^\psi} - L_{t_k}^{\pi^\psi}\right).$$

Note that, by using the martingale orthogonality condition, we need at least $L_\xi + L_\psi$ equations to fully determine the parameters (ξ, ψ). Here, we choose the test functions in the conventional sense that

$$\iota_t = e^{-\rho t} \frac{\partial J^\xi}{\partial \xi} \left(Y_t^{\pi^\psi}\right), \quad \varsigma_t = e^{-\rho t} \frac{\partial q^\psi}{\partial \psi} \left(Y_s^{\pi^\psi}, \theta_s^{\pi^\psi}\right).$$

Based on the above updating rules, we then present the pseudo-code of the offline q-learning algorithm in Algorithm 1.

9.4 Numerical Examples

In this section, we consider our auxiliary control problem with reflections as an example beyond the LQ control framework to illustrate the proposed q-learning algorithm. In particular, we will first choose the temperature parameter $\gamma = \rho/d$, where we recall that ρ is the discount factor and d is number of risky assets. We also note that, in this case, Theorem 9.1-(iii) asserts that the map $\mathcal{I}$ has a fixed point π^* given by (9.34). However, it remains to verify the transversality condition (9.33) to check the optimality of π^*.

We next derive the explicit solution of the exploratory HJB equation (9.52) and shows that π^* is indeed the optimal policy. Plugging (9.31) into the exploratory HJB equation (9.29), we obtain

$$\frac{1}{2}\sigma_Z^2 \kappa^2 (y+1)^2 v_{yy}(y) + \sqrt{1-\kappa^2}\zeta(y+1)v_y(y) + \gamma \ln\left(\sqrt{-\frac{(2\pi\gamma)^d}{|\sigma\sigma^\top|(v_{yy}(y))^d}}\right)$$

Reinforcement Learning for Optimal Tracking Portfolio in Incomplete Markets   289

Algorithm 1 Offline q-Learning Algorithm
Input: Initial state y_0, horizon T, time step Δt, number of episodes N, number of mesh grids K, initial learning rates $\alpha_\xi(\cdot), \alpha_\psi(\cdot)$ (a function of the number of episodes), functional forms of parameterized value function $J^\xi(\cdot)$, q-function $q^\psi(\cdot)$, policy $\boldsymbol{\pi}^\psi(\cdot \mid \cdot)$ and temperature parameter γ.
Required Program: an environment simulator $(y', L') = \text{Environment}_{\Delta t}(t, y, L, \tilde{\theta})$ that takes current time-state pair (t, y, L) and action θ as inputs and generates state y' and L' at time $t + \Delta t$ as outputs.
Learning Procedure:
1: Initialize ξ, ψ and $i = 1$.
2: **while** $i < N$ **do**
3:     Initialize $j = 0$. Observe initial state y_0 and store $y_{t_0} \leftarrow y_0$.
4:     **while** $j < K$ **do**
5:         Generate action $\tilde{\theta}_{t_j} \sim \boldsymbol{\pi}^\psi\left(\cdot \mid y_{t_j}\right)$.
6:         Apply $\tilde{\theta}_{t_j}$ to environment simulator $(y, L) = \text{Environment}_{\Delta t}(t_j, y_{t_j}, L_{t_j}, \tilde{\theta}_{t_j})$.
7:         Observe new state y and L as output. Store $y_{t_{j+1}} \leftarrow y$, and $L_{t_{j+1}} \leftarrow L$.
8:     **end while**
9:     For every $k = 0, 1, \ldots, K-1$, compute $$G_k = J^\xi\left(y_{t_{k+1}}\right) - J^\xi\left(y_{t_k}\right) - q^\psi\left(y_{t_k}, \tilde{\theta}_{t_k}\right)\Delta t - \left(L_{t_{k+1}} - L_{t_k}\right) - \rho J^\xi\left(y_{t_k}\right)\Delta t.$$
10:    Update ξ and ψ by $$\xi \leftarrow \xi + \alpha_\xi(i) \sum_{k=0}^{K-1} e^{-\rho t_k} \frac{\partial J^\xi}{\partial \xi}\left(y_{t_k}\right) G_k,$$ $$\psi \leftarrow \psi + \alpha_\psi(i) \sum_{k=0}^{K-1} e^{-\rho t_k} \frac{\partial q^\psi}{\partial \psi}\left(y_{t_k}, \tilde{\theta}_{t_k}\right) G_k.$$
11:    Update $i \leftarrow i + 1$.
12: **end while**

$$-\alpha \frac{(v_y(y))^2}{v_{yy}(y)} = \rho v(y) \tag{9.52}$$

with the Neumann boundary condition $v_y(0) = 1$. We also recall that $\alpha = \frac{1}{2}\mu^\top(\sigma\sigma^\top)^{-1}\mu$ and $\zeta = \sigma_Z \eta^\top \sigma^{-1}\mu$. We conjecture that $v(y)$ for $y \geq 0$ satisfies the form of

$$v(y) = A\ln(1+y) + B, \tag{9.53}$$

for some constants A and B. By plugging the expression into (9.52), (9.53) and using the choice of $\gamma = \rho/d$, we obtain that

$$A = 1, \quad B = \frac{1}{\rho}\left(-\frac{1}{2}\sigma_Z^2\kappa^2 + \sqrt{1-\kappa^2}\zeta + \frac{\gamma}{2}\ln\left(\frac{(2\pi\gamma)^d}{|\sigma\sigma^T|}\right) + \alpha\right).$$

It then follows that

$$v(y) = \ln(1+y) \qquad (9.54)$$
$$+ \frac{1}{\rho}\left(-\frac{1}{2}\sigma_Z^2\kappa^2 + \sqrt{1-\kappa^2}\zeta + \frac{\gamma}{2}\ln\left(\frac{(2\pi\gamma)^d}{|\sigma\sigma^T|}\right) + \alpha\right)$$

is a classical solution of Eq. (9.52). For the general temperature parameter $\gamma \neq \rho/d$, the classical solution to the exploratory HJB equation (9.52), if it exists, is not in the simple form of (9.54) and actually does not admit an explicit expression, which shows the significant influence of the entropy regularizer in our stochastic control problem when it is not the LQ type control.

Back to the special choice of $\gamma = \rho/d$, as a result of (9.31) and (9.54), the candidate optimal policy is given by the following Gaussian policy

$$\boldsymbol{\pi}^*(\cdot|y) \qquad (9.55)$$
$$= \mathcal{N}\left(\tilde{\theta} \mid (\sigma\sigma^T)^{-1}(\mu + \sigma_z\sqrt{1-\kappa^2}\sigma\eta)(1+y), (\sigma\sigma^T)^{-1}\gamma(1+y)^2\right),$$

which is also the fixed point as shown in Theorem 9.1-(iii). It is notable that both the mean and the variance of the Gaussian policy $\boldsymbol{\pi}^*(\cdot|y)$ depend on the state variable y, and in particular, its variance is increasing in y. That is, for the larger state process $Y = (Y_t)_{t\geq 0}$ or the state process $X = (X_t)_{t\geq 0}$, the agent needs to implement the more random policies with the larger variance for the purpose of learning.

The next verification theorem shows that the classical solution of the exploratory HJB equation (9.29) coincides with the value function and provides the optimal policy.

Theorem 9.4 (Verification Theorem). *Consider the classical solution v on $\overline{\mathbb{R}}_+$ of the exploratory HJB equation (9.52) given by (9.54), the policy $\boldsymbol{\pi}^*$ given by (9.55), and the controlled state process $Y^* = (Y_t^*)_{t\geq 0}$ that obeys the following reflected SDE, for all $t \geq 0$,*

$$dY_t^* = (2\alpha + \sqrt{1-\kappa^2}\zeta)(1+Y_t^*)dt \qquad (9.56)$$
$$+ \sqrt{\alpha + \frac{d}{2}\gamma + \frac{1}{2}\kappa^2\sigma_Z^2 + \sqrt{1-\kappa^2}\zeta}(1+Y_t^*)dB_t + dL_t^*,$$

where $L^* = (L_t^*)_{t\geq 0}$ is the local time process for the process $Y^* = (Y_t^*)_{t\geq 0}$ at the reflecting boundary 0, namely, $L^* = (L_t^*)_{t\geq 0}$ is a continuous non-decreasing process satisfying $\int_0^t \mathbf{1}_{Y_s^*=0} dL_s^* = L_t^*$. Then, π^* is an optimal policy and $v(y)$ is the value function. That is, for all admissible $\pi \in \Pi$, we have $J(y;\pi) \leq v(y)$ for all $y \in \mathbb{R}_+$, where the equality holds when $\pi = \pi^*$.

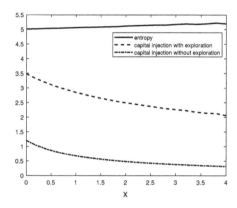

Fig. 9.1   Expected entropy and expected capital injection.

Remark 9.1. For the case where model parameters are known, we present in Figure 9.1 the numerical comparison between the expectation of the discounted capital injection with exploration and the expectation of the discounted capital injection without exploration, illustrating the influence by the additional entropy regularizer or the policy exploration. As shown in Figure 9.1, both expectations of the discounted capital injection with exploration and without exploration are decreasing with respect to the variable y, which can be explained by the fact that the larger value of y indicates the larger initial distance between the portfolio process V^θ and the benchmark process Z, thereby it will have smaller chances for the capital injection. More importantly, Figure 9.1 shows that if the fund manager employs the exploratory policy to learn the optimal policy in the unknown environment, the randomized actions will inevitably cause the underlying controlled processes to hit the reflection boundary 0 more often comparing with the case using the strict control, thereby leading to the higher expected capital injection.

Let us denote some parameters $(\xi^*, \psi_1^*, \psi_2^*, \psi_3^*)$ by

$$\begin{cases} \xi^* = \dfrac{1}{\rho}\left(-\dfrac{1}{2}\sigma_Z^2 \kappa^2 + \sqrt{1-\kappa^2}\zeta + \dfrac{\gamma}{2}\ln\left(\dfrac{(2\pi\gamma)^d}{|\sigma\sigma^\top|}\right) + \alpha\right), \\ \psi_1^* = \tilde{\mu} + \sigma_z\sqrt{1-\kappa^2}\sigma\eta, \quad \psi_2^* = \sigma, \\ \psi_3^* = \dfrac{1}{2}\sigma_Z^2(\kappa^2-1) - \alpha - \sqrt{1-\kappa^2}\zeta - \dfrac{\gamma}{2}\ln\left(\dfrac{(2\pi\gamma)^d}{|\sigma\sigma^\top|}\right). \end{cases} \quad (9.57)$$

Using Definition 9.2, the q-function can be expressed by, for all $(y, \tilde{\theta}) \in \overline{\mathbb{R}}_+ \times \mathcal{A}$,

$$q(y, \tilde{\theta}) = \dfrac{(\psi_1^*)^\top \tilde{\theta}}{1+y} - \dfrac{\tilde{\theta}^\top \psi_2^* (\psi_2^*)^\top \tilde{\theta}}{2(1+y)^2} - \rho\ln(1+y) + \psi_3^*, \quad (9.58)$$

and the value function $v(y)$ for $y \geq 0$ can be written by

$$v(y) = \ln(1+y) + \xi^*, \quad \forall y \geq 0. \quad (9.59)$$

Based on (9.58)–(9.59), for all $(y, \tilde{\theta}) \in \overline{\mathbb{R}}_+ \times \mathcal{A}$, we can parameterize the optimal value function and the optimal q-function in the exact form as:

$$\begin{cases} J^\xi(y) = \ln(1+y) + \xi, \\ q^\psi(y, \tilde{\theta}) = \dfrac{\psi_1^\top \tilde{\theta}}{1+y} - \dfrac{\tilde{\theta}^\top \psi_2 \psi_2^\top \tilde{\theta}^\top}{2(1+y)^2} - \rho\ln(1+y) + \psi_3 \end{cases} \quad (9.60)$$

with the parameters $(\xi, \psi_1, \psi_2) \in \mathbb{R} \times \mathbb{R}^d \times \mathbb{R}^{d \times d}$ to be learnt, the parameter ψ_3 satisfying

$$\psi_3 = -\dfrac{1}{2}\psi_1^\top (\psi_2 \psi_2^\top)^{-1}\psi_1 - \dfrac{\gamma}{2}\ln\left(\dfrac{(2\pi\gamma)^d}{|\psi_2 \psi_2^\top|}\right),$$

and the optimal policy can be parameterized by the distribution $\pi^\psi(\theta \mid y) = \dfrac{\exp\{\frac{1}{\gamma}q^\psi(y,\theta)\}}{\int_{\mathcal{A}}\exp\{\frac{1}{\gamma}q^\psi(y,\theta)\}d\theta}$. We can verify that the parameterized value function and q-function satisfy assumptions $(\mathbf{A}_\xi)$ and $(\mathbf{A}_\psi)$.

We consider the model with one risky asset (i.e., $d = 1$), and we set the coefficients of the simulator to be $\tilde{\mu} = 0.2$, $\sigma = 1$, $\sigma_Z = 0.2$ and $\kappa = 0.5$. Furthermore, we set $\gamma = \rho = 0.2$ and the truncated horizon $T = 12$, the time step $\Delta t = 0.005$ and the number of episodes $N = 2 \times 10^4$. The learning rates $(\alpha_\psi, \alpha_\xi)$ are chosen by

$$\alpha_\psi(i) = \begin{cases} \left(\dfrac{0.1}{i^{0.61}}, \dfrac{0.01}{i^{0.61}}\right), & 1 \leq i \leq 1 \cdot 10^4, \\ \left(\dfrac{0.05}{i^{0.81}}, \dfrac{0.005}{i^{0.81}}\right), & 1 \cdot 10^4 < i \leq 2 \cdot 10^4, \end{cases}$$

$$\alpha_\xi(i) = \begin{cases} \dfrac{0.015}{i^{0.61}}, & 1 \leq i \leq 1 \cdot 10^4, \\ \dfrac{0.005}{i^{0.81}}, & 1 \cdot 10^4 < i \leq 2 \cdot 10^4, \end{cases}$$

which decay along the iterations. Based on Algorithm 1, we plot in Figure 9.2 the numerical results on the convergence of iterations for parameters $(\psi_1, \psi_2, \psi_3, \xi)$ in the optimal value function and the optimal q-function. The learnt parameters for the optimal value function and the optimal q-function are summarized in Table 9.1.

	ψ_1	ψ_2	ψ_3	ξ
True value	0.3732	1.0000	-0.0050	0.3624
Learnt value	0.2520	1.2410	-0.0002	0.3326

For the general choice of temperature parameter $\gamma \neq \rho/d$, one needs to apply neural networks as the parameterized approximations of the optimal value function and the optimal q-function. In what follows, we implement our proposed q-learning algorithm together with neural networks. More precisely, we consider the parameterized value function $J^\xi(y)$ that satisfies $J_y^\xi(0) = 1$ together with the transversality condition and parameterized q-function given by

$$q^\psi(y, \tilde{\theta}) = -a^\psi(y)(\tilde{\theta} - b^\psi(y))^2 + c^\psi(y)$$

with $c^\psi(y) = -\frac{\gamma}{2} \ln(\frac{(2\pi\gamma)^d}{|b^\psi(y)|})$ for $(y, \tilde{\theta}) \in \mathbb{R}_+ \times \mathcal{A}$. Thus, we adopt three two-layer fully connected neural networks with Sigmoid activation functions to train the functions $J^\xi(y)$ and $(a^\psi(y), b^\psi(y))$. Consider the model with one risky asset (i.e., $d = 1$), and still set the coefficients of the simulator to be $\tilde{\mu} = 0.2$, $\sigma = 1$, $\sigma_Z = 0.2$, $\kappa = 0.5$ and $\rho = 0.2$. Here, we choose different temperature parameters $\gamma = 0.05, 0.1, 0.3$. The truncated horizon is $T = 12$ and the time step is $\Delta t = 0.1$. The learning rates are given by $\alpha_\psi(i) = \alpha_\xi(i) = \frac{0.001}{i^{0.61}}$. We plot in Figure 9.3 the learnt value function, the mean and variance functions of the learnt policy by using Algorithm 1 together with the neural networks after $N = 100$ iterations of policy update. We next also compare the performance of the proposed q-learning algorithm with the classical stochastic control approach using maximum likelihood estimation (MLE) method for model parameters. In particular, we implement two methods based on real market data. We choose the S&P 500 index as the benchmark process and select the Amazon.com Inc

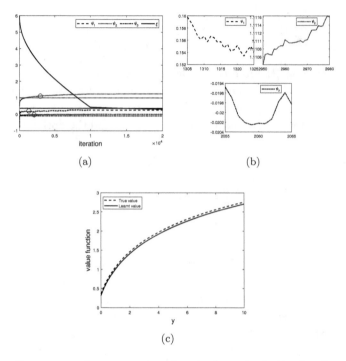

Fig. 9.2 Convergence of parameter iterations in the optimal value function and the optimal q-function using Algorithm 1. (a): paths of learnt parameters for the optimal value function and the optimal q-function vs optimal parameters shown in the dashed line; (b): "zoom-in" paths of learnt parameters for the hollow circles selected in panel (a) to illustrate the fluctuations of the local path; (c): the learnt value function vs the optimal value function.

(AMZN) as the risky asset. We set the discount factor $\rho = 0.1$, time step $\Delta t = 1$ (Day) and initial wealth v = 3281.1 (10 higher then initial value of the benchmark process). The daily returns from January 1, 2000 through July 30, 2020 are used as the training set while daily returns from July 31, 2020 through July 30, 2024 are used as the test set. For the MLE method, we first use the training set to obtain the estimated values of the parameters as $\hat{\mu} = 0.0012$, $\hat{\sigma} = 0.0324$ and $\hat{\sigma}_Z = 0.0126$. Suppose that the two Brownian motions W^κ and W^η are independent, i.e., $\kappa = 1$. We then compute the optimal portfolio strategy given by (9.21) on the test set. On the other hand, for the reinforcement learning approach, we choose the temperature parameter $\gamma = \rho/d = 0.1$, and train $N = 500$ times with

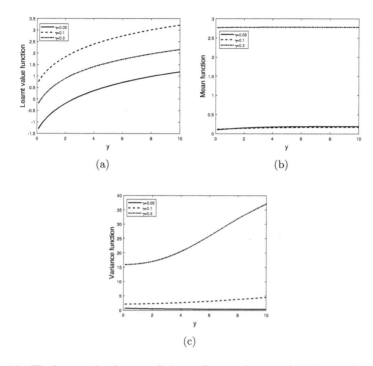

Fig. 9.3 The learnt value function (left panel), mean function (middle panel) and variance function (right panel) of the learnt policy by implementing Algorithm 1 together with the neural networks for $\gamma \neq \rho/d$.

learning rates given by

$$\alpha_\psi(i) = \left(\frac{0.005}{i^{0.71}}, \frac{0.005}{i^{0.71}}\right), \quad \alpha_\xi = \frac{0.005}{i^{0.71}}, \quad 1 \leq i \leq 500$$

on the training set. We then implement the q-learning algorithm on the test set with the learned value function and q-function. We plot the total wealth (including capital injection) process of the agent and the cumulative capital injection process in Figure 9.4. It shows that the wealth process using the q-learning algorithm outperforms the one using the MLE method. The total injected capital required by the agent using the reinforcement learning method is 2243.46, which is 5.86% lower than the total capital injection of 2383.21 using the MLE method. These plots can show the effectiveness and robustness of the q-learning algorithm for some real market data.

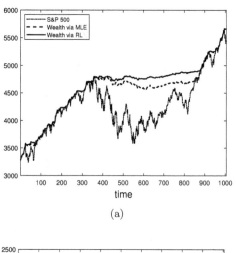

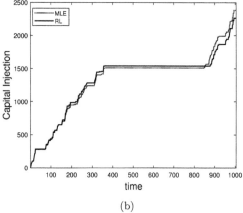

Fig. 9.4 (a): The price process of the S&P500, the total wealth (including capital injection) process under the MLE method and under the RL approach from July 31, 2020 through July 30, 2024; (b): The cumulative capital injection process under the MLE method and under the RL approach from July 31, 2020 through July 30, 2024.

9.5 Proofs of Main Results

This section collects all proofs of main results presented in previous sections.

Proof of Theorem 9.1. We first prove item (i). For $T > 0$ and $y \in \overline{\mathbb{R}}_+$, applying Itô's formula to the process $e^{-\rho t}J(\tilde{Y}_t^{\pi'};\pi)$ from 0 to T, we can obtain

$$e^{-\rho T} J(\tilde{Y}_T^{\pi'}; \boldsymbol{\pi}) - J(y; \boldsymbol{\pi}) - \int_0^T e^{-\rho t} dL_t^{\pi'}$$

$$- \gamma \int_0^T \int_{\mathcal{A}} e^{-\rho t} \ln\left(\boldsymbol{\pi}'(\tilde{\theta}; \tilde{Y}_t^{\pi'})\right) \boldsymbol{\pi}'(\tilde{\theta}; \tilde{Y}_t^{\pi'}) d\tilde{\theta} dt$$

$$= \int_0^T e^{-\rho t} \int_{\mathcal{A}} \left(\left[H\left(\tilde{\theta}, J_y(\tilde{Y}_t^{\pi'}; \boldsymbol{\pi}), J_{yy}(\tilde{Y}_t^{\pi'}; \boldsymbol{\pi})\right) - \rho J(\tilde{Y}_t^{\pi'}; \boldsymbol{\pi}) \right] \right.$$

$$\left. + \frac{\sigma_Z^2 (\tilde{Y}_t^{\pi'} + 1)^2}{2} J_{yy}(\tilde{Y}_t^{\pi'}; \boldsymbol{\pi}) - \gamma \ln \boldsymbol{\pi}'(\tilde{\theta}; \tilde{Y}_t^{\pi'}) \right) \boldsymbol{\pi}'(\tilde{\theta}; \tilde{Y}_t^{\pi'}) d\tilde{\theta} dt$$

$$+ \int_0^T e^{-\rho t} \left(J_y(\tilde{Y}_t^{\pi'}; \boldsymbol{\pi}) - 1 \right) dL_t^{\pi'}$$

$$+ \int_0^T e^{-\rho t} \tilde{\sigma}(\tilde{Y}_t^{\pi'}, \boldsymbol{\pi}'(\cdot \mid \tilde{Y}_t^{\pi'})) J_y(\tilde{Y}_t^{\pi'}; \boldsymbol{\pi}) dB_t. \quad (9.61)$$

As $J(\cdot; \boldsymbol{\pi}) \in C^2(\overline{\mathbb{R}}_+)$ satisfies the PDE (9.28), one has

$$\int_0^T e^{-\rho t} \left(J_y(\tilde{Y}_t^{\pi'}; \boldsymbol{\pi}) - 1 \right) dL_t^{\pi'}$$

$$= \int_0^T e^{-\rho t} (J_y(0; \boldsymbol{\pi}) - 1) dL_t^{\pi'} = 0. \quad (9.62)$$

Moreover, it follows from Lemma 2 in Jia and Zhou (2023) that, $\forall y \in \overline{\mathbb{R}}_+$,

$$\int_{\mathcal{A}} \left(\left[H\left(\tilde{\theta}, J_y(y; \boldsymbol{\pi}), J_{yy}(y; \boldsymbol{\pi})\right) - \rho J(y; \boldsymbol{\pi}) \right] + \frac{\sigma_Z^2 (y+1)^2}{2} J_{yy}(y; \boldsymbol{\pi}) \right.$$

$$\left. - \gamma \ln \boldsymbol{\pi}'(\tilde{\theta}; y) \right) \boldsymbol{\pi}'(\tilde{\theta}; y) d\tilde{\theta} \quad (9.63)$$

$$\geq \int_{\mathcal{A}} \left(\left[H\left(\tilde{\theta}, J_y(y; \boldsymbol{\pi}), J_{yy}(y; \boldsymbol{\pi})\right) - \rho J(y; \boldsymbol{\pi}) \right] + \frac{\sigma_Z^2 (y+1)^2}{2} J_{yy}(y; \boldsymbol{\pi}) \right.$$

$$\left. - \gamma \ln \boldsymbol{\pi}(\tilde{\theta}; y) \right) \boldsymbol{\pi}(\tilde{\theta}; y) d\tilde{\theta} = 0.$$

From (9.61), (9.62) and (9.63), we deduce that

$$e^{-\rho T} J(\tilde{Y}_T^{\pi'}; \boldsymbol{\pi}) - J(y; \boldsymbol{\pi}) - \int_0^T e^{-\rho t} dL_t^{\pi'}$$

$$- \gamma \int_0^T \int_{\mathcal{A}} e^{-\rho t} \ln\left(\boldsymbol{\pi}'(\tilde{\theta}; \tilde{Y}_t^{\pi'})\right) \boldsymbol{\pi}'(\tilde{\theta}; \tilde{Y}_t^{\pi'}) d\tilde{\theta} dt$$

$$\geq \int_0^T e^{-\rho t} \tilde{\sigma}(\tilde{Y}_t^{\pi'}, \pi'(\cdot \mid \tilde{Y}_t^{\pi'})) J_y(\tilde{Y}_t^{\pi'}; \pi) dB_t. \tag{9.64}$$

Taking the expectation on both side of (9.64) and letting $T \to \infty$, by (9.32) and MCT, we conclude that for any $y \in \overline{\mathbb{R}}_+$,

$$J(y; \pi) \leq -\gamma \lim_{T \to \infty} \mathbb{E}^{\mathbb{Q}^W} \left[\int_0^T \int_{\mathcal{A}} e^{-\rho t} \ln \left(\pi'(\tilde{\theta}; \tilde{Y}_t^{\pi'}) \right) \pi'(\tilde{\theta}; \tilde{Y}_t^{\pi'}) d\tilde{\theta} dt \right]$$

$$- \lim_{T \to \infty} \mathbb{E}^{\mathbb{Q}^W} \left[\int_0^T e^{-\rho t} dL_t^{\pi'} \right] + \lim_{T \to \infty} \mathbb{E}^{\mathbb{Q}^W} \left[e^{-\rho T} J(\tilde{Y}_T^{\pi'}; \pi) \right]$$

$$= \mathbb{E}^{\mathbb{Q}^W} \left[- \int_0^\infty e^{-\rho t} dL_t^{\pi'} - \gamma \int_0^\infty \int_{\mathcal{A}} e^{-\rho t} \ln \left(\pi'(\tilde{\theta}; \tilde{Y}_t^{\pi'}) \right) \pi'(\tilde{\theta}; \tilde{Y}_t^{\pi'}) d\tilde{\theta} dt \right]$$

$$= J(y; \pi').$$

Next, we handle item (ii). Denote by $J^*(y) = J(y; \pi^*)$. Thus, $J^*(y)$ satisfies Eq. (9.28), and Lemma 2 in Jia and Zhou (2023) shows that

$$\int_{\mathcal{A}} \left(\left[H\left(\tilde{\theta}, y, J_y^*(y), J_{yy}^*(y)\right) - \rho J^*(y) \right] + \frac{\sigma_Z^2 (y+1)^2}{2} J_{yy}^*(y) \right.$$

$$\left. - \gamma \ln \pi^*(\tilde{\theta}; y) \right) \pi^*(\tilde{\theta}; y) d\tilde{\theta}$$

$$= \sup_{\pi \in \mathcal{P}(\mathcal{A})} \int_{\mathcal{A}} \left(\left[H\left(\tilde{\theta}, y, J_y^*(y), J_{yy}^*(y)\right) - \rho J^*(y) \right] + \frac{\sigma_Z^2 (y+1)^2}{2} J_{yy}^*(y) \right.$$

$$\left. - \gamma \ln \pi(\tilde{\theta}; y) \right) \pi(\tilde{\theta}; y) d\tilde{\theta} = 0.$$

Fix $(T, y) \in \mathbb{R}_+ \times \overline{\mathbb{R}}_+$ and $\pi \in \Pi$. By applying Itô's rule to $e^{-\rho T} J^*(\tilde{Y}_T^{\pi^*})$, we arrive at

$$e^{-\rho T} J^*(\tilde{Y}_T^\pi)$$

$$= J^*(y) + \int_0^T e^{-\rho s} J_y^*(\tilde{Y}_s^\pi) dL_s^Y + \int_0^T e^{-\rho s} (\mathcal{L}^{\pi(\cdot \mid \tilde{Y}_s^\pi)} J^* - \rho J^*)(\tilde{Y}_s^\pi) ds$$

$$+ \int_0^T e^{-\rho s} J_y^*(\tilde{Y}_s^\pi) \tilde{\sigma}(\tilde{Y}_s^\pi, \pi(\cdot \mid \tilde{Y}_s^\pi)) dB_s, \tag{9.65}$$

where, for any $\pi \in \Pi$, the operator $\mathcal{L}^\pi$ acting on $C^2(\overline{\mathbb{R}}_+)$ is defined by

$$\mathcal{L}^\pi g(y) := \tilde{b}(y, \pi) g'(y) + \frac{1}{2} \tilde{\sigma}(y, \pi) g''(y), \quad \forall y \in \overline{\mathbb{R}}_+.$$

Taking the expectation on both sides of (9.65), we deduce from the Neumann boundary condition $J_y^*(0) = 1$ that

$$\mathbb{E}^{\mathbb{Q}^W}\left[-\int_0^T e^{-\rho s}dL_s^Y\right]$$

$$= J^*(y) - \mathbb{E}^{\mathbb{Q}^W}\left[e^{-\rho T}v(\tilde{Y}_T^\pi)\right] + \mathbb{E}^{\mathbb{Q}^W}\left[\int_0^T e^{-\rho s}(\mathcal{L}^{\pi(\cdot|\tilde{Y}_s^\pi)}v - \rho v)(\tilde{Y}_s)ds\right]$$

$$\leq J^*(y) - \mathbb{E}^{\mathbb{Q}^W}\left[e^{-\rho T}J^*\left(\tilde{Y}_T^\pi\right)\right]. \tag{9.66}$$

Here, the last inequality in (9.66) holds true due to $(\mathcal{L}^\pi v - \rho v)(y) \leq 0$ for all $y \in \mathbb{R}_+$ and $\pi \in \mathcal{P}(\mathcal{A})$. Toward this end, letting $T \to \infty$ in (9.66), we obtain from (9.33) and MCT that, for all $\pi \in \Pi$,

$$\mathbb{E}^{\mathbb{Q}^W}\left[-\int_0^\infty e^{-\rho s}dL_s^Y\right] \leq v(y), \quad \forall y \geq 0,$$

where the equality holds when $\pi = \pi^*$. This implies that $J^*(y) = v(y)$ for $y \geq 0$ is the optimal value function and π^* is the optimal policy.

Next, we move to item (iii). To find the fix point, we use the iteration method and start with a special Gaussian policy: $\pi_0(\cdot|y) \sim \mathcal{N}\left(\cdot\,|c_1(1+y), c_2(1+y)^2\right)$, with vector $c_1 \in \mathbb{R}^d$ and positive matrix $c_2 \in \mathbb{R}^{d\times d}$. Then, the resulting value function $J^{\pi_0}(y)$ satisfies that

$$\begin{cases} \int_{\mathcal{A}}\left[H(\theta, y, J_y^{\pi_0}(y), J_{yy}^{\pi_0}(y)) - \gamma \log \pi_0(\theta|y)\right]\pi_0(\theta|y)d\theta \\ \qquad + \dfrac{\sigma_Z^2}{2}(y+1)^2 J_{yy}^{\pi_0}(y) = \rho J^{\pi_0}(y), \\ J_y^{\pi_0}(y) = 1. \end{cases} \tag{9.67}$$

By using standard arguments, we can show that the classical solution of (9.67) is given by

$$J^{\pi_0}(y) = \ln(1+y) + C, \quad \forall y \in \overline{\mathbb{R}}_+,$$

where C is a constant defined by

$$C := \frac{1}{\rho}\left(\tilde{\mu}^\top c_1 + \sqrt{1-\kappa^2}\sigma_Z c_1^\top \sigma\eta - \frac{1}{2}c_1^\top \sigma\sigma^\top c_1 - \frac{1}{2}\mathrm{tr}(\sigma\sigma^\top c_2)\right.$$
$$\left. - \frac{1}{2}\sigma_Z^2 + \frac{1}{2}\gamma \ln\left((2\pi)^d|c_2|\right) + \frac{d}{2}\gamma\right),$$

and for a matrix $A \in \mathbb{R}^{d\times d}$, we denote $|A| := \det(A)$. Then, using once iteration, we get that

$$\pi_1(\cdot|y) = \mathcal{I}(\pi_0) \tag{9.68}$$

$$\sim \mathcal{N}\left(\cdot \left|(\sigma\sigma^\top)^{-1}(\tilde{\mu} + \sigma_z\sqrt{1-\kappa^2}\sigma\eta)(1+y), (\sigma\sigma^\top)^{-1}\gamma(1+y)^2\right.\right).$$

Again, we can calculate the corresponding reward function as

$$J^{\pi_1}(y) = \ln(1+y) \qquad (9.69)$$
$$+ \frac{1}{\rho}\left(-\frac{1}{2}\sigma_Z^2\kappa^2 + \sqrt{1-\kappa^2}\zeta + \frac{\gamma}{2}\ln\left(\frac{(2\pi\gamma)^d}{|\sigma\sigma^\top|}\right) + \alpha\right)$$

Then, the iteration is applicable again, which yields the improved policy π_2 as exactly the Gaussian policy π_1 given in (9.68), together with the reward function in (9.69). Then we find that π^* give in (9.34) is a fixed point of $\mathcal{I}$. Thus, we complete the proof. □

Proof of Proposition 9.2. For $T > 0$ and $y \in \mathbb{R}_+$, applying Itô's formula to the process $e^{-\rho s}J(Y_s^\pi; \pi)$ from 0 to t, we can obtain that

$$e^{-\rho t}J(Y_t^\pi; \pi) - J(y; \pi) - \int_0^t e^{-\rho s}\hat{q}\left(Y_s^\pi, \tilde{\theta}_s^\pi\right)ds - \int_0^t e^{-\rho s}dL_s^\pi$$
$$= \int_0^t e^{-\rho s}\left(H\left(\tilde{\theta}^\pi, Y_s^\pi, J_y(Y_s^\pi; \pi), J_{yy}(Y_s^\pi; \pi)\right) - \rho J(Y_s^\pi; \pi)\right.$$
$$+ \left.\frac{\sigma_Z^2(Y_s^\pi+1)^2}{2}J_{yy}(Y_s^\pi;\pi)\right)ds$$
$$+ \int_0^t e^{-\rho t}(J_y(Y_s^\pi;\pi) - 1)dL_s^\pi + \int_0^t e^{-\rho s}(\tilde{\theta}_s^\pi)^\top \sigma J_y(Y_s^\pi;\pi)d\tilde{W}_s$$
$$- \int_0^t e^{-\rho s}\sigma_Z(Y_s^\pi+1)J_y(Y_s^\pi;\pi)d\tilde{W}_s^\kappa$$
$$= \int_0^t e^{-\rho s}\left(q(Y_s^\pi,\tilde{\theta}_s^\pi;\pi) - \hat{q}\left(Y_s^\pi,\tilde{\theta}_s^\pi\right)\right)ds + \int_0^t e^{-\rho s}(\tilde{\theta}_s^\pi)^\top \sigma J_y(Y_s^\pi;\pi)d\tilde{W}_s$$
$$- \int_0^t e^{-\rho s}\sigma_Z(Y_s^\pi+1)J_y(Y_s^\pi;\pi)d\tilde{W}_s^\kappa. \qquad (9.70)$$

It follows from (9.70) that, if $\hat{q}(y, \tilde{\theta}) = q(y, \tilde{\theta}; \pi)$ for all $(y, \tilde{\theta}) \in \overline{\mathbb{R}}_+ \times \mathcal{A}$, then the above process, and hence the process defined by (9.36) is an $((\mathcal{F}_t)_{t\geq 0}, \mathbb{Q})$-martingale.

On the other hand, if the process (9.36) is an $((\mathcal{F}_t)_{t\geq 0}, \mathbb{Q})$-martingale, we next prove that $\hat{q}(y, \tilde{\theta}) = q(y, \tilde{\theta}; \pi)$ for $(y, \tilde{\theta}) \in \overline{\mathbb{R}}_+ \times \mathcal{A}$. By using (9.70), $\int_0^t e^{-\rho s}(q(Y_s^\pi, \tilde{\theta}_s^\pi; \pi) - \hat{q}(Y_s^\pi, \tilde{\theta}_s^\pi))ds$ for $t \geq 0$ is a continuous local martingale with finite variation, and hence zero quadratic variation. Therefore, it follows that $\int_0^t e^{-\tilde{\rho}s}(q(Y_s^\pi, \tilde{\theta}_s^\pi; \pi) - \hat{q}(Y_s^\pi, \tilde{\theta}_s^\pi))ds = 0$ for all $t \geq 0$, $\mathbb{Q}$-a.s. (see Chapter 1, Exercise 5.21 in Karatzas and Shreve (1991)). For $y \in \mathbb{R}_+$

and $a \in \mathcal{A}$, denote by $h(y, \tilde{\theta}) = q(y, \tilde{\theta}; \boldsymbol{\pi}) - \hat{q}(y, \tilde{\theta})$, which is a continuous function that maps $\overline{\mathbb{R}}_+ \times \mathcal{A}$ to $\overline{\mathbb{R}}_+$. Next, we argue by contradiction. Suppose there exists a pair $(y_0, \tilde{\theta}_0) \in \overline{\mathbb{R}}_+ \times \mathcal{A}$ and $\epsilon > 0$ such that $h(y_0, \tilde{\theta}_0) > \epsilon$. By the continuity of h, there exists $\delta > 0$ such that $h(y, \tilde{\theta}) > \epsilon/2$ for all $(y, \tilde{\theta})$ with $\max\{|y - y_0|, |\tilde{\theta} - \tilde{\theta}_0|\} < \delta$.

Now, consider the state process, still denoted by $Y^{\boldsymbol{\pi}}$, starting from $(y_0, \tilde{\theta}_0)$, namely, $(Y_t^{\boldsymbol{\pi}})_{t \geq 0}$ follows (6) with $Y_0^{\boldsymbol{\pi}} = y_0$ and $\tilde{\theta}_0^{\boldsymbol{\pi}} = \tilde{\theta}_0$. Define the stopping time τ by $\tau := \inf\{t \geq 0; |Y_t^{\boldsymbol{\pi}} - y_0| > \delta\}$. We have already shown that there exists $\Omega_0 \in \mathcal{F}$ with $\mathbb{Q}(\Omega_0) = 0$ such that for all $\omega \in \Omega \setminus \Omega_0$, $\int_0^t e^{-\rho s}(q(Y_s^{\boldsymbol{\pi}}, \tilde{\theta}_s^{\boldsymbol{\pi}}; \boldsymbol{\pi}) - \hat{q}(Y_s^{\boldsymbol{\pi}}, \tilde{\theta}_s^{\boldsymbol{\pi}}))ds = 0$ for all $t \geq 0$. It follows from Lebesgue's differentiation theorem that, for any $\omega \in \Omega \setminus \Omega_0$, $h(Y_t^{\boldsymbol{\pi}}(\omega), \tilde{\theta}_t^{\boldsymbol{\pi}}(\omega)) = 0$, a.e., for $t \in [0, \tau(\omega)]$. Consider the set $\mathcal{O}(\omega) = \{t \in [0, \tau(\omega)]; |\tilde{\theta}_t^{\boldsymbol{\pi}}(\omega) - \tilde{\theta}_0| < \delta\} \subset [0, \tau(\omega)]$. Because $h(Y_t^{\boldsymbol{\pi}}(\omega), \tilde{\theta}_t^{\boldsymbol{\pi}}(\omega)) > \frac{\epsilon}{2}$ when $t \in \mathcal{O}(\omega)$, we conclude that $\mathcal{O}(\omega)$ has Lebesgue measure zero for any $\omega \in \Omega \setminus \Omega_0$. That is

$$\int_0^\infty \mathbf{1}_{\{t \leq \tau(\omega)\}} \mathbf{1}_{\{|\tilde{\theta}_t^{\boldsymbol{\pi}}(\omega) - \tilde{\theta}_0| < \delta\}} dt = 0.$$

Integrating ω with respect to $\mathbb{Q}$ and applying Fubini's theorem, we obtain that

$$\begin{aligned}
0 &= \int_\Omega \int_0^\infty \mathbf{1}_{\{t \leq \tau(\omega)\}} \mathbf{1}_{\{|\tilde{\theta}_t^{\boldsymbol{\pi}}(\omega) - \tilde{\theta}_0| < \delta\}} dt \mathbb{Q}(d\omega) \int_0^\infty \mathbb{E}^{\mathbb{Q}}\left[\mathbf{1}_{\{t \leq \tau\}} \mathbf{1}_{\{|\tilde{\theta}_t^{\boldsymbol{\pi}} - \tilde{\theta}_0| < \delta\}}\right] dt \\
&= \int_0^\infty \mathbb{E}\left[\mathbf{1}_{\{t \leq \tau\}} \mathbb{Q}^{\mathbb{Q}}\left(|\tilde{\theta}_t^{\boldsymbol{\pi}} - \tilde{\theta}_0| < \delta | \mathcal{F}_t\right)\right] dt \\
&= \int_0^\infty \mathbb{E}\left[\mathbf{1}_{\{t \leq \tau\}} \int_{|\tilde{\theta} - \tilde{\theta}_0| < \delta} \boldsymbol{\pi}\left(\tilde{\theta}|Y_t^{\boldsymbol{\pi}}\right) d\tilde{\theta}\right] dt \\
&\geq \min_{|y - y_0| < \delta} \left\{ \int_{|\tilde{\theta} - \tilde{\theta}_0| < \delta} \boldsymbol{\pi}\left(\tilde{\theta}|y\right) d\tilde{\theta} \right\} \int_0^\infty \mathbb{E}^{\mathbb{Q}}\left[\mathbf{1}_{\{t \leq \tau\}}\right] dt \\
&= \min_{|y - y_0| < \delta} \left\{ \int_{|\tilde{\theta} - \tilde{\theta}_0| < \delta} \boldsymbol{\pi}\left(\tilde{\theta}|y\right) d\tilde{\theta} \right\} \mathbb{E}^{\mathbb{Q}}[\tau] \geq 0.
\end{aligned}$$

The above yields that $\min_{|y - y_0| < \delta}\{\int_{|\tilde{\theta} - \tilde{\theta}_0| < \delta} \boldsymbol{\pi}(\tilde{\theta}|y)d\tilde{\theta}\} = 0$. However, this contradicts Definition 9.1 about an admissible policy. In fact, Definition 9.1-(i) stipulates $\operatorname{supp} \boldsymbol{\pi}(\cdot|y) = \mathcal{A}$ for any $y \in \overline{\mathbb{R}}_+$; hence $\int_{|\tilde{\theta} - \tilde{\theta}_0| < \delta} \boldsymbol{\pi}(\tilde{\theta}|y)d\tilde{\theta} > 0$. Then, the continuity in Definition 9.1-(iii) yields $\min_{|y - y_0| < \delta}\{\int_{|\tilde{\theta} - \tilde{\theta}_0| < \delta} \boldsymbol{\pi}(\tilde{\theta}|y)d\tilde{\theta}\} > 0$, which is a contradiction. Hence, we conclude that $q(y, \tilde{\theta}; \boldsymbol{\pi}) = \hat{q}(y, \tilde{\theta})$ for every $(y, \tilde{\theta}) \in \overline{\mathbb{R}}_+ \times \mathcal{A}$. □

Proof of Theorem 9.2. If $\hat{J}$ and $\hat{q}$ are respectively the value function and the q-function associated with the policy π, it follows from the same argument as in the proof of Proposition 9.2 that
$$e^{-\rho t}\hat{J}(Y_t^\pi) - \int_0^t e^{-\rho s}\hat{q}\left(Y_s^\pi, \tilde{\theta}_s^\pi\right) ds - \int_0^t e^{-\rho s} dL_s^\pi, \quad t \geq 0$$
is an $((\mathcal{F}_t)_{t\geq 0}, \mathbb{Q})$-martingale. On the other hand, assume that $e^{-\rho t}\hat{J}(Y_t^\pi) - \int_0^t e^{-\rho s}\hat{q}(Y_s^\pi, \tilde{\theta}_s^\pi) ds - \int_0^t e^{-\rho s} dL_s^\pi$ is an $((\mathcal{F}_t)_{t\geq 0}, \mathbb{Q})$ martingale. It then holds that, for any $T > 0$,

$$\hat{J}(y) \tag{9.71}$$
$$= \mathbb{E}^{\mathbb{Q}}\left[e^{-\rho T}\hat{J}(Y_T^\pi) - \int_0^T e^{-\rho s}\hat{q}\left(Y_s^\pi, \tilde{\theta}_s^\pi\right) ds - \int_0^T e^{-\rho s} dL_s^\pi\right].$$

Integrating over the action randomization with respect to the policy π on the side of Eq. (9.71), we have

$$\hat{J}(y, \pi) = \mathbb{E}^{\mathbb{Q}^W}\left[e^{-\rho T}\hat{J}(\tilde{Y}_T^\pi; \pi) - \int_0^T e^{-\rho s}\int_{\mathcal{A}} \hat{q}\left(\tilde{Y}_s^\pi, \tilde{\theta}\right)\pi(\tilde{\theta}|\tilde{Y}_s^\pi)d\tilde{\theta}ds \right.$$
$$\left. - \int_0^T e^{-\rho s} dL_s^\pi\right]. \tag{9.72}$$

In view that $\int_{\mathcal{A}}[\hat{q}(y, \tilde{\theta}) - \gamma \ln \pi(\tilde{\theta}|y)]\pi(\tilde{\theta}|y)d\tilde{\theta} = 0$, we obtain

$$\hat{J}(y) = \mathbb{E}^{\mathbb{Q}^W}\left[e^{-\rho T}\hat{J}(\tilde{Y}_T^\pi) - \gamma\int_0^T e^{-\rho s}\int_{\mathcal{A}} \ln\pi(\tilde{\theta}|\tilde{Y}_s^\pi)\pi(\tilde{\theta}|\tilde{Y}_s^\pi)d\tilde{\theta}ds \right.$$
$$\left. - \int_0^T e^{-\rho s} dL_s^\pi\right]. \tag{9.73}$$

Letting T go to infinity on both side of (9.73), we deduce from MCT and (9.37) that

$$\hat{J}(y) = \mathbb{E}^{\mathbb{Q}^W}\left[-\gamma\int_0^\infty e^{-\rho s}\int_{\mathcal{A}} \ln\pi(\tilde{\theta}|\tilde{Y}_s^\pi)\pi(\tilde{\theta}|\tilde{Y}_s^\pi)d\tilde{\theta}ds \right.$$
$$\left. - \int_0^\infty e^{-\rho s} dL_s^\pi\right]. \tag{9.74}$$

This implies that $\hat{J}(y) = J(y; \pi)$ for all $y \in \mathbb{R}_+$ by virtue of (9.28). Furthermore, based on Proposition 9.2, the martingale condition implies that $\hat{q}(y, \tilde{\theta}) = q(y, \tilde{\theta}; \pi)$ for all $(y, \tilde{\theta}) \in \overline{\mathbb{R}}_+ \times \mathcal{A}$.

Finally, if $\pi(\tilde{\theta}|y) = \frac{\exp\{\frac{1}{\gamma}\hat{q}(y,\tilde{\theta})\}}{\int_{\mathcal{A}} \exp\{\frac{1}{\gamma}\hat{q}(y,\tilde{\theta})\}d\tilde{\theta}}$ satisfies the condition (9.39), then $\pi = \mathcal{I}(\pi)$ where $\mathcal{I}$ is the map defined in Theorem 9.1. This in turn implies that π is the optimal policy and $\hat{J}$ is the optimal value function. $\square$

Proof of Theorem 9.3. To prove the theorem, we need the following auxiliary lemma:

Lemma 9.2 (Lemma 8 in Jia and Zhou (2022b)). Let $f_h(x) = f(x) + r_h(x)$, where f is a continuous function and r_h converges to 0 uniformly on any compact set as $h \to 0$. Assume that $f_h(x_h^*) = 0$ and $\lim_{h \to 0} x_h^* = x^*$. Then $f(x^*) = 0$.

We apply Lemma 9.2 to prove item (i). Take $f(x) = \mathbb{E}^{\mathbb{Q}}[\int_0^\infty \varsigma_t dM_t^{\xi,\psi}]$ and $f_h(x) = \mathbb{E}^{\mathbb{Q}}[\int_0^T \varsigma_t dM_t^{\xi,\psi}]$ with $h = 1/T$ and $x = (\xi, \psi)$. Thus, we need to prove that

$$\mathbb{E}^{\mathbb{Q}}\left[\int_0^\infty \varsigma_t dM_t^{\xi,\psi}\right] - \mathbb{E}^{\mathbb{Q}}\left[\int_0^T \varsigma_t dM_t^{\xi,\psi}\right] = \mathbb{E}^{\mathbb{Q}}\left[\int_T^\infty \varsigma_t dM_t^{\xi,\psi}\right] \to 0, \quad (9.75)$$

as $T \to \infty$ uniformly on any compact subset of $\Theta \times \Psi$. Let $\mathcal{C} \subset \Theta \times \Psi$ be a compact set, then $\mathcal{C}$ is a bounded closed set. By applying Proposition 9.3 and assumptions $(\mathbf{A}_\xi)$ and $(\mathbf{A}_\psi)$, we have

$$\mathbb{E}^{\mathbb{Q}}\left[\int_T^\infty \varsigma_t dM_t^{\xi,\psi}\right]$$

$$= \mathbb{E}^{\mathbb{Q}}\left[\int_T^\infty e^{-\rho t} \varsigma_t \left(\mathcal{L}^{\tilde{\theta}_t^\pi} J^\xi(Y_t^\pi) - \rho J^\xi(Y_t^\pi) - q^\psi(Y_t^\pi, \tilde{\theta}_t^\pi)\right) dt\right]$$

$$+ \mathbb{E}^{\mathbb{Q}}\left[\int_T^\infty e^{-\rho t} \varsigma_t dL_t^\pi\right]$$

$$= \mathbb{E}^{\mathbb{Q}}\left[\int_T^\infty e^{-\rho t} \varsigma_t \left((\tilde{\theta}_t^\pi)^\top \tilde{\mu} J_y^\xi(Y_t^\pi) + \frac{1}{2}(\tilde{\theta}_t^\pi)^\top \sigma \sigma^\top \tilde{\theta}_t^\pi J_{yy}^\xi(Y_t^\pi)\right.\right.$$

$$- \sqrt{1-\kappa^2}\sigma_Z(Y_t^\pi + 1)(\tilde{\theta}_t^\pi)^\top \sigma \eta J_{yy}^\xi(Y_t^\pi) + \frac{1}{2}\sigma_Z^2(Y_t^\pi + 1)^2 J_{yy}^\xi(Y_t^\pi)$$

$$\left.\left. - \rho J^\xi(Y_t^\pi) - q^\psi(Y_t^\pi, \tilde{\theta}_t^\pi)\right) dt\right] + \mathbb{E}^{\mathbb{Q}}\left[\int_T^\infty e^{-\rho t} \varsigma_t dL_t^\pi\right]$$

$$\leq (G_J(\xi) + G_q(\psi))\mathbb{E}^{\mathbb{Q}}\left[\int_T^\infty e^{-\rho t} \varsigma_t \left(\left(\left|(\tilde{\theta}_t^\pi)^\top \tilde{\mu}\right| + \frac{1}{2}(\tilde{\theta}_t^\pi)^\top \sigma \sigma^\top \tilde{\theta}_t^\pi\right.\right.\right.$$

$$\left.+ \left|\sqrt{1-\kappa^2}\sigma_Z(Y_t^\pi + 1)(\tilde{\theta}_t^\pi)^\top \sigma \eta\right| + \frac{1}{2}\sigma_Z^2(Y_t^\pi + 1)^2 + \rho\right) \tilde{J}(Y_t^\pi)$$

$$\left.\left.+ \tilde{q}(Y_t^\pi, \tilde{\theta}_t^\pi)\right) dt\right] + \mathbb{E}^{\mathbb{Q}}\left[\int_T^\infty e^{-\rho t} \varsigma_t dL_t^\pi\right]. \quad (9.76)$$

This yields that as $T \to \infty$, $\mathbb{E}^{\mathbb{Q}}[\int_T^\infty \varsigma_t dM_t^{\xi,\psi}]$ converges to 0 uniformly in $(\xi, \psi) \in \mathcal{C}$. By using Lemma 9.2, we get the desired result.

Next, we deal with item (ii). By applying Lemma 9.2 again, it is sufficient to show that

$$\mathbb{E}^{\mathbb{Q}}\left[\int_0^\infty \varsigma_t dM_t^{\xi,\psi}\right] - \mathbb{E}^{\mathbb{Q}}\left[\sum_{k=0}^{K-1} \varsigma_{t_k}\left(M_{t_{k+1}}^{\xi,\psi} - M_{t_k}^{\xi,\psi}\right)\right] \to 0, \qquad (9.77)$$

as $\Delta t \to \infty$ uniformly on any compact subset of $\Theta \times \Psi$. Let $\mathcal{C} \subset \Theta \times \Psi$ be a compact set, which is also a bounded closed set. It follows from that

$$\mathbb{E}^{\mathbb{Q}}\left[\int_0^\infty \varsigma_t dM_t^{\xi,\psi}\right] - \mathbb{E}^{\mathbb{Q}}\left[\sum_{k=0}^{K-1} \varsigma_{t_k}\left(M_{t_{k+1}}^{\xi,\psi} - M_{t_k}^{\xi,\psi}\right)\right]$$

$$= \mathbb{E}^{\mathbb{Q}}\left[\int_0^\infty \varsigma_t dM_t^{\xi,\psi}\right] - \mathbb{E}^{\mathbb{Q}}\left[\sum_{k=0}^{K-1} \int_{t_k}^{t_{k+1}} \varsigma_{t_k} dM_t^{\xi,\psi}\right]$$

$$= \mathbb{E}^{\mathbb{Q}}\left[\sum_{k=0}^{K-1} \int_{t_k}^{t_{k+1}} e^{-\rho t}(\varsigma_t - \varsigma_{t_k})\left(\mathcal{L}^{\tilde{\theta}_t^\pi} J^\xi(Y_t^\pi) - \rho J^\xi(Y_t^\pi) - q^\psi(Y_t^\pi, \tilde{\theta}_t^\pi)\right) dt\right]$$

$$+ \mathbb{E}^{\mathbb{Q}}\left[\int_0^T e^{-\rho t}\varsigma_t dL_t^\pi - \sum_{k=0}^{K-1}\int_{t_k}^{t_{k+1}} e^{-\rho t}\varsigma_{t_k} dL_t^\pi\right].$$

Note that

$$\mathbb{E}^{\mathbb{Q}}\left[\int_0^T e^{-\rho t}\varsigma_t dL_t^\pi - \sum_{k=0}^{K-1}\int_{t_k}^{t_{k+1}} e^{-\rho t}\varsigma_{t_k} dL_t^\pi\right]$$

$$= \mathbb{E}^{\mathbb{Q}}\left[\int_0^T e^{-\rho t}\varsigma_t dL_t^\pi - \sum_{k=0}^{K-1} e^{-\rho t_k}\varsigma_{t_k}\left(L_{t_{k+1}}^\pi - L_{t_k}^\pi\right)\right]$$

$$+ \mathbb{E}^{\mathbb{Q}}\left[\sum_{k=0}^{K-1}\int_{t_k}^{t_{k+1}}(e^{-\rho t} - e^{-\rho t_k})\varsigma_{t_k} dL_t^\pi\right]$$

$$\leq \mathbb{E}^{\mathbb{Q}}\left[\int_0^T e^{-\rho t}\varsigma_t dL_t^\pi - \sum_{k=0}^{K-1} e^{-\rho t_k}\varsigma_{t_k}\left(L_{t_{k+1}}^\pi - L_{t_k}^\pi\right)\right]$$

$$+ \sup_{|t-s|\leq \Delta t, t, s \in [0,T]} (e^{-\rho t} - e^{-\rho s})\mathbb{E}^{\mathbb{Q}}\left[\sum_{k=0}^{K-1}\int_{t_k}^{t_{k+1}} \varsigma_{t_k} dL_t^\pi\right],$$

where the 1st term in above inequality converges to 0 as $\Delta t \to 0$ by using the definition of Riemann-Stieltjes integral and the 2nd term converges to 0 as $\Delta t \to 0$ since $\sup_{|t-s|\leq \Delta t, t, s \in [0,T]}(e^{-\rho t} - e^{-\rho s}) \to 0$. On the other hand, we have

$$\mathbb{E}^{\mathbb{Q}}\left[\sum_{k=0}^{K-1}\int_{t_k}^{t_{k+1}} e^{-\rho t}(\varsigma_t - \varsigma_{t_k})\left(\mathcal{L}^{\tilde{\theta}_t^\pi} J^\xi(Y_t^\pi) - \rho J^\xi(Y_t^\pi) - q^\psi(Y_t^\pi, \tilde{\theta}_t^\pi)\right) dt\right]$$

$$\leq \sum_{k=0}^{K-1} \left(\int_{t_k}^{t_{k+1}} \mathbb{E}^{\mathbb{Q}} \left[(\varsigma_t - \varsigma_{t_k})^2 \right] dt \right)^{\frac{1}{2}}$$

$$\times \left(\int_{t_k}^{t_{k+1}} \mathbb{E}^{\mathbb{Q}} \left[\left(\mathcal{L}^{\tilde{\theta}_t^\pi} J^\xi(Y_t^\pi) - \rho J^\xi(Y_t^\pi) - q^\psi(Y_t^\pi, \tilde{\theta}_t^\pi) \right)^2 \right] dt \right)^{\frac{1}{2}}$$

$$\leq d(\ell(\cdot), \Delta t)\sqrt{\Delta t}$$

$$\times \sum_{k=0}^{K-1} \left(\int_{t_k}^{t_{k+1}} \mathbb{E}^{\mathbb{Q}} \left[\left(\mathcal{L}^{\tilde{\theta}_t^\pi} J^\xi(Y_t^\pi) - \rho J^\xi(Y_t^\pi) - q^\psi(Y_t^\pi, \tilde{\theta}_t^\pi) \right)^2 \right] dt \right)^{\frac{1}{2}}$$

$$\leq d(\ell(\cdot), \Delta t)\sqrt{\Delta t}$$

$$\times \sqrt{K} \left(\sum_{k=0}^{K-1} \int_{t_k}^{t_{k+1}} \mathbb{E}^{\mathbb{Q}} \left[\left(\mathcal{L}^{\tilde{\theta}_t^\pi} J^\xi(Y_t^\pi) - \rho J^\xi(Y_t^\pi) - q^\psi(Y_t^\pi, \tilde{\theta}_t^\pi) \right)^2 \right] dt \right)^{\frac{1}{2}}$$

$$\leq d(\ell(\cdot), \Delta t)\sqrt{T} \left(\mathbb{E}^{\mathbb{Q}} \left[\int_0^T \left(\mathcal{L}^{\tilde{\theta}_t^\pi} J^\xi(Y_t^\pi) - \rho J^\xi(Y_t^\pi) - q^\psi(Y_t^\pi, \tilde{\theta}_t^\pi) \right)^2 dt \right] \right)^{\frac{1}{2}}$$

$$\leq d(\ell(\cdot), \Delta t)\sqrt{T}(G_J^2(\xi) + G_q^2(\psi))^{\frac{1}{2}}$$

$$\times \mathbb{E}^{\mathbb{Q}} \left[\int_0^T \left(\left(|(\tilde{\theta}_t^\pi)^\top \mu| + \frac{1}{2}(\tilde{\theta}_t^\pi)^\top \sigma\sigma^\top \tilde{\theta}_t^\pi + \left| \sqrt{1-\kappa^2} \sigma_Z(Y_t^\pi + 1)(\tilde{\theta}_t^\pi)^\top \sigma\eta \right| \right. \right.$$

$$\left. \left. + \frac{1}{2}\sigma_Z^2(Y_t^\pi + 1)^2 + \rho \right) \tilde{J}(Y_t^\pi) + \tilde{q}(Y_t^\pi, \tilde{\theta}_t^\pi) \right)^2 dt \right]^{\frac{1}{2}},$$

where $\ell(t) := \varsigma_t$ and $d(\ell(\cdot), \Delta t) = \sup_{|t-s| \leq \Delta t, t, s \in [0,T]} \mathbb{E}^{\mathbb{Q}}[|\ell(t) - \ell(s)|^2]^{\frac{1}{2}}$ is the modulus of continuity of $\ell(\cdot)$ in $\mathbb{L}^2(\Omega)$. Therefore, $d^2(\ell(\cdot), \Delta t) \to 0$ as $\Delta t \to 0$ and we get the desired result (9.77). □

Proof of Theorem 9.4. By using Theorem 9.1, it suffices to verify the transversality conditions in (9.33). Note that $y \to v(y)$ is non-decreasing, we have

$$v(y) = \ln(1+y) + \frac{1}{\rho}\left(-\frac{1}{2}\sigma_Z^2\kappa^2 + \sqrt{1-\kappa^2}\varsigma + \frac{\gamma}{2}\ln\left(\frac{(2\pi\gamma)^d}{|\sigma\sigma^\top|}\right) + \alpha\right)$$

$$\geq \frac{1}{\rho}\left(-\frac{1}{2}\sigma_Z^2\kappa^2 + \sqrt{1-\kappa^2}\varsigma + \frac{\gamma}{2}\ln\left(\frac{(2\pi\gamma)^d}{|\sigma\sigma^\top|}\right) + \alpha\right).$$

This yields that, for any $\boldsymbol{\pi} \in \Pi$,

$$\limsup_{T \to \infty} \mathbb{E}^{\mathbb{Q}^W}\left[e^{-\rho T}v(\tilde{Y}_T^\pi)\right] \geq 0. \qquad (9.78)$$

Next, we check the validity of the second transversality condition in (9.33), that is

$$\lim_{T\to\infty} \mathbb{E}^{\mathbb{Q}^W}\left[e^{-\rho T} v(Y_T^*)\right] = 0. \tag{9.79}$$

It follows from (9.25) and (9.54) that the controlled state process $Y^* = (Y_t^*)_{t\geq 0}$ with policy $\boldsymbol{\pi}^*$ obeys the following reflected SDE, for $t > 0$,

$$dY_t^* = (2\alpha + \sqrt{1-\kappa^2}\zeta)(1+Y_t^*)dt$$
$$+ \sqrt{\alpha + \frac{d}{2}\gamma + \frac{1}{2}\kappa^2\sigma_Z^2 + \sqrt{1-\kappa^2}\zeta}(1+Y_t^*)dB_t + dL_t^*.$$

Introduce the processes $H_t := \ln(1+Y_t^*)$ and $K_t := \int_0^t \frac{dL_s^*}{1+Y_s^*}$ for $t \geq 0$. Then, $t \to H_t$ is a non-negative process with $H_0 = h = \ln(1+y)$. Moreover, $t \to K_t$ is a non-decreasing process and satisfies that, a.s.

$$\int_0^t \mathbf{1}_{H_s=0} dK_s = \int_0^t \mathbf{1}_{Y_s^*=0} \frac{dL_s^*}{1+Y_s^*} = \int_0^t \frac{dL_s^*}{1+Y_s^*} = K_t, \quad t \geq 0$$

This implies that $K = (K_t)_{t\geq 0}$ is the local time of the process $H = (H_t)_{t\geq 0}$ at the reflecting boundary 0. By applying Itô's formula to $H_t = \ln(1+Y_t^*)$, we arrive at

$$dH_t = \hat{\mu} dt + \hat{\sigma} dB_t + \frac{dL_t^*}{1+Y_t^*} = \hat{\mu} dt + \hat{\sigma} dB_t + dK_t, \quad t > 0, \tag{9.80}$$

where constants $\hat{\mu}$ and $\hat{\sigma}$ are defined by

$$\hat{\mu} := 2\alpha + \sqrt{1-\kappa^2}\zeta - \frac{1}{2}\left(\alpha + \frac{d}{2}\gamma + \frac{1}{2}\kappa^2\sigma_Z^2 + \sqrt{1-\kappa^2}\zeta\right),$$

$$\hat{\sigma} := \sqrt{\alpha + \frac{d}{2}\gamma + \frac{1}{2}\kappa^2\sigma_Z^2 + \sqrt{1-\kappa^2}\zeta}.$$

From the solution representation of the "Skorokhod problem", it follows that, for any $y \geq 0$,

$$K_t = 0 \vee \left\{-h + \max_{s\in[0,t]} (-\hat{\mu}s - \hat{\sigma}B_s)\right\} \leq h + |\hat{\mu}|t + |\hat{\sigma}| \max_{s\in[0,t]} B_s. \tag{9.81}$$

Thus, we deduce from (9.80) and (9.81) that, for $T > 0$,

$$e^{-\rho T}\mathbb{E}^{\mathbb{Q}^W}\left[\ln(1+Y_T^*)\right] = e^{-\rho T}\mathbb{E}^{\mathbb{Q}^W}\left[K_T\right]$$
$$\leq e^{-\rho T}\left(2h + 2|\hat{\mu}|T + |\hat{\sigma}|\sqrt{\frac{2T}{\pi}}\right) \to 0, \text{ as } T \to \infty.$$

This yields the desired transversality condition (9.79) and completes the proof. $\square$

Bibliography

Abraham, R. (2000): Reflecting Brownian snake and a Neumann-Dirichlet problem. *Stoch. Process. Appl.* **89**(2), 239-260.

Aliprantis, C.D., and K.C. Border (1999): *Infinite-Dimensional Analysis: A Hitchhiker's Guide.* Springer-Verlag, Berlin.

Andruszkiewicz, G., H.A. Davis, and S. LIeo (2016): Risk-sensitive investment in a finite-factor model. *Stochastics.* 89, 89-114.

Ang, A., and G. Bekaert (2002): International asset allocation with regime shifts. *Rev. Finan. Stud.* 15, 1137-1187.

Ang, A., and A. Timmermann (2012): Regime changes and financial markets. *Ann. Rev. Finan. Econ.* 4, 313-337.

Ankirchner, S., C. Blanchet-Scalliet, and A. Eyeraud-Loisel (2010): Credit risk premia and quadratic BSDEs with a single jump. *Inter. J. Theor. Appl. Finan.* 13, 1103-1129.

Antonelli, F., and C. Mancini (2016): Solutions of BSDEs with jumps and quadratic/locally Lipschitz generator *Stoch. Process. Appl.* 126, 3124-3144.

Avellaneda, M., A. Lévy, and A. Parás (1995): Pricing and hedging derivative securities in markets with uncertain volatilities. *Appl. Math. Finan.* 2, 73-88.

Bai, L., T. Gamage, J. Ma and P. Xie (2023): Reinforcement learning for optimal dividend problem under diffusion model. Preprint, available at arXiv:2309.10242.

Bank, P., and F. Riedel (2001): Optimal consumption choice with intertemporal substitution. *Ann. Appl. Probab.* 11, 750-788.

Barles, G., C. Daher and M. Romano (1994): Optimal control on the L^∞ norm of a diffusion process. *SIAM J. Contr. Optim.* 32, 612-634.

Barron, E.N. (1993): The Bellman equation for control of the running max of a diffusion and applications to look-back options. *Appl. Anal.* 48, 205-222.

Barron, E.N., and H. Ishii (1989): The Bellman equation for minimizing the maximum cost. *Nonlinear Anal.: Theor. Meth. Appl.* 13, 1067-1090.

Bass, R.F. (2010): The measurability of hitting times. *Electron. Commun. Probab.* 15, 99-105.

Bäuerle, N., and U. Rieder (2007): Portfolio optimization with jumps and unobservable intensity process. *Math. Finan.* 17, 205-224.

Bayraktar, E., and A. Cohen (2018): Risk sensitive control of the lifetime ruin problem. *Appl. Math. Optim.* 77, 229-252.

Bayraktar, E., and M. Egami (2008): An analysis of monotone follower problems for diffusion processes. *Math. Oper. Res.* 33, 336-350.

Bayraktar, E., I. Karatzas, and S. Yao (2010): Optimal stopping for dynamic convex risk measures. *Illinois J. Math.* 54(3), 1025-1067.

Bayraktar, E., and S. Yao (2011): Optimal stopping for non-linear expectations–Part I. *Stoch. Process. Appl.* 121(2), 185-211.

Bayraktar, E., and S. Yao (2011): Optimal stopping for non-linear expectations–Part II. *Stoch. Process. Appl.* 121(2), 212-264.

Bayraktar, E., and S. Yao (2013): A weak dynamic programming principle for zero-sum stochastic differential games with unbounded controls. *SIAM J. Contr. Optim.* 51, 2036-2080.

Bayraktar, E., and S. Yao (2014): On the robust optimal stopping problem. *SIAM J. Contr. Optim.* 52, 3135-3175.

Bayraktar, E., and S. Yao (2017): Optimal stopping with random maturity under nonlinear expectation. *Stoch. Process. Appl.* 127, 2586-2629.

Bayraktar, E., J. Zhang, and Z. Zhou (2021): Equilibrium concepts for time-inconsistent stopping problems in continuous time. *Math. Finan.* 31(1), 508-530.

Beissner, P., Q. Lin, and F. Riedel (2020): Dynamically consistent alpha-maxmin expected utility. *Math. Finan.* 30(3), 1073-1102.

Belomestny, D., and V. Krätschmer (2016): Optimal stopping under model uncertainty: randomized stopping times approach. *Ann. Appl. Probab.* 26, 1260-1295.

Bertsekas, D.P., and S.E. Shreve (1978): *Stochastic Optimal Control–The Discrete Time Case.* Academic Press Inc., New York.

Benedetti, G., and L. Campi (2012): Multivariate utility maximization with proportional transaction costs and random endowment. *SIAM J. Contr. Optim.* 50(3), 1283-1308.

Bhidé, A. (1999): *The Origins and Evolution of New Businesses.* Oxford University Press, Oxford.

Biagini, S., B. Bouchard, C. Kardaras, and M. Nutz (2017): Robust fundamental theorem for continuous processes. *Math. Finan.* 27(4), 963-987.

Bielecki, T.R., and I. Jang (2006): Portfolio optimization with a defaultable security *Asia-Pacific Finan. Markets* 13, 113-127.

Bielecki, T.R., and M. Rutkowski (2002): *Credit Risk: Modeling, Valuation and Hedging.* Springer-Verlag, New York.

Bielecki, T.R., and S.R. Pliska (1999): Risk-sensitive dynamic asset management. *Appl. Math. Optim.* 39, 337-360.

Birge, J.R., L. Bo, and A. Capponi (2018): Risk-sensitive asset management and cascading defaults. *Math. Opers. Res.* 43, 1-28.

Bismut, J.M. (1973): Conjugate convex functions in optimal stochastic control. *J. Math. Anal. Appl.* 44, 384-404.

Bismut, J.M. (1976): Linear quadratic optimal stochastic control with random coefficients. *SIAM J. Contr. Optim.* 14(3), 419-444.

Bismut, J.M. (1978): An introductory approach to duality in optimal stochastic control. *SIAM Rev.* 20(1), 62-78.

Björk, T., and A. Murgoci (2014): A theory of Markovian time-inconsistent stochastic control in continuous time. *Finan. Stoch.* 18, 545-592.

Bo, L., and A. Capponi (2016): Optimal investment in credit derivatives portfolio under contagion risk. *Math. Finan.* 26, 785-834.

Bo, L., and A. Capponi (2017): Optimal investment under information driven contagious distress. *SIAM J. Contr. Optim.* 55, 1020-1068.

Bo, L., and A. Capponi (2018): Portfolio choice with market-credit risk dependencies. *SIAM J. Contr. Optim.* 56, 3050-3091.

Bo, L., A. Capponi, and P.C. Chen (2019): Credit portfolio selection with decaying contagion intensities. *Math. Finan.* 29, 137-173.

Bo, L., Y. Huang, and X. Yu (2024): Stochastic control problems with state-reflections arising from relaxed benchmark tracking. *Math. Oper. Res.* Forthcoming, DOI: 10.1287/moor.2023.0265.

Bo, L., Y. Huang, and X. Yu (2024): A decomposition-homogenization method for Robin boundary problems on the nonnegative orthant. *Elect. J. Probab.* 29, 1-25.

Bo, L., Y. Huang, and X. Yu (2025): On optimal tracking portfolio in incomplete markets: The reinforcement learning approach. *SIAM J. Contr. Optim.* 63, 321-348.

Bo, L., H. Liao, and Y. Wang (2019): Optimal credit investment and risk control for an insurer with regime-switching. *Math. Finan. Econ.* 13, 147-172.

Bo, L., H. Liao, and X. Yu (2019): Risk sensitive portfolio optimization with default contagion and regime-switching. *SIAM J. Contr. Optim.* 57(1), 366-401.

Bo, L., H. Liao, and X. Yu (2021): Optimal tracking portfolio with a ratcheting capital benchmark. *SIAM J. Contr. Optim.* 59(3), 2346-2380.

Bo, L., H. Liao, and X. Yu (2022): Risk-sensitive credit portfolio optimization under partial information and contagion risk. *Ann. Appl. Probab.* 32(4), 2355-2399.

Bokanowski, O., A. Picarelli, and H. Zidani (2015): Dynamic programming and error estimates for stochastic control problems with maximum cost. *Appl. Math. Optim.* 71, 125-163.

Borodin, A.N., and P. Salminen (2002): *Handbook of Brownian Motion–Facts and Formulae*. Probability and its Applications, second edn., Birkhäuser Verlag, Basel.

Bouchard, B. (2007): *Introduction to Stochastic Control of Mixed Diffusion Processes, Viscosity Solutions and Applications in Finance and Insurance.* Lecture Notes. Available at `http://www.ceremade.dauphine.fr/~bouchard/pdf/PolyContSto.pdf`.

Bouchard, B., R. Elie, and C. Imbert (2010): Optimal control under stochastic target constraints. *SIAM J. Contr. Optim.* 48, 3501-3531.

Bouchard, B., and H. Pham (2004): Wealth-path dependent utility maximization in incomplete markets. *Finan. Stoch.* 8, 579-603.

Branger, N., H. Kraft, and C. Meinerding (2014): Partial information about contagion risk, self-exciting processes and portfolio optimization. *J. Econ. Dyn. Contr.* 39, 18-36.

Brandao, L.E., J.S. Dyer, and W.J. Hahn (2005): Using binomial decision trees to solve real-option valuation problems. *Dec. Anal.* 2(2), 69-88.

Brannath, W., and W. Schachermayer (1999): A bipolar theorem for $L^0_+(\Omega, \mathcal{F}, \mathbf{P})$. *Séminaire de Probabilités, XXXIII, Lecture Notes in Math.* 1709, 349-354. Springer, Berlin.

Browne, S. (1999a): Reaching goals by a deadline: Digital options and continuous-time active portfolio management. *Adv. Appl. Probab.* 31, 551-577.

Browne, S. (1999b): Beating a moving target: Optimal portfolio strategies for outperforming a stochastic benchmark. *Finan. Stoch.* 3, 275-294.

Browne, S. (2000): Risk-constrained dynamic active portfolio management. *Manag. Sci.* 46, 1188-1199.

Brunick, G., and S. Shreve (2013): Mimicking an Itô process by a solution of a stochastic differential equation. *Ann. Appl. Probab.* 23(4), 1584-1628.

Callegaro, G., M. Jeanblanc, and W. Runggaldier (2012): Portfolio optimization in a defaultable market under incomplete information. *Dec. Econ. Finan.* 35, 91-111.

Campbell, J.Y., and J.H. Cochrane (1999): By force of habit: A consumption-based explanation of aggregate stock market behavior. *J. Polit. Econ.* 107, 205-251.

Campi, L., and W. Schachermayer (2006): A super-replication theorem in Kabanov's model of transaction costs. *Finan. Stoch.* 10(4), 579-596.

Capponi, A., and J.E. Figueroa-López (2014): Dynamic portfolio optimization with a defaultable security and regime-switching. *Math. Finan.* 24, 207-249.

Capponi, A., J.E. Figueroa-López, and A. Pascucci (2015): Dynamic credit investment in partially observed markets. *Finan. Stoch.* 19, 891-939.

Capponi, A., J.E. Figueroa-López, and J. Nisen (2014): Pricing and semimartingale representations of vulnerable contingent claims in regime-switching markets. *Math. Finan.* 24, 250-288.

Carbone, R., B. Ferrario, and M. Santacroce (2008): Backward stochastic differential equations driven by càdlàg martingales. *Theor. Probab. Appl.* 52(2), '304-314.

Castaing, C., and M. Valadier (1977): *Convex Analysis and Measurable Multifunctions.* Lecture Notes in Math., Vol. 580, Springer-Verlag, New York.

Cerrai, S. (2001): *Second Order PDEs in Finite and Infinite Dimension: A Probabilistic Approach.* Lecture Notes in Math. 1762, Springer-Verlag, New York.

Chateauneuf, A., J. Eichberger, and S. Grant (2007): Choice under uncertainty with the best and worst in mind: neo-additive capacities. *J. Econ. Theor.* 137, 538-567.

Cheng, X., and F. Riedel (2013): Optimal stopping under ambiguity in continuous time. *Math. Finan. Econ.* 7, 29-68.

Cheng, X., Y. Dong, P. Bartlett, and M. Jordan (2020): Stochastic gradient and Langevin processes. *Proceedings of the 37th International Conference on Machine Learning*, PMLR 119, 1810-1819.

Christensen, S., and K. Lindensjö (2018): On finding equilibrium stopping times for time-inconsistent Markovian problems. *SIAM J. Contr. Optim.* 56(6), 4228-4255.

Chow, Y., X. Yu, and C. Zhou (2020): On dynamic programming principle for stochastic control under expectation constraints. *J. Optim. Theor. Appl.* 185, 803-818.

Cohen, S., S. Ji, and S. Peng (2011): Sublinear expectations and martingales in discrete time. Preprint, Available at https://arxiv.org/abs/1104.5390.

Cohn, L. (1993): *Measure Theory*. Birkhäuser Boston, Inc., Boston, MA.

Cox, J.C. (1996): The constant elasticity of variance option pricing model. *J. Portfolio Manag.* A Tribute to Fischer Black. 23, 15-17.

Cox, J.C., J.E. Ingersoll, and S.A. Ross (1985): A theory of the term structure of interest rates. *Econometrica* 53, 385-408.

Cox, J.C., and C. Huang (1989): Optimal consumption and portfolio policies when asset prices follow a diffusion process. *J. Econ. Theor.* 49, 33-83.

Cox, J.C., and C. Huang (1991): A variational problem arising in financial economics. *J. Math. Econ.* 20, 465-487.

Crauel, H. (2002): *Random Probability Measures on Polish Spaces*. Vol. 11 of Stochastics Monographs, Taylor & Francis, London.

Curley, S., and J. Yates (1989): An empirical evaluation of descriptive models of ambiguity reactions in choice situations. *J. Math. Psych.* 33, 397-427.

Cvitanić, J., W. Schachermayer, and H. Wang (2001): Utility maximization in incomplete markets with random endowment. *Finan. Stoch.* 5, 259-272.

Das, M.K., A. Goswami, and N. Rana (2018): Risk sensitive portfolio optimization in a jump diffusion model with regimes. *SIAM J. Contr. Optim.* 56, 1550-1576.

Davis, M., and F. Lischka (2002): Convertible bonds with market risk and credit risk, in Applied Probability, R. Chan et al., eds. Studies in Advanced Mathematics. Providence, RI: American Mathematical Society/International Press, 45-58.

Davis, M., and S. Lleo (2011): Jump-diffusion risk-sensitive asset management I: Diffusion factor model. *SIAM J. Finan. Math.* 2, 22-54.

Davis, M., and S. Lleo (2013): Jump-diffusion risk-sensitive asset management II: Jump-diffusion factor model. *SIAM J. Contr. Optim.* 51, 1441-1480.

Davis, M.H., and S. Lleo (2014): *Risk-Sensitive Investment Management*. World Scientific, Singapore.

De Figueiredo, D.G. (1991): Lectures on the Ekeland variational principle with applications and detours. *Acta. Appl. Math.* 24, 195-196.

Delbaen, F., and W. Schachermayer (1994): A general version of the fundamental theorem of asset pricing. *Math. Ann.* 300(3), 463-520.

Delbaen, F., and W. Schachermayer (1997): The Banach space of workable contingent claims in arbitrage theory. *Ann. Inst. Henri Poincaré, Probab. Stats.* 33(1), 113-144.

Delbaen, F., and W. Schachermayer (1998): The fundamental theorem of asset pricing for unbounded stochastic processes. *Math. Ann.* 312(2), 215-250.

Delong, Ł., and C. Klüppelberg (2008): Optimal investment and consumption in a Black-Scholes market with Lévy-driven stochastic coefficients. *Ann. Appl. Probab.* 18(3), 879-908.

Denis, L., M. Hu, and S. Peng (2011): Function spaces and capacity related to a sublinear expectation: application to G-Brownian motion paths. *Potential Anal.* 34(2), 139-161.

Deng, S., X. Li, H. Pham, and X. Yu (2022): Optimal consumption with reference to past spending maximum. *Finan. Stoch.* 26, 217-266.

Detemple, J.B., and I. Karatzas (2003): Non-addictive habits: optimal consumption-portfolio policies. *J. Econ. Theor.* 113(2), 265-285.

Detemple, J.B., and F. Zapatero (1991): Asset prices in an exchange economy with habit formation. *Econometrica* 59(6), 1633-1657.

Detemple, J.B., and F. Zapatero (1992): Optimal consumption-portfolio policies With habit formation. *Math. Finan.* 2(4), 251-274.

Di Giacinto, M., S. Federico, and F. Gozzi (2011): Pension funds with a minimum guarantee: a stochastic control approach. *Finan. Stoch.* 15, 297-342.

Di Giacinto, M., S. Federico, F. Gozzi, and E. Vigna (2014): Income drawdown option with minimum guarantee. *Euro. J. Oper. Res.* 234, 610-624.

Dixit, A., and R. Pindyck (1994): Investment under uncertainty. Princeton University Press, Princeton, NJ.

Duffie, D., D. Filipović, and W. Schachermayer (2003): Affine processes and applications in finance. *Ann. Appl. Probab.* 13, 984-1053.

Egglezos, N., and I. Karatzas (2009): Utility maximization with habit formation: dynamic programming and stochastic PDEs. *SIAM J. Contr. Optim.* 48(2), 481-520.

Ekeland, I., and A. Lazrak (2006): Being serious about non-commitment: subgame perfect equilibrium in continuous time. Preprint, available at http://arxiv.org/abs/math/0604264.

Ekren, I., N. Touzi, and J. Zhang (2014): Optimal stopping under nonlinear expectation. *Stoch. Process. Appl.* 124, 3277-3311.

El Karoui, N., and S. Hamadène (2003): BSDEs and risk-sensitive control, zero-sum and nonzero-sum game problems of stochastic functional differential equations. *Stoch. Process. Appl.* 107, 145-169.

El Karoui, N., M. Jeanblanc, and V. Lacoste (2005): Optimal portfolio manangement with American capital guarantee. *J. Econ. Dyn. Contr.* 29, 449-468.

El Karoui, N., and A. Meziou (2006): Constrained optimization with respect to stochastic dominance: application to portfolio insurance. *Math. Finan.* 16, 103-117.

Epstein, L., and M. Schneider (2003): Recursive multiple-priors. *J. Econ. Theor.* 113(1), 1-31.

Feller, W. (1951): Two singular diffusion problems. *Ann. Math.* 173-182.

Fleming, W. (2006): Risk sensitive stochastic control and differential games. *Commun. Inf. Syst.* 6, 161-177.

Fleming, W.H., and R.W. Rishel (1975): *Deterministic and Stochastic Optimal Control*. Springer-Verlag, New York.

Frey, R., and J. Backhaus (2008): Pricing and hedging of portfolio credit derivatives with interacting default intensities. *Inter. J. Theor. Appl. Finan.* 11, 611-634.

Frey, R., and W. Runggaldier (2010): Pricing credit derivatives under incomplete information: a nonlinear-filtering approach. *Finan. Stoch.* 14, 495-526.

Frey, R., and T. Schmidt (2012): Pricing and hedging of credit derivatives via the innovations approach to nonlinear filtering. *Finan. Stoch.* 16, 105-133.

Friedman, A. (1975): *Stochastic Differential Equations and Applications*. Dover Publications, Reprint of the Academic Press, Inc., New York.

Gaivoronski, A., S. Krylov, and N. Wijst (2005): Optimal portfolio selection and dynamic benchmark tracking. *Euro. J. Oper. Res.* 163, 115-131.

Ghirardato, P., F. Maccheroni, and M. Marinacci (2004): Differentiating ambiguity and ambiguity attitude. *J. Econ. Theor.* 118, 133-173.

Giesecke, K., F. Longstaff, S. Schaefer, and I. Strebulaev (2011): Corporate bond default risk: A 150-year perspective. *J. Finan. Econ.* 102, 233-250.

Giesecke, K. (2011): Default and information. *J. Econ. Dyn. Contr.* 30, 2281-2303.

Graves, L.M. (1946): *The Theory of Functions of Real Variables*. McGraw-Hill, New York.

Guasoni, P., G. Huberman, and D. Ren (2020): Shortfall aversion. *Math. Finan.* 30(3), 869-920.

Guasoni, P., G. Huberman, and Z.Y. Wang (2011): Performance maximization of actively managed funds. *J. Finan. Econ.* 101, 574-595.

Guasoni, P., E. Lépinette, and M. Rásonyi (2012): The fundamental theorem of asset pricing under transaction costs. *Finan. Stoch.* 16(4), 741-777.

Guasoni, P. and M. Rásonyi, and W. Schachermayer (2010): The fundamental theorem of asset pricing for continuous processes under small transaction costs. *Ann. Finan.* 6(2), 157-191.

Hansen, L., and T. Sargent (2007): Recursive robust estimation and control without commitment. *J. Econ. Theor.* 136, 1-27.

Hansen, L., T. Sargent, G. Turmuhambetova, and N. Williams (2006): Robust control and model misspecification. *J. Econ. Theor.* 128, 45-90.

Harrison, M. (1985): *Brownian Motion and Stochastic Flow Systems*. John Wiley and Son, New York.

Heath, C., and A. Tversky (1991): Preference and belief: Ambiguity and competence in choice under uncertainty. *J. Risk Uncertain.* 4, 5-28.

Heath, D., and M. Schweizer (2000): Martingales versus PDEs in finance: An equivalence result with examples. *J. Appl. Probab.* 37, 947-957.

Hicks, J. (1965): *Capital and Growth*. Oxford University Press, New York.

Hiriart-Urruty, J.B., and C. Lemaréchal (2001): *Fundamentals of Convex Analysis*. Grundlehren Text Editions, Springer-Verlag, New York.

Howard, R.A., and J.E. Matheson (1972): Risk-sensitive markov decision processes. *Manag. Sci.* 18, 356-369.

Huang, Y.J., and A. Nguyen-Huu (2018): Time-consistent stopping under decreasing impatience. *Finan. Stoch.* 22(1), 69-95.

Huang, Y.J., A. Nguyen-Huu, and X.Y. Zhou (2020): General stopping behaviors of naive and non-committed sophisticated agents, with applications to probability distortion. *Math. Finan.* 30(1), 310-340.

Huang, Y.J., and Z. Zhou (2021): Strong and weak equilibria for time-inconsistent stochastic control in continuous time. *Math. Opers. Res.* 46(2), 405-833.

Huang, Y.J., and Z. Zhou (2019): The optimal equilibrium for time-inconsistent stopping problems–the discrete-time case. *SIAM J. Contr. Optim.* 57(1), 590-609.

Huang, Y.J., and Z. Zhou (2020): Optimal equilibria for time-inconsistent stopping problems in continuous time. *Math. Finan.* 30(3), 1103-1134.

Hugonnier, J., and D. Kramkov (2004): Optimal investment with random endowments in incomplete markets. *Ann. Appl. Probab.* 14(2), 845-864.

Hugonnier, J., and E. Morellec (2007): Corporate control and real investment in incomplete markets. *J. Econ. Dyn. Contr.* 31(5), 1781-1800.

Ikeda, N., and S. Watanabe (1989): *Stochastic Differential Equations and Diffusion Processes, 2nd ed.* North-Holland Math. Libr., Vol. 24, North-Holland, Amsterdam.

Itô, K., and S. Watanabe (1978): Introduction to stochastic differential equations. *Proc. Intern. Symp. on Stoch. Diff. Eqn., Kyoto 1976* (K. Itô, ed.), i-xxx. Kinokuniya, Tokyo.

Jacka, S., and A. Berkaoui (2007): On the density of properly maximal claims in financial markets with transaction costs. *Ann. Appl. Probab.* 17(2), 716-740.

Jacobson, D. (1973): Optimal stochastic linear systems with exponential performance criteria and relation to deterministic differential games. *IEEE Trans. Autom. Contr.* 18, 124-131

Jacod, J., and A.N. Shiryaev (2003): *Limit Theorems for Stochastic Processes. 2nd Ed.* Springer-Verlag, Berlin Heidelberg.

Jia, Y., and X.Y. Zhou (2022a): Policy gradient and actor-critic learning in continuous time and space: Theory and algorithms. *J. Machine Learning Res.* 23, 1-50.

Jia, Y., and X.Y. Zhou (2022b): Policy evaluation and temporal-difference learning in continuous time and space: A martingale approach. *J. Machine Learning Res.* 23, 1-55.

Jia, Y., and X.Y. Zhou (2023): q-Learning in continuous time. *J. Machine Learning Res.* 24, 1-61.

Jiao, Y., I. Kharroubi, and H. Pham (2013): Optimal investment under multiple defaults risk: A BSDE-decomposition approach. *Ann. Appl. Probab.* 23(2), 455-491.

Kabanov, Y.M., and G. Last (2002): Hedging under transaction costs in currency markets: a continuous-time model. *Math. Finan.* 12(1), 63-70.

Kabanov, Y.M., and C. Stricker (2002): *Hedging of Contingent Claims under Transaction Costs.* Advances in Finance and Stochastics, 125-136, Springer-Verlag, New York.

Karatzas, I., J. Lehoczky, S.E. Shreve, and G. Xu (1991): Martingale and duality methods for utility maximization in an incomplete market. *SIAM J. Contr. Optim.* 29, 702-730.

Karatzas, I., and S.E. Shreve (1984): Connections between optimal stopping and singular stochastic control: I. monotone follower problems. *SIAM J. Contr. Optim.* 22, 856-877.

Karatzas, I., and S.E. Shreve (1991): *Borwonian motion and Stochastic Calculus.* Springer-Verlag, New York.

Karatzas, I., and S.E. Shreve (1998): *Methods of Mathematical Finance.* Springer-Verlag, New York.

Karatzas, I., and G. Žitković (2003): Optimal consumption from investment and random endowment in incomplete semimartingale markets. *Ann. Probab.* 31(4), 1821-1858.

Kardaras, C., and E. Platen (2011): On the semimartingale property of discounted asset-price processes. *Stoch. Process. Appl.* 121(11), 2678-2691.

Kauppila, H. (2010): *Convex Duality in Singular Control: Optimal Consumption with Intertemporal Substitution and Optimal Investment in Incomplete Markets.* Thesis (Ph.D.), Columbia University.

Kazamaki, N. (1994): *Continuous Exponential Martingales and BMO.* Springer-Verlag, New York.

Kazi-Tani, N., D. Possamai, and C. Zhou (2015): Quadratic BSDEs with jumps: a fixed point approach. *Electron. J. Probab.* 66, 1-28.

Kharroubi, I., and T. Lim (2014): Progressive enlargement of filtrations and backward stochastic differential equations with jump. *J. Theor. Probab.* 27, 683-724.

Klibanoff, P., M. Marinacci, and S. Mukerji (2005): A smooth model of decision making under ambiguity. *Econometrica* 73, 1849-1892.

Kobylanski, M. (2000): Backward stochastic differential equations and partial differential equations with quadratic growth. *Ann. Probab.* 28, 558-602.

Konno H., S.R. Pliska, and K.I. Suzuki (1993): Optimal portfolios with asymptotic criteria. *Ann. Oper. Res.* 45, 184-204.

Korn, R. (1997): *Optimal Portfolios: Stochastic Models for Optimal Investment and Risk Management in Continuous Time.* World Scientific Pub. Co. Inc., Singapore.

Korn, R., and E. Korn (2001): *Option Pricing and Portfolio Optimization: Modern Methods of Financial Mathematics.* Graduate Studies in Mathematics. American Mathematical Society, Washington.

Kraft, H., and M. Steffensen (2009): Asset allocation with contagion and explicit bankruptcy procedures. *J. Math. Econ.* 45, 147-167.

Kramkov, D. (1996): Optional decomposition of supermartingales and hedging contingent claims in incomplete security markets. *Probab. Theor. Rel. Field* 105(4), 459-479.

Kramkov, D., and W. Schachermayer (1999): The asymptotic elasticity of utility functions and optimal investment in incomplete markets. *Ann. Appl. Probab.* 9(3), 904-950.

Kramkov, D., and W. Schachermayer (2003): Necessary and sufficient conditions in the problem of optimal investment in incomplete markets. *Ann. Appl. Probab.* 13(4), 1504-1516.

Kramkov, D., and M. Sîrbu (2006): Sensitivity analysis of utility-based prices and risk-tolerance wealth processes. *Ann. Appl. Probab.* 16(4), 2140-2194.

Kröner, A., A. Picarelli and H. Zidani (2018): Infinite horizon stochastic optimal control problems with running maximum cost. *SIAM J. Contr. Optim.* 56, 3296-3319.

Kumar, K.S., and C. Pal (2013): Risk-sensitive control of pure jump process on countable space with near monotone cost. *Appl. Math. Optim.* 68, 311-331.

Kunita, H. (2019): *Stochastic Flows and Jump-Diffusions.* Springer-Verlag, New York.

Kushner, H.J. (1998): Existence of optimal controls for variance control. In McEneaney, W., G. Yin and Q. Zhang (Eds.), *Stochastic analysis, control, optimization and applications: a volume in honor of W.H. Fleming.* Boston: Birkhäuser.

Kushner, H.J., and P. Dupuis (1991): *Numerical Methods for Stochastic Control Problems in Continuous Time.* Springer-Verlag, New York.

Łaukajtys, W., and L. Słomiński (2013): Penalization methods for the Skorokhod problem and reflecting SDEs with jumps. *Bernoulli* 19(5A), 1750-1775.

Lewis, A. (2000): *Option Valuation Under Stochastic Volatility: With Mathematica Code.* Finance Press, Newport Beach, California.

Lim, T., and M.C. Quenez (2015): Portfolio optimization in a default model under full/partial information. *Probab. Engine. Inf. Sci.* 29, pp. 565-587.

Linetsky, V. (2000): Pricing equity derivatives subject to bankruptcy. *Math. Finan.* 16(2), 255-282.

Lyons, T.J. (1995): Uncertain volatility and the risk-free synthesis of derivatives. *Appl. Math. Finan.* 2, 117-133.

McDonald, R., and D. Siegel (1986): The value of waiting to invest. *Quart. J. Econ.* 101(4), 707-727.

Mehra, R., and E.C. Prescott (1985): The equity premium: A puzzle. *J. Monet. Econ.* 15(2), 145-161.

Merton, R.C. (1969): Lifetime portfolio selection under uncertainty: The continuous-time model. *Rev. Econ. Stats.* 51, 247-257.

Merton, R.C. (1971): Optimum consumption and portfolio rules in a continuous-time model. *J. Econ. Theor.* 3, 373-413.

Miao, J., and N. Wang (2011): Risk, uncertainty and option exercise. *J. Econ. Dyn. Contr.* 35, 442-461.

Morlais, M.A. (2009): Utility maximization in a jump market model. *Stochastics.* 81, 1-27.

Morlais, M.A. (2010): A new existence result for quadratic BSDEs with jumps with application to the utility maximization problem. *Stoch. Process. Appl.* 120, 1966-1995.

Munk, C. (2008): Portfolio and consumption choice with stochastic investment opportunities and habit formation in preferences. *J. Econ. Dyn. Contr.* 32(11), 3560-3589.

Myers, C. (1977): Determinants of corporate borrowing. *J. Finan. Econ.* 5, 147-175.

Nagai, H., and S. Peng (2002): Risk-sensitive dynamic portfolio optimization with partial information on infinite time horizon. *Ann. Appl. Probab.* 12, 173-195.

Neufeld, A., and M. Nutz (2013): Superreplication under volatility uncertainty for measurable claims. *Electron. J. Probab.* 18, 1-14.

Ni, C., Y. Li, P. Forsyth and R. Carroll (2022): Optimal asset allocation for outperforming a stochastic benchmark target. *Quant. Finance* 22(9), 1595-1626.

Nishimura, K.G., and H. Ozaki (2007): Irreversible investment and knightian uncertainty. *J. Econ. Theor.* 136(1), 668-694.

Nguyen, D.H., and G. Yin (2016): Modeling and analysis of switching diffusion systems: Past-dependent switching with a countable state space. *SIAM J. Contr. Optim.* 54, 2450-2477.

Nutz, M., and R. van Handel (2013): Constructing sublinear expectations on path space. *Stoch. Process. Appl.* 123(8), 3100-3121.

Nutz, M., and J. Zhang (2015): Optimal stopping under adverse nonlinear expectation and related games. *Ann. Appl. Probab.* 25, 2503-2534.

Otroka, C., B. Ravikumar, and C.H. Whiteman (2002): Habit formation: a resolution of the equity premium puzzle? *J. Monet. Econ.* 49, 1261-1288.

A. Papanicolaou (2019): Backward SDEs for control with partial information. *Math. Finan.* 29, 208-248.

Peng, S.G. (1990): A general stochastic maximum principle for optimal control problems. *SIAM J. Contr. Optim.* 28(4), 966-979.

Pham, H., and M.C. Quenez (2001): Optimal portfolio in partially observed stochastic volatility models. *Ann. Appl. Probab.* 11, 210-238.

Pham, H. (2009): *Continuous-time Stochastic Control and Optimization with Financial Applications.* Stochastic Modelling and Applied Probability (SMAP, volume 61), Springer-Verlag, New York.

Pontryagin, L.S., V.G. Boltyanskii, R.V. Gamkrelidze, and E.F. Mishchenko (1962): *The Mathematical Theory of Optimal Processes* International Series of Monographs in Pure and Appl. Math. Interscience, New York.

Protter, P. (2005): *Stochastic Integration and Differential Equations, 2nd ed.* Springer-Verlag, New York.

Rásonyi, M. (2003): A remark on the superhedging theorem under transaction costs. In Jacques Azma, Michel mery, Michel Ledoux, and Marc Yor, editors, Sminaire de Probabilits XXXVII, volume 1832 of Lecture Notes in Math. 394-398, Springer-Verlag, New York.

Revuz, D., and M. Yor (1999): *Continuous Martingales and Brownian Motion.* Springer-Verlag, New York.

Riedel, F. (2009): Optimal stopping with multiple priors. *Econometrica* 77, 857-908.

Robbins, H., and S. Monro (1951): A stochastic approximation method. *Ann. Math. Statist.* 22(3), 400-407.

Rockafellar, R.T. (1970): *Convex Analysis.* Princeton University Press, Princeton.

Rogers, L.C.G., and D. Williams (2000): *Diffusions, Markov processes, and Martingales*. Vol. 2, Cambridge Mathematical Library, Cambridge University Press, Cambridge.

Ryder, H.E., and G.M. Heal (1973): Optimal growth with intertemporally dependent preferences. *Rev. Econ. Stud.* 40, 1-33.

Samuelson, P.A. (1969): Lifetime portfolio selection by dynamic stochastic programming. *Rev. Econ. Stats.* 51(3), 239-246.

Sass, J., and U. G. Haussmann (2004): Optimizing the terminal wealth under partial information: The drift process as a continuous time Markov chain. *Finan. Stoch.* 8, 553-577.

Schachermayer, W. (2004): *Portfolio Optimization in Incomplete Financial Markets*. Scuola Normale Superiore, Classe di Scienze, Pisa.

Schachermayer, M. (2004): The fundamental theorem of asset pricing under proportional transaction costs in finite discrete time. *Math. Finan.* 14(1), 1467-9965.

Schroder, M., and C. Skiadas (2002): An isomorphism between asset pricing models with and without linear habit formation. *Rev. Finan. Stud.* 15(4), 1189-1221.

Schröder, D. (2011): Investment under ambiguity with the best and worst in mind. *Math. Finan. Econ.* 4, 107-133.

Schwartz, E. (2013): The real options approach to valuation: challenges and opportunities. *Latin Amer. J. Econ.* 50(2), 163-177.

Sekine, J. (2012): Long-term optimal portfolios with floor. *Finan. Stoch.* 16, 369-401.

Smith, H.L. (2008): *Monotone Dynamical Systems: An Introduction to the Theory of Competitive and Cooperative Systems*. Mathematical Surveys and Monographs, Vol. 41, AMS.

Smith, J.E., and R.F. Nau (1995): Valuing risky projects: option pricing theory and decision analysis. *Manag. Sci.* 41(5), 795-816.

Stroock, D.W., and S.R.S. Varadhan (2006): *Multidimensional Diffusion Processes*. Springer-Verlag, Berlin.

Strotz, R.H. (1955): Myopia and inconsistency in dynamic utility maximization. *Rev. Econ. Stud.* 23(3), 165-180.

Strub, O., and P. Baumann (2018): Optimal construction and rebalancing of index-tracking portfolios. *Euro. J. Oper. Res.* 264, 370-387.

Sun, Y. (2006): The exact law of large numbers via Fubini extension and characterization of insurable risks. *J. Econ. Theor.* 126(1), 31-69.

Sutton, R.S. (1988): Learning to predict by the methods of temporal differences. *Mach. Learn.* 3, 9-44.

Tang, W.P., Y.M. Zhang and X.Y. Zhou (2022): Exploratory HJB equation and their convergence. *SIAM J. Contr. Optim.* 60, 3191-3216.

L. Teplá (2001): Optimal investment with minimum performance constraints. *J. Econ. Dyn. Contr.* **25**, 1629-1645.

Trigeorgis, L. (1991): A log-transformed binomial analysis method for valuing complex multi-option investments. *Adv. Futures Opers. Res.* 26(3), 309-326.

Trojanowska, M., and P. Kort (2010): The worst case for real options. *J. Optim. Theor. Appl.* 146, 709-734.

Trutnau, G. (2010): Weak existence of the squared Bessel and CIR processes with skew reflection on a deterministic time-dependent curve. *Stoch. Process. Appl.* 120, 381-402.

van den Broek, B., W. Wiegerinck, and B. Kappen (2010): Risk sensitive path integral control. UAI'10: Proceedings of the Twenty-Sixth Conference on Uncertainty in Artificial Intelligence, 615-622.

D. Veestraeten (2004): The conditional probability density function for a reflected Brownian motion. *Comput. Econ.* **24**(2), 185-207-2380.

Wang, H., T. Zariphopoulou, and X.Y. Zhou (2020): Reinforcement learning in continuous time and space: A stochastic control approach. *J. Machine Learning Res.* 21, 1-34.

Wang, B., X. Gao, and L. Li (2023): Reinforcement learning for continuous-time optimal execution: Actor-critic algorithm and error analysis. Preprint, available at SSRN:4378950.

Weerasinghe, A., and C. Zhu (2016): Optimal inventory control with path-dependent cost criteria. *Stoch. Process. Appl.* 126, 1585-1621.

X. Wei, and X. Yu (2023): Continuous-time q-learning for mean-field control problems. Preprint, available at arXiv:2306.16208.

Xi, F.B., and C. Zhu (2017): On Feller and strong Feller properties and exponential ergodicity of regime-switching jump diffusion processes with countable regimes. *SIAM J. Contr. Optim.* 55(3), 1789-1818.

Xiong, J., and X.Y. Zhou (2007): Mean-variance portfolio selection under partial information. *SIAM J. Contr. Optim.* 46, 156-175

Yao, D., S. Zhang, and X. Zhou (2006): Tracking a financial benchmark using a few assets. *Oper. Res.* 54, 232-246.

Yamada, T., and S. Watanabe (1971): On the uniqueness of solutions of stochastic differential equations. *J. Math. Kyoto Univ.* 11, 155-167.

Yu, X. (2012): *Utility Maximization with Consumption Habit Formation in Incomplete Markets*. Ph.D. Thesis, The University of Texas at Austin.

Yu, X. (2015): Utility maximization with addictive consumption habit formation in incomplete semimartingale markets. *Ann. Appl. Probab.* 25(3), 1383-1419.

Zhang, J.F. (2017): *Backward Stochastic Differential Equations*. Part of the Probability Theory and Stochastic Modelling book series (PTSM, volume 86), 79-99.

Zhao H.Y., W.P. Tang, and D. Yao (2023): Policy optimization for continuous reinforcement learning. *37th Adv. Neur. Inf. Process. Syst.* NeurIPS 2023, 1-27.

Žitković, G. (2002): A filtered version of the bipolar theorem of Brannath and Schachermayer. *J. Theor. Probab.* 15, 41-61.

Žitković, G. (2005): Utility maximization with a stochastic clock and an unbounded random endowment. *Ann. Appl. Probab.* 15(1B), 748-777.

Žitković, G. (2009): Convex-compactness and its applications. *Math. Finan. Econ.* 3(1), 1-12.

Index

ϵ-optimal control, 45

adjoint equation, 59, 60
admissible control, 43, 51, 126
affine process, 24

backward recursively, 80
Bellman equation, 48
benchmark growth rate, 200
benchmark process, 9, 201, 235
Bernoulli equation, 3
Black–Scholes model, 1, 6, 8
BSDEs, 38

calculus of variations, 54
capital injection, 9
Cauchy problem, 35
CEV model, 23
change of measure, 73, 74, 126, 277
change of numéraire, 8, 200, 201
CIR process
   Yamada–Wanatabe SDE, 23
classical solution, 3, 101, 207
Cole–Hopf transformation, 64, 76
conjugate duality, 183
constant relative risk aversion, 2
consumption habit, 161, 165
convergence rate, 105
convex duality, 162
convex perturbation, 58
convexly compact, 177

Countably Infinite State, 91
credit portfolio, 67
credit risk, 30

default contagion, 30, 67
default indicator process, 4, 30
default intensity, 31
defaultable bond, 4
defaultable security, 68
defaultable stock, 32
Doléans–Dade exponential, 18, 74
DPP, 49, 51, 63
   dynamic program, 2, 7, 47, 48
dual problem, 172
dynamic floor constraint, 201
dynamic optimal tracking, 8

equivalent local martingale measure, 165
Euler–Lagrange equation, 55
European call option, 38
European put option, 37
exploratory HJB equation, 281
extended Merton's problem, 235

FBSDE, 41
Feller diffusion process
   Yamada–Wanatabe SDE, 23
Feller's condition, 24
Fenchel–Legendre transform, 3, 162, 187, 191, 207

Feynman–Kac's formula, 33, 35, 65, 209
filter process, 115, 121
financeable consumption process, 166
FPK equation, 65

Gâteaux derivative, 60
GARCH model, 21
generalized Hamiltonian, 59
geometric Brownian motion, 20, 200

habit formation, 161
Hamiltonian, 3, 54, 56
HJB equation, 2, 48, 53
homogenization method, 253
hybrid diffusion system, 24, 30

incomplete information, 112
incomplete market, 162
information-induced contagion, 111
initial-boundary problem, 35
interacting default intensity, 67
interlacing procedure, 25
Itô SDE, 15

Jacod–Yor theorem, 125
jump-to-default, 32

Langevin equation
    OU process, 19
linear addictive habit, 161
linear BSDE, 41
linear dual PDE, 208
linear SDE, 17
logarithm utility, 167

market isomorphism, 173
Markovian feedback control, 51
martingale optimality condition, 206
martingale orthogonality condition, 285
martingale representation theorem, 40, 112, 117
maximum principle, 56
mean-reverting, 19
Minimax theorem, 183, 184

Neumann boundary, 240
Neumann boundary condition, 199, 206, 218, 232

offline q-learning algorithm, 288
optimal feedback strategy
    Markovian feedback control, 104
optimal tracking portfolio, 235
optimal tracking problem, 201, 229
optional decomposition theorem, 166
OU process, 19

partial information, 114
path dependent optimization problem, 162
path integral method, 64
pathwise uniqueness, 22
physical contagion, 111
Poisson random measure, 29
policy improvement theorem, 282
policy space, 80
Pontryagin's maximum principle, 57, 58
post-default, 7
power utility, 167
pre-default, 7, 32, 71

q-function, 283
quadratic BSDE, 112
quadratic BSDE with jumps, 130

ratcheting benchmark process, 198
reasonable asymptotic elasticity, 164, 167
recursive HJB equation, 8, 75, 94
reflected Brownian motion, 209
reflected process, 10
reflected state process, 229
regime-switching, 24, 67, 113
reinforcement learning, 271, 279
relaxed benchmark tracking, 235
risk sensitive control, 61, 69
riskless bond, 4
running function, 44
running maximum process, 204

self-exciting default, 8
self-financing, 1, 6, 9, 20, 165
Shannon's entropy, 280
singular control, 9, 239
Skorokhod problem, 205
SMP, 56
solid hull, 162
stationary HJB equation, 11, 232
stochastic approximation, 288
stochastic flow, 207
stochastic volatility model, 37
strong solution, 15

terminal function, 44

transversality condition, 266, 283
truncated BSDE, 132
type K condition, 83, 85, 98

uniformly elliptic, 35, 102
utility function, 166

verification result, 101
viscosity solution, 42

weak solution, 15

Yamada–Watanabe SDE, 22

www.ingramcontent.com/pod-product-compliance
Lightning Source LLC
Chambersburg PA
CBHW050609010725
28797CB00004B/37